## Discrete Random Variables (continued)

**Negative Binomial Random Variable**

$S_X = \{r, r+1, \ldots\}$ where $r$ is a positive integer

$$p_k = \binom{k-1}{r-1} p^r (1-p)^{k-r} \qquad k = r, r+1, \ldots$$

$$E[X] = \frac{r}{p} \qquad \text{VAR}[X] = \frac{r(1-p)}{p^2}$$

$$G_X(z) = \left(\frac{pz}{1-qz}\right)^r$$

*Remarks*: $X$ is the number of trials until the $r$th success in a sequence of independent Bernoulli trials.

**Poisson Random Variable**

$S_X = \{0, 1, 2, \ldots\}$

$$p_k = \frac{\alpha^k}{k!} e^{-\alpha} \qquad k = 0, 1, \ldots \quad \text{and} \quad \alpha > 0$$

$$E[X] = \alpha \qquad \text{VAR}[X] = \alpha$$

$$G_X(z) = e^{\alpha(z-1)}$$

*Remarks*: $X$ is the number of events that occur in one time unit when the time between events is exponentially distributed with mean $1/\alpha$.

W9-BIB-939

# Probability and Random Processes for Electrical Engineering

# Probability and Random Processes for Electrical Engineering

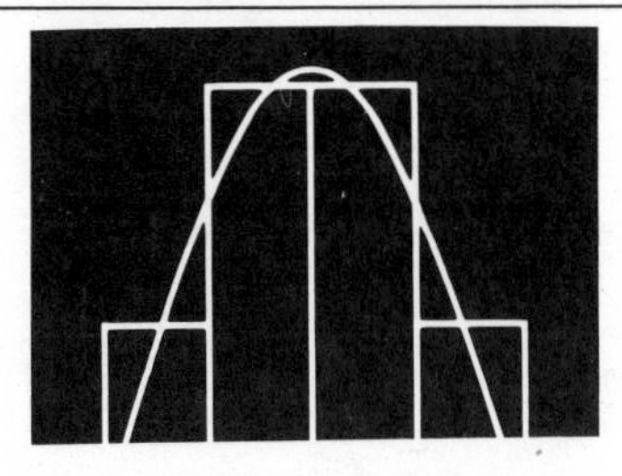


**Alberto Leon-Garcia**

University of Toronto

**ADDISON-WESLEY PUBLISHING COMPANY**

Reading, Massachusetts • Menlo Park, California • New York
Don Mills, Ontario • Wokingham, England • Amsterdam • Bonn
Sydney • Singapore • Tokyo • Madrid • San Juan

This book is in the **Addison-Wesley Series in Electrical and Computer Engineering**

Barbara Rifkind and Tom Robbins, *Sponsoring Editors*
Bette J. Aaronson, *Production Supervisor*
Patricia Steele, *Copy Editor*
Beth Anderson, *Text Designer*
Long Associates, *Illustrator*
Joseph Vetere, *Technical Art Consultant*
Sheila Bendikian, *Production Coordinator*
Marshall Henrichs, *Cover Designer*
Hugh Crawford, *Manufacturing Supervisor*

**Library of Congress Cataloging-in-Publication Data**

Leon-Garcia, Alberto
Probability and random processes for electrical engineering / by Alberto Leon-Garcia.
p. cm. —(Addison-Wesley series in electrical engineering)
Includes bibliographies and index.
ISBN 0-201-12906-X: ISBN 0-201-12907-8 (solutions manual)
1. Electric engineering—Mathematics. 2. Probabilities. 3. Stochastic processes. I. Title. II. Series.
TK153.L425 1989
621.3—dc 19
88-19374
CIP

Copyright © 1989 by the Addison-Wesley Publishing Company.

All rights reserved. No part of this publication may be reproduced, stored in a retrieval system, or transmitted, in any form or by any means, electronic, mechanical, photocopying, recording, or otherwise, without the prior written permission of the publisher. Printed in the United States of America. Published simultaneously in Canada.

ABCDEFGHIJ-AL-898

**To my mother**
**and to the memory of my father**

# Preface

This book is intended for introductory courses on probability theory and random processes taught in electrical and computer engineering programs. Chapters 1 through 5 are designed for a one-semester junior/senior level course on probability theory. Chapters 6 through 8 are designed for a one-semester senior or first-year graduate course on random processes including Markov chains. Chapters 6, 8, and 9 can be used for a one-semester introduction to queueing theory.

The complexity of the systems encountered in electrical and computer engineering calls for an understanding of probability concepts and a facility in the use of probability tools from an increasing number of B.S. degree graduates. The introductory course should therefore teach the student not only the basic theoretical concepts, but also how to solve problems that arise in engineering practice. This course requires that the student develop problem-solving skills and understand how to make the transition from a real problem to a probability model for that problem.

*Probability and Random Processes for Electrical Engineering* presents a carefully motivated, accessible, and interesting introduction to probability. It is designed to allow the instructor maximum flexibility in the selection of topics. In addition to the standard topics taught in introductory courses on probability, random variables, and random processes, the book includes sections on modelling, basic statistical techniques, computer simulation, reliability, and concise but relatively complete introductions to Markov chains and queueing theory.

### Relevance to Engineering Practice

A major problem in introductory probability courses is in motivating the students. One needs to show the student the relevance of the material to engineering practice. This book addresses this problem in Chapter 1 by discussing the role of probability models in engineering design. Applications from various areas of electrical and computer engineering are used to show how averages and relative frequencies provide the right tools for handling the design of systems that involve randomness. These application areas are used in examples throughout the text.

### From Problems to Probability Models

How the transition is made from real problems to probability models is shown in several ways. First, important concepts usually are motivated by presenting real data or computer simulation data. Second, sections on basic statistical techniques have been integrated into the text. These sections show how statistical methods provide the link between theory and the real world. Finally, the important random variables and random processes are developed using model-building arguments that range from the simple to the complex. For example, in Chapters 2 and 3 we proceed from coin tossing to Bernoulli trials, and then to the binomial and geometric distributions, and finally via limiting arguments to the Poisson, exponential, and Gaussian distributions.

### Examples and Problems

Numerous examples are included in every section. Examples are used to demonstrate analytical and problem-solving techniques, to motivate concepts using simplified cases, and to illustrate applications. The book includes over 550 problems, which are identified by section to help the instructor in selecting homework problems. Answers to selected problems are included at the end of the book, and a solutions manual is available to the instructor.

### Computer Methods

The development of an intuition for randomness can be aided by the use of computer exercises. Simple computer programs can be used to generate random numbers and then random variables of various types. The resulting data can then be analyzed using the statistical methods introduced in the book. The sections on computer methods have been integrated into the text rather than isolated in a separate chapter because I feel that the learning of basic probability concepts will be assisted by performing the computer exercises. It should be noted that the computer methods introduced in Sections 2.7, 3.11, and 4.9 do not necessarily require entirely new lectures. The transformation method in Section 3.11 can be incorporated into the discussion on functions of a random variable, and similarly the material in Section 4.9 can be incorporated into the discussion on transformations of random vectors.

### Random Variables and Continuous-time Random Processes

Discrete-time random processes provide a crucial "bridge" in going from random variables to continuous-time random processes. Care is taken in the first five chapters to lay the proper groundwork for this transition. Thus sequences of dependent experiments are discussed in Chapter 2 as a preview of Markov chains. In Chapter 4 I emphasize how a joint distribution generates a consistent family of marginal distributions. Chapter 5 introduces sequences of independent identically distributed (iid) random variables, and Chapter 6 considers the sum of an iid sequence to produce important examples of random processes. Throughout Chapters 6 and 7, a concise development of the concepts is achieved by developing discrete-time and continuous-time results in parallel.

### Markov Chains and Queueing Theory

Markov chains and queueing theory have become essential tools in communication network and computer system modelling. In the introductory course on probability only a few changes need to be made to accommodate these new requirements. The treatment of conditional probability and conditional expectation needs to be modified, and the Poisson and gamma random variables need to be given greater prominence. In an introductory course on random processes a new balance needs to be struck between the traditional discussion of wide-sense stationary processes and linear systems, and the discussion of Markov chains and queueing theory. The "optimum" balance between these two needs will surely vary from instructor to instructor, so I have provided more material than can be covered in one semester in order to give the instructor leeway in striking a balance.

### Suggested Syllabuses

The first five chapters form the basis for a one-semester introduction to probability. In addition to the optional sections on computer methods, these chapters also contain optional sections on combinatorics, reliability, confidence intervals, and basic results from renewal theory. In a one-semester course, it is possible to provide an introduction to random processes by omitting all the starred sections in the first five chapters and covering instead the first part of Chapter 6. The material in the first five chapters has been used at the University of Toronto in an introductory junior-level required course for electrical engineers.

A one-semester course on random processes with Markov chains can be taught using Chapters 6 through 8. A quick introduction to Markov chains and queueing theory is possible by covering only the first three sections of Chapter 8 and then proceeding to the first few sections in Chapter 9. A one-semester introduction to queueing theory can be taught from Chapters 6, 8, and 9.

### Acknowledgments

I would like to acknowledge the help of several individuals. During his brief but brilliant career, my ex-student and colleague Gilbert Williams exemplified the engineer who can transform the insights that result from probabilistic reasoning into tangible improvements in the performance of complex systems. The origin of many of the ideas in this book lie in discussions I had with Gil over the years.

During the initial phases of this project I received excellent advice on how to write a book from Paul Shields, Adel Sedra, and Safwat Zaky. At Addison-Wesley, my sponsoring editors, Barbara Rifkind and Tom Robbins, enthusiastically supported this project from the beginning and it was a pleasure to work with Bette Aaronson, the production supervisor. In addition, I would like to thank the many students who read the manuscript, especially Renos Melas, Massoud Khansari, and Peter Lau

The comments of the many reviewers who read various parts of the manuscript were beneficial. The reviewers include: Richard H. Williams, University of New Mexico; Shih-Chun Chang, George Mason University; Kai-Bor Yu, Virginia Polytechnic Institute and State University; John N. Daigle, Virginia Polytechnic Institute and State University; Jonathan D. Cryer, University of Iowa; James H. Stapleton, Michigan State University; Richard A. Roberts, University of Colorado (Boulder); Demissie Alemayehu, Western Michigan University; Steven A. Tretter, University of Maryland (College Park); Arthur Werbner, Wentworth Institute of Technology and HH Aerospace Design Co., Inc.; Richard E. Mortensen, University of California, Los Angeles; Ronald A. Iltis, University of California, Santa Barbara; and Ken C. Sercik, University of Toronto.

My colleague, Pas Pasupathy, provided constant encouragement and constructive criticism, and countless (!) valuable suggestions. Most of all I would like to thank my wife, Karen Carlyle, for not only putting up with this project while completing her own Ph.D., but also giving generously of her time whenever help was needed.

*Toronto* A. L-G

# Contents

## Chapter 1



## Chapter 2

## Chapter 5

## Chapter 6



## Chapter 7

## Chapter 8



## Chapter 9

## Appendixes

# CHAPTER 1

# Probability Models in Electrical and Computer Engineering

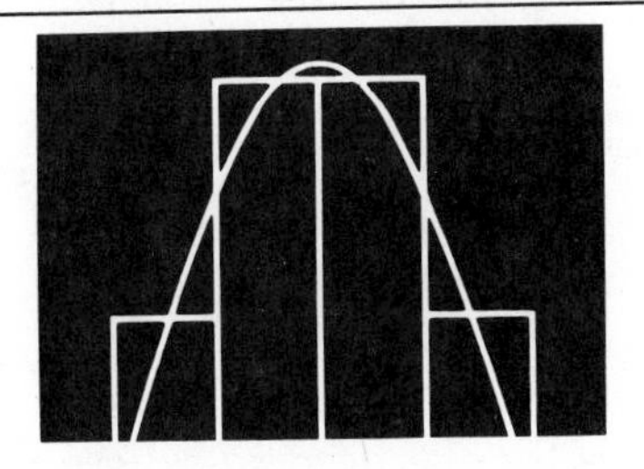

Designers today must often build systems that work in a chaotic environment:

- A large computer system must satisfy the diverse and fluctuating processing demands of the community it serves.
- Communication networks must be continually ready to meet the irregular demands of the customer for voice and data "information pipelines."
- Communication systems must provide continuous and error-free communication over channels that are subject to interference and noise.
- Word recognition systems must decode speaker inputs with high reliability.

Probability models are one of the tools that enable the designer to make sense out of the chaos and to successfully build systems that are efficient, reliable, and cost-effective. This book is an introduction to the theory underlying probability models as well as to the basic techniques used in the development of such models.

This chapter introduces probability models and shows how they differ from the deterministic models that are pervasive in engineering. The key properties of the notion of probability are developed, and various examples from electrical and computer engineering, where probability models play a key role, are presented. Section 1.6 gives an overview of the book.

## 1.1 MATHEMATICAL MODELS AS TOOLS IN ANALYSIS AND DESIGN

The design or modification of any complex system involves the making of choices from various feasible alternatives. Choices are made on the basis of criteria such as cost, reliability, and performance. The quantitative evaluation of these criteria is seldom made through the actual implementation and experimental evaluation of the alternative configurations. Instead, decisions are made based on estimates that are obtained using models of the alternatives.

A **model** is an approximate representation of a physical situation. A model attempts to explain observed behavior using a set of simple and understandable rules. These rules can be used to predict the outcome of experiments involving the given physical situation. A useful model explains all relevant aspects of a given situation. Such models can therefore be used instead of experiments to answer questions regarding the given situation. Models therefore allow the engineer to avoid the costs of experimentation, namely, labor, equipment, and time.

**Mathematical models** are used when the observational phenomenon has measurable properties. A mathematical model consists of a set of assumptions about how a system or physical process works. These assumptions are stated in the form of mathematical relations involving the important parameters and variables of the system. The conditions under which an experiment involving the system is carried out determine the "givens" in the mathematical relations, and the solution of these relations allows us to predict the measurements that would be obtained if the experiment were performed.

Mathematical models are used extensively by engineers in guiding system design and modification decisions. Intuition and rules-of-thumb are not always reliable in predicting the performance of complex and novel systems, and experimentation is not possible during the initial phases of a system design. Furthermore, the cost of extensive experimentation in existing systems frequently proves to be prohibitive. The availability of adequate models for the components of a complex system combined with a knowledge of their interactions allows the scientist and engineer to develop an overall mathematical model for the system. It is then possible to quickly and inexpensively answer questions about the performance of complex systems. Indeed computer programs for obtaining the solution of mathematical models form the basis of many computer-aided analysis and design systems.

In order to be useful, a model must fit the facts of a given situation. Therefore the process of developing and validating a model necessarily consists of a series of experiments and model modifications as shown in Fig. 1.1. Each experiment investigates a certain aspect of the phenomenon under investigation and involves the taking of observations and measurements under a specified set of conditions. The model is used to predict the outcome of the experiment, and these predictions are compared with the actual observations that result when the experiment is carried out. If there is a significant discrepancy, the model is then modified to account for it. The modeling process continues until the investigator is satisfied that the behavior of all relevant aspects of the phenomenon can be predicted to within a desired accuracy. It should be emphasized that the decision of when to stop the modeling process depends on the immediate objectives of the investigator. Thus a model that is adequate for one application may prove to be completely inadequate in another setting.

The predictions of a mathematical model should be treated as hypothetical until the model has been validated through a comparison with experimental measurements. A dilemma arises in a system design situation: The model cannot be validated experimentally because the real system does not exist. Computer simulation models play a useful role in this situation by presenting an alternative means of predicting system behavior, and thus as a means of checking the predictions made by a

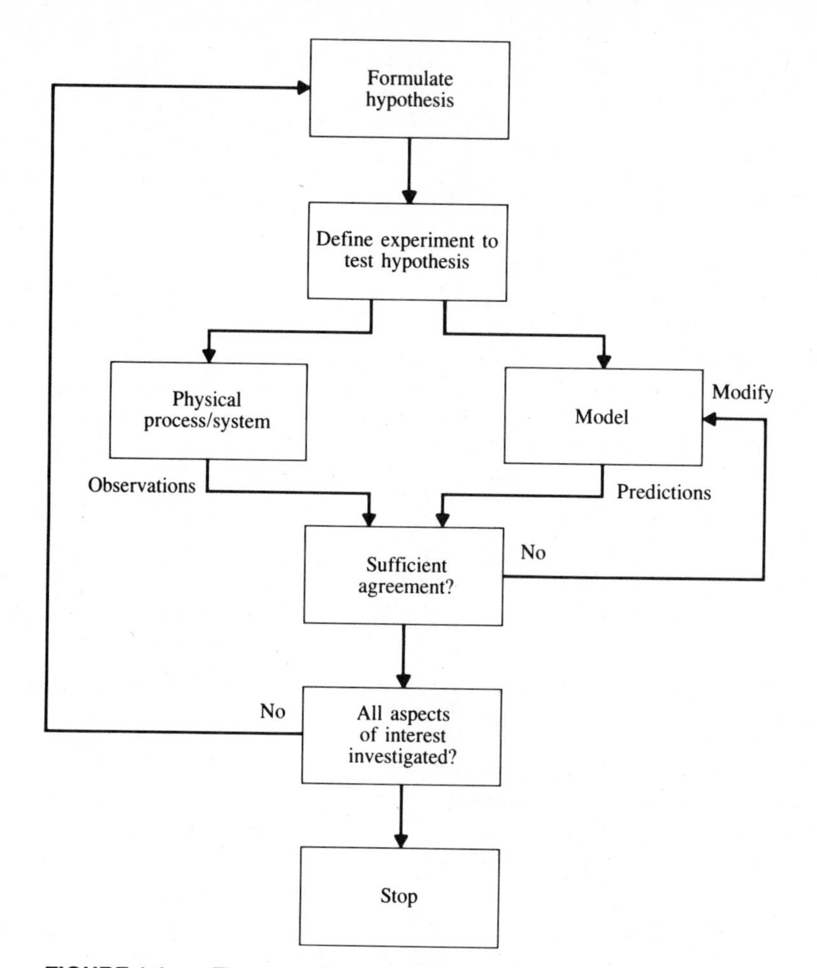


**FIGURE 1.1** The modeling process.

mathematical model. A **computer simulation model** consists of a computer program that simulates or mimics the dynamics of a system. Incorporated into the program are instructions that "measure" the relevant performance parameters. In general, simulation models are capable of representing systems in greater detail than mathematical models. However, they tend to be less flexible and usually require more computation time than mathematical models.

In the following two sections we discuss the two basic types of mathematical models, deterministic models and probability models.

## 1.2

## DETERMINISTIC MODELS

In **deterministic models** the conditions under which an experiment is carried out determine the exact outcome of the experiment. In deterministic mathematical models, the solution of a set of mathematical equations specifies the exact outcome of the experiment. Circuit theory is an example of a deterministic mathematical model.

Circuit theory models the interconnection of electronic devices by ideal circuits that consist of discrete components with idealized voltage-current characteristics. The theory assumes that the interaction between these idealized components is completely described by Kirchhoff's voltage and current laws. For example, Ohm's law states that the voltage-current characteristic of a resistor is $I = V/R$. The voltages and currents in any circuit consisting of an interconnection of batteries and resistors can be found by solving a system of simultaneous linear equations that is found by applying Kirchhoff's laws and Ohm's law.

If an experiment involving the measurement of a set of voltages is repeated a number of times under the same conditions, circuit theory predicts that the observations will always be exactly the same. In practice there will be some variation in the observations due to measurement errors and uncontrolled factors. Nevertheless, this deterministic model will be adequate as long as the deviation about the predicted values remains small.

## 1.3

## PROBABILITY MODELS

Many systems of interest involve phenomena that exhibit unpredictable variation and randomness. We will define a **random experiment** to be an experiment in which the outcome varies in an unpredictable fashion when the experiment is repeated under the same conditions. Deterministic models are not appropriate for random experiments since they predict the same outcome for each repetition of an experiment. In this section we introduce probability models that are intended for random experiments.

As an example of a random experiment, suppose a ball is selected from an urn containing three identical balls, labeled 0, 1, and 2. The urn is first shaken to randomize the position of the balls, and a ball is then selected. The number of the ball is noted, and the ball is then returned to the urn. The **outcome** of this experiment is a number from the set $S = \{0, 1, 2\}$. We call the set $S$ of all possible outcomes the **sample space.** Figure 1.2

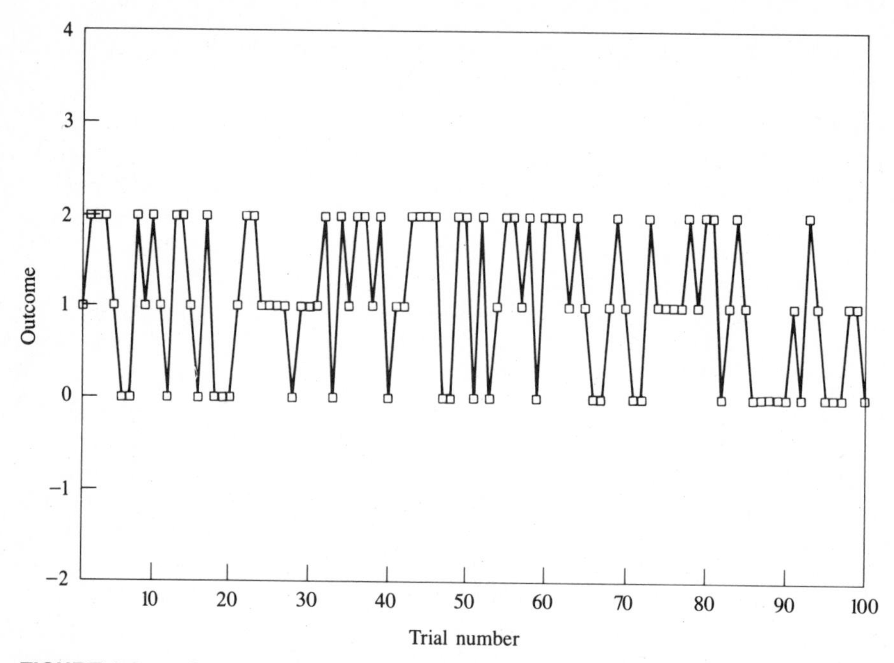


**FIGURE 1.2** Outcomes of urn experiment.

shows the outcomes in 100 repetitions (trials) of a computer simulation of this urn experiment. It is clear that the outcome of this experiment cannot consistently be predicted correctly.

### Statistical Regularity

In order to be useful, a model must enable us to make predictions about the future behavior of a system, and in order to be predictable, a phenomenon must exhibit regularity in its behavior. Many probability models in engineering are based on the fact that averages obtained in long sequences of repetitions (trials) of random experiments consistently yield approximately the same value. This property is called **statistical regularity**.

Suppose that the above urn experiment is repeated $n$ times under identical conditions. Let $N_0(n)$, $N_1(n)$, and $N_2(n)$ be the number of times in which the outcomes are balls 0, 1, and 2, respectively, and let the **relative frequency** of outcome $k$ be defined by

$$f_k(n) = \frac{N_k(n)}{n}. \tag{1.1}$$

By statistical regularity we mean that $f_k(n)$ varies less and less about a constant value as $n$ is made large, that is,

$$\lim_{n\to\infty} f_k(n) = p_k. \tag{1.2}$$

The constant $p_k$ is called the **probability** of the outcome $k$. Equation (1.2) states that the probability of an outcome is the long-term proportion of times it arises in a long sequence of trials. We will see throughout the book that Eq. (1.2) provides the key connection in going from the measurement of physical quantities to the probability models discussed in this book.

Figures 1.3 and 1.4 show the relative frequencies for the three outcomes in the above urn experiment as the number of trials $n$ is increased. It is clear that all the relative frequencies are converging to the value 1/3. This is in agreement with our intuition that the three outcomes are equiprobable.

Suppose we alter the above urn experiment by placing in the urn a fourth identical ball with the number 0. The probability of the outcome 0 is now 2/4 since two of the four balls in the urn have the number 0. The probabilities of the outcomes 1 and 2 would be reduced to 1/4 each. This demonstrates a key property of probability models, namely, *the conditions under which a random experiment is performed determine the probabilities of the outcomes of an experiment.*

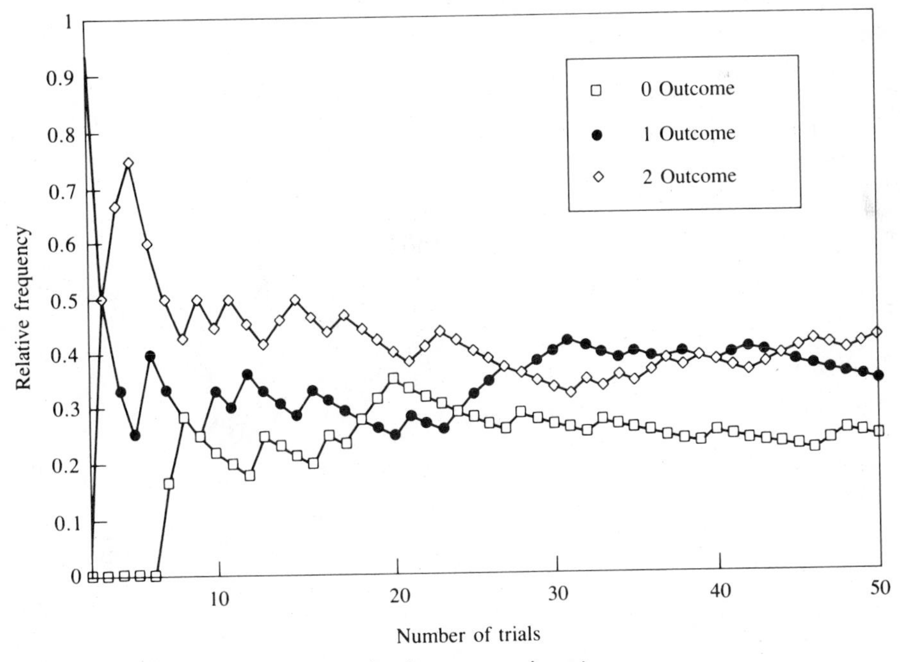


**FIGURE 1.3** Relative frequencies in urn experiment.

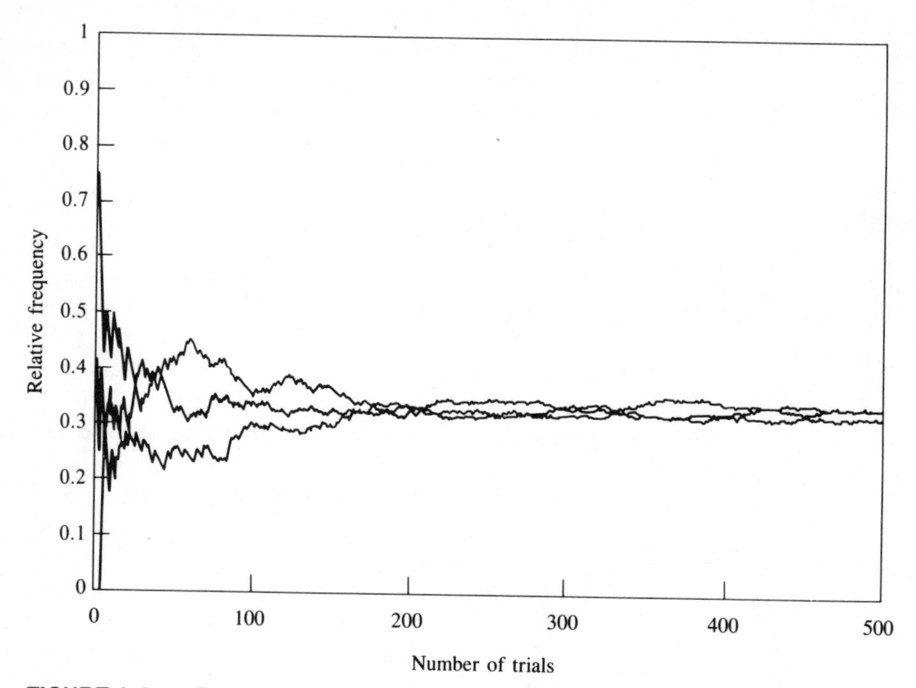


**FIGURE 1.4** Relative frequencies in urn experiment.

### Properties of Relative Frequency

We will now present several properties of relative frequency. Suppose that a random experiment has $K$ possible outcomes, that is, $S = \{1, 2, \ldots, K\}$. Since the number of occurrences of any outcome in $n$ trials is a number between zero and $n$, we have that

$$0 \leq N_k(n) \leq n \qquad \text{for } k = 1, 2, \ldots, K,$$

and thus dividing the above equation by $n$, we find that the relative frequencies are a number between zero and one:

$$0 \leq f_k(n) \leq 1 \qquad \text{for } k = 1, 2, \ldots, K. \tag{1.3}$$

The sum of the number of occurrences of all possible outcomes must be $n$:

$$\sum_{k=1}^{K} N_k(n) = n.$$

If we divide both sides of the above equation by $n$, we find that the sum of all the relative frequencies equals one:

$$\sum_{k=1}^{K} f_k(n) = 1. \tag{1.4}$$

Sometimes we are interested in the occurrence of **events** associated with the outcomes of an experiment. For example, consider the event "an even-numbered ball is selected" in the above urn experiment. What is the relative frequency of this event? The event will occur whenever the number of the ball is 0 or 2. The number of experiments in which the outcome is an even-numbered ball is therefore $N_E(n) = N_0(n) + N_2(n)$. The relative frequency of the event is thus

$$f_E(n) = \frac{N_e(n)}{n} = \frac{N_0(n) + N_2(n)}{n} = f_0(n) + f_2(n).$$

This example shows that the relative frequency of an event is the sum of the relative frequencies of the associated outcomes. More generally, let $C$ be the event "$A$ or $B$ occurs," where $A$ and $B$ are two events that cannot occur simultaneously, then the number of times when $C$ occurs is $N_C(n) = N_A(n) + N_B(n)$, so

$$f_C(n) = f_A(n) + f_B(n). \tag{1.5}$$

Equations (1.3), (1.4), and (1.5) are the three basic properties of relative frequency from which we can derive many other useful results.

### The Axiomatic Approach to a Theory of Probability

Equation (1.2) suggests that we define the probability of an event by its long-term relative frequency. There are problems with using this definition of probability to develop a mathematical theory of probability. First of all, it is not clear when and in what mathematical sense the limit in Eq. (1.2) exists. Second, we can never perform an experiment an infinite number of times so we can never know the probabilities $p_k$ exactly. Finally, the use of relative frequency to define probability would rule out the applicability of probability theory to situations in which an experiment cannot be repeated. Thus it makes practical sense to develop a mathematical theory of probability that is not tied to any particular application or to any particular notion of what probability means. On the other hand, we must insist that, when appropriate, the theory should allow us to use our intuition and interpret probability as relative frequency.

In order to be consistent with the relative frequency interpretation, any definition of "probability of an event" must satisfy the properties in Eqs. (1.3) through (1.5). The modern theory of probability begins with a construction of a set of axioms that specify that probability assignments must satisfy these properties. It supposes that: (1) A random experiment has been defined, and a set $S$ of all possible outcomes has been identified; (2) a class of subsets of $S$ called events has been specified; and (3) each event $A$ has been assigned a number, $P[A]$, in such a way that the following axioms are satisfied:

1. $0 \leq P[A] \leq 1$.
2. $P[S] = 1$.
3. If $A$ and $B$ are events that cannot occur simultaneously, then $P[A \text{ or } B] = P[A] + P[B]$.

The correspondence between the three axioms and the properties of relative frequency stated in Eqs. (1.3) through (1.5) are apparent. These three axioms lead to many useful and powerful results. Indeed, we will spend the remainder of this book developing many of these results.

Note that the theory of probability does not concern itself with how the probabilities are obtained or with what they mean. Any assignment of probabilities to events that satisfies the above axioms is legitimate. It is up to the user of the theory, the model builder, to determine what the probability assignment should be and what interpretation of probability makes sense in any given application.

### Building a Probability Model

Let us consider how we proceed from a real-world problem that involves randomness to a **probability model** for the problem. The theory requires that we identify the elements in the above axioms. This involves (1) defining the random experiment inherent in the application, (2) specifying the set $S$ of all possible outcomes and the events of interest, and (3) specifying a probability assignment from which the probabilities of all events of interest can be computed. The challenge is to develop the simplest model that explains all the relevant aspects of the real-world problem.

As an example, suppose that we test a telephone conversation to determine whether a speaker is currently speaking or silent. We know that on the average the typical speaker is active only 1/3 of the time; the rest of the time he is listening to the other party or pausing between words and phrases. We can model this physical situation as an urn experiment in which we select a ball from an urn containing two white balls (silence) and one black ball (active speech). We are making a great simplification here, not all speakers are the same, not all languages have the same silence-activity behavior, and so forth. The usefulness and power of this simplification becomes apparent when we begin asking questions that arise in system design such as: What is the probability that more than 24 speakers out of 48 independent speakers are active at the same time? This question is equivalent to: What is the probability that more than 24 black balls are selected in 48 independent repetitions of the above urn experiment? By the end of Chapter 2 you will be able to answer the latter question *and* all the real-world problems that can be reduced to it!

## 1.4

## A DETAILED EXAMPLE: A PACKET VOICE TRANSMISSION SYSTEM

In the beginning of this chapter we claimed that probability models provide a tool that enables the designer to successfully design systems that must operate in a random environment, but that nevertheless are efficient, reliable, and cost-effective. In this section, we present a detailed example of such a system. Our objective here is to convince you of the power and usefulness of probability theory. The presentation intentionally draws upon your intuition. Many of the derivation steps that may appear nonrigorous now will be made precise later in the book.

Suppose that a communication system is required to transmit 48 simultaneous conversations from city $A$ to city $B$ using "packets" of voice information. The speech of each speaker is converted into voltage waveforms that are first digitized (i.e., converted into a sequence of binary numbers) and then bundled into packets of information that correspond to 10-millisecond (ms) segments of speech. A source and destination address is appended to each voice packet before it is transmitted (see Fig. 1.5).

The simplest design for the communication system would transmit 48 packets every 10 ms in each direction. This is an inefficient design,

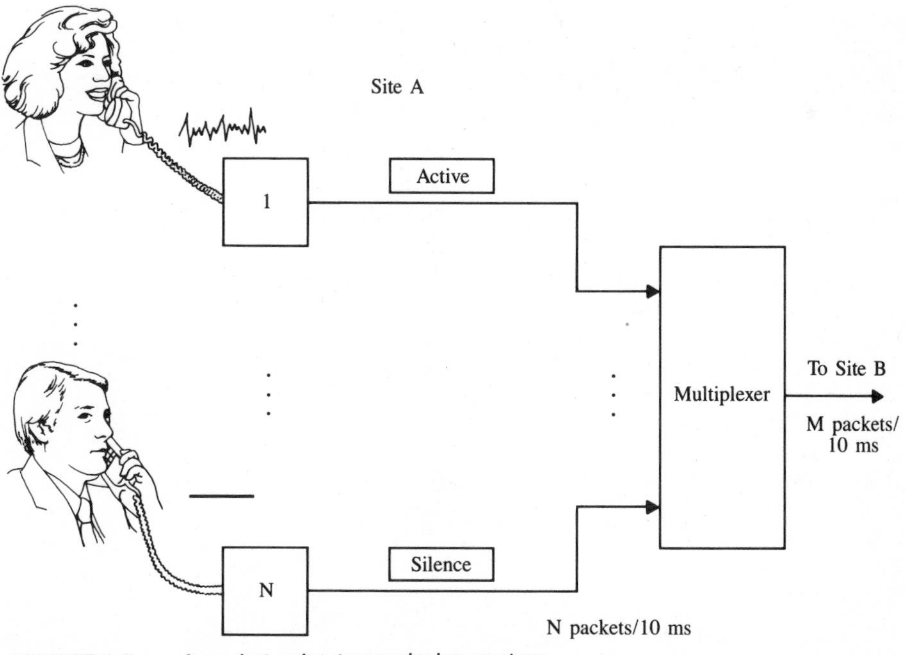


**FIGURE 1.5** A packet voice transmission system.

however, since it is known that on the average about 2/3 of all packets contain silence and hence no speech information. In other words, on the average the 48 speakers only produce about $48/3 = 16$ active (nonsilence) packets per 10 ms period. We therefore consider another system that transmits only $M < 48$ packets every 10 ms.

Every 10 ms, the new system determines which speakers have produced packets with active speech. Let the outcome of this random experiment be $A$, the number of active packets produced in a given 10 ms segment. The quantity $A$ takes on values in the range from 0 (all speakers silent) to 48 (all speakers active). If $A \leq M$, then all the active packets are transmitted. However if $A > M$, then the system is unable to transmit all the active packets, so $A - M$ of the active packets are selected at random and discarded. The discarding of active packets results in the loss of speech, so we would like to keep the fraction of discarded active packets at a level that the speakers do not find objectionable.

First consider the relative frequencies of $A$. Every 10 ms, the above experiment is repeated. Let $A(n)$ be the outcome in the $n$th trial. Let $N_k(n)$ be the number of trials in which the number of active packets is $k$. The relative frequency of the outcome $k$ in the first $n$ trials is then $f_k(n) = N_k(n)/n$, which we suppose converges to a probability $p_k$:

$$\lim_{n\to\infty} f_k(n) = p_k \qquad 0 \leq k \leq 48. \tag{1.6}$$

Next consider the rate at which active packets are produced. The average number of active packets produced per 10 ms interval is given by the **sample mean** of the number of active packets:

$$\langle A \rangle_n = \frac{1}{n}\sum_{j=1}^{n} A(j) \tag{1.7}$$

$$= \frac{1}{n}\sum_{k=0}^{48} kN_k(n). \tag{1.8}$$

The first expression adds the number of active packets produced in the first $n$ trials in the order in which the observations were recorded. The second expression counts how many of these observations had $k$ active packets for each possible value of $k$, and then computes the total.[1] As $n$ gets large, the ratio $N_k(n)/n$ in the second expression approaches $p_k$. Thus the average number of active packets produced per 10 ms segment

1. Suppose you pull out the following change from your pocket: 1 quarter, 1 dime, 1 quarter, 1 nickel. Equation (1.7) says your total is 25 + 10 + 25 + 5 = 65 cents. Equation (1.8) says your total is (1)5 + (1)10 + (2)(25) = 65 cents.

approaches

$$\langle A \rangle_n \rightarrow \sum_{k=0}^{48} kp_k \triangleq E[A]. \tag{1.9}$$

The expression on the right-hand side is called the **expected value of *A*** and is completely determined by the probabilities $p_k$.

The fraction of active packets that are discarded by the system in $n$ trials is

$$\frac{\text{number of active packets discarded}}{\text{number of active packets produced}} = \frac{\sum_{k=M+1}^{48} (k - M) N_k(n)}{\sum_{k=0}^{48} k N_k(n)}.$$

The term $(k - M)$ in the numerator is the number of packets that are discarded when $k > M$ active packets are produced. If we divide the numerator and denominator by $n$, and let $n$ get large, we again obtain an expression in terms of the probabilities $p_k$:

$$\frac{\sum_{k=M+1}^{48} (k - M) N_k(n)/n}{\sum_{k=0}^{48} k N_k(n)/n} \rightarrow \frac{\sum_{k=M+1}^{48} (k - M) p_k}{\sum_{k=0}^{48} k p_k}. \tag{1.10}$$

The expression on the right side is the long-term fraction of active packets that are discarded. We have thus found that *all of the performance measures of interest in this problem can be evaluated if we know the probabilities $p_k$ for the number of active packets produced in a 10 ms segment.*

In general, it might be necessary to perform the experiment a large number of times to compute relative frequencies, and use these as estimates of the probabilities $p_k$. In this particular problem, however, it turns out that this is not necessary because probability theory allows us to derive the probabilities, $p_k$, in terms of the probabilities of a simpler experiment. In the next chapter we will see that the $p_k$'s are given by the Binomial Distribution (Eq. (2.32) in Chapter 2). We will thus obtain a closed-form expression for all of the performance measures of interest in this packet voice transmission system.

Figure 1.6 shows the long-term fraction of active packets that are discarded in the 48 speaker system under consideration. It can be seen that this fraction decreases as $M$ is increased. If we assume that the customers will tolerate the loss of 1% of active packets, then Fig. 1.6 shows that the required value of $M$ is 24. Thus we are able to provide voice transmission for all 48 speakers with only half the transmission speed

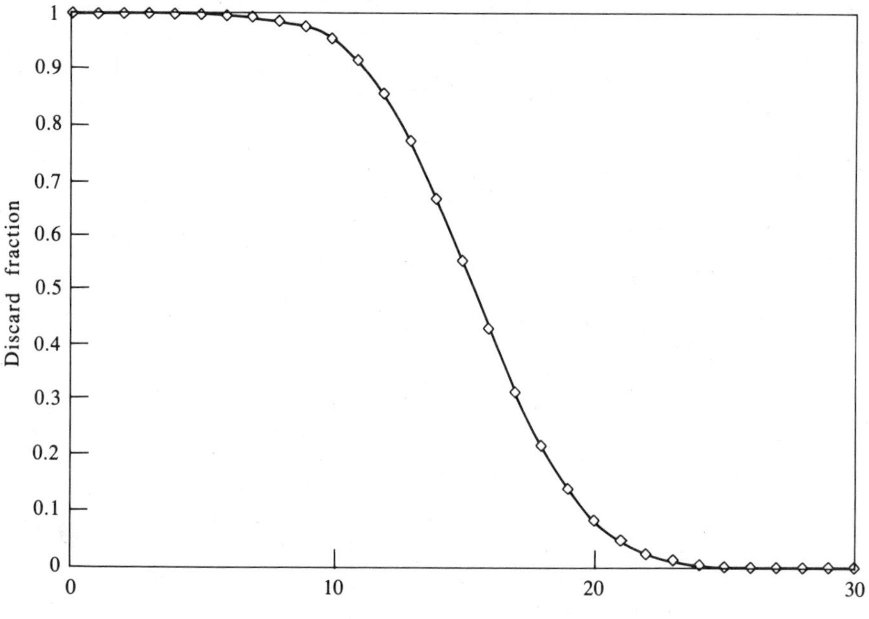


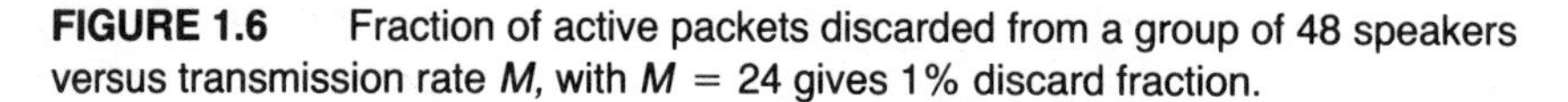

**FIGURE 1.6** Fraction of active packets discarded from a group of 48 speakers versus transmission rate $M$, with $M = 24$ gives 1% discard fraction.

required in the brute force approach. If we are dealing with expensive long-distance telephone calls, the new system could result in significant savings.

Let us summarize what we have done in this section. We have presented an example in which the system behavior is intrinsically random, and in which the system performance measures are stated in terms of long-term averages. We have shown how these long-term measures lead to expressions involving the probabilities of the various outcomes. Finally we have indicated that, in some cases, probability theory allows us to derive these probabilities. We are then able to predict the long-term averages of various quantities of interest and proceed with the system design.

## 1.5 OTHER EXAMPLES

In this section we present further examples from electrical and computer engineering where probability models are used to design systems that work in a random environment. Our intention here is to show how probabilities and long-term averages arise naturally as performance measures in many systems. We hasten to add, however, that this book is intended to present the basic concepts of probability theory and not detailed applications. For the interested reader, references for further reading are provided at the end of this and other chapters.

### Communication over Unreliable Channels

Many communication systems operate in the following way. Every $T$ seconds, the transmitter accepts a binary input, namely, a 0 or a 1, and transmits a corresponding signal. At the end of the $T$ seconds, the receiver makes a decision as to what the input was, based on the signal it has received. Most communications systems are unreliable in the sense that the decision of the receiver is not always the same as the transmitter input. Figure 1.7(a) models systems in which transmission errors occur at random with probability $\varepsilon$. As indicated in the figure, the output is not equal to the input with probability $\varepsilon$. Thus $\varepsilon$ is the long-term proportion of bits delivered in error by the receiver. In situations where this error rate is not acceptable, error control techniques are introduced to reduce the error rate in the delivered information.

One method for reducing the error rate in the delivered information is to use error-correcting codes as shown in Fig. 1.7(b). As a simple example, consider a repetition code where each information bit is transmitted three times:

$$0 \rightarrow 000$$

$$1 \rightarrow 111.$$

If we suppose that the decoder makes a decision on the information bit by

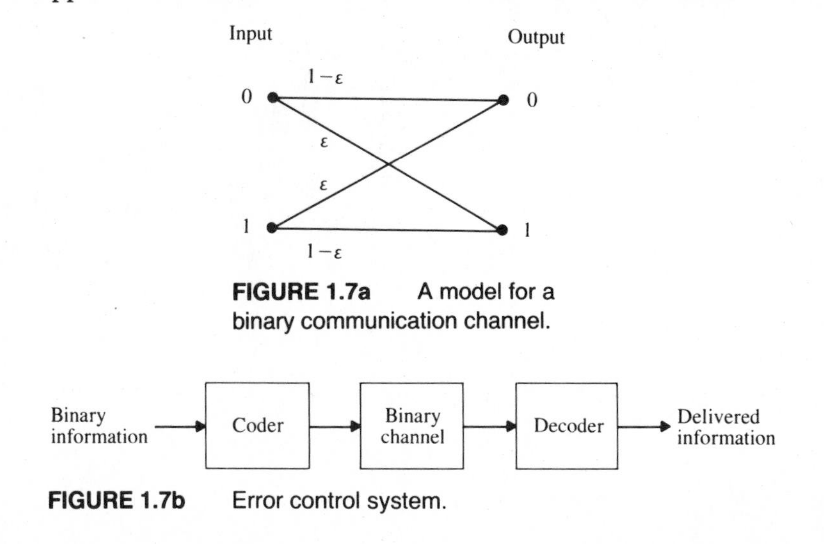


**FIGURE 1.7a** A model for a binary communication channel.



**FIGURE 1.7b** Error control system.

taking a majority vote of the three bits output by the receiver, then the decoder will make the wrong decision only if two or three of the bits are in error. In Example 2.37, we show that this occurs with probability $3\varepsilon^2 + \varepsilon^3$. Thus if the bit error rate of the channel without coding is $10^{-3}$, then the delivered bit error with the above simple code will be $3 \times 10^{-6}$, a reduction of three orders of magnitude! This improvement is obtained at a cost however: The rate of transmission of information has been slowed down to 1 bit every $3T$ seconds. By going to longer, more complicated codes, it is possible to obtain reductions in error rate without the drastic reduction in transmission rate of this simple example.

### Processing of Random Signals

The outcome of a random experiment need not be a single number, but can also be an entire function of time. For example, the outcome of an experiment could be a voltage waveform corresponding to speech or music. In these situations we are interested in the properties of a signal and of processed versions of the signal.

As a specific example, suppose we are given an observed voltage waveform, $Y(t)$, which is the sum of a voltage waveform of interest, $S(t)$ (signal), and an unwanted voltage waveform, $N(t)$ (noise). For example in

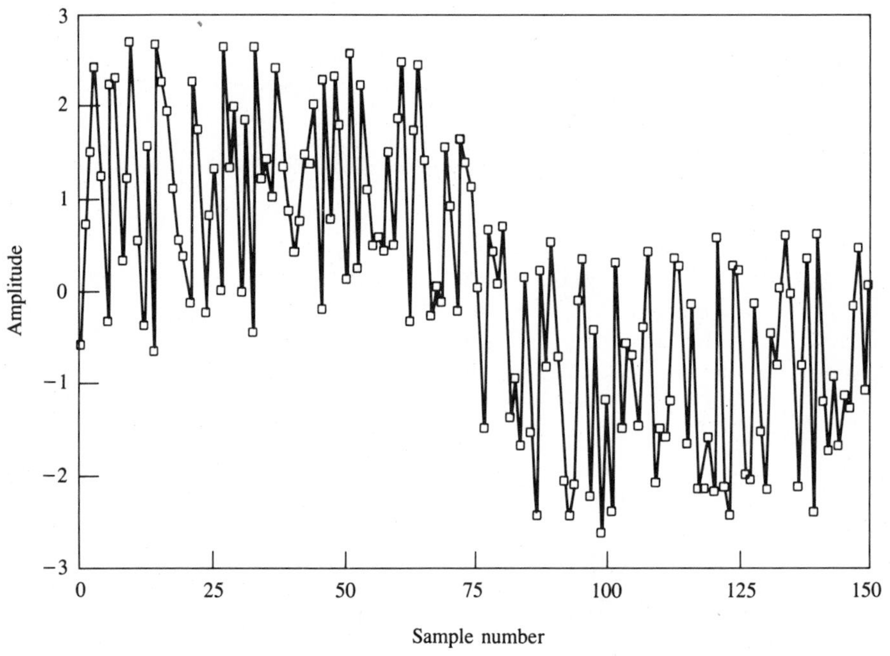


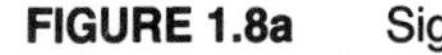
**FIGURE 1.8a** Signal plus noise.

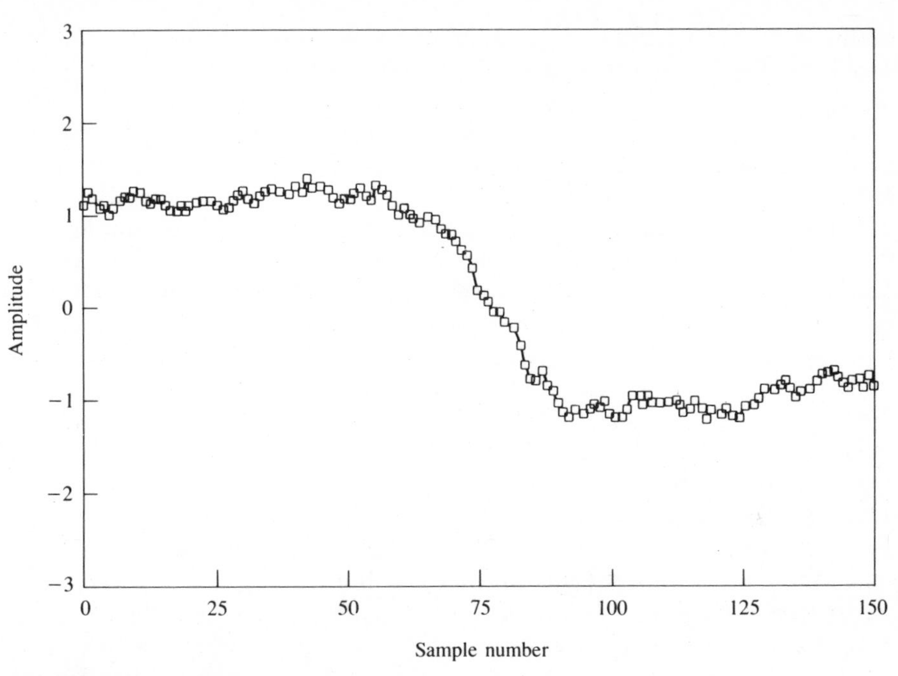


**FIGURE 1.8b** Filtered signal plus noise.

magnetic recording, the signal may be a voltage waveform corresponding to music, and the noise may be the hiss inherent in the magnetic medium. The measure of quality in these types of systems is the *signal-to-noise ratio* (SNR), which is defined as the ratio of the average power in the signal and the average power in the noise. The quality of the observed signal improves as the SNR increases, because the noise then produces smaller perturbations about the desired signal.

Figure 1.8(a) depicts an example in which the observed waveform is a square wave signal plus a noise waveform. We will see in Chapter 7 that the observed waveform can be "filtered" to produce a waveform that has higher SNR. Figure 1.8(b) shows the results of filtering the waveform in Fig. 1.8(a). Clearly, filtering works.

### Resource Sharing Systems

In many applications, expensive resources such as computers and communication lines are subject to unsteady and random demand. Users intersperse demands for short periods of service between long idle periods. The demands of the users can be met by dedicating sufficient resources to each individual user. This approach however is extremely wasteful because these dedicated resources go unused when the user is idle. The challenge to the designer is to configure the system in an efficient and economic way, where the demands of the users are met through the dynamic sharing of resources.

Multi-user computer systems are an example of a resource sharing system. An entire computer system can be viewed as a single resource. If only a single user is allowed to use the system, the system state alternates between periods where the computer is idle waiting for a command from the user, and periods during which the computer is active and the user is waiting for the response to his command. In typical applications, the computer spends a very large proportion of the time in the idle state. For this reason, computers are usually set up to be shared among a group of users. A computer system could be made to handle a number of users by providing a queue (line) in which commands wait service from the computer as shown in Fig. 1.9. At any given time instant, a number of users are in the process of preparing a command (job), and the rest are awaiting a response from the system.

The performance measures of interest here are the average response time that elapses from the instant a user submits a job to the instant when the response is received, and the average rate at which the computer completes jobs (throughput). These measures can be predicted using the queueing model discussed in Section 9.5. Figures 1.10(a) and 1.10(b) show the average response time and the average throughput as the number of users in the system is increased. The results are as expected: As the number of users increases, the computer system is busy more of the time, and hence completes more jobs per second, but the line of jobs waiting for service increases as well, resulting in higher average response times and increased user dissatisfaction.

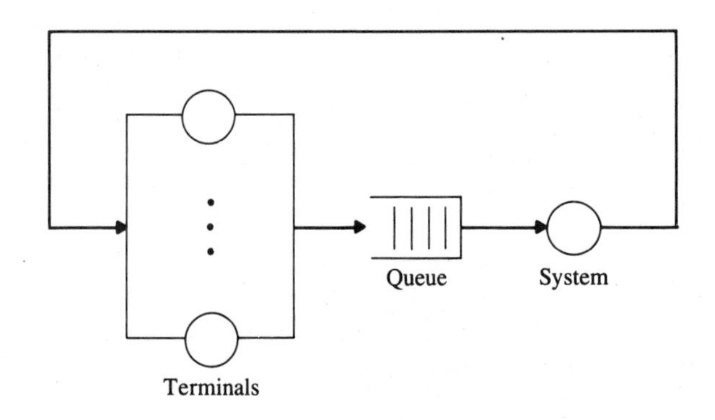


**FIGURE 1.9** Simple model for a multi-user computer system.

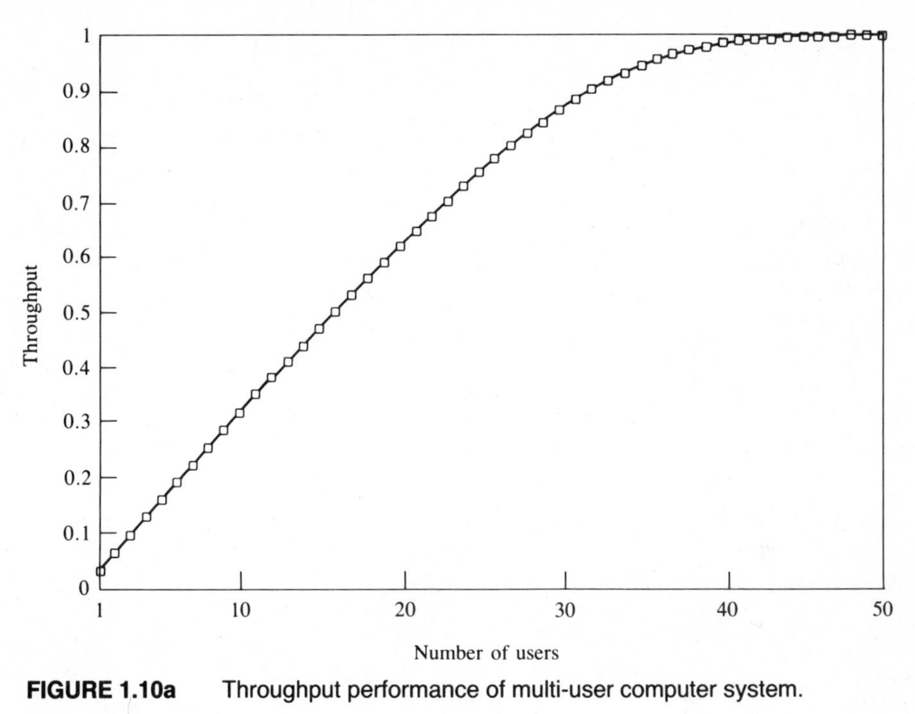


**FIGURE 1.10a** Throughput performance of multi-user computer system.

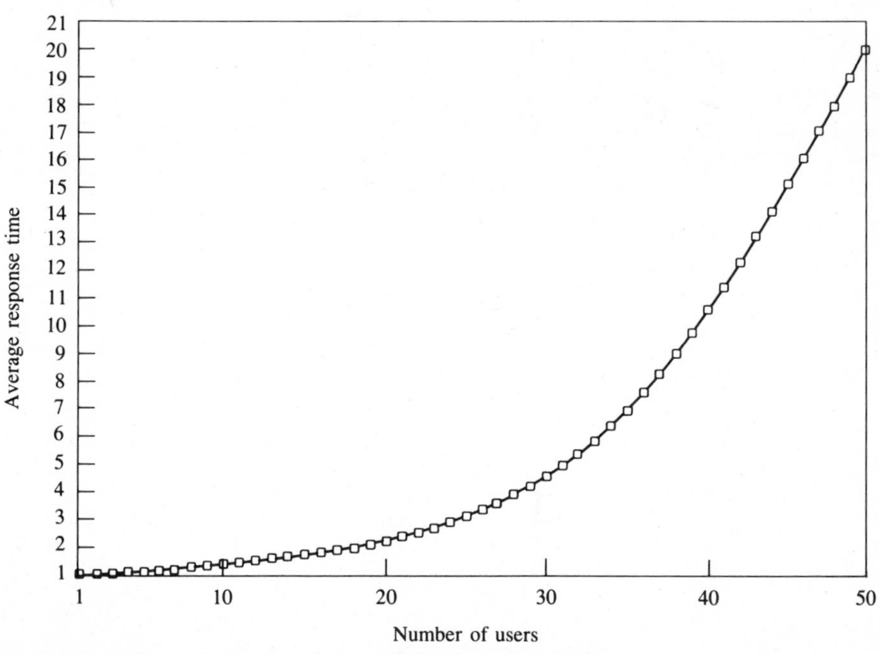


**FIGURE 1.10b** Response time performance of multi-user computer system.

### Reliability of Systems

Reliability is a major concern in the design of modern systems. A prime example of this is the system of computers and communication networks that support the electronic transfer of funds between banks. It is of critical importance that this system continues operating even in the face of subsystem failures. The key question is, How does one build reliable systems from unreliable components? Probability models provide us with the tools to address this question in a quantitative way.

The operation of a system requires the operation of some or all of its components. For example, Fig. 1.11(a) shows a system that functions only when all of its components are functioning, and Fig. 1.11(b) shows a system that functions as long as at least one of its components is functioning. More complex systems can be obtained as combinations of these two basic configurations.

We all know from experience that it is not possible to predict exactly when a component will fail. Probability theory allows us to evaluate measures of reliability such as the average *time to failure* and the probability that a component is still functioning after a certain time has elapsed. Furthermore, we will see in Chapters 2 and 3 that probability theory enables us to determine these averages and probabilities for an entire system in terms of the probabilities and averages of its components.

(a) Series configuration of components.

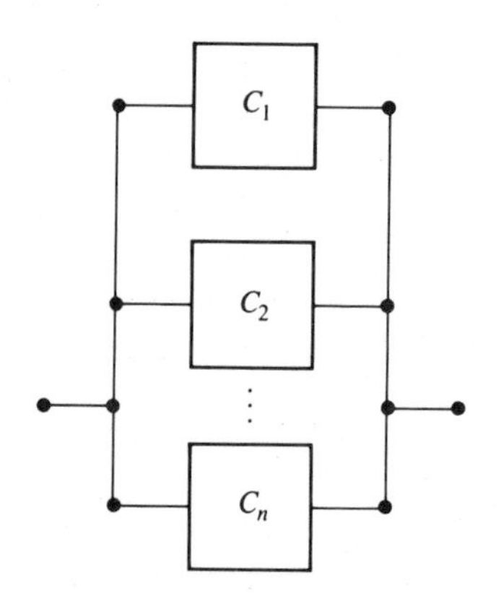


(b) Parallel configuration of components.

**FIGURE 1.11** Systems with $n$ components.

This allows us to evaluate system configurations in terms of their reliability, and thus to select system designs that are reliable.

## 1.6 OVERVIEW OF BOOK

In this chapter we have discussed the important role that probability models play in the design of systems that involve randomness. *The principal objective of this book is to introduce the student to the basic concepts of probability theory that are required to understand probability models that are used in electrical and computer engineering.* The book is not intended to cover applications per se; there are far too many applications, with each one requiring its own detailed discussion. On the other hand, we do attempt to keep the examples relevant to the intended audience by drawing from relevant application areas.

*Another objective of the book is to present some of the basic techniques required to develop probability models.* The discussion in this chapter has made it clear that the probabilities used in a model must be determined experimentally. *Statistical techniques* are required to do this, so we have included a discussion of a few of the basic but essential statistical techniques. We have also alluded to the usefulness of *computer simulation models* in validating probability models. Most chapters include a section that presents some useful computer method. These sections are marked by an asterisk, and can be skipped without loss of continuity. However, the student is encouraged to explore these techniques. They are fun to play with, and they will provide insight into the nature of randomness.

The remainder of the book is organized as follows:

- Chapter 2 presents the basic concepts of probability theory. We begin with the axioms of probability that were stated in Section 1.3, and we discuss their implications. Several basic probability models are introduced in Chapter 2.

- In general, probability theory does not require that the outcomes of random experiments be numbers. Thus the outcomes can be objects (e.g., black or white balls) or conditions (e.g., computer system up or down). However, we are usually interested in experiments where the outcomes are numbers. The notion of a random variable addresses this situation. Chapters 3 and 4 discuss experiments where the outcome is a single number, and a vector of numbers, respectively. In these two chapters we develop several extremely useful problem-solving techniques.

- Chapter 5 presents mathematical results (limit theorems) that answer the question of what happens in a very long sequence of independent repetitions of an experiment. The results presented will justify our extensive use of relative frequency to motivate the notion of probability.

■ Chapter 6 introduces the notion of a random or stochastic process, which is simply an experiment in which the outcome is a function of time.

■ Chapter 7 introduces the notion of the power spectral density and its use in the analysis and processing of random signals.

■ Chapter 8 discusses Markov chains, which are random processes that allow us to model sequences of nonindependent experiments.

■ Chapter 9 presents an introduction to queueing theory and various applications.

## SUMMARY

■ Mathematical models relate important system parameters and variables using mathematical relations. They allow system designers to predict system performance by using equations when experimentation is not feasible or too costly.

■ Computer simulation models are an alternative means of predicting system performance. They can be used to validate mathematical models.

■ In deterministic models the conditions under which an experiment is performed determine the *exact* outcome. The equations in deterministic models predict an exact outcome.

■ In probability models the conditions under which a random experiment is performed determine the *probabilities* of the possible outcomes. The solution of the equations in probability models yields the probabilities of outcomes and events as well as various types of averages.

■ The probabilities and averages for a random experiment can be found experimentally by computing relative frequencies and sample averages in a large number of repetitions of a random experiment.

■ The performance measures in many systems of practical interest involve relative frequencies and long-term averages. Probability models are used in the design of these systems.

## CHECKLIST OF IMPORTANT TERMS

Deterministic model
Event
Expected value
Probability
Probability model
Random experiment
Relative frequency
Sample mean
Sample space
Statistical regularity

## ANNOTATED REFERENCES

Section 1.8 in reference [1] discusses the process of modelling in the context of circuit theory. References [2] through [7] discuss probability models in an engineering context. References [8] and [9] are classic works, and they contain excellent discussions on the foundations of probability models. References [10] and [11] are introductions to error control. Reference [12] discusses random signal analysis in the context of communication systems, and references [13] and [14] discuss various aspects of random signal analysis. References [15] through [17] are introductions to performance aspects of computer communications.


1. M. E. Van Valkenburg, *Network Analysis,* Prentice-Hall, Englewood Cliffs, N.J., 1974.
2. L. Breiman, *Probability and Stochastic Processes: With A View Toward Applications,* Houghton Mifflin, Boston, 1969.
3. P. L. Meyer, *Introductory Probability and Statistical Applications,* Addison-Wesley, Reading, Mass., 1970.
4. W. B. Davenport, *Probability and Random Processes: An Introduction for Applied Scientists and Engineers,* McGraw-Hill, New York, 1970.
5. A. Papoulis, *Probability, Random Variables, and Stochastic Processes,* McGraw-Hill, New York, 1965.
6. A. B. Clarke and R. L. Disney, *Probability and Random Processes: A First Course with Applications,* Wiley, New York, 1985.
7. C. W. Helstrom, *Probability and Stochastic Processes for Engineers,* Macmillan, New York, 1984.
8. H. Cramer, *Mathematical Methods of Statistics,* Princeton University Press, Princeton, N.J., 1946.
9. W. Feller, *An Introduction to Probability Theory and Its Applications,* Wiley, New York, 1968.
10. G. C. Clark and J. B. Cain, *Error-Correction Coding for Digital Communications,* Plenum Press, New York, 1981.
11. S. Lin and R. Costello, *Error Control Coding: Fundamentals and Applications,* Prentice-Hall, Englewood Cliffs, N.J., 1983.
12. S. Haykin, *Communication Systems,* Wiley, New York, 1983.
13. N. Jayant and P. Noll, *Digital Coding of Waveforms,* Prentice-Hall, Englewood Cliffs, N.J., 1984.
14. A. V. Oppenheim and R. W. Schafer, *Digital Signal Processing,* Prentice-Hall, Englewood Cliffs, N.J., 1975.
15. M. Schwartz, *Telecommunication Networks: Protocols, Modeling, and Analysis,* Addison-Wesley, Reading, Mass., 1987.

16. D. Bertsekas and R. G. Gallager, *Data Networks,* Prentice-Hall, Englewood Cliffs, N.J., 1987.
17. J. F. Hayes, *Modeling and Analysis of Communications Networks,* Plenum Press, New York, 1984.

## PROBLEMS

1. A random experiment consists of selecting two balls in succession from an urn containing two black balls and and one white ball.
   a. Specify the sample space for this experiment.
   b. Suppose that the experiment is modified so that the ball is immediately put back into the urn after the first selection. What is the sample space now?
   c. What is the relative frequency of the outcome (white, white) in a large number of repetitions of the experiment in Part a? In Part b?
   d. Does the outcome of the second draw from the urn depend in any way on the outcome in the first draw in either of these experiments?
2. Let $A$ be an event associated with outcomes of a random experiment, and let the event $B$ be defined as "event $A$ does not occur." Show that $f_B(n) = 1 - f_A(n)$.
3. Let $A$, $B$, and $C$ be events that cannot occur simultaneously as pairs or triplets, and let $D$ be the event "$A$ or $B$ or $C$ occurs." Show that

$$f_D(n) = f_A(n) + f_B(n) + f_C(n).$$

4. Consider a random experiment with sample space $S = \{1, 2, \ldots, N\}$. Suppose we are given the following probability assignment:

$$P[\text{outcome is } k] = \frac{1}{N} \qquad \text{for } k = 1, \ldots, N.$$

   Verify that the three axioms of probability are satisfied.
5. The *sample mean* for a series of numerical outcomes $X(1)$, $X(2), \ldots, X(n)$ of a sequence of random experiments is defined by

$$\langle X \rangle_n = \frac{1}{n} \sum_{j=1}^{n} X(j).$$

   Show that the sample mean satisfies the recursion formula:

$$\langle X \rangle_n = \langle X \rangle_{n-1} + \frac{X(n) - \langle X \rangle_{n-1}}{n}, \qquad \langle X \rangle_0 = 0.$$

6. The *sample mean-squared value* of the numerical outcomes $X(1), X(2), \ldots, X(n)$ of a series of $n$ repetitions of an experiment is defined by

$$\langle X^2 \rangle_n = \frac{1}{n} \sum_{j=1}^{n} X^2(j).$$

   a. What would you expect this expression to converge to as the number of repetitions $n$ becomes very large?
   b. Find a recursion formula for $\langle X^2 \rangle_n$ similar to the one found in Problem 5.

7. The *sample variance* is defined as the mean-squared value of the variation of the samples about the sample mean:

$$\langle V^2 \rangle_n = \frac{1}{n} \sum_{j=1}^{n} \{X(j) - \langle X \rangle_n\}^2.$$

   Note that the $\langle X \rangle_n$ also depends on the sample values. (It is customary to replace the $n$ in the denominator by $n - 1$ for technical reasons that will be discussed in Chapter 5. For now we will use the above definition.)
   a. Show that the sample variance is also given by the following expression:

$$\langle V^2 \rangle_n = \langle X^2 \rangle_n - \langle X \rangle_n^2.$$

   b. Show that the sample variance satisfies the following recursion formula:

$$\langle V^2 \rangle_n = \left(1 - \frac{1}{n}\right)\langle V^2 \rangle_{n-1} + \frac{1}{n}\left(1 - \frac{1}{n}\right)(X(n) - \langle X \rangle_n)^2,$$

   with $\langle V^2 \rangle_0 = 0$.

8. The following data is obtained by sampling a voltage waveform: 7, 3, −9, 4, 7, −2, −8, 4, 3, 4, −5, 5, 4, 1, −6, 3, −7, 1, −9, 0.
   a. Find the relative frequency of the event "voltage is positive."
   b. Find the sample mean and variance of the data.
   c. Sketch the empirical distribution function, which is defined by

$$F(x) = \frac{\text{number of outcomes less than } x}{\text{total number of outcomes}}, \qquad -\infty < x < \infty.$$

9. Consider the following data for job interarrival times (in milliseconds) to a computer system: 14, 3, 11, 4, 12, 10, 2, 3, 7, 8, 14,

a. Find the sample mean and variance of the data.
b. Find the relative frequency of the event "the interarrival time is greater than 10 ms."
c. Sketch the empirical distribution function defined in the previous problem.

CHAPTER 2

# Basic Concepts of Probability Theory

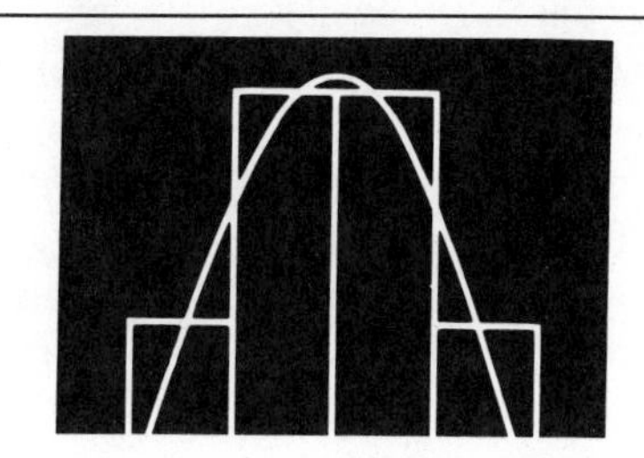

This chapter presents the basic concepts of probability theory. In the remainder of the book, we will usually be further developing or elaborating the basic concepts presented here. You will be well-prepared to deal with the rest of the book if you have a good understanding of these basic concepts when you complete the chapter.

These are the concepts that will be developed. First, set theory is used to specify the sample space and the events of a random experiment. Second, the axioms of probability specify rules for computing the probabilities of events. Third, the notion of conditional probability allows us to determine how partial information about the outcome of an experiment affects the probabilities of events. Conditional probability also allows us to formulate the notion of "independence" of events and of experiments. Finally, we consider "sequential" random experiments that consist of performing a sequence of simple random subexperiments. We show how the probabilities of events in these experiments can be derived from the probabilities of the simpler subexperiments. Throughout the book it is shown that complex random experiments can be analyzed by decomposing them into simple subexperiments.

## 2.1 SPECIFYING RANDOM EXPERIMENTS

A random experiment is an experiment in which the outcome varies in an unpredictable fashion when the experiment is repeated under the same conditions. *A random experiment is specified by stating an experimental procedure and a set of one or more measurements or observations.*

### Example 2.1

*Experiment $E_1$:* Select a ball from an urn containing balls numbered 1 to 50. Note the number of the ball.

*Experiment $E_2$:* Select a ball from an urn containing balls numbered 1 to 4. Suppose that balls 1 and 2 are black and that balls 3 and 4 are white. Note the number and color of the ball.

*Experiment $E_3$:* Toss a coin three times and note the sequence of heads and tails.

*Experiment $E_4$:* Toss a coin three times and note the number of heads.

*Experiment $E_5$:* Count the number of voice packets containing only silence produced from a group of $N$ speakers in a 10 ms period.

*Experiment $E_6$:* A block of information is transmitted repeatedly over a noisy channel until an error-free block arrives at the receiver. Count the number of transmissions required.

*Experiment* $E_7$*:* Pick a number at random between zero and one.

*Experiment* $E_8$*:* Measure the time between two message arrivals at a message center.

*Experiment* $E_9$*:* Measure the lifetime of a given computer memory chip in a specified environment.

*Experiment* $E_{10}$*:* Determine the value of a voltage waveform at time $t_1$.

*Experiment* $E_{11}$*:* Determine the values of a voltage waveform at times $t_1$ and $t_2$.

*Experiment* $E_{12}$*:* Pick two numbers at random between zero and one.

*Experiment* $E_{13}$*:* Pick a number $X$ at random between zero and one, then pick a number $Y$ at random between $X$ and one.

*Experiment* $E_{14}$*:* A system component is installed at time $t = 0$. For $t \geq 0$ let $X(t) = 1$ as long as the component is functioning, and let $X(t) = 0$ after the component fails. ■■

---

The specification of a random experiment must include an unambiguous statement of exactly what is measured or observed. For example, random experiments may consist of the same procedure but differ in the observations made, as illustrated by $E_3$ and $E_4$.

A random experiment may involve more than one measurement or observation, as illustrated by $E_2$, $E_3$, $E_{11}$, $E_{12}$, and $E_{13}$. A random experiment may even involve a continuum of measurements, as shown by $E_{14}$.

Experiments $E_3$, $E_4$, $E_5$, $E_6$, $E_{12}$, and $E_{13}$ are examples of sequential experiments that can be viewed as consisting of a sequence of simple subexperiments. Can you identify the subexperiments in each of these? Note that in $E_{13}$ the second subexperiment depends on the outcome of the first subexperiment.

### The Sample Space

Since random experiments do not consistently yield the same result, it is necessary to determine what the set of possible results can be. We define an **outcome** or **sample point** of a random experiment as a result that is "finest grain" in the sense that it cannot be decomposed into other results. When we perform a random experiment, one and only one outcome occurs. Thus outcomes are mutually exclusive in the sense that they cannot occur simultaneously. The **sample space** $S$ of a random experiment is defined as the set of all possible outcomes.

We will denote an outcome of an experiment by $\zeta$, where $\zeta$ is an element or point in $S$. Each performance of a random experiment can then be viewed as the selection at random of a single point (outcome) from $S$.

The sample space $S$ can be specified compactly by using set notation. It can be visualized by drawing tables, diagrams, intervals of the real line, or regions of the plane.

### ■■ Example 2.2

The sample spaces corresponding to the experiments in Example 2.1 are given below using set notation:

$S_1 = \{1, 2, \ldots, 50\}$

$S_2 = \{(1, b), (2, b), (3, w), (4, w)\}$

$S_3 = \{\text{HHH, HHT, HTH, THH, TTH, THT, HTT, TTT}\}$

$S_4 = \{0, 1, 2, 3\}$

$S_5 = \{0, 1, 2, \ldots, N\}$

$S_6 = \{1, 2, 3, \ldots\}$

$S_7 = \{x : 0 \leq x \leq 1\} = [0, 1]$ See Fig. 2.1(a).

$S_8 = \{t : t \geq 0\} = [0, \infty)$

$S_9 = \{t : t \geq 0\} = [0, \infty)$ See Fig. 2.1(b).

$S_{10} = \{v : -\infty < v < \infty\} = (-\infty, \infty)$

$S_{11} = \{(v_1, v_2) : -\infty < v_1 < \infty \text{ and } -\infty < v_2 < \infty\}$

$S_{12} = \{(x, y) : 0 \leq x \leq 1 \text{ and } 0 \leq y \leq 1\}$ See Fig. 2.1(c).

$S_{13} = \{(x, y) : 0 \leq y \leq x \leq 1\}$ See Fig. 2.1(d).

$S_{14}$ = set of functions $X(t)$ for which $X(t) = 1$ for $0 \leq t < t_0$ and $X(t) = 0$ for $t \geq t_0$, where $t_0 > 0$ is the time when the component fails. ■■

---

Random experiments involving the same experimental procedure may have different sample spaces as shown by Experiments $E_3$ and $E_4$. Thus the purpose of an experiment affects the choice of sample space.

There are three possibilities for the number of outcomes in a sample space. A sample space can be finite, countably infinite, or uncountably infinite. We will call $S$ a **discrete sample space** if $S$ is countable; that is, its outcomes can be put into one-to-one correspondence with the positive integers. We will call $S$ a **continuous sample space** if $S$ is not countable. Experiments $E_1$, $E_2$, $E_3$, $E_4$, and $E_5$ have finite discrete sample spaces.

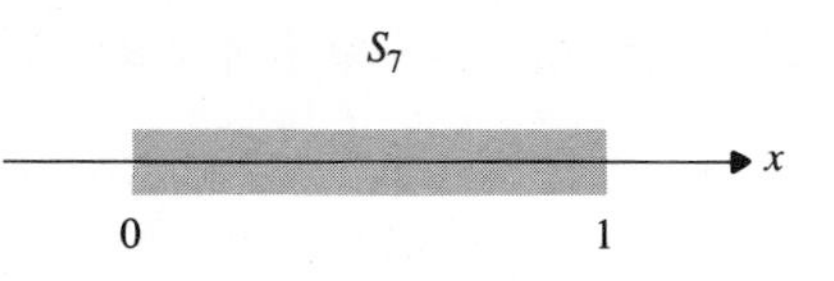


(a) Sample space for Experiment 7.

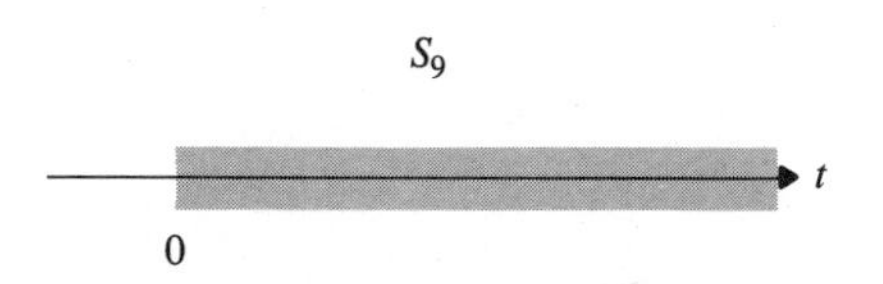


(b) Sample space for Experiment 9.

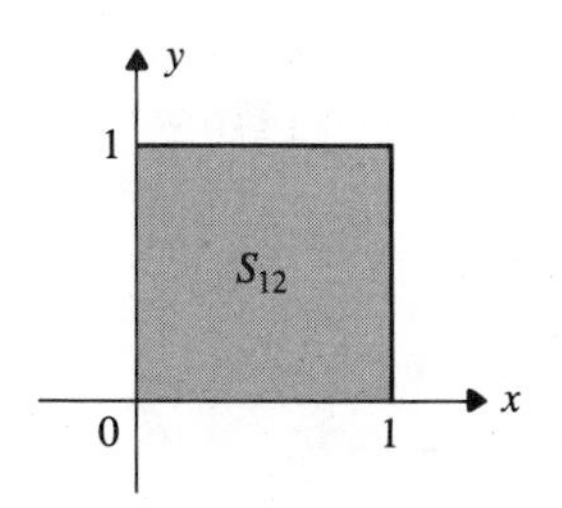


(c) Sample space for Experiment 12.

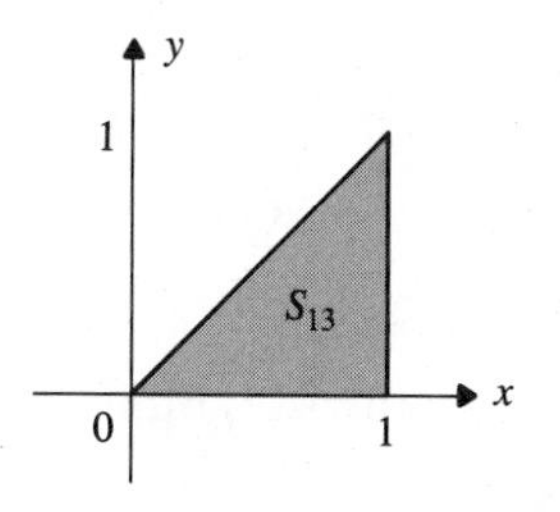


(d) Sample space for Experiment 13.

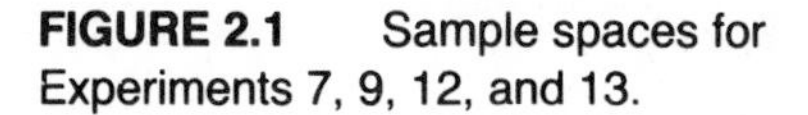

**FIGURE 2.1** Sample spaces for Experiments 7, 9, 12, and 13.

Experiment $E_6$ has a countably infinite discrete sample space. Experiments $E_7$ through $E_{13}$ have continuous sample spaces.

Since an outcome of an experiment can consist of one or more observations or measurements, the sample space $S$ can be multidimensional. For example, the outcomes in Experiments $E_2$, $E_{11}$, $E_{12}$, and $E_{13}$ are two-dimensional, and those in Experiment $E_3$ are three-dimensional. In some instances, the sample space can be written as the Cartesian product of other sets.[1] For example, $S_{11} = R \times R$, where $R$ is the real line, and $S_3 = S \times S \times S$, where $S = \{\text{H}, \text{T}\}$.

It is sometimes convenient to let the sample space include outcomes that are impossible. For example, in Experiment $E_9$ it is convenient to define the sample space as the positive real line, even though a device cannot have an infinite lifetime.

### Events

We are usually not interested in the occurrence of specific outcomes, but rather on the occurrence of some event (i.e., whether the outcome satisfies certain conditions). For example, in Experiment $E_{10}$, which involves the measurement of a voltage, we might be interested in the event "voltage is negative." The conditions of interest define a subset of the sample space, namely the set of points $\zeta$ from $S$ that satisfy the given conditions. For example, "voltage is negative" corresponds to the set $\{\zeta: -\infty < \zeta < 0\}$. The event occurs if and only if the outcome of the experiment $\zeta$ is in this subset. For this reason we define an **event** as a subset of $S$.

Two events of special interest are the **certain event**, $S$, which consists of all outcomes and hence always occurs, and the **impossible** or **null event**, $\varnothing$, which contains no outcomes and hence never occurs.

### ■■ Example 2.3

In the following examples, $A_k$ refers to an event corresponding to experiment $E_k$ in Example 2.1.

$E_1$: "An even-numbered ball is selected," $A_1 = \{2, 4, \ldots, 48, 50\}$.

$E_2$: "The ball is white and even numbered," $A_2 = \{(4, w)\}$.

$E_3$: "The three tosses give the same outcome," $A_3 = \{\text{HHH}, \text{TTT}\}$.

$E_4$: "The number of heads equals the number of tails," $A_4 = \varnothing$.

$E_5$: "No active packets are produced," $A_5 = \{0\}$.

$E_6$: "Fewer than 10 transmissions are required," $A_6 = \{1, \ldots, 9\}$.

$E_7$: "The number selected is nonnegative," $A_7 = S_7$.

$E_8$: "Less than $t_0$ seconds elapse between message arrivals," $A_8 = \{t: 0 < t < t_0\} = [0, t_0)$.

$E_9$: "The chip lasts more than 1000 hours but less than 1500 hours," $A_9 = \{t: 1000 < t < 1500\} = (1000, 1500)$.

1. The Cartesian product of the sets $A$ and $B$ consists of the set of all ordered pairs $(a, b)$, where the first element is taken from $A$ and the second from $B$.

$E_{10}$: "The absolute value of the voltage is less than 1 volt," $A_{10} = \{v: -1 < v < 1\} = (-1, 1)$.

$E_{11}$: "The two voltages have opposite polarities," $A_{11} = \{(v_1, v_2):(v_1 < 0$ and $v_2 > 0)$ or $(v_1 > 0$ and $v_2 < 0)\}$.

$E_{12}$: "The two numbers differ by less than 1/10," $A_{12} = \{(x, y):(x, y)$ in $S_{12}$ and $|x - y| < 1/10\}$.

$E_{13}$: "The two numbers differ by less than 1/10," $A_{13} = \{(x, y):(x, y)$ in $S_{13}$ and $|x - y| < 1/10\}$.

$E_{14}$: "The system is functioning at time $t_1$," $A_{14}$ = subset of $S_{14}$ for which $X(t_1) = 1$. ■■

---

An event may consist of a single outcome, as in $A_2$ and $A_5$. An event from a *discrete* sample space that consists of a single outcome is called an **elementary event**. Events $A_2$ and $A_5$ are elementary events. An event may also consist of the entire sample space, as in $A_7$. The null event, $\varnothing$, arises when none of the outcomes satisfy the conditions that specify a given event, as in $A_4$.

### Set Operations

We can combine events using **set operations** to obtain other events. We can also express complicated events as combinations of simple events.

The **union** of two events $A$ and $B$ is denoted by $A \cup B$ and is defined as the set of outcomes that are either in $A$ or in $B$, or both. The event $A \cup B$ occurs if either $A$, or $B$, or both $A$ and $B$ occur.

The **intersection** of two events $A$ and $B$ is denoted by $A \cap B$ and is defined as the set of outcomes that are in both $A$ and $B$. The event $A \cap B$ occurs when both $A$ and $B$ occur. Two events are said to be **mutually exclusive** if their intersection is the null event, $A \cap B = \varnothing$. Mutually exclusive events cannot occur simultaneously.

The **complement** of an event $A$ is denoted by $A^c$ and is defined as the set of all outcomes not in $A$. The event $A^c$ occurs when the event $A$ does not occur and vice versa.

Figures 2.2(a), 2.2(b), and 2.2(c) show the basic set operations using Venn diagrams. In these diagrams the rectangle represents the sample space $S$, and the shaded regions represent the various events. Figure 2.2(d) shows two mutually exclusive events.

If an event $A$ is a subset of an event $B$, that is $A \subset B$, then event $B$ will occur whenever event $A$ occurs because all the outcomes in $A$ are also in $B$ (see Fig. 2.2e). For this reason we say that event $A$ **implies** event $B$. It should be clear to you from the diagrams in Figs. 2.2(a) and 2.2(b) that $A \cap B$ implies both $A$ and $B$, and that $A$ and $B$ each imply $A \cup B$.

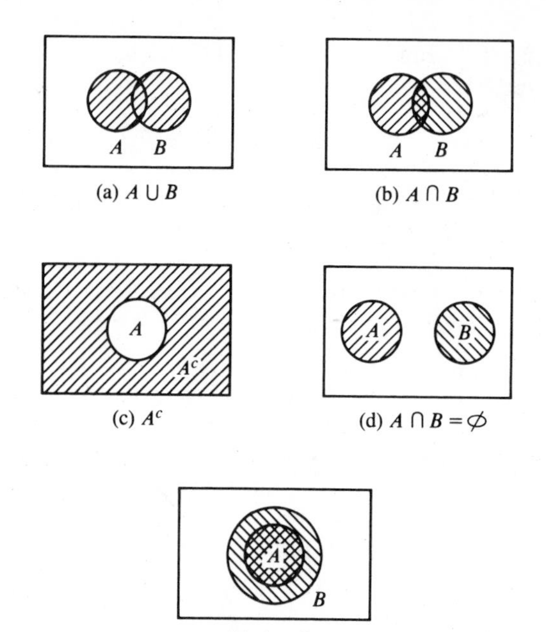


**FIGURE 2.2** Set operations and set relations.

The events $A$ and $B$ are **equal**, $A = B$, if they contain the same outcomes.

The three basic operations can be combined to form other events. The following properties of set operations and their combinations are useful:

*Commutative Properties:*

$$A \cup B = B \cup A \quad \text{and} \quad A \cap B = B \cap A. \tag{2.1}$$

*Associative Properties:*

$$A \cup (B \cup C) = (A \cup B) \cup C \quad \text{and} \quad A \cap (B \cap C) = (A \cap B) \cap C. \tag{2.2}$$

*Distributive Properties:*

$$A \cup (B \cap C) = (A \cup B) \cap (A \cup C) \quad \text{and}$$
$$A \cap (B \cup C) = (A \cap B) \cup (A \cap C). \tag{2.3}$$

*DeMorgan's Rules:*

$$(A \cap B)^c = A^c \cup B^c \quad \text{and} \quad (A \cup B)^c = A^c \cap B^c. \tag{2.4}$$

### ■■ Example 2.4

For Experiment $E_{10}$, let the events $A$, $B$, and $C$ be defined by

$A = \{v : |v| > 10\}$, "magnitude of $v$ is greater than 10 volts,"

$B = \{v : v < -5\}$, "$v$ is less than $-5$ volts," and

$C = \{v : v > 0\}$, "$v$ is positive."

You should then verify that

$$A \cup B = \{v : v < -5 \text{ or } v > 10\},$$
$$A \cap B = \{v : v < -5\},$$
$$C^c = \{v : v \leq 0\},$$
$$(A \cup B) \cap C = \{v : v > 10\},$$
$$A \cap B \cap C = \varnothing, \text{ and}$$
$$(A \cup B)^c = \{v : -5 \leq v \leq 10\}.$$ ■■

### ■■ Example 2.5
System Reliability

Figure 2.3 shows three systems of three components. Figure 2.3(a) is a "series" system in which the system is functioning only if all three components are functioning. Figure 2.3(b) is a "parallel" system in which the system is functioning as long as at least one of the three components is functioning. Figure 2.3(c) is a "two-out-of-three" system in which the system is functioning as long as at least two components are functioning. Let $A_k$ be the event "component $k$ is functioning." For each of the three system configurations, express the event "system is functioning" in terms of the events $A_k$.

The series system is functioning if and only if all components are functioning. Thus the event $D_a$ "system $a$ is functioning" is given by

$$D_a = A_1 \cap A_2 \cap A_3.$$

The parallel system is functioning as long as at least one of the components is functioning, that is, if component 1 or component 2 or component 3 or any combination thereof are functioning. Thus the event $D_b$, "system $b$ is functioning" is given by

$$D_b = A_1 \cup A_2 \cup A_3.$$

Finally, the two-out-of-three system is functioning as long as no more than one component has failed. Thus the event $D_c$; "system $c$ is functioning" is given by

$$D_c = (A_1 \cap A_2 \cap A_3) \cup (A_1^c \cap A_2 \cap A_3) \cup (A_1 \cap A_2^c \cap A_3) \cup (A_1 \cap A_2 \cap A_3^c)$$ ■■

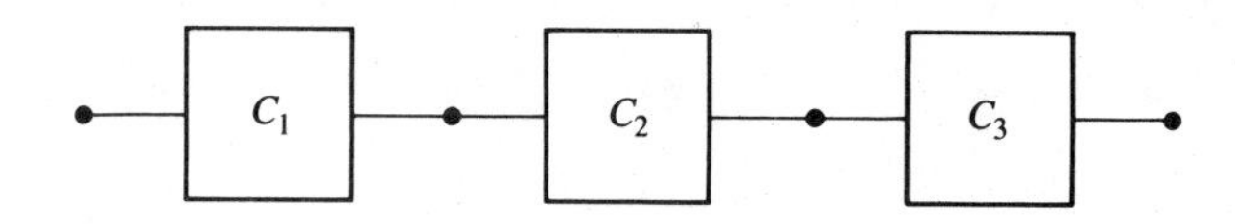


(a) Series system

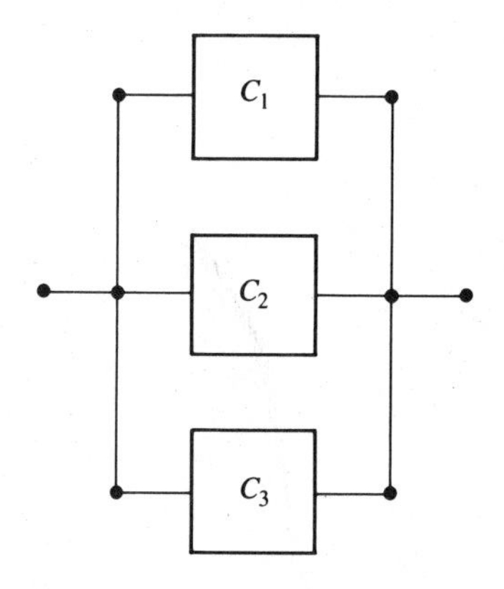


(b) Parallel system

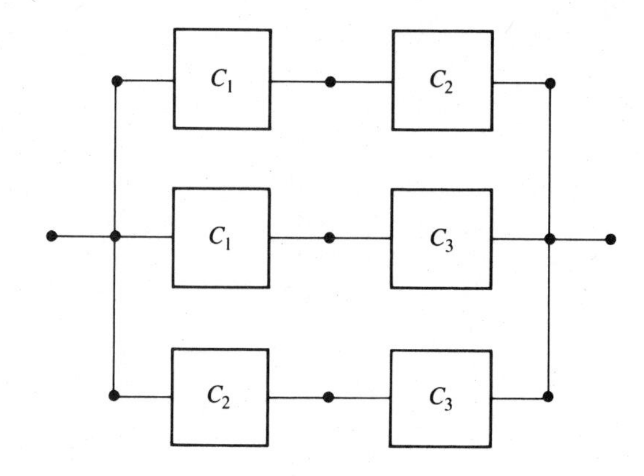


(c) Two-out-of-three system

**FIGURE 2.3** Configurations of a three-component system. When a component fails it is removed from the diagram. The system is functioning as long as there is a path from the right end to the left end.

The union and intersection operations can be repeated for an arbitrary number of events. Thus the event

$$\bigcup_{k=1}^{n} A_k = A_1 \cup A_2 \cup \cdots \cup A_n$$

occurs if one or more of the events $A_k$ occur. The event

$$\bigcap_{k=1}^{n} A_k = A_1 \cap A_2 \cap \cdots \cap A_n$$

occurs when *all* of the events $A_1, \ldots, A_n$ occur. The operations can also be applied to a countably infinite sequence of events. Thus we can also have events of the form

$$\bigcup_{k=1}^{\infty} A_k \quad \text{and} \quad \bigcap_{k=1}^{\infty} A_k.$$

## 2.2 THE AXIOMS OF PROBABILITY

Probabilities are numbers assigned to events that indicate how "likely" it is that the events will occur when an experiment is performed. A **probability law** for a random experiment is a rule that assigns probabilities to the events of the experiment. Thus a probability law is a function that assigns a number to sets (events). In Section 1.3 we found a number of properties of relative frequency that any definition of probability should satisfy. The axioms of probability formally state that a probability law must satisfy these properties. In this section, we develop a number of results that follow from this set of axioms.

Let $E$ be a random experiment with sample space $S$. A *probability law* for the experiment $E$ is a rule that assigns to each event $A$ a number $P[A]$, called the **probability of $A$**, that satisfies the following axioms:

*Axiom I* $\quad 0 \le P[A]$

*Axiom II* $\quad P[S] = 1$

*Axiom III* $\quad$ If $A \cap B = \varnothing$, then $P[A \cup B] = P[A] + P[B]$

*Axiom III′* $\quad$ If $A_1, A_2, \ldots$ is a sequence of events such that $A_i \cap A_j = \varnothing$ for $i \neq j$, then

$$P\left[\bigcup_{k=1}^{\infty} A_k\right] = \sum_{k=1}^{\infty} P[A_k].$$

Axioms I, II, and III are enough to deal with experiments with finite sample spaces. In order to handle experiments with infinite sample spaces,

Axiom III needs to be replaced by Axiom III′. Note that Axiom III′ includes Axiom III as a special case, by letting $A_k = \varnothing$ for $k \geq 3$. Thus we really only need Axioms I, II, and III′. Nevertheless we will gain greater insight by starting with Axioms I, II, and III.

The axioms allow us to view events as objects that possess a property (i.e., their probability) that has attributes similar to physical mass. Axiom I states that the probability (mass) is nonnegative, and Axiom II states that there is a fixed total amount of probability (mass), namely 1 unit. Axiom III states that the total probability (mass) in two disjoint objects is the sum of the individual probabilities (masses).

The axioms provide us with a set of consistency rules that any valid probability assignment must satisfy. We will now develop several properties that stem from the axioms that are useful in the computation of probabilities.

The first result states that if we partition the sample space into two mutually exclusive events, $A$ and $A^c$, then the probabilities of these two events add up to one.

**Corollary 1.** $P[A^c] = 1 - P[A]$
*Proof:* Since an event $A$ and its complement $A^c$ are mutually exclusive, $A \cap A^c = \varnothing$, we have from Axiom III that

$$P[A \cup A^c] = P[A] + P[A^c].$$

Since $S = A \cup A^c$, by Axiom II,

$$1 = P[S] = P[A \cup A^c] = P[A] + P[A^c].$$

The corollary follows after solving for $P[A^c]$.

The next corollary states that the probability of an event is always less than or equal to one. Corollary 2 combined with Axiom I provide good checks in problem solving: If your probabilities are negative or are greater than one, you have made a mistake somewhere!

**Corollary 2.** $P[A] \leq 1$
*Proof:* From Corollary 1,

$$P[A] = 1 - P[A^c] \leq 1,$$

since $P[A^c] \geq 0$.

Corollary 3 states that the impossible event has probability zero.

**Corollary 3.** $P[\varnothing] = 0$
*Proof*: Let $A = S$ and $A^c = \varnothing$ in Corollary 1:

$$P[\varnothing] = 1 - P[S] = 0.$$

Corollary 4 provides us with the standard method for computing the probability of a complicated event $A$. The method involves decomposing the event $A$ into the union of disjoint events $A_1, A_2, \ldots, A_n$. The probability of $A$ is the sum of the probabilities of the $A_k$'s.

**Corollary 4.** If $A_1, A_2, \ldots, A_n$ are pairwise mutually exclusive, then

$$P\left[\bigcup_{k=1}^{n} A_k\right] = \sum_{k=1}^{n} P[A_k] \qquad \text{for } n \geq 2.$$

*Proof:* We use mathematical induction. Axiom III implies that the result if true for $n = 2$. Next we need to show that if the result is true for some $n$, then it is also true for $n + 1$. This combined with the fact that the result is true for $n = 2$, implies that the result is true for $n \geq 2$.

Suppose that the result is true for $n$, that is,

$$P\left[\bigcup_{k=1}^{n} A_k\right] = \sum_{k=1}^{n} P[A_k], \tag{2.5}$$

and consider the $n + 1$ case

$$P\left[\bigcup_{k=1}^{n+1} A_k\right] = P\left[\left\{\bigcup_{k=1}^{n} A_k\right\} \cup A_{n+1}\right] = P\left[\bigcup_{k=1}^{n} A_k\right] + P[A_{n+1}], \tag{2.6}$$

where we have applied Axiom III to to the second expression after noting that the union of events $A_1$ to $A_n$ is mutually exclusive with $A_{n+1}$:

$$\left\{\bigcup_{k=1}^{n} A_k\right\} \cap A_{n+1} = \bigcup_{k=1}^{n} \{A_k \cap A_{n+1}\} = \bigcup_{k=1}^{n} \varnothing = \varnothing.$$

Substitution of Eq. (2.5) into Eq. (2.6) gives the $n + 1$ case

$$P\left[\bigcup_{k=1}^{n+1} A_k\right] = \sum_{k=1}^{n+1} P[A_k].$$

Corollary 5 gives an expression for the union of two events that are not necessarily mutually exclusive.

**Corollary 5.** $P[A \cup B] = P[A] + P[B] - P[A \cap B]$
*Proof:* First we decompose $A \cup B$, $A$, and $B$ as unions of disjoint events. From the Venn diagram in Fig. 2.4,

$$P[A \cup B] = P[A \cap B^c] + P[B \cap A^c] + P[A \cap B]$$

$$P[A] = P[A \cap B^c] + P[A \cap B]$$

$$P[B] = P[B \cap A^c] + P[A \cap B]$$

By substituting $P[A \cap B^c]$ and $P[B \cap A^c]$ from the two lower equations into the top equation, we obtain the corollary.

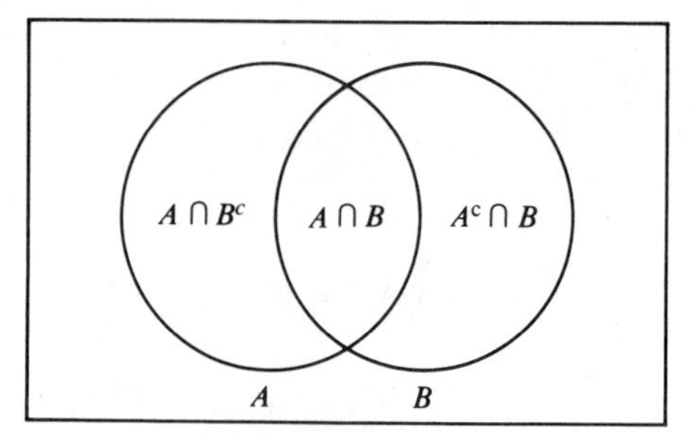


**FIGURE 2.4** Decomposition of $A \cup B$ into three disjoint sets.

By looking at the Venn diagram in Fig. 2.4, you will see that the sum $P[A] + P[B]$ counts the probability (mass) of the set $A \cap B$ twice. The expression in Corollary 5 makes the appropriate correction.

Corollary 5 is easily generalized to three events,

$$P[A \cup B \cup C] = P[A] + P[B] + P[C] - P[A \cap B] - P[A \cap C] - P[B \cap C] + P[A \cap B \cap C], \quad (2.7)$$

and in general to $n$ events, as shown in Corollary 6.

**Corollary 6.**

$$P\left[\bigcup_{k=1}^{n} A_k\right] = \sum_{j=1}^{n} P[A_j] - \sum_{j<k} P[A_j \cap A_k] + \cdots + (-1)^{n+1} P[A_1 \cap \cdots \cap A_n].$$

*Proof* is by induction (see Problems 18 and 19).

Since probabilities are nonnegative, Corollary 5 implies that the probability of the union of two events is no greater than the sum of the individual event probabilities

$$P[A \cup B] \leq P[A] + P[B]. \quad (2.8)$$

The above inequality is a special case of the fact that a subset of another set must have smaller probability. This result is frequently used to obtain upper bounds for probabilities of interest. In the typical situation, we are interested in an event $A$ whose probability is difficult to find; so we find an event $B$ for which the probability can be found and that includes $A$ as a subset.

**Corollary 7.** If $A \subset B$, then $P[A] \leq P[B]$.

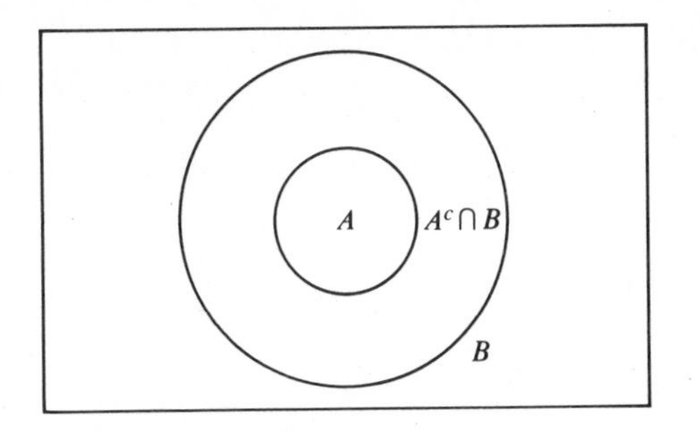


**FIGURE 2.5** If $A \subset B$, then $P(A) \leq P(B)$.

and $A^c \cap B$, thus

$$P[B] = P[A] + P[A^c \cap B] \geq P[A],$$

since $P[A^c \cap B] \geq 0$.

The axioms together with the corollaries provide us with a set of rules for computing the probability of certain events in terms of other events. However, we still need an initial probability assignment for some basic set of events from which the probability of all other events can be computed. This problem is dealt with in the next two subsections.

### Discrete Sample Spaces

In this section we will show that *the probability law for an experiment with a countable sample space can be specified by giving the probabilities of the elementary events.* First, suppose that the sample space is finite, $S = \{a_1, a_2, \ldots, a_n\}$. All distinct elementary events are mutually exclusive, so by Corollary 4 the probability of any event $B = \{a_{k_1}, a_{k_2}, \ldots, a_{k_m}\}$ is given by

$$\begin{aligned} P[B] &= P[\{a_{k_1}, a_{k_2}, \ldots, a_{k_m}\}] \\ &= P[\{a_{k_1}\}] + P[\{a_{k_2}\}] + \cdots + P[\{a_{k_m}\}], \end{aligned} \tag{2.9}$$

that is, the probability of an event is equal to the sum of the probabilities of the outcomes in the event. If $S$ is countably infinite, then Axiom III′ implies that the probability of an event such as $D = \{b_{k_1}, b_{k_2}, \ldots\}$ is given by

$$P[D] = P[\{b_{k_1}\}] + P[\{b_{k_2}\}] + \cdots. \tag{2.10}$$

Again, the probability of an event is determined from the probabilities of its outcomes. Thus we conclude that the probability law for a random experiment with a discrete sample space is specified by giving the probabilities of the elementary events.

If the sample space has $n$ elements, $S = \{a_1, \ldots, a_n\}$, a probability assignment of particular interest is the case of **equally likely outcomes**. The probability of the elementary events is

$$P[\{a_1\}] = P[\{a_2\}] = \cdots = P[\{a_n\}] = \frac{1}{n}. \tag{2.11}$$

The probability of any event that consists of $k$ outcomes, say $B = \{a_{j_1}, \ldots, a_{j_k}\}$, is

$$P[B] = P[\{a_{j_1}\}] + \cdots + P[\{a_{j_k}\}] = \frac{k}{n}. \tag{2.12}$$

Thus *if outcomes are equally likely, then the probability of an event is equal to the number of outcomes in the event divided by the total number of outcomes in the sample space.* Section 2.3 discusses counting methods that are useful in finding probabilities in experiments that have equally likely outcomes.

### ■■ Example 2.6

An urn contains 10 identical balls numbered $0, 1, \ldots, 9$. A random experiment involves selecting a ball from the urn and noting the number of the ball. Find the probability of the following events:

$A$ = "number of ball selected is odd,"

$B$ = "number of ball selected is a multiple of three,"

$C$ = "number of ball selected is less than 5,"

and of $A \cup B$ and $A \cup B \cup C$.

The sample space is $S = \{0, 1, \ldots, 9\}$, so the sets of outcomes corresponding to the above events are

$$A = \{1, 3, 5, 7, 9\}, \qquad B = \{3, 6, 9\}, \qquad \text{and} \qquad C = \{0, 1, 2, 3, 4\}.$$

If we assume that the outcomes are equally likely, then

$$P[A] = P[\{1\}] + P[\{3\}] + P[\{5\}] + P[\{7\}] + P[\{9\}] = \frac{5}{10}.$$

$$P[B] = P[\{3\}] + P[\{6\}] + P[\{9\}] = \frac{3}{10}.$$

$$P[C] = P[\{0\}] + P[\{1\}] + P[\{2\}] + P[\{3\}] + P[\{4\}] = \frac{5}{10}.$$

From Corollary 5

$$P[A \cup B] = P[A] + P[B] - P[A \cap B] = \frac{5}{10} + \frac{3}{10} - \frac{2}{10} = \frac{6}{10},$$

where we have used the fact that $A \cap B = \{3, 9\}$, so $P[A \cap B] = 2/10$. From Corollary 6

$$\begin{aligned} P[A \cup B \cup C] &= P[A] + P[B] + P[C] - P[A \cap B] \\ &\quad - P[A \cap C] - P[B \cap C] + P[A \cap B \cap C] \\ &= \frac{5}{10} + \frac{3}{10} + \frac{5}{10} - \frac{2}{10} - \frac{2}{10} - \frac{1}{10} + \frac{1}{10} \\ &= \frac{9}{10}. \end{aligned}$$

You should verify the answers for $P[A \cup B]$ and $P[A \cup B \cup C]$ by enumerating the outcomes in the events. ■■

---

Many probability models can be devised for the same sample space and events by varying the probability assignment; in the case of finite sample spaces all we need to do is come up with $n$ nonnegative numbers that add up to one for the probabilities of the elementary events. Of course, in any particular situation, the probability assignment should be selected to reflect experimental observations to the extent possible. The following example shows that situations can arise where there is more than one "reasonable" probability assignment and where experimental evidence is required to decide on the appropriate assignment.

### ■■ Example 2.7

Suppose that a coin is tossed three times. If we observe the sequence of heads and tails, then there are eight possible outcomes $S_3 = \{\text{HHH, HHT, HTH, THH, TTH, THT, HTT, TTT}\}$. If we assume that the outcomes of $S_3$ are equiprobable, then the probability of each of the eight elementary events is 1/8. This probability assignment implies that the probability of obtaining two heads in three tosses is by Corollary 3

$$\begin{aligned} P[\text{“2 heads in 3 tosses”}] &= P[\{\text{HHT, HTH, THH}\}] \\ &= P[\{\text{HHT}\}] + P[\{\text{HTH}\}] + P[\{\text{THH}\}] = \frac{3}{8}. \end{aligned}$$

Now suppose that we toss a coin three times but we count the number of heads in three tosses instead of observing the sequence of heads and tails. The sample space is now $S_4 = \{0, 1, 2, 3\}$. If we assume the outcomes of $S_4$ to be equiprobable, then each of the elementary events of $S_4$ has probability 1/4. This second probability assignment predicts that the probability of obtaining two heads in three tosses is

$$P[\text{“2 heads in 3 tosses”}] = P[\{2\}] = \frac{1}{4}.$$

The first probability assignment implies that the probability of two heads in three tosses is 3/8, and the second probability assignment predicts that the probability is 1/4. Thus the two assignments are not consistent with each other. As far as the theory is concerned, either one of the assignments is acceptable. It is up to us to decide which assignment is more appropriate. Later in the chapter we will see that only the first assignment is consistent with the assumption that the coin is fair and that the tosses are "independent." This assignment correctly predicts the relative frequencies that would be observed in an actual coin tossing experiment. ■■

---

Finally we consider an example with a countably infinite sample space.

### ■■ Example 2.8

A fair coin is tossed repeatedly until the first head shows up; the outcome of the experiment is the number of tosses required until the first head occurs. Find a probability law for this experiment.

It is conceivable that an arbitrarily large number of tosses will be required until a head occurs, so the sample space is $S = \{1, 2, 3, \ldots\}$. Suppose the experiment is repeated $n$ times. Let $N_j$ be the number of trials in which the $j$th toss results in the first head. If $n$ is very large, we expect $N_1$ to be approximately $n/2$ since the coin is fair. This implies that a second toss is necessarily about $n - N_1 \approx n/2$ times, and again we expect that about half of these, that is $n/4$, will result in heads, and so on as

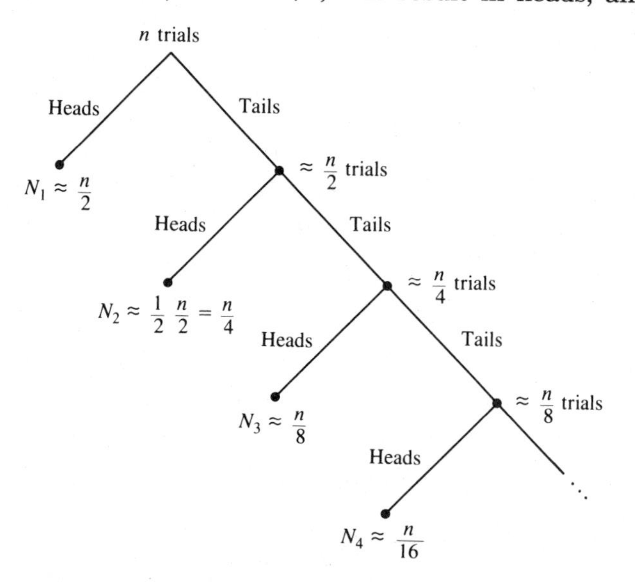


**FIGURE 2.6** In $n$ trials "heads" comes up in the first toss approximately $n/2$ times; in the second toss approximately $n/4$ times; and so on.

shown in Fig. 2.6. Thus for large $n$, the relative frequencies are

$$f_j \approx \frac{N_j}{n} = \left(\frac{1}{2}\right)^j \qquad j = 1, 2, \ldots.$$

We therefore conclude that a reasonable probability law for this experiment is

$$P[j \text{ tosses till first head}] = \left(\frac{1}{2}\right)^j \qquad j = 1, 2, \ldots. \tag{2.13}$$

We can verify that these probabilities add up to one by using the geometric series with $\alpha = 1/2$

$$\sum_{j=1}^{\infty} \alpha^j = \frac{\alpha}{1-\alpha}\bigg|_{\alpha=1/2} = 1.$$

■■

---

### Continuous Sample Spaces

Continuous sample spaces arise in experiments in which the outcomes are numbers. The events of interest in these experiments consist of intervals of the real line, and of complements, unions, and intersections of intervals. For this reason, *probability laws in experiments with continuous sample spaces specify a rule for assigning numbers to intervals of the real line.*

#### ■■ Example 2.9

Consider the random experiment "pick a number $x$ at random between zero and one." The sample space $S$ for this experiment is the unit interval $[0, 1]$, which is uncountably infinite. If we suppose that all the outcomes $S$ are equally likely to be selected, then we would guess that the probability that the outcome is in the interval $[0, 1/2]$ is the same as the probability that the outcome is in the interval $[1/2, 1]$. We would also guess that the probability of the outcome being exactly equal to $1/2$ would be zero since there are an uncountably infinite number of equally likely outcomes.

Consider the following probability law: "The probability that the outcome falls in a subinterval of $S$ is equal to the length of the subinterval," that is,

$$P[[a, b]] = (b - a) \qquad \text{for } 0 \le a \le b \le 1, \tag{2.14}$$

where by $P[[a, b]]$ we mean the probability of the event corresponding to the interval $[a, b]$. Clearly, Axiom I is satisfied since $b \ge a \ge 0$. Axiom II follows from $S = [a, b]$ with $a = 0$ and $b = 1$.

We now show that the probability law is consistent with the previous guesses about the probabilities of the events $[0, 1/2]$, $[1/2, 1]$, and $\{1/2\}$:

$$P[[0, 0.5]] = 0.5 - 0 = .5$$

$$P[[0.5, 1]] = 1 - 0.5 = .5$$

In addition, if $x_0$ is any point in $S$, then $P[[x_0, x_0]] = 0$ since individual points have zero width.

Now suppose that we are interested in an event that is the union of several intervals, for example, "the outcome is at least 0.3 away from the center of the unit interval," that is $A = [0, 0.2] \cup [0.8, 1]$. Since the two intervals are disjoint, we have by Axiom III

$$P[A] = P[[0, 0.2]] + P[[0.8, 1]] = .4.$$

■■

The next example shows that an initial probability assignment that specifies the probability of semi-infinite intervals also suffices to specify the probabilities of all events of interest.

### ■■ Example 2.10

Suppose that the lifetime of a computer memory chip is measured, and we find that "the proportion of chips whose lifetime exceeds $t$ decreases exponentially at a rate $\alpha$." Find an appropriate probability law.

Let the sample space in this experiment be $S = (0, \infty)$. If we interpret the above finding as "the probability that a chip's lifetime exceeds $t$ decreases exponentially at a rate $\alpha$," we then obtain the following assignment of probabilities to events of the form $(t, \infty)$:

$$P[(t, \infty)] = e^{-\alpha t}, \qquad \text{for } t > 0, \tag{2.15}$$

where $\alpha > 0$. Note that the exponential is a number between 0 and 1 for $t > 0$, so Axiom I is satisfied. Axiom II is satisfied since

$$P[S] = P[(0, \infty)] = 1.$$

The probability that the lifetime is in the interval $[r, s]$ is found by noting in Fig. 2.7 that $(r, s] \cup (s, \infty) = (r, \infty)$, thus by Axiom III,

$$P[(r, \infty)] = P[(r, s]] + P[(s, \infty)].$$

By rearranging the above equation we obtain

$$P[(r, s]] = P[(r, \infty)] - P[(s, \infty)] = e^{-\alpha r} - e^{-\alpha s}.$$

We thus obtain the probability of arbitrary intervals in $S$.

■■

In both Example 2.9 and Example 2.10, the probability that the outcome takes on a specific value is zero. You may ask: If an outcome (or event) has probability zero, doesn't that mean that it cannot occur? And

r s

FIGURE 2.7 $(r, \infty) = (r, s] \cup (s, \infty)$.

you may then ask: How can all the outcomes in a sample space have probability zero? We can explain this paradox by using the relative frequency interpretation of probability. An event that occurs only once in an infinite number of trials will have relative frequency zero. Hence the fact that an event or outcome has relative frequency zero, does not imply that it cannot occur, but rather that it occurs *very infrequently*. In the case of continuous sample spaces, the set of possible outcomes is so rich that all outcomes occur infrequently enough that their relative frequencies are zero.

We end this section with an example where the events are regions in the plane.

### ■■ Example 2.11

Consider experiment $E_{12}$ where we picked two numbers $x$ and $y$ at random between zero and one. The sample space is then the unit square shown in Fig. 2.8(a). If we suppose that all pairs of numbers in the unit square are equally likely to be selected, then it is reasonable to use a probability assignment in which the probability of any region $R$ inside the unit square is equal to the area of $R$. Find the probability of the following events: $A = \{x > 0.5\}$, $B = \{y > 0.5\}$, and $C = \{x > y\}$.

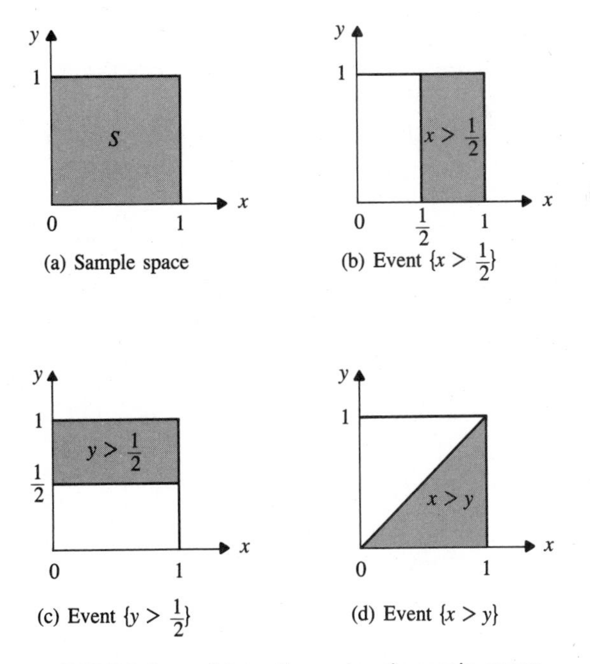


**FIGURE 2.8** A two-dimensional sample space and three events.

Figures 2.8(b) through 2.8(c) show the regions corresponding to the events $A$, $B$, and $C$. Clearly each of these regions has area 1/2. Thus

$$P[A] = \frac{1}{2}, \qquad P[B] = \frac{1}{2}, \qquad P[C] = \frac{1}{2}.$$

■■

## *2.3

## COMPUTING PROBABILITIES USING COUNTING METHODS[2]

In many experiments with finite sample spaces, the outcomes can be assumed to be equiprobable. The probability of an event is then the ratio of the number of outcomes in the event of interest to the total number of outcomes in the sample space (Eq. 2.12). The calculation of probabilities reduces to counting the number outcomes in an event. In this section, we develop several useful counting (combinatorial) formulas.

Suppose that a multiple-choice test has $k$ questions and that for question $i$ the student must select one of $n_i$ possible answers. What is the total number of ways of answering the entire test? The answer to question $i$ can be viewed as specifying the $i$th component of a $k$-tuple, so the above question is equivalent to: How many distinct ordered $k$-tuples $(x_1, \ldots, x_k)$ are possible if $x_i$ is an element from a set with $n_i$ distinct elements?

Consider the $k = 2$ case. If we arrange all possible choices for $x_1$ and for $x_2$ along the sides of a table as shown in Fig. 2.9, we see that there are $n_1 n_2$ distinct ordered pairs. For triplets we could arrange the $n_1 n_2$ possible pairs $(x_1, x_2)$ along the vertical side of the table and the $n_3$ choices for $x_3$ along the horizontal side. Clearly, the number of possible triplets is $n_1 n_2 n_3$.

In general, *the number of distinct ordered $k$-tuples $(x_1, \ldots, x_k)$ with components $x_i$ from a set with $n_i$ distinct elements is*

$$\text{number of distinct ordered } k\text{-tuples} = n_1 n_2 \ldots n_k. \tag{2.16}$$

Many counting problems can be posed as sampling problems where we select "balls" from "urns" or "objects" from "populations." We will now use Eq. (2.16) to develop combinatorial formulas for various types of sampling.

### Sampling with Replacement and with Ordering

Suppose we choose $k$ objects from a set $A$ that has $n$ distinct objects, with replacement, that is, after selecting an object and noting its identity in an ordered list, the object is placed back in the set before the next choice is

2. This section and all sections marked with asterisks may be omitted without loss of continuity.

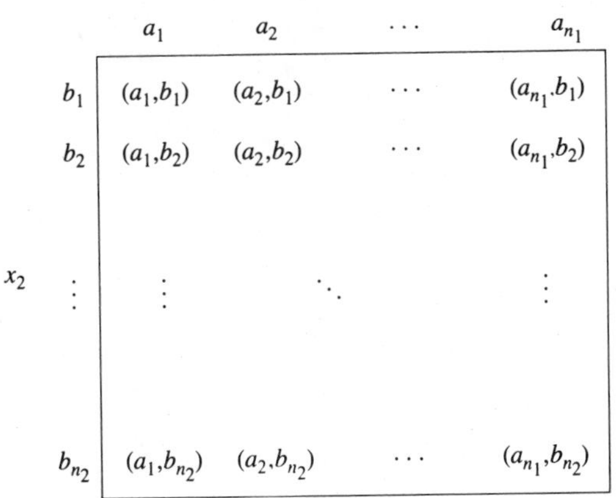


**FIGURE 2.9** If there are $n_1$ distinct choices for $x_1$ and $n_2$ distinct choices for $x_2$, then there are $n_1n_2$ distinct ordered pairs $(x_1, x_2)$.

made. We will refer to the set $A$ as the "population." The experiment produces an ordered $k$-tuple

$$(x_1, \ldots, x_k),$$

where $x_i \in A$ and $i = 1, \ldots, k$. Equation (2.16) with $n_1 = n_2 = \cdots = n_k = n$ implies that

$$\text{number of distinct ordered } k\text{-tuples} = n^k. \tag{2.17}$$

### ■■ Example 2.12

An urn contains five balls numbered 1 to 5. Suppose we select two balls from the urn with replacement. How many possible distinct *ordered* pairs are possible? What is the probability that the two draws yield the same number?

Equation (2.17) states that the number of ordered pairs is $5^2 = 25$. Figure 2.10(a) shows the 25 possible pairs. Five of the 25 outcomes have the two draws yielding the same number; if we suppose that all pairs are equiprobable, then the probability that the two draws yield the same number is $5/25 = .2$. ■■

### Sampling without Replacement and with Ordering

Suppose we choose $k$ objects in succession without replacement from a population $A$ of $n$ distinct objects. Clearly, $k \leq n$. The number of possible outcomes in the first draw is $n_1 = n$; the number of possible outcomes in

| | | | | |
|---|---|---|---|---|
| (1, 1) | (1, 2) | (1, 3) | (1, 4) | (1, 5) |
| (2, 1) | (2, 2) | (2, 3) | (2, 4) | (2, 5) |
| (3, 1) | (3, 2) | (3, 3) | (3, 4) | (3, 5) |
| (4, 1) | (4, 2) | (4, 3) | (4, 4) | (4, 5) |
| (5, 1) | (5, 2) | (5, 3) | (5, 4) | (5, 5) |

(a) Ordered pairs for sampling with replacement.

| | | | | |
|---|---|---|---|---|
| | (1, 2) | (1, 3) | (1, 4) | (1, 5) |
| (2, 1) | | (2, 3) | (2, 4) | (2, 5) |
| (3, 1) | (3, 2) | | (3, 4) | (3, 5) |
| (4, 1) | (4, 2) | (4, 3) | | (4, 5) |
| (5, 1) | (5, 2) | (5, 3) | (5, 4) | |

(b) Ordered pairs for sampling without replacement.

| | | | | |
|---|---|---|---|---|
| | (1, 2) | (1, 3) | (1, 4) | (1, 5) |
| | | (2, 3) | (2, 4) | (2, 5) |
| | | | (3, 4) | (3, 5) |
| | | | | (4, 5) |

(c) Pairs for sampling without replacement or ordering.

**FIGURE 2.10** Enumeration of possible outcomes in various types of sampling of two balls from an urn containing five distinct balls.

the second draw is $n_2 = n - 1$, namely all $n$ objects except the one selected in the first draw; and so on up to $n_k = n - (k - 1)$ in the final draw. Equation (2.16) then gives

$$\text{number of distinct ordered } k\text{-tuples} = n(n-1)\dots(n-k+1). \tag{2.18}$$

## ■■ Example 2.13

An urn contains five balls numbered 1 to 5. Suppose we select two balls in succession without replacement. How many possible distinct *ordered* pairs are possible? What is the probability that the first ball has a number larger than that of the second ball?

Equation (2.18) states that the number of ordered pairs is $5(4) = 20$. The 20 possible ordered pairs are shown in Fig. 2.10(b). Ten ordered pairs in Fig. 2.10(b) have the first number larger than the second number; thus the probability of this event is $10/20 = 1/2$. ■■

## ■■ Example 2.14

An urn contains five balls numbered $1, 2, \dots, 5$. Suppose we draw three balls with replacement. What is the probability that all three balls are different?

From Eq. (2.17) there are $5^3 = 125$ possible outcomes, which we will suppose are equiprobable. The number of these outcomes for which the three draws are different is given by Eq. (2.18): $5(4)(3) = 60$. Thus the probability that all three balls are different is $60/125 = .48$. ■■

### Permutations of $n$ Distinct Objects

Consider sampling without replacement with $k = n$. This is simply drawing objects from an urn containing $n$ distinct objects until the urn is empty. Thus, the number of possible orderings (arrangements, permutations) of $n$ distinct objects is equal to the number of ordered $n$-tuples in sampling without replacement with $k = n$. From Eq. (2.18), we have

$$\text{number of permutations of } n \text{ objects} = n(n-1)\ldots(2)(1) \triangleq n!. \quad (2.19)$$

We refer to $n!$ as **$n$ factorial.**

We will see that $n!$ appears in many of the combinatorial formulas. For large $n$, **Stirling's formula** is very useful:

$$n! \sim \sqrt{2\pi}\, n^{n+1/2}e^{-n}, \quad (2.20)$$

where the sign $\sim$ indicates that the ratio of the two sides tends to unity as $n \to \infty$ (Feller, p. 52).

### ■■ Example 2.15

Find the number of permutations of three distinct objects $\{1, 2, 3\}$. Equation (2.19) gives $3! = 3(2)(1) = 6$. The six permutations are

$$123 \quad 312 \quad 231 \quad 132 \quad 213 \quad 321.$$

■■

### ■■ Example 2.16

Suppose that 12 balls are placed at random into 12 cells, where more than 1 ball is allowed to occupy a cell. What is the probability that all cells are occupied?

The placement of each ball into a cell can be viewed as the selection of a cell number between 1 and 12. Equation (2.17) implies that there are $12^{12}$ possible placements of the 12 balls in the 12 cells. In order for all cells to be occupied, the first ball selects from any of the 12 cells, the second ball from the remaining 11 cells, and so on. Thus the number of placements that occupy all cells is 12!. If we suppose that all $12^{12}$ possible placements are equiprobable, we find that the probability that all cells are occupied is

$$\frac{12!}{12^{12}} = \left(\frac{12}{12}\right)\left(\frac{11}{12}\right)\cdots\left(\frac{1}{12}\right) = 5.37(10^{-5}).$$

This answer is surprising if we reinterpret the question as follows.

Given that 12 airplane crashes occur at random in a year, what is the probability that there is exactly 1 crash each month? The above result shows that this probability is very small. Thus a model that assumes that crashes occur randomly in time does *not* predict that they tend to occur uniformly over time (Feller, p. 32). ■■

### Sampling without Replacement and without Ordering

Suppose we pick $k$ objects from a set (population) of $n$ distinct objects without replacement and that we record the result without regard to order. (You can imagine putting each selected object into another jar, so that when the $k$ selections are completed we have no record of the order in which the selection was done.) We call the resulting subset of $k$ selected objects a "subpopulation of size $k$."

From Eq. (2.19), there are $k!$ possible orders in which the $k$ objects in the second jar could have been selected. Thus if $C_k^n$ denotes the number of subpopulations of size $k$ from a set of size $n$, then $C_k^n k!$ must be the total number of distinct ordered samples of $k$ objects, which is given by Eq. (2.18). Thus

$$C_k^n k! = n(n-1)\dots(n-k+1), \tag{2.21}$$

and the *number of subpopulations of size $k$ from a population of size $n$, $k \leq n$, is*

$$C_k^n = \frac{n(n-1)\dots(n-k+1)}{k!} = \frac{n!}{k!\,(n-k)!} \triangleq \binom{n}{k}. \tag{2.22}$$

The expression $\binom{n}{k}$ is called a **binomial coefficient.**

### ■■ Example 2.17

Find the number of ways of selecting two objects from $A = \{1, 2, 3, 4, 5\}$ without regard to order.

Equation (2.22) gives

$$\binom{5}{2} = \frac{5!}{2!\,3!} = 10.$$

Figure 2.10(c) gives the 10 pairs. ■■

### ■■ Example 2.18

Find the number of distinct permutations of $k$ white balls and $n - k$ black balls.

This problem is equivalent to the following sampling problem: Put $n$ tokens numbered 1 to $n$ in an urn, where each token represents a position in the arrangement of balls; pick a subpopulation of $k$ tokens and put the $k$ white balls in the corresponding positions. Each subpopulation of size $k$ leads to a distinct arrangement (permutation) of $k$ white balls and $n - k$ black balls. Thus the number of distinct permutations of $k$ white balls and $n - k$ black balls is $C_k^n$.

As a specific example let $n = 4$ and $k = 2$. The number of subpopulations of size 2 from a set of four distinct objects is

$$\binom{4}{2} = \frac{4!}{2!\,2!} = \frac{4(3)}{2(1)} = 6.$$

The 6 distinct permutations with 2 whites (zeros) and 2 blacks (ones) are

1100 0110 0011 1001 1010 0101. ■■

---

**■■ Example 2.19**
Quality Control

A batch of 50 items contains 10 defective items. Suppose 10 items are selected at random and tested. What is the probability that exactly 5 of the items tested are defective?

The number of ways of selecting 10 items out of a batch of 50 is the number of subpopulations of size 10 from a set of 50 objects:

$$\binom{50}{10} = \frac{50!}{10!\,40!}.$$

The number of ways of selecting 5 defective and 5 nondefective items from the batch of 50 is the product $N_1 N_2$, where $N_1$ is the number of ways of selecting the 5 items from the set of 10 defective items, and $N_2$ is the number of ways of selecting 5 items from the 40 nondefective items. Thus the probability that exactly 5 tested items are defective is

$$\frac{\binom{10}{5}\binom{40}{5}}{\binom{50}{10}} = \frac{10!\,40!\,10!\,40!}{5!\,5!\,35!\,5!\,50!} = .016.$$

■■

---

Example 2.18 shows that sampling without replacement and without ordering is equivalent to partitioning the set of $n$ distinct objects into two sets: $B$, containing the $k$ items that are picked from the urn, and $B^c$, containing the $n - k$ left behind. Suppose we partition a set of $n$ distinct objects into $J$ subsets $B_1, B_2, \ldots, B_J$, where $B_J$ is assigned $k_J$ elements and $k_1 + k_2 + \cdots + k_J = n$.

In Problem 43, it is shown that the number of distinct partitions is

$$\frac{n!}{k_1!\,k_2!\ldots k_J!}. \tag{2.23}$$

Equation (2.23) is called the **multinomial coefficient.** The binomial coefficient is the $J = 2$ case of the multinomial coefficient.

**■■ Example 2.20**

A six-sided die is tossed 12 times. How many distinct sequences of faces (numbers from the set $\{1, 2, 3, 4, 5, 6\}$) have each number appearing exactly twice? What is the probability of obtaining such a sequence?

The number of distinct sequences in which each face of the die appears exactly twice is the same as the number of partitions of the set $\{1, 2, \ldots, 12\}$ into 6 subsets of size 2, namely

$$\frac{12!}{2!\,2!\,2!\,2!\,2!\,2!} = \frac{12!}{2^6} = 7{,}484{,}400$$

From Eq. (2.17) we have that there are $6^{12}$ possible outcomes in twelve tosses of a die. If we suppose that all of these have equal probabilities, then the probability of obtaining a sequence in which each face appears exactly twice is

$$\frac{12!/2^6}{6^{12}} = \frac{7{,}484{,}400}{2{,}176{,}782{,}336} \simeq 3.4(10^{-3}).$$ ■■

## 2.4

## CONDITIONAL PROBABILITY

Quite often we are interested in determining whether two events, $A$ and $B$, are related in the sense that knowledge about the occurrence of one, say $B$, alters the likelihood of occurrence of the other, $A$. This requires that we find the **conditional probability**, $P[A \mid B]$, of event $A$ given that event $B$ has occurred. The conditional probability is defined by

$$P[A \mid B] = \frac{P[A \cap B]}{P[B]} \qquad \text{for } P[B] > 0. \tag{2.24}$$

Knowledge that event $B$ has occurred implies that the outcome of the experiment is in the set $B$. In computing $P[A \mid B]$ we can therefore view the experiment as now having the reduced sample space $B$ as shown in Fig. 2.11. The event $A$ occurs in the reduced sample space if and only if the outcome $\zeta$ is in $A \cap B$. Equation (2.24) simply renormalizes the probabil-

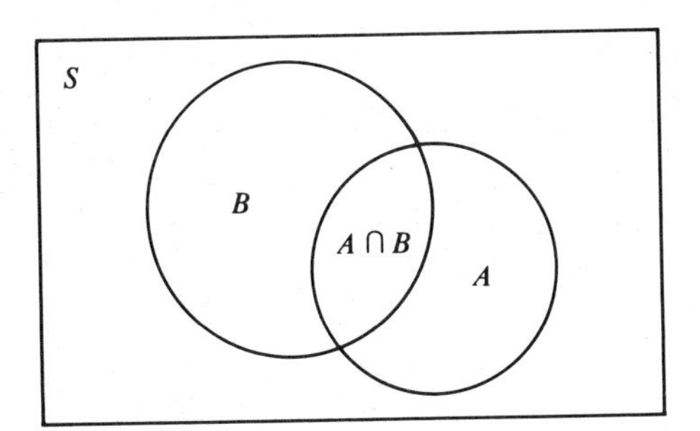


**FIGURE 2.11** If $B$ is known to have occurred, then $A$ can occur only if $A \cap B$ occurs.

ity of events that occur jointly with $B$. Thus if we let $A = B$, Eq. (2.24) gives $P[B \mid B] = 1$, as required.

If we interpret probability as relative frequency, then $P[A \mid B]$ should be the relative frequency of the event $A \cap B$ in experiments where $B$ occurred. Suppose that the experiment is performed $n$ times, and suppose that event $B$ occurs $n_B$ times, and event $A \cap B$, $n_{A \cap B}$ times. The relative frequency of interest is then

$$\frac{n_{A \cap B}}{n_B} = \frac{n_{A \cap B}/n}{n_B/n} \to \frac{P[A \cap B]}{P[B]},$$

where we have implicitly assumed that $P[B] > 0$. This is in agreement with Eq. (2.24).

### ■■ Example 2.21

A ball is selected from an urn containing two black balls, numbered 1 and 2, and two white balls, numbered 3 and 4. The number and color of the ball is noted, so the sample space is $\{(1, b), (2, b), (3, w), (4, w)\}$. Assuming that the four outcomes are equally likely, find $P[A \mid B]$ and $P[A \mid C]$, where $A$, $B$, and $C$ are the following events:

$A = \{(1, b), (2, b)\}$, "black ball selected,"

$B = \{(2, b), (4, w)\}$, "even-numbered ball selected," and

$C = \{(3, w), (4, w)\}$, "number of ball is greater than 2."

Since $P[A \cap B] = P[(2, b)]$ and $P[A \cap C] = P[\varnothing] = 0$, Eq. (2.21) gives

$$P[A \mid B] = \frac{P[A \cap B]}{P[B]} = \frac{.25}{.5} = .5 = P[A]$$

$$P[A \mid C] = \frac{P[A \cap C]}{P[C]} = \frac{0}{.5} = 0 \neq P[A].$$

In the first case, knowledge of $B$ did not alter the probability of $A$. In the second case, knowledge of $C$ implied that $A$ had not occurred. ■■

If we multiply both sides of the definition of $P[A \mid B]$ by $P[B]$ we obtain

$$P[A \cap B] = P[A \mid B]P[B]. \tag{2.25a}$$

Similarly we also have that

$$P[A \cap B] = P[B \mid A]P[A]. \tag{2.25b}$$

In the next example we show how this equation is useful in finding probabilities in sequential experiments. The example also introduces a **tree diagram** that facilitates the calculation of probabilities.

### ■■ Example 2.22

An urn contains two black balls and three white balls. Two balls are selected at random from the urn without replacement and the sequence of colors is noted. Find the probability that both balls are black.

This experiment consists of a sequence of two subexperiments. We can imagine working our way down the tree shown in Fig. 2.12 from the topmost node to one of the bottom nodes. We reach node 1 in the tree if the outcome of the first draw is a black ball; then the next subexperiment consists of selecting a ball from an urn containing one black ball and three white balls. On the other hand, if the outcome of the first draw is white, then we reach node 2 in the tree and the second subexperiment consists of selecting a ball from an urn that contains two black balls and two white balls. Thus if we know which node is reached after the first draw, then we can state the probabilities of the outcome in the next subexperiment.

Let $B_1$ and $B_2$ be the events that the outcome is a black ball in the first

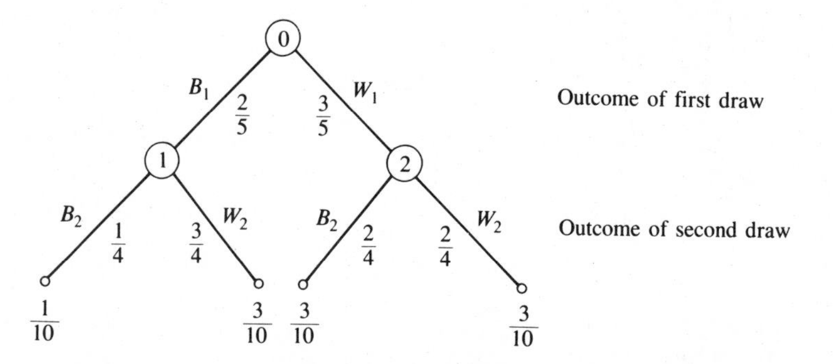


**FIGURE 2.12** The paths from the top node to a bottom node correspond to the possible outcomes in the drawing of two balls from an urn without replacement. The probability of a path is the product of the probabilities in the associated transitions.

and second draw respectively. From Eq. (2.25b) we have

$$P[B_1 \cap B_2] = P[B_2 \mid B_1]P[B_1].$$

In terms of the tree diagram in Fig. 2.12, $P[B_1]$ is the probability of reaching node 1 and $P[B_2 \mid B_1]$ is the probability of reaching the leftmost bottom node from node 1. Now $P[B_1] = 2/5$ since the first draw is from an urn containing two black balls and three white balls; $P[B_2 \mid B_1] = 1/4$ since, given $B_1$, the second draw is from an urn containing one black ball and three white balls. Thus

$$P[B_1 \cap B_2] = \frac{1}{4}\frac{2}{5} = \frac{1}{10}.$$

In general, the probability of any sequence of colors is obtained by multiplying the probabilities corresponding to the node transitions in the tree in Fig. 2.12. ■■

## ■■ Example 2.23

Many communication systems can be modeled in the following way. First, the user inputs a 0 or a 1 into the system, and a corresponding signal is transmitted. Second, the receiver makes a decision about what was the input to the system, based on the signal it received. Suppose that the user sends 0s with probability $1 - p$ and 1s with probability $p$, and suppose that the receiver makes random decision errors with probability $\varepsilon$. For $i = 0, 1$ let $A_i$ be the event "input was $i$," and let $B_i$ be the event "receiver decision was $i$." Find the probabilities $P[A_i \cap B_j]$ for $i = 0, 1$ and $j = 0, 1$.

The tree diagram for this experiment is shown in Fig. 2.13. We then readily obtain the desired probabilities

$$P[A_0 \cap B_0] = (1 - p)(1 - \varepsilon),$$
$$P[A_0 \cap B_1] = (1 - p)\varepsilon,$$
$$P[A_1 \cap B_0] = p\varepsilon, \text{ and}$$
$$P[A_1 \cap B_1] = p(1 - \varepsilon).$$

■■

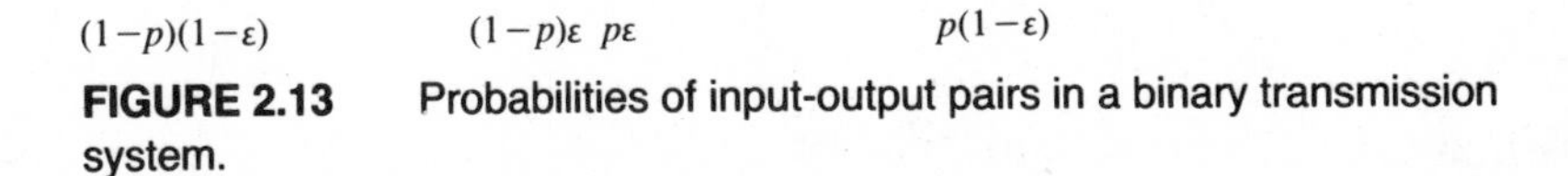


**FIGURE 2.13** Probabilities of input-output pairs in a binary transmission system.

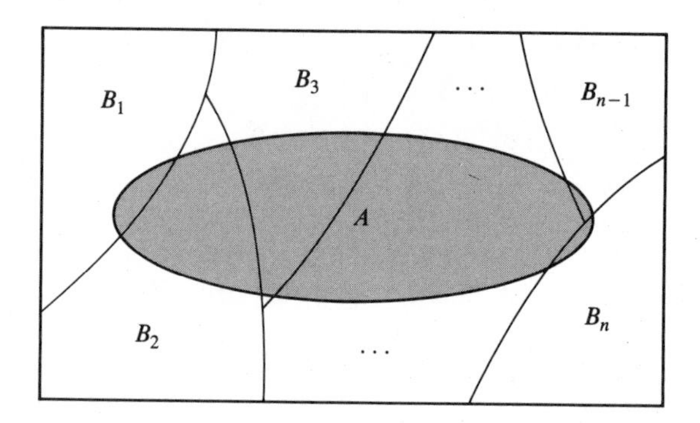


**FIGURE 2.14** A partition of $S$ into $n$ disjoint sets.

Let $B_1, B_2, \ldots, B_n$ be mutually exclusive events whose union equals the sample space $S$ as shown in Fig. 2.14. We refer to these sets as a **partition** of $S$. Any event $A$ can be represented as the union of mutually exclusive events in the following way:

$$A = A \cap S = A \cap (B_1 \cup B_2 \cup \cdots \cup B_n)$$
$$= (A \cap B_1) \cup (A \cap B_2) \cup \cdots \cup (A \cap B_n).$$

See Fig. 2.14. By Corollary 4, the probability of $A$ is

$$P[A] = P[A \cap B_1] + P[A \cap B_2] + \cdots + P[A \cap B_n].$$

By applying Eq. (2.25a) to each of the terms on the right-hand side, we obtain the **theorem on total probability**:

$$P[A] = P[A \mid B_1]P[B_1] + P[A \mid B_2]P[B_2] + \cdots + P[A \mid B_n]P[B_n]. \tag{2.26}$$

This result is particularly useful when the experiments can be viewed as consisting of a sequence of two subexperiments as shown in the tree diagram in Fig. 2.12.

### ■■ Example 2.24

In the experiment discussed in Example 2.22, find the probability of the event $W_2$ that the second ball is white.

The events $B_1 = \{(b, b), (b, w)\}$ and $W_1 = \{(w, b), (w, w)\}$ form a partition of the sample space, so applying Eq. 2.26 we have

$$P[W_2] = P[W_2 \mid B_1]P[B_1] + P[W_2 \mid W_1]P[W_1]$$
$$= \frac{3}{4}\frac{2}{5} + \frac{1}{2}\frac{3}{5} = \frac{3}{5}.$$

It is interesting to note that this is the same as the probability of selecting a white ball in the first draw. The result makes sense because we are computing the probability of a white ball in the second draw under the assumption that we have no knowledge of the outcome of the first draw. ■■

---

### ■■ Example 2.25

A manufacturing process produces a mix of "good" memory chips and "bad" memory chips. The lifetime of good chips follow the exponential law introduced in Example 2.10, with a rate of failure $\alpha$. The lifetime of bad chips also follows the exponential law but the rate of failure is $1000\alpha$. Suppose that the fraction of good chips is $1 - p$ and of bad chips, $p$. Find the probability that a randomly selected chip is still functioning after $t$ seconds.

Let $C$ be the event, "chip still functioning after $t$ seconds," and let $G$ be the event "the chip is good" and $B$ be the event "the chip is bad." By the theorem of total probability we have

$$\begin{aligned} P[C] &= P[C \mid G]P[G] + P[C \mid B]P[B] \\ &= P[C \mid G](1 - p) + P[C \mid B]p \\ &= (1 - p)e^{-\alpha t} + pe^{-1000\alpha t}, \end{aligned}$$

where we used the fact that $P[C \mid G] = e^{-\alpha t}$ and $P[C \mid B] = e^{-1000\alpha t}$. ■■

---

### Bayes' Rule

Let $B_1, B_2, \ldots, B_n$ be a partition of a sample space $S$. Suppose that event $A$ occurs, what is the probability of event $B_j$? By the definition of conditional probability we have

$$P[B_j \mid A] = \frac{P[A \cap B_j]}{P[A]} = \frac{P[A \mid B_j]P[B_j]}{\sum_{k=1}^{n} P[A \mid B_k]P[B_k]}, \qquad (2.27)$$

where we used the theorem on total probability to replace $P[A]$. Equation (2.27) is called **Bayes' rule**.

Bayes' rule is often applied in the following situation. We have some random experiment in which the events of interest form a partition. The "a priori probabilities" of these events, $P[B_j]$, are the probabilities of the events before the experiment is performed. Now suppose that the experiment is performed, and we are informed that event $A$ occurred; the "a posteriori probabilities" are the probabilities of the events in the partition, $P[B_j \mid A]$, given this additional information. The following two examples illustrate this situation.

### Example 2.26

In the binary communication system in Example 2.23, find which input is more probable given that the receiver has output a 1. Assume that, a priori, the input is equally likely 0 or 1.

Let $A_k$ be the event that the input was $k$, $k = 0, 1$, then $A_0$ and $A_1$ are a partition of the sample space of input-output pairs. Let $B_1$ be the event receiver output was a 1. The probability of $B_1$ is

$$P[B_1] = P[B_1 \mid A_0]P[A_0] + P[B_1 \mid A_1]P[A_1]$$
$$= \varepsilon\left(\frac{1}{2}\right) + (1 - \varepsilon)\left(\frac{1}{2}\right) = \frac{1}{2}.$$

Applying Bayes' rule, we obtain the a posteriori probabilities

$$P[A_0 \mid B_1] = \frac{P[B_1 \mid A_0]P[A_0]}{P[B_1]} = \frac{\varepsilon/2}{1/2} = \varepsilon$$

$$P[A_1 \mid B_1] = \frac{P[B_1 \mid A_1]P[A_1]}{P[B_1]} = \frac{(1-\varepsilon)/2}{1/2} = (1 - \varepsilon).$$

Thus, if $\varepsilon$ is less than 1/2, then input 1 is more likely than input 0 when a 1 is observed at the output of the channel.

### Example 2.27
Quality Control

Consider the memory chips discussed in Example 2.25. Recall that a fraction $p$ of the chips are bad and tend to fail much more quickly than good chips. Suppose that in order to "weed out" the bad chips, every chip is tested for $t$ seconds prior to leaving the factory. The chips that fail are discarded and the remaining chips are sent out to customers. Find the value of $t$ for which 99% of the chips sent out to customers are good.

Let $C$ be the event "chip still functioning after $t$ seconds," and let $G$ be the event "chip is good," and $B$ the event "chip is bad." The problem requires that we find the value of $t$ for which

$$P[G \mid C] = .99.$$

We find $P[G \mid C]$ by applying Bayes' rule:

$$P[G \mid C] = \frac{P[C \mid G]P[G]}{P[C \mid G]P[G] + P[C \mid B]P[B]}$$
$$= \frac{(1-p)e^{-\alpha t}}{(1-p)e^{-\alpha t} + pe^{-\alpha 1000t}}$$
$$= \frac{1}{1 + \dfrac{pe^{-\alpha 1000t}}{(1-p)e^{-\alpha t}}} = .99.$$

The above equation can then be solved for $t$:

$$t = \frac{1}{999\alpha} \ln\left(\frac{99p}{1-p}\right).$$

For example if $1/\alpha = 20{,}000$ hours and $p = .10$, then $t = 48$ hours. ■■

## 2.5

## INDEPENDENCE OF EVENTS

If knowledge of the occurrence of an event $B$ does not alter the probability of some other event $A$, then it would be natural to say that event $A$ is independent of $B$. In terms of probabilities this situation occurs when

$$P[A] = P[A \mid B] = \frac{P[A \cap B]}{P[B]}.$$

The above equation has the problem that the right-hand side is not defined when $P[B] = 0$.

We will define two events $A$ and $B$ to be **independent** if

$$P[A \cap B] = P[A]P[B]. \tag{2.28}$$

Equation (2.28) then implies both

$$P[A \mid B] = P[A] \tag{2.29a}$$

and

$$P[B \mid A] = P[B] \tag{2.29b}$$

Note also that Eq. (2.29a) implies Eq. (2.28) when $P[B] \neq 0$ and Eq. (2.29b) implies Eq. (2.28) when $P[A] \neq 0$.

### ■■ Example 2.28

A ball is selected from an urn containing two black balls, numbered 1 and 2, and two white balls, numbered 3 and 4. Let the events $A$, $B$, and $D$ be defined as follows:

$A = \{(1, b), (2, b)\}$, "black ball selected";

$B = \{(2, b), (4, w)\}$, "even-numbered ball selected"; and

$C = \{(3, w), (4, w)\}$, "number of ball is greater than 2."

Are events $A$ and $B$ independent? Are events $A$ and $C$ independent?

First, consider events $A$ and $B$. The probabilities required by Eq. (2.28)

are

$$P[A] = P[B] = \frac{1}{2}, \qquad \text{and}$$

$$P[A \cap B] = P[\{(2, b)\}] = \frac{1}{4}.$$

Thus

$$P[A \cap B] = \frac{1}{4} = P[A]P[B],$$

and the events $A$ and $B$ are independent. Equation (2.29b) gives more insight into the meaning of independence:

$$P[A \mid B] = \frac{P[A \cap B]}{P[B]} = \frac{P[\{(2, b)\}]}{P[\{(2, b), (4, w)\}]} = \frac{1/4}{1/2} = \frac{1}{2}$$

$$P[A] = \frac{P[A]}{P[S]} = \frac{P[\{(1, b), (2, b)\}]}{P[\{(1, b), (2, b), (3, w), (4, w)\}]} = \frac{1/2}{1}.$$

These two equations imply that $P[A] = P[A \mid B]$ because the proportion of outcomes in $S$ that lead to the occurrence of $A$ is equal to the proportion of outcomes in $B$ that lead to $A$. Thus knowledge of the occurrence of $B$ does not alter the probability of the occurrence of $A$.

Events $A$ and $C$ are not independent since $P[A \cap C] = P[\varnothing] = 0$ so

$$P[A \mid C] = 0 \neq P[A] = .5.$$

In fact, $A$ and $C$ are mutually exclusive since $A \cap C = \varnothing$, so the occurrence of $C$ implies that $A$ has definitely not occurred. ■■

---

In general if two events have nonzero probability and are mutually exclusive, then they cannot be independent. For suppose they were independent and mutually exclusive, then

$$0 = P[A \cap B] = P[A]P[B],$$

which implies that at least one of the events must have zero probability.

### ■■ Example 2.29

Two numbers $x$ and $y$ are selected at random between zero and one. Let the events $A$, $B$, and $C$ be defined as follows:

$$A = \{x > 0.5\}, \qquad B = \{y > 0.5\}, \qquad \text{and } C = \{x > y\}.$$

Are the events $A$ and $B$ independent? Are $A$ and $C$ independent?

Figure 2.15(a and b) shows the regions of the unit square that

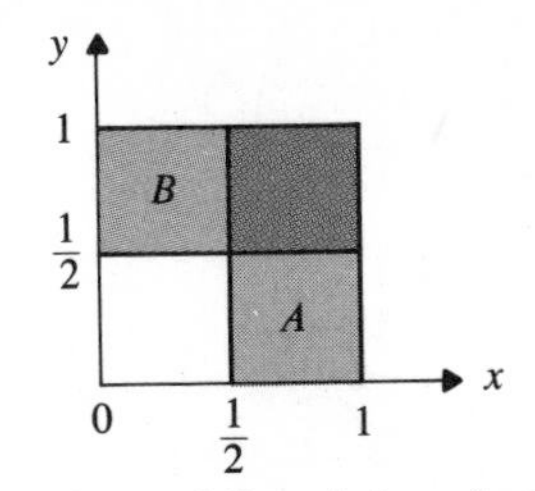

(a) Events $A$ and $B$ are independent.

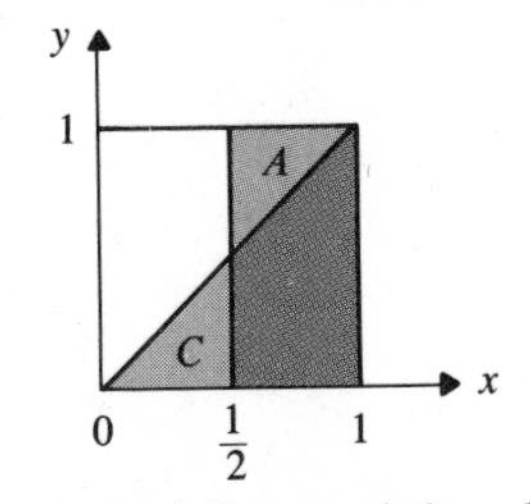


(b) Events $A$ and $C$ are not independent.

**FIGURE 2.15** Examples of independent and nonindependent events.

correspond to the above events. Using Eq. (2.29b), we have

$$P[A \mid B] = \frac{P[A \cap B]}{P[B]} = \frac{1/4}{1/2} = \frac{1}{2} = P[A],$$

so events $A$ and $B$ are independent. Again we have that the "proportion" of outcomes in $S$ leading to $A$ is equal to the "proportion" in $B$ that lead to $A$.

Using Eq. (2.29a), we have

$$P[A \mid C] = \frac{P[A \cap C]}{P[C]} = \frac{3/8}{1/2} = \frac{3}{4} \neq \frac{1}{2} = P[A],$$

so events $A$ and $C$ are not independent. Indeed from Fig. 2.15(b) we can see that knowledge of the fact that $x$ is greater than $y$ increases the probability that $x$ is greater than 0.5. ■■

---

What conditions should three events $A$, $B$, and $C$ satisfy in order for them to be independent? First, they should be pairwise independent, that is,

$$P[A \cap B] = P[A]P[B], \qquad P[A \cap C] = P[A]P[C], \text{ and}$$
$$P[B \cap C] = P[B]P[C].$$

In addition, knowledge of the joint occurrence of any two, say $A$ and $B$, should not affect the probability of the third, that is,

$$P[C \mid A \cap B] = P[C].$$

In order for this to hold, we must have

$$P[C \mid A \cap B] = \frac{P[A \cap B \cap C]}{P[A \cap B]} = P[C].$$

This in turn implies that we must have

$$P[A \cap B \cap C] = P[A \cap B]P[C] = P[A]P[B]P[C],$$

where we have used the fact that $A$ and $B$ are pairwise independent. Thus we conclude that *three events A, B, and C are independent if the probability of the intersection of any pair or triplet of events is equal to the product of the probabilities of the individual events.*

The following example shows that if three events are pairwise independent, it does not necessarily follow that $P[A \cap B \cap C] = P[A]P[B]P[C]$.

### ■■ Example 2.30

Consider the experiment discussed in Example 2.29 where two numbers are selected at random from the unit interval. Let the events $B$, $D$, and $F$

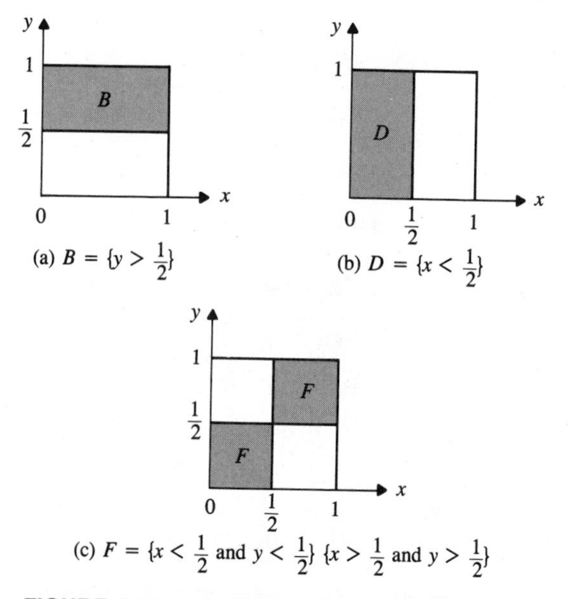


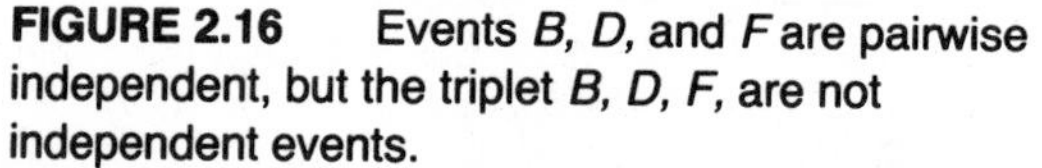

**FIGURE 2.16** Events *B, D,* and *F* are pairwise independent, but the triplet *B, D, F,* are not independent events.

be defined as follows:

$$B = \left\{y > \frac{1}{2}\right\}, \qquad D = \left\{x < \frac{1}{2}\right\}$$

$$F = \left\{x < \frac{1}{2} \text{ and } y < \frac{1}{2}\right\} \cup \left\{x > \frac{1}{2} \text{ and } y > \frac{1}{2}\right\}.$$

The three events are shown in Fig. 2.16. It can be easily verified that any pair of these events is independent:

$$P[B \cap D] = \frac{1}{4} = P[B]P[D],$$

$$P[B \cap F] = \frac{1}{4} = P[B]P[F], \text{ and}$$

$$P[D \cap F] = \frac{1}{4} = P[D]P[F].$$

However the three events are not independent, since $B \cap D \cap F = \varnothing$, so

$$P[B \cap D \cap F] = P[\varnothing] = 0 \neq P[B]P[D]P[F] = \frac{1}{8}.$$

■■

In order for a set of $n$ events to be independent, the probability of an event should be unchanged when we are given the joint occurrence of any subset of the other events. This requirement naturally leads to the following definition of independence. *The events $A_1, A_2, \ldots, A_n$ are said to be* **independent** if for $k = 2, \ldots, n$,

$$P[A_{i_1} \cap A_{i_2} \cap \cdots \cap A_{i_k}] = P[A_{i_1}]P[A_{i_2}] \ldots P[A_{i_k}], \tag{2.30}$$

where $1 \le i_1 < i_2 < \cdots < i_k \le n$. For a set of $n$ events we need to verify that all $2^n - n - 1$ possible intersections factor in the right way.

The above definition of independence appears quite cumbersome because it requires that so many conditions be verified. However, the most common application of the independence concept is in making the assumption that the outcomes of separate experiments are independent. For example it is common to assume that the outcome of a coin toss is independent of the outcomes of all prior and all subsequent coin tosses.

### ■■ Example 2.31

Suppose a fair coin is tossed three times and we observe the resulting sequence of heads and tails. Find the probability of the elementary events.

The sample space of this experiment is $S = \{\text{HHH, HHT, HTH, THH, TTH, THT, HTT, TTT}\}$. The assumption that the coin is fair means that the outcomes of a single toss are equiprobable, that is, $P[\text{H}] = P[\text{T}] = 1/2$.

If we assume that the outcomes of the coin tosses are independent, then

$$P[\{\text{HHH}\}] = P[\{\text{H}\}]P[\{\text{H}\}]P[\{\text{H}\}] = \frac{1}{8},$$

$$P[\{\text{HHT}\}] = P[\{\text{H}\}]P[\{\text{H}\}]P[\{\text{T}\}] = \frac{1}{8},$$

$$P[\{\text{HTH}\}] = P[\{\text{H}\}]P[\{\text{T}\}]P[\{\text{H}\}] = \frac{1}{8},$$

$$P[\{\text{THH}\}] = P[\{\text{T}\}]P[\{\text{H}\}]P[\{\text{H}\}] = \frac{1}{8},$$

$$P[\{\text{TTH}\}] = P[\{\text{T}\}]P[\{\text{T}\}]P[\{\text{H}\}] = \frac{1}{8},$$

$$P[\{\text{THT}\}] = P[\{\text{T}\}]P[\{\text{H}\}]P[\{\text{T}\}] = \frac{1}{8},$$

$$P[\{\text{HTT}\}] = P[\{\text{H}\}]P[\{\text{T}\}]P[\{\text{T}\}] = \frac{1}{8}, \text{ and}$$

$$P[\{\text{TTT}\}] = P[\{\text{T}\}]P[\{\text{T}\}]P[\{\text{T}\}] = \frac{1}{8}.$$

### Example 2.32
System Reliability

A system consists of a controller and three peripheral units. The system is said to be "up" if the controller and at least two of the peripherals are functioning. Find the probability that the system is up assuming that all components fail independently.

Define the following events: $A$ is "controller is functioning"; and $B_i$ is "peripheral $i$ is functioning"; where $i = 1, 2, 3$. The event $F$, "two or more peripheral units are functioning," occurs if all three units are functioning or if exactly two units are functioning. Thus

$$F = [B_1 \cap B_2 \cap B_3^c) \cup (B_1 \cap B_2^c \cap B_3) \\ \cup (B_1^c \cap B_2 \cap B_3) \cup (B_1 \cap B_2 \cap B_3).$$

Note that the events in the above union are mutually exclusive. Thus

$$\begin{aligned} P[F] &= P[B_1]P[B_2]P[B_3^c] + P[B_1]P[B_2^c]P[B_3] \\ &\quad + P[B_1^c]P[B_2]P[B_3] + P[B_1]P[B_2]P[B_3]\} \\ &= 3(1 - a)^2 a + (1 - a)^3, \end{aligned}$$

where we have assumed that each peripheral fails with probability $a$, so that $P[B_i] = 1 - a$ and $P[B_i^c] = a$.

The event "system is up" is then $A \cap F$. If we assume that the controller fails with probability $p$, then

$$\begin{aligned} P[\text{"system up"}] &= P[A \cap F] = P[A]P[F] \\ &= (1-p)P[F] \\ &= (1-p)\{3(1-a)^2a + (1-a)^3\}. \end{aligned}$$

Let $a = 10\%$, then all three peripherals are functioning $(1-a)^3 = 72.9\%$ of the time and two are functioning and one "down" $3(1-a)^2a = 24.3\%$ of the time. Thus two or more peripherals are functioning 97.2% of the time. Suppose that the controller is not very reliable, say $p = 20\%$, then the system is up only 77.8% of the time, mostly because of controller failures.

Suppose a second identical controller with $p = 20\%$ is added to the system, and that the system is "up" if at least one of the controllers is functioning and if two or more of the peripherals are functioning. In Problem 65, you are asked to show that at least one of the controllers is functioning 96% of the time, and that the system is up 93.3% of the time. This is an increase of 16% over the system with a single controller. ■■

## 2.6 SEQUENTIAL EXPERIMENTS

Many random experiments can be viewed as sequential experiments that consist of a sequence of simpler subexperiments. These subexperiments may or may not be independent. In this section we discuss methods for obtaining the probabilities of events in sequential experiments.

### Sequences of Independent Experiments

Suppose that a random experiment consists of performing experiments $E_1, E_2, \ldots, E_n$. The outcome of this experiment will then be an $n$-tuple $s = (s_1, \ldots, s_n)$, where $s_k$ is the outcome of the $k$th subexperiment. The sample space of the sequential experiment is defined as the set that contains the above $n$-tuples and is denoted by the Cartesian product of the individual sample spaces $S_1 \times S_2 \times \cdots \times S_n$.

We can usually determine, because of physical considerations, when the subexperiments are independent, in the sense that the outcome of any given subexperiment cannot affect the outcomes of the other subexperiments. Let $A_1, A_2, \ldots, A_n$ be events such that $A_k$ concerns only the outcomes of the $k$th subexperiment. If the subexperiments are independent, then it is reasonable to assume that the above events $A_1, A_2, \ldots, A_n$ are independent. Thus

$$P[A_1 \cap A_2 \cap \cdots \cap A_n] = P[A_1]P[A_2]\ldots P[A_n]. \tag{2.31}$$

This expression allows us to compute all probabilities of events of the sequential experiment.

### ■■ Example 2.33

Suppose that 10 numbers are selected at random from the interval [0, 1]. Find the probability that the first 5 numbers are less than 1/4 and the last 5 numbers are greater than 1/2. Let $x_1, x_2, \ldots, x_{10}$ be the sequence of 10 numbers, then the events of interest are:

$$A_k = \left\{x_k < \frac{1}{4}\right\} \qquad \text{for } k = 1, \ldots, 5$$

$$A_k = \left\{x_k > \frac{1}{2}\right\} \qquad \text{for } k = 6, \ldots, 10.$$

If we assume that each selection of a number is independent of the other selections, then

$$P[A_1 \cap A_2 \cap \cdots \cap A_{10}] = P[A_1]P[A_2]\ldots P[A_{10}]$$
$$= \left(\frac{1}{4}\right)^5 \left(\frac{1}{2}\right)^5.$$

■■

---

We will now derive several important models for experiments that consist of sequences of independent subexperiments.

### The Binomial Probability Law

A **Bernoulli trial** involves performing an experiment once and noting whether a particular event $A$ occurs. The outcome of the Bernoulli trial is said to be a "success" if $A$ occurs and a "failure" otherwise. In this section we are interested in finding the probability of $k$ successes in $n$ independent repetitions of a Bernoulli trial.

We can view the outcome of a single Bernoulli trial as the outcome of a toss of a coin for which the probability of heads (success) is $p = P[A]$. The probability of $k$ successes in $n$ Bernoulli trials is then equal to the probability of $k$ heads in $n$ tosses of the coin.

### ■■ Example 2.34

Suppose that a coin is tossed three times. If we assume that the *tosses are independent* and the probability of heads is $p$, then the probability for the sequences of heads and tails is

$$P[\{\text{HHH}\}] = P[\{\text{H}\}]P[\{\text{H}\}]P[\{\text{H}\}] = p^3,$$

$$P[\{\text{HHT}\}] = P[\{\text{H}\}]P[\{\text{H}\}]P[\{\text{T}\}] = p^2(1-p),$$

$$P[\{\text{HTH}\}] = P[\{\text{H}\}]P[\{\text{T}\}]P[\{\text{H}\}] = p^2(1-p),$$
$$P[\{\text{THH}\}] = P[\{\text{T}\}]P[\{\text{H}\}]P[\{\text{H}\}] = p^2(1-p),$$
$$P[\{\text{TTH}\}] = P[\{\text{T}\}]P[\{\text{T}\}]P[\{\text{H}\}] = p(1-p)^2,$$
$$P[\{\text{THT}\}] = P[\{T\}]P[\{\text{H}\}]P[\{\text{T}\}] = p(1-p)^2,$$
$$P[\{\text{HTT}\}] = P[\{\text{H}\}]P[\{\text{T}\}]P[\{\text{T}\}] = p(1-p)^2, \text{ and}$$
$$P[\{\text{TTT}\}] = P[\{\text{T}\}]P[\{\text{T}\}]P[\{\text{T}\}] = (1-p)^3,$$

where we used the fact that the tosses are independent. Let $k$ be the number of heads in three trials, then

$$P[k=0] = P[\{\text{TTT}\}] = (1-p)^3,$$
$$P[k=1] = P[\{\text{TTH, THT, HTT}\}] = 3p(1-p)^2,$$
$$P[k=2] = P[\{\text{HHT, HTH, THH}\}] = 3p^2(1-p), \text{ and}$$
$$P[k=3] = P[\{\text{HHH}\}] = p^3.$$

■■

The result in Example 2.34 is the $n = 3$ case of the binomial probability law.

**THEOREM**

Let $k$ be the number of successes in $n$ independent Bernoulli trials, then the probabilities of $k$ are given by the **binomial probability law**:

$$p_n(k) = \binom{n}{k} p^k (1-p)^{n-k} \qquad \text{for} \qquad k = 0, \ldots, n, \tag{2.32}$$

where $p_n(k)$ is the probability of $k$ successes in $n$ trials, and

$$\binom{n}{k} = \frac{n!}{k!\,(n-k)!} \tag{2.33}$$

is the Binomial coefficient.

The term $n!$ in Eq. 2.33 is called $n$ factorial and is defined by $n! = n(n-1)\ldots(2)(1)$. By definition 0! is equal to 1.

We now prove the above theorem. In Example 2.34 we saw that each of the sequences with $k$ successes and $n - k$ failures has the same probability, namely $p^k(1-p)^{n-k}$. Let $N_n(k)$ be the number of distinct sequences that have $k$ successes and $n - k$ failures, then

$$p_n(k) = N_n(k) p^k (1-p)^{n-k}. \tag{2.34}$$

The expression $N_n(k)$ is the number of ways of picking $k$ positions out of $n$

for the successes. It can be shown that[3]

$$N_n(k) = \binom{n}{k}. \tag{2.35}$$

The theorem follows by substituting Eq. (2.35) into Eq. (2.34).

### ■■ Example 2.35

Verify that Eq. (2.32) gives the probabilities found in Example 2.34.

In Example 2.34, let "toss results in heads" correspond to a "success," then

$$p_3(0) = \frac{3!}{0!\,3!}p^0(1-p)^3 = (1-p)^3,$$

$$p_3(1) = \frac{3!}{1!\,2!}p^1(1-p)^2 = 3p(1-p)^2,$$

$$p_3(2) = \frac{3!}{2!\,1!}p^2(1-p)^1 = 3p^2(1-p), \text{ and}$$

$$p_3(3) = \frac{3!}{0!\,3!}p^3(1-p)^0 = p^3,$$

which are in agreement with our previous results. ■■

---

You were introduced to the binomial coefficient in an introductory calculus course when the **binomial theorem** was discussed:

$$(a+b)^n = \sum_{k=0}^{n} \binom{n}{k} a^k b^{n-k}. \tag{2.36}$$

If we let $a = b = 1$, then

$$2^n = \sum_{k=0}^{n} \binom{n}{k} = \sum_{k=0}^{n} N_n(k),$$

which is in agreement with the fact that there are $2^n$ distinct possible sequences of successes and failures in $n$ trials. If we let $a = p$ and $b = 1 - p$ in Eq. (2.36) we then obtain

$$1 = \sum_{k=0}^{n} \binom{n}{k} p^k (1-p)^{n-k} = \sum_{k=0}^{n} p_n(k),$$

which confirms that the probabilities of the binomial probabilities sum to 1.

The term $n!$ grows very quickly with $n$, so numerical problems are

---

3. See Example 2.18.

encountered for relatively small values of $n$ if one attempts to compute $p_n(k)$ directly using Eq. (2.32). The following recursive formula extends the range of $n$ for which $p_n(k)$ can be computed before encountering numerical difficulties:

$$p_n(k+1) = \frac{(n-k)p}{(k+1)(1-p)} p_n(k). \tag{2.37}$$

Later in the book, we present two approximations for the binomial probabilities for the case when $n$ is large.

### ■■ Example 2.36

Let $k$ be the number of active (nonsilent) speakers in a group of eight noninteracting (i.e., independent) speakers. Suppose that a speaker is active with probability 1/3. Find the probability that the number of active speakers is greater than six.

For $i = 1, \ldots, 8$, let $A_i$ denote the event "$i$th speaker is active." The number of active speakers is then the number of successes in eight Bernoulli trials with $p = 1/3$. Thus the probability that more than six speakers are active is

$$P[k=7] + P[k=8] = \binom{8}{7}\left(\frac{1}{3}\right)^7\left(\frac{2}{3}\right) + \binom{8}{8}\left(\frac{1}{3}\right)^8$$
$$= .00244 + .00015 = .00259.$$

■■

---

### ■■ Example 2.37

A communication system transmits binary information over a channel that introduces random bit errors with probability $\varepsilon = 10^{-3}$. The transmitter transmits each information bit three times, and a decoder takes a majority vote of the received bits to decide on what the transmitted bit was. Find the probability that the receiver will make an incorrect decision.

The receiver will make the wrong decision if the channel introduces two or more errors. If we view each transmission as a Bernoulli trial in which a "success" corresponds to the introduction of an error, then the probability of two or more errors in three Bernoulli trials is

$$P[k \geq 2] = \binom{3}{2}(.001)^2(.999) + \binom{3}{3}(.001)^3 \simeq 3(10^{-6})$$

■■

---

### The Multinomial Probability Law

The binomial probability law can be generalized to the case where we note the occurrence of more than one event. Let $B_1, B_2, \ldots, B_M$ be a partition of the sample space $S$ of some random experiment and let $P[B_j] = p_j$. The events are mutually exclusive, so

$$p_1 + p_2 + \cdots + p_M = 1.$$

Suppose that $n$ independent repetitions of the experiment are performed. Let $k_j$ be the number of times event $B_j$ occurs, then the vector $(k_1, k_2, \ldots, k_M)$ specifies the number of times each of the events $B_j$'s occur. The probability of the vector $(k_1, \ldots, k_M)$ satisfies the **multinomial probability law:**

$$P[(k_1, k_2, \ldots, k_M)] = \frac{n!}{k_1!\, k_2! \ldots k_M!} p_1^{k_1} p_2^{k_2} \ldots p_M^{k_M}, \tag{2.38}$$

where $k_1 + k_2 + \cdots + k_M = n$. The binomial probability law is the $M = 2$ case of the multinomial probability law.

### ■■ Example 2.38

A dart is thrown nine times at a target consisting of three areas. Each throw has a probability of .2, .3, and .5 of landing in areas 1, 2, and 3, respectively. Find the probability that exactly three darts land in each of the areas.

This experiment consists of nine independent repetitions of a subexperiment that has three possible outcomes. The probability for the number of occurrences of each outcome is given by the multinomial probabilities with parameters $n = 9$ and $p_1 = .2$, $p_2 = .3$, and $p_3 = .5$:

$$P[(3, 3, 3)] = \frac{9!}{3!\, 3!\, 3!} (.2)^3 (.3)^3 (.5)^3 = .04536.$$

■■

### ■■ Example 2.39

Suppose we pick 10 telephone numbers at random from a telephone book and note the last digit in each of the numbers. What is the probability that we obtain each of the integers from 0 to 9 only once?

The probabilities for the number of occurrences of the integers is given by the multinomial probabilities with parameters $M = 10$, $n = 10$, and $p_j = 1/10$ if we assume that the 10 integers in the range 0 to 9 are equiprobable. The probability of obtaining each integer once in 10 draws is then

$$\frac{10!}{1!\, 1! \ldots 1!} (.1)^{10} \simeq 3.6(10^{-4}).$$

■■

### The Geometric Probability Law

Consider a sequential experiment in which we repeat independent Bernoulli trials until the occurrence of the first success. Let the outcome of this experiment be $m$, the number of trials carried out until the occurrence of the first success. The sample space for this experiment is the set of positive integers. The probability, $p(m)$, that $m$ trials are required is found by noting that this can only happen if the first $m-1$ trials result in failures and the $m$th trial in success.[4] The probability of this event is

$$p(m) = P[A_1^c A_2^c \ldots A_{m-1}^c A_m] = (1-p)^{m-1}p \qquad m = 1, 2, \ldots, \tag{2.39}$$

where $A_i$ is the event "success in $i$th trial." The probability assignment specified by Eq. (2.39) is called the **geometric probability law**.

The probabilities in Eq. (2.39) sum 1:

$$\sum_{m=1}^{\infty} p(m) = p \sum_{m=1}^{\infty} q^{m-1} = p\frac{1}{1-q} = 1,$$

where $q = 1 - p$, and where we have used the formula for the summation of a geometric series. The probability that more than $K$ trials are required before a success occurs has a simple form:

$$\begin{aligned} P[\{m > K\}] &= p \sum_{m=K+1}^{\infty} q^{m-1} = pq^K \sum_{j=0}^{\infty} q^j \\ &= pq^K \frac{1}{1-q} \\ &= q^K. \end{aligned} \tag{2.40}$$

### ■■ Example 2.40

Computer $A$ sends a message to computer $B$ over an unreliable telephone line. The message is encoded so that $B$ can detect when errors have been introduced into the message during transmission. If $B$ detects an error it requests $A$ to retransmit it. If the probability of a message transmission error is $q = .1$, what is the probability that a message needs to be transmitted more than two times?

Each transmission of a message is a Bernoulli trial with probability of success $p = 1 - q$. The Bernoulli trials are repeated until the first success (error-free transmission). The probability that more than two transmissions are required is given by Eq. (2.40):

$$P[m > 2] = q^2 = 10^{-2}.$$ ■■

## Sequences of Dependent Experiments

In this section we consider a sequence or "chain" of subexperiments in which the outcome of a given subexperiment determines which subexperiment is performed next. We first give a simple example of such an

4. See Example 2.8 in Section 2.2 for a relative frequency interpretation of how the geometric probability law comes about.

experiment and show how diagrams can be used to specify the sample space.

### ■■ Example 2.41

A sequential experiment involves repeatedly drawing a ball from one of two urns, noting the number on the ball, and replacing the ball in its urn. Urn 0 contains a ball with the number 0 and two balls with the number 1, and urn 1 contains five balls with the number 0 and one ball with the number 1. The urn from which the first draw is made is selected at random by flipping a fair coin. Urn 0 is used if the outcome is heads and urn 1 if the outcome is tails. Thereafter the urn used in a subexperiment corresponds to the number on the ball selected in the previous subexperiment.

The sample space of this experiment consists of sequences of 0s and 1s. Each possible sequence corresponds to a path through the "trellis" diagram shown in Fig. 2.17(a). The nodes in the diagram denote the urn used in the $n$th subexperiment, and the labels in the branches denote the outcome of a subexperiment. Thus the path 0011 corresponds to the

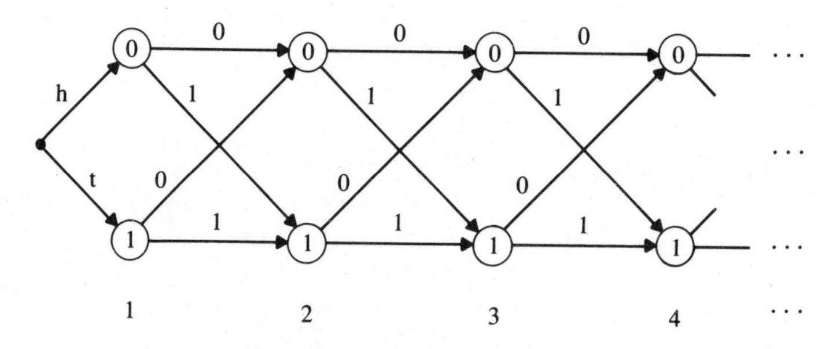


(a) Each sequence of outcomes corresponds to a path through this trellis diagram.

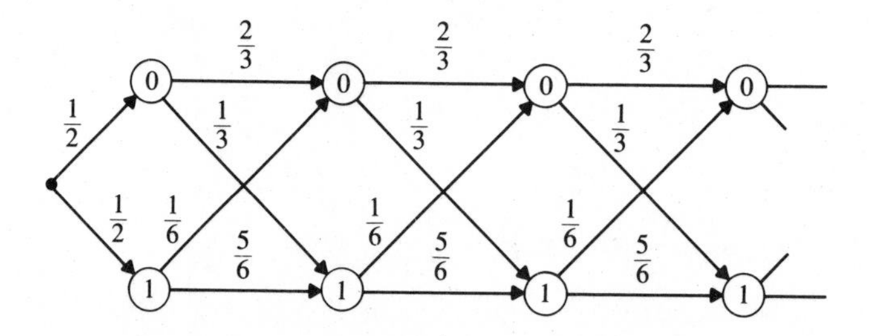


(b) The probability of a sequence of outcomes is the product of the probabilities along the associated path.

**FIGURE 2.17** Trellis diagram for a Markov chain.

sequence: The coin toss was heads so the first draw was from urn 0; the outcome of the first draw was 0, so the second draw was from urn 0; the outcome of second draw was 1, so the third draw was from urn 1; and the outcome from the third draw was 1, so the fourth draw is from urn 1. ■■

Now suppose that we want to compute the probability of a particular sequence of outcomes, say $s_0$, $s_1$, $s_2$. Denote this probability by $P[\{s_0\} \cap \{s_1\} \cap \{s_2\}]$. Let $A = \{s_2\}$ and $B = \{s_0\} \cap \{s_1\}$, then since $P[A \cap B] = P[A \mid B]P[B]$ we have

$$\begin{aligned} P[\{s_0\} \cap \{s_1\} \cap \{s_2\}] &= P[\{s_2\} \mid \{s_0\} \cap \{s_1\}]P[\{s_0\} \cap \{s_1\}] \\ &= P[\{s_2\} \mid \{s_0\} \cap \{s_1\}]P[\{s_1\} \mid \{s_0\}]P[\{s_0\}]. \end{aligned} \tag{2.41}$$

Now note that in the above urn example the probability $P[\{s_n\} \mid \{s_0\} \cap \cdots \cap \{s_{n-1}\}]$ depends only on $\{s_{n-1}\}$ since the most recent outcome determines which subexperiment is performed:

$$P[\{s_n\} \mid \{s_0\} \cap \cdots \cap \{s_{n-1}\}] = P[\{s_n\} \mid \{s_{n-1}\}]. \tag{2.42}$$

Therefore for the sequence of interest we have that

$$P[\{s_0\} \cap \{s_1\} \cap \{s_2\}] = P[\{s_2\} \mid \{s_1\}]P[\{s_1\} \mid \{s_0\}]P[\{s_0\}]. \tag{2.43}$$

Sequential experiments that satisfy Eq. (2.42) are called **Markov chains**. For these experiments, the probability of a sequence $s_0, s_1, \ldots, s_n$ is given by

$$P[s_0, s_1, \ldots, s_n] = P[s_n \mid s_{n-1}]P[s_{n-1} \mid s_{n-2}] \ldots P[s_1 \mid s_0]P[s_0]. \tag{2.44}$$

Thus the probability of the sequence $s_0, \ldots, s_n$ is given by the product of the probability of the first outcome $s_0$ and the probabilities of all subsequent transitions, $s_0$ to $s_1$, $s_1$ to $s_2$, and so on. Chapter 8 deals with Markov chains.

## ■■ Example 2.42

Find the probability of the sequence 0011 for the urn experiment introduced in Example 2.41.

Recall that urn 0 contains two balls with label 0 and one ball with label 1, and that urn 1 contains five balls with label 1 and one ball with label 0. We can readily compute the probabilities of sequences of outcomes by labeling the branches in the trellis diagram with the probability of the corresponding transition as shown in Fig. 2.17(b). Thus the probability of the sequence 0011 is given by

$$P[0011] = P[1 \mid 1]P[1 \mid 0]P[0 \mid 0]P[0],$$

where the transition probabilities are given by

$$P[1 \mid 0] = \frac{1}{3} \quad \text{and} \quad P[0 \mid 0] = \frac{2}{3}$$

$$P[1 \mid 1] = \frac{5}{6} \quad \text{and} \quad P[0 \mid 1] = \frac{1}{6},$$

and the initial probabilities are given by

$$P(0) = \frac{1}{2} = P[1].$$

If we substitute these values into the expression for $P[0011]$, we obtain

$$P[0011] = \left(\frac{5}{6}\right)\left(\frac{1}{3}\right)\left(\frac{2}{3}\right)\left(\frac{1}{2}\right) = \frac{5}{54}.$$

■■

The two-urn experiment in Examples 2.41 and 2.42 is the simplest example of the Markov chain models that are discussed in Chapter 8. The two-urn experiment discussed here is used to model situations in which there are only two outcomes, and in which the outcomes tend to occur in bursts. For example, the two-urn model has been used to model the "bursty" behavior of the voice packets generated by a single speaker where bursts of active packets are separated by relatively long periods of silence. The model has also been used for the sequence of black and white dots that result from scanning a black and white image line-by-line.

## *2.7

## A COMPUTER METHOD FOR SYNTHESIZING RANDOMNESS: RANDOM NUMBER GENERATORS

This section introduces the basic method for generating sequences of "random" numbers using a computer. Any computer simulation of a system that involves randomness must include a method for generating sequences of random numbers. These random numbers must satisfy long-term average properties of the processes they are simulating. In this section we focus on the problem of generating random numbers that are "uniformly distributed" in the interval $[0, 1]$. In the next chapter we will show how these random numbers can be used to generate numbers with arbitrary probability laws.

The first problem we must confront in generating a random number in the interval $[0, 1]$ is the fact that there are an uncountably infinite number of points in the interval, but the computer is limited to representing numbers with finite precision only. We must therefore be content with generating equiprobable numbers from some finite set, say $\{0, 1, \ldots, M - 1\}$ or $\{1, 2, \ldots, M\}$. By dividing these numbers by $M$, we obtain numbers in the unit interval. These numbers can be made increasingly dense in the unit interval by making $M$ very large.

The next step involves finding a mechanism for generating random numbers. The direct approach involves performing random experiments. For example, we can generate integers in the range 0 to $2^m - 1$ by flipping a fair coin $m$ times and replacing the sequence of heads and tails by 0s and 1s to obtain the binary representation of an integer. Another example would involve drawing a ball from an urn containing balls numbered 1 to $M$. Computer simulations involve the generation of long sequences of random numbers. If we were to use the above mechanisms to generate random numbers, we would have to perform the experiments a large number of times and store the outcomes in computer storage for access by the simulation program. It is clear that this approach is cumbersome and quickly becomes impractical.

The preferred approach for the computer generation of random numbers involves the use of recursive formulas that can be implemented easily and quickly. We will discuss the **power residue method**, which involves the following recursive formula:

$$Z_k = \alpha Z_{k-1} \bmod M, \tag{2.45}$$

where $\alpha$ is a carefully chosen integer between 1 and $M$, and $M$ is a prime number $p$ or an integer power of a prime number $p^m$. Equation (2.45) involves taking the product of $\alpha$ and $Z_{k-1}$, dividing it by $M$, and letting $Z_k$ be the remainder of the division. The resulting number is in the range 0 to $M - 1$.

### ■■ Example 2.43

Find the sequences yielded by Eq. (2.45) for: $M = 11$, $\alpha = 7$, $Z_0 = 1$; $M = 11$, $\alpha = 3$, $Z_0 = 1$; $M = 2^2$, $\alpha = 2$, $Z_0 = 1$.

For $M = 11$, $\alpha = 7$, and $Z_0 = 1$, we have

$$Z_1 = \text{remainder of } \frac{(7 \times 1)}{11} = 7,$$

$$Z_2 = \text{remainder of } \frac{(7 \times Z_1)}{11} = \text{remainder of } \frac{49}{11} = 5,$$

and so on. You should verify that the resulting sequence is

$$1, 7, 5, 2, 3, 10, 4, 6, 9, 8, 1, 7, 5, 2, 3, 10, 4, 6, 9, 8, 1, \ldots.$$

Note that the sequence cycles through all the integers in the range 1 to 10, and then repeats the sequence indefinitely.

For $M = 11$, $\alpha = 3$ and $Z_0 = 1$, the sequence yielded by Eq. (2.45) is

$$1, 3, 9, 5, 4, 1, 3, 9, 5, 4, 1, \ldots.$$

This sequence does not cycle through all the integers in the range 1 to 10 before it begins repeating.

For $M = 2^2 = 4$, $\alpha = 2$, $Z_0 = 1$, the sequence yielded by Eq. (2.45) is

$$1, 2, 0, 0, \ldots.$$

■■

---

Note that if $\alpha$ divides $M$ evenly, then the sequence generated by Eq. (2.45) eventually yields all zeros. Otherwise, the sequence is periodic with maximum period $M - 1$. In order for the sequence to have the maximum possible length, $\alpha$ must be a "primitive root of $M$." We do not discuss how such primitive roots are found. We refer the interested reader to reference [6] at the end of the chapter for further reading.

Example 2.43 shows that the sequences produced by Eq. (2.45) are periodic and not truly random. Equation (2.45) implies that the entire sequence will begin repeating as soon as a number appears for the second time in the sequence. For this reason the sequences produced by Eq. (2.45) are called **pseudo-random**.

If $M$ is made extremely large, then the numbers in the sequence do not repeat during the course of a simulation. The key question then becomes whether the sequence of numbers *appear* to be random. And how do we check this? One way is to compute relative frequencies. If we examine the sequence of numbers normalized by $M$, single numbers should appear to be uniformly distributed in the unit interval, pairs of numbers should appear to be uniformly distributed in the unit square, triplets in the unit cube, and so on. Chapter 3 presents statistical tests for determining whether observed relative frequencies are consistent with specified distributions.

The statistical properties of the sequences generated by recursive formulas of the type in Eq. (2.45) have been studied extensively. Only a few sequences have been identified with good statistical properties. The following parameters are recommended for the power residue method:

$$Z_i = 7^5 Z_{i-1} \bmod (2^{31} - 1). \tag{2.46}$$

This generator with multiplier $\alpha = 7^5 = 16{,}807$ produces a sequence of length $M - 1 = 2^{31} - 2 = 2{,}147{,}483{,}646$. The choice of $Z_0$ is called the "seed" of the random number generator, and it determines the point at which the sequence is begun.

You should have no problem writing a program to implement Eq. (2.46). The most time consuming operation in such a program is the division by $M$. This becomes an important consideration in simulations that involve the generation of very large numbers of random numbers. A

method for implementing Eq. (2.46), called **simulated division,** has been devised that executes substantially faster than conventional division. References [6] and [7] explain simulated division and reference [7] presents FORTRAN programs that implement the method.

## SUMMARY

- A probability model is specified by identifying the sample space $S$, the events of interest, and an initial probability assignment, a "probability law," from which the probability of all events can be computed.

- The sample space $S$ specifies the set of all possible outcomes. If it has a finite or countable number of elements, $S$ is discrete; $S$ is continuous otherwise.

- Events are subsets of $S$ that result by specifying conditions that are of interest in the particular experiment. When $S$ is discrete, events consist of the union of elementary events. When $S$ is continuous, events consist of the union or intersection of intervals in the real line.

- The axioms of probability specify a set of properties that must be satisfied by the probabilities of events. The corollaries that follow from the axioms provide rules for computing the probabilities of events in terms of the probabilities of other related events.

- An initial probability assignment that specifies the probability of certain events must be determined by the person doing the modelling. If $S$ is discrete, it suffices to specify the probabilities of the elementary events. If $S$ is continuous, it suffices to specify the probabilities of intervals or of semi-infinite intervals.

- Combinatorial formulas are used to evaluate probabilities in experiments that have an equiprobable, finite number of outcomes.

- A conditional probability quantifies the effect of partial knowledge about the outcome of an experiment on the probabilities of events. It is particularly useful in sequential experiments where the outcomes of subexperiments constitute the "partial knowledge."

- Bayes' rule gives the a posteriori probability of an event given that another event has been observed. It can be used to synthesize decision rules that attempt to determine the most probable "cause" in light of an observation.

- Two events are independent if knowledge of the occurrence of one does not alter the probability of the other. Two experiments are independent if all of their respective events are independent. The notion of independence is useful for computing probabilities in experiments that involve noninteracting subexperiments.

- Many experiments can be viewed as consisting of a sequence of independent subexperiments. In this chapter we presented the binomial, the multinomial, and the geometric probability laws as models that arise in this context.
- A Markov chain consists of a sequence of subexperiments in which the outcome of a subexperiment determines which subexperiment is performed next. The probability of a sequence of outcomes in a Markov chain is given by the product of the probability of the first outcome and the probabilities of all subsequent transitions.
- Computer simulation models use recursive equations to generate sequences of pseudo-random numbers.

## CHECKLIST OF IMPORTANT TERMS

Bayes' rule
Bernoulli trial
Binomial coefficient
Binomial theorem
Conditional probability
Continuous sample space
Discrete sample space
Elementary event
Event
Independent events
Independent experiments
Initial probability assignment
Markov chain
Partition
Probability law
Sample space
Set operations
Theorem on total probability
Tree diagram
Trellis diagram

## ANNOTATED REFERENCES

There are dozens of introductory books on probability and statistics. References [1] through [5] are my favorites. They start from the very beginning, they draw on intuition, they point out where mysterious complications lie below the surface, and they are fun to read! References [6] and [7] contain excellent introductions to computer simulation methods of random systems.

1. Y. A. Rozanov, *Probability Theory: A Concise Course,* Dover Publications, New York, 1969.
2. P. L. Meyer, *Introductory Probability and Statistical Applications,* Addison-Wesley, Reading, Mass. 1970.
3. K. L. Chung, *Elementary Probability Theory,* Springer-Verlag, New York, 1974.
4. A. B. Clarke and R. L. Disney, *Probability and Random Processes,* 2d ed., Wiley, New York, 1985.

5. L. Breiman, *Probability and Stochastic Processes,* Houghton Mifflin, Boston, 1969.
6. H. Kobayashi, *Modeling and Analysis: An Introduction to System Performance Evaluation Methods,* Addison-Wesley, Reading, Mass. 1978.
7. A. M. Law and W. D. Kelton, *Simulation Modeling and Analysis,* McGraw-Hill, New York, 1982.
8. W. Feller, *An Introduction to Probability Theory and Its Applications,* 3d ed., Wiley, New York, 1968.

## PROBLEMS

### Section 2.1 Specifying Random Experiments

1. A die is tossed and the number of dots facing up is counted and noted.
   a. What is the sample space?
   b. What is the set $A$ corresponding to the event "even number of dots are facing up"?
   c. Find the set $A^c$ and describe the corresponding event in words.
2. A die is tossed twice and the number of dots facing up is counted and noted in the order of occurrence.
   a. Find the sample space.
   b. Find the set $A$ corresponding to the event, "the total number of dots showing is even."
   c. Find the set $B$ corresponding to the event, "both dice are even."
   d. Does $A$ imply $B$ or does $B$ imply $A$?
   e. Find $A \cap B^c$ and describe this event in words.
   f. Let $C$ correspond to the event "the number of dots in the die differ by 1." Find $A \cap C$.
3. Two dice are tossed and the total number of dots facing up is counted and noted.
   a. Find the sample space.
   b. Find the set $A$ corresponding to the event "the total number of dots showing is even."
   c. Express each of the elementary events in this experiment as the union of elementary events from Problem 2.
4. A die is tossed and the number $N_1$ of dots facing up is counted and noted; an integer $N_2$ is then selected at random from the range 1 to $N_1$.
   a. Find the sample space.

b. Find the set of outcomes corresponding to the event "the die shows four dots facing up."
c. Find the set of outcomes corresponding to the event "$N_2 = 3$."
d. Find the set of outcomes corresponding to the event "$N_2 = 6$."

5. A desk drawer contains five pens, three of which are dry.
   a. The pens are selected at random one by one until a good pen is found. The sequence of test results are noted. What is the sample space?
   b. Suppose that only the number, and not the sequence, of pens tested in Part a is noted. Specify the sample space.
   c. Suppose that the pens are selected one by one and tested until both good pens have been identified, and that sequence of test results is noted. What is the sample space?
   d. Specify the sample space in Part c if only the number of pens tested is noted.
6. Two components in a system, $C_1$ and $C_2$, are tested and declared to be in one of three possible states: $F$, functioning; $R$, not functioning but repairable; and $K$, kaput.
   a. What is the sample space in this experiment?
   b. What is the set corresponding to the event "none of the components is kaput."
7. The three balls numbered 1 to 3 in an urn are drawn at random one at a time until the urn is empty. The sequence of the ball numbers is noted.
   a. Find the sample space.
   b. Find the sets $A_k$ corresponding to the events "ball number $k$ is selected in the $k$th draw," for $k = 1, 2, 3$.
   c. Find the set $A_1 \cap A_2 \cap A_3$ and describe the event in words.
   d. Find the set $A_1 \cup A_2 \cup A_3$ and describe the event in words.
   e. Find the set $(A_1 \cup A_2 \cup A_3)^c$ and describe the event in words.
8. The sample space of an experiment is the real line. Let the events $A$ and $B$ correspond to the following subsets of the real line: $A = (-\infty, r]$ and $B = (-\infty, s]$, where $r \leq s$. Find an expression for the event $C = (r, s]$ in terms of $A$ and $B$.
9. In experiment $E_9$ in Example 2.3, the lifetime of a chip is measured. Let the events $A$, $B$, and $C$ be defined by: $A = (5, \infty)$, $B = (7, \infty)$, and $C = (0, 3]$. Describe these events in words. Find the events $A \cap B$, $A \cap C$, and $A \cup B$ and describe these in words.
10. Use Venn diagrams to verify the set identities given in Eqs. (2.2) to (2.4). You will need to use different colors or different shadings to denote the various regions clearly.

11. Use Venn diagrams to verify that:
    a. If event $A$ implies $B$, and $B$ implies $C$, then $A$ implies $C$.
    b. If event $A$ implies $B$, then $B^c$ implies $A^c$.
12. Let $A$ and $B$ be events. Find an expression for the event "exactly one of the events $A$ and $B$ occurs."
13. Let $A$, $B$, and $C$ be events. Find expressions for the following events:
    a. Exactly one of the three events occurs.
    b. Exactly two of the events occur.
    c. One or more of the events occur.
    d. Two or more of the events occur.
    e. None of the events occur.
14. A system consists of three key subsystems that are duplicated for redundancy. The system is "up" if at least one unit in each of the subsystems is functioning.
    a. Let $A_{jk}$ correspond to the event "unit $k$ in subsystem $j$ is functioning," for $j = 1, 2, 3$ and $k = 1, 2$. Explain why the above problem is equivalent to the problem of having a connection in the network of switches shown in Fig. P2.1.
    b. Find a network that corresponds to a system that is "up" if both units in the first subsystem are functioning and at least one of the units in each of the other subsystems is functioning.

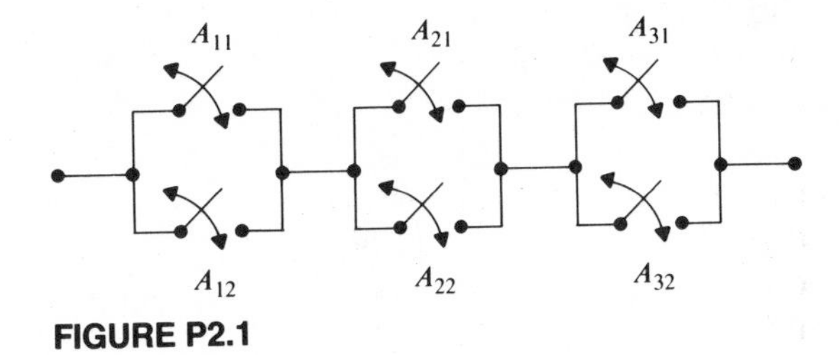


**FIGURE P2.1**

15. In a specified 24 hour period, a student wakes up at time $T_1$ and goes to sleep at some later time $T_2$.
    a. Find the sample space and sketch it on the $x$-$y$ plane if the outcome of this experiment consists of the pair $(T_1, T_2)$.
    b. Specify the set $A$ and sketch the region on the plane corresponding to the event "the student is awake at 9 o'clock."
    c. Specify the set $B$ and sketch the region on the plane corresponding to the event "the student is asleep more time than he is awake."
    d. Sketch the region corresponding to $A^c \cap B$ and describe the corresponding event in words.

## Section 2.2
## The Axioms of Probability

16. A die is tossed and the number of dots facing up is noted.
    a. Find the probability of the elementary events under the assumption that all faces of the die are equally likely to be facing up after a toss.
    b. Find the probability of the elementary events under the assumption that the face with a single dot is twice as likely to end up facing up as the rest of the faces.
    c. Find the probability that the outcome of a toss is even under the assumptions in Parts a and b of this problem.
17. A die is tossed twice and the number of dots facing up noted in the order of occurrence. Assuming that all outcomes are equally likely to occur, find the probabilities of the following events:
    a. $A_k$: the sum of the two outcomes is $k$, for $k = 2, \ldots, 12$.
    b. $B$: the outcomes of the two tosses are different.
18. Show that

$$P[A \cup B \cup C] = P[A] + P[B] + P[C] - P[A \cap B] - P[A \cap C] - P[B \cap C] + P[A \cap B \cap C].$$

19. Use the argument from Problem 18 to prove Corollary 6 by induction.
20. A fair coin is tossed four times. Let $A_i$ be the event "$i$th toss results in heads." Find the probabilities of the following events: $A_2$, $A_1 \cap A_3$, $A_1 \cap A_2 \cap A_3 \cap A_4$, and $A_1 \cup A_2 \cup A_3 \cup A_4$.
21. A fair coin is tossed until the first head comes up. Let $k$ be the number of the toss in which the first head occurs. Let $A$ be the event "$k > 5$," and let $B$ be the event "$k > 10$." Find the probabilities of $A$, $B$, $B^c$, $A \cap B$, $A \cup B$.
22. Use Corollary 7 to prove the following:
    a. $P[A \cup B \cup C] \le P[A] + P[B] + P[C]$.
    b. $P\left[\bigcup_{k=1}^{n} A_k\right] \le \sum_{k=1}^{n} P[A_k]$.

    The second expression is called the **union bound**.
23. A document consisting of $n$ characters is typed into a computer. The probability that a single character is typed incorrectly is $p$. Use the union bound to obtain an upper bound to the probability of there being any errors in the typed document.
24. A number $x$ is selected at random in the interval $[-1, 1]$. Let the events $A = \{x < 0\}$, $B = \{|x - 0.5| < 1\}$ and $C = \{x > 0.75\}$.

a. Find the probabilities of $B$, $A \cap B$, and $A \cap C$.
b. Find the probabilities of $A \cup B$, $A \cup C$, and $A \cup B \cup C$ first, by directly evaluating the sets and then their probabilities, and second, by use of the appropriate axioms or corollaries.

25. A number $x$ is selected at random in the interval $[-1, 1]$. Numbers from the subinterval $[0, 1]$ occur twice as frequently as those from $[-1, 0)$.
    a. Find the probability assignment for an interval completely within $[-1, 0)$; completely within $[0, 1]$; and partly in each of the above intervals.
    b. Repeat Problem 24 with this probability assignment.

26. The lifetime of a device behaves according to the exponential probability law introduced in Example 2.10, with $\alpha = 1$. Let $A$ be the event "lifetime is greater than 5," and $B$ be the event "lifetime is greater than 10."
    a. Find the probability of $A \cap B$, and $A \cup B$.
    b. Find the probability of the event "lifetime is greater than 5 but less than or equal to 10."

27. Consider an experiment for which the sample space is the real line. A probability law assigns probabilities to subsets of the form $(-\infty, r]$.
    a. Show that we must have $P[(-\infty, r]] \leq P[(-\infty, s]]$ when $r < s$.
    b. Find an expression for $P[(r, s]]$ in terms of $P[(-\infty, r]]$ and $P[(-\infty, s]]$.

28. Two numbers are selected at random from the interval $[0, 1]$. Find the probability that they differ by more than 1/2.

### Section *2.3
### Computing Probabilities Using Counting Methods

29. The combination to a lock is given by three numbers from the set $\{0, 1, 2, \ldots, 59\}$. Find the number of combinations possible.

30. A six-sided die is tossed, a coin is flipped, and a card is selected at random from a deck of 52 distinct cards. Find the number of possible outcomes.

31. A student has four different pairs of shoes and he never wears the same pair on two consecutive days. In how many ways can he wear shoes in 5 days?

32. How many seven-digit telephone numbers are possible if the first number is not allowed to be 0 or 1?

33. The "deluxe" pizza requires that you pick four choices from 15 available toppings. How many pizza combinations are possible if items can be repeated? If items may not be repeated?

34. In how many ways can 10 students occupy 10 desks? 12 desks?
35. A toddler pulls three volumes of an encyclopedia from a bookshelf and after being scolded places them back in random order. What is the probability that the books are in the correct order?
36. A deck of cards contains 10 red cards numbered 1 to 10 and 10 black cards numbered 1 to 10. How many ways are there of arranging the 20 cards in a row? Suppose we draw the cards at random and lay them in a row. What is the probability that red and black cards alternate?
37. A fast food counter provides onions, relish, mustard, catsup, and hot peppers for your hot dog. How many combinations are possible using one ingredient? Two ingredients? None, some, or all of the above ingredients?
38. A lot of 100 items contains $k$ defective items. $M$ items are chosen at random and tested. What is the probability that $m$ are found defective?
39. A park has $N$ raccoons of which 10 were previously captured and tagged. Suppose that 20 raccoons are captured and that 5 of these are found to be tagged. Find the probability of this happening. Denote this probability by $p(N)$. Find the value of $N$ that maximizes this probability. *Hint:* Compare this ratio $p(N)/p(N-1)$ to unity.
40. You win a lottery if you correctly predict the numbers of six balls drawn from an urn containing balls numbered $1, 2, \ldots, 49$, without replacement and without regard to ordering. What is the probability of winning if you buy one ticket?
41. How many distinct permutations are there of four red balls, two white balls, and three black balls?
42. Find the probability that the sum of the outcomes of three tosses of a die is 7.
43. In this problem we derive the multinomial coefficient. Suppose we partition a set of $n$ distinct objects into $J$ subsets $B_1, B_2, \ldots, B_J$ of size $k_1, \ldots, k_J$, respectively, where $k_i \geq 0$, and $k_1 + k_2 + \cdots + k_J = n$.
    a. Let $N_i$ denote the number of possible outcomes when the $i$th subset is selected. Show that

$$N_1 = \binom{n}{k_1}, \qquad N_2 = \binom{n-k_1}{k_2}, \ldots, N_{J-1} = \binom{n-k_1-\ldots-k_{J-2}}{k_{J-1}}.$$

    b. Show that the number of partitions is then

$$N_1 N_2 \ldots N_{J-1} = \frac{n!}{k_2!\, k_2! \ldots k_J!}.$$

## Section 2.4 Conditional Probability

44. Find $P[A \mid B]$ if $A \cap B = \varnothing$, if $A \subset B$, and if $B \subset A$.
45. Show that if $P[A \mid B] > P[A]$, then $P[B \mid A] > P[B]$. Comment.
46. Show that $P[A \cap B \cap C] = P[A \mid B \cap C][P[B \mid C]P[C]$.
47. A die is tossed twice and the number of dots facing up is counted and noted in order of occurrence. Let $A$ be the event "total number of dots is even," and let $B$ be the event "both tosses had an even number of dots." Find $P[A \mid B]$ and $P[B \mid A]$.
48. A number $x$ is selected at random in the interval $[-1, 1]$. Let $B$ the event $\{|x - 1/2| < 1\}$ and let $C$ be the event $\{x > 3/4\}$. Find $P[B \mid C]$ and $P[C \mid B]$.
49. In each lot of 100 items 5 items are tested, and the lot is rejected if any of the tested items is found defective.
    a. Find the probability of accepting a lot with 5 defective items. Repeat for 10 defective items.
    b. Recompute the probabilities in Part a if a lot is accepted if at most 1 of the tested items is found defective.
50. Find the probability that 2 or more students in a class of 20 students have the same birthday. *Hint:* Use Corollary 1.
51. Use conditional probabilities and tree diagrams to find the probabilities for the elementary events in the random experiments defined in Parts a to d of Problem 2.5.
52. The arrival time of a professor to his office is uniformly distributed in the interval between 8 and 9 AM Find the probability that the professor will arrive during the next minute given that he has not arrived by 9:30. Repeat for 9:50. Comment on the results.
53. A nonsymmetric binary communications channel is shown in Fig. P2.2. Assume the inputs are equiprobable.
    a. Find the probability that the output is 0.
    b. Find the probability that the input was 0 given that the output is

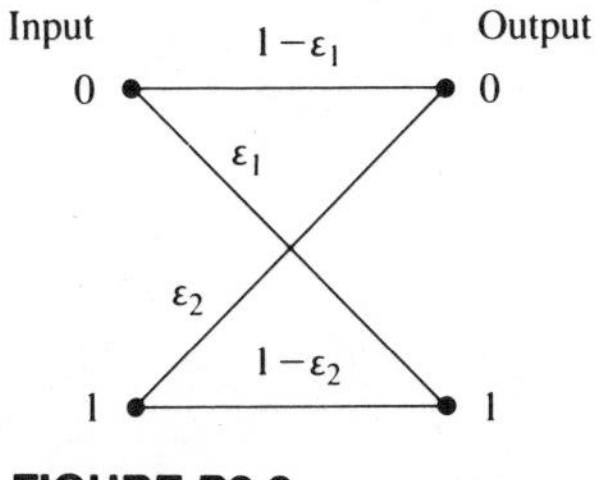

FIGURE P2.2

1. Find the probability that the input was 1 given the output is a 1. Which input is more probable?

54. A die is tossed and the number of dots $N_1$ is noted; an integer $N_2$ is then selected at random from $\{1, \ldots, N_1\}$.
    a. Use a tree diagram to specify the sample space.
    b. Find the probability of the event $\{N_2 = 3\}$.
    c. Find the probability of the event $\{N_1 = 4\}$ given $\{N_2 = 3\}$.
    d. Find the probability of $\{N_1 = 4\}$ given $\{N_2 = 5\}$.

55. One of two coins is selected at random and tossed. The first coin comes up heads with probability $p_1$ and the second coin with probability $p_2$.
    a. What is the probability that the outcome of the toss is heads?
    b. What is the probability that coin 2 was used given that a "heads" occurred?

56. A ternary communication channel is shown in Fig. P2.3. Suppose that the input symbols 0, 1, 2 occur with probability 1/2, 1/4, and 1/4, respectively.
    a. Find the probabilities of the output symbols.
    b. Suppose that a 1 is observed as an output. What is the probability that the input was 0, 1, 2?

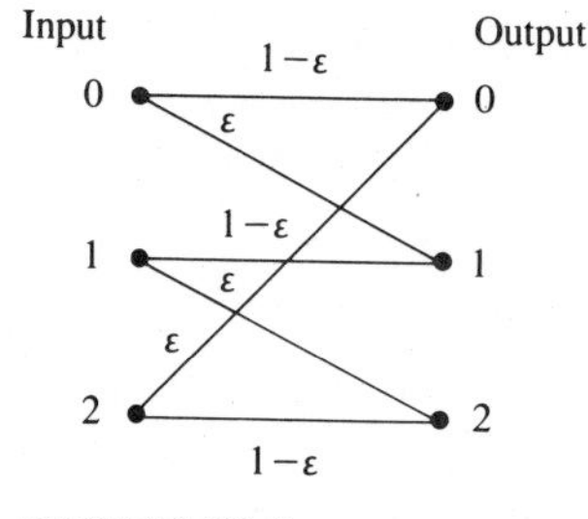


**FIGURE P2.3**

57. A computer manufacturer uses chips from three sources. Chips from sources $A$, $B$, and $C$ are defective with probabilities, .001, .005, and .01. If a randomly selected chip is found to be defective, find the probability that the manufacturer was $A$; $C$.

## Section 2.5
## Independence of Events

58. Show that if $A$ and $B$ are independent events, then the pairs $A$ and $B^c$, $A^c$ and $B$, and $A^c$ and $B^c$ are also independent.

59. Show that events $A$ and $B$ are independent if $P[A \mid B] = P[A \mid B^c]$.

60. Let $A$ and $B$ be events with probabilities $P[A]$ and $P[B]$.

a. Find $P[A \cup B]$ if $A$ and $B$ are independent.
b. Find $P[A \cup B]$ if $A$ and $B$ are mutually exclusive.

61. An experiment consists of picking one of two urns at random and then selecting a ball from the urn and noting its color (black or white). Let $A$ be the event "urn 1 is selected" and $B$ the event "a black ball is observed." Under what conditions are $A$ and $B$ independent?
62. Find the probabilities in Problem 2.13 assuming that events $A$, $B$, and $C$ are independent.
63. Find the probabilities that the systems are "up" in Problem 2.14, Parts a and b. Assume that all units in the system fail independently and that a type $k$ unit fails with probability $p_k$.
64. A random experiment is repeated a large number of times and the occurrence of events $A$ and $B$ is noted. How would you test whether events $A$ and $B$ are independent?
65. Compute the probability of the system in Example 2.32 being "up" when a second controller is added to the system.

### Section 2.6
### Sequential Experiments

66. A block of 100 bits is transmitted over a binary communications channel with probability of bit error $p = 10^{-3}$. Find the probability that the block contains three or more errors.
67. Ten percent of items from a certain production line are defective. What is the probability that there is more than one defective item in a batch of $n$ items?
68. A student needs 10 chips of a certain type in a circuit. It is known that 5% of these chips are defective. How many chips should he buy for there to be a greater than 90% probability of having enough chips for the circuit?
69. Let $k$ be the number of active speakers in a group of $n$ noninteracting (i.e., independent) speakers. Write a computer program to compute $P[k]$ for $k = 0, \ldots, n$ and $n = 8$, 24, and 48. Use your results to recompute Fig. 1.6 in Chapter 1.
70. A system contains 10 chips. The lifetime of each chip has an exponential probability law with parameter $\alpha$. Find the probability that at least half of the chips are functioning after $1/\alpha$ seconds.
71. One of two coins is selected at random and tossed three times. The coins are known to have probabilities of "heads," $p_1$ and $p_2$, respectively, with $p_1 > p_2$.

    a. Find the probability that coin 1 was tossed given that $k$ heads were observed, for $k = 0, 1, 2, 3$.

b. In Part a, which coin is more probable when $k$ heads have been observed?
c. Generalize the solution in Part b to the case where the selected coin is tossed $m$ times. In particular, find a threshold value $T$ such that for $k > T$ heads are observed, coin 1 is more probable, and when $k \leq T$, coin 2 is more probable.

72. A machine makes errors in a certain operation with probability $p$. There are two types of errors: type 1 errors occur with probability $a$ and type 2 with probability $1 - a$.
    a. What is the probability of $k$ errors in $n$ operations?
    b. What is the probability of $k_1$ type 1 errors in $n$ operations?
    c. What is the probability of $k_2$ type 2 errors in $n$ operations?
    d. What is the joint probability of $k_1$ and $k_2$ type 1 and 2 errors, respectively, in $n$ operations?
73. Four types of messages arrive at a message center. A type $k$ message arrives with probability $p_k$. Find the probability that $k$ out of the $N$ messages are type 1 or type 2.
74. A runlength coder segments a binary information sequence into strings that consist of either a "run" of $k$ "zeros" punctuated by a "one", for $k = 0, \ldots, m - 1$ or a string of $m$ "zeros." The $m = 3$ case is:

| *string* | *runlength k* |
|---|---|
| 1 | 0 |
| 01 | 1 |
| 001 | 2 |
| 000 | 3 |

Suppose that the information sequence is produced by a sequence of independent Bernoulli trials with $P[\text{"one"}] = P[\text{success}] = p$.
    a. Find the probability of runlength $k$ in the $m = 3$ case.
    b. Find the probability of runlength $k$ for general $m$.
75. The time spent by cars in a parking lot follows an exponential probability law with parameter 1. The charge for parking in the lot is $1 for each 1/2 hour or less.
    a. Find the probability that a car pays $k$ dollars.
    b. Suppose that there is a maximum charge of $5. Find the probability that a car pays $k$ dollars.
76. A biased coin is tossed repeatedly until "heads" has occurred twice. Find the probability that $k$ tosses are required. *Hint:* Let $A$ be the event "$k$ tosses are required" and $B$ be "$k$th toss is heads," and find $P[A \mid B]$.

77. An urn initially contains two black balls and two white balls. The following experiment is repeated indefinitely: A ball is drawn from the urn; if the ball is white it is put back in the urn, otherwise it is left out.
    a. Draw the trellis diagram for this experiment and label the branches by the transition probabilities.
    b. Find the probabilities for the sequences www, bww, bbw, and bbwww.
    c. Find the probability that the urn contains no black balls after three draws.

78. In Example 2.42, let $p_0(n)$ and $p_1(n)$ be the probabilities that urn 0 or urn 1 are used in the $n$th subexperiment.
    a. Find $p_0(1)$ and $p_1(1)$.
    b. Express $p_0(n + 1)$ and $p_1(n + 1)$ in terms of $p_0(n)$ and $p_1(n)$.
    c. Find the solution to the recursion in Part b with the initial conditions given in Part a.
    d. What are the urn probabilities as $n$ approaches infinity?

### Section *2.7 A Computer Method for Synthesizing Randomness: Random Number Generators

79. Write a computer program to implement the random number generator specified by Eq. 2.32.
    a. To check your program find $Z_{1000}$ with the initial seed $Z_0 = 1$, it should be 522,329,230.
    b. Generate 10,000 random numbers in the unit interval and truncate these numbers to a single decimal place. Compute histograms for single numbers, pairs of numbers, and triplets of numbers. Note that each number will be involved in the count of two pairs of numbers, once as the first component and once as the second component, and similarly for triplets of numbers. Do the histograms correspond to the results expected?

80. Suppose you have a program that gives you numbers $U_n$ that are uniformly distributed in the interval $[0, 1]$. Let $Y_n = \alpha U_n + \beta$. Find $\alpha$ and $\beta$ so that $Y_n$ is uniformly distributed in the interval $[a, b]$. Let $a = -5$ and $b = 15$. Write a computer program to generate $Y_n$ and to compute the sample mean and sample variance in 1000 repetitions. Compare the sample mean and variance to $(a + b)/2$ and $(b - a)^2/12$, respectively.

81. Suppose you have a program that gives you numbers $U_n$ that are uniformly distributed in the interval $[0, 1]$.
    a. Suppose we generate a sequence $B_n$ of *binary* numbers in the

following manner: if $U_n \leq 1/2$, then $B_n = 0$; otherwise $B_n = 1$. In what way does the sequence of $B_n$'s simulate the flipping of a fair coin?

b. How would you modify the procedure in Part a to simulate a sequence of Bernoulli trials?

c. How would you modify the procedure in Part a to simulate the tossing of a fair die?

d. How would you modify the procedure in Part a to simulate *any* random experiment that has a finite number of outcomes? Can the method handle countable number of outcomes?

82. Write a computer program to simulate the urn experiment discussed in Section 1.3. Compute the relative frequencies of the outcomes in 1000 draws from the urn.

83. Find a method for using a sequence $U_n$ of uniformly distributed random numbers in $[0, 1]$ to generate a sequence of integers that obey the geometric probability law. Write a program to simulate the coin tossing experiment discussed in Example 2.8. Repeat the experiment 500 times and compute the tree diagram shown in Fig. 2.6.

84. Find a method for using a sequence $U_n$ of uniformly distributed random numbers in $[0, 1]$ to generate a sequence of integers that obey the binomial probability law. Modify the method to generate vectors $(k_1, \ldots, k_M)$ that obey the multinomial probability law.

CHAPTER 3

# Random Variables

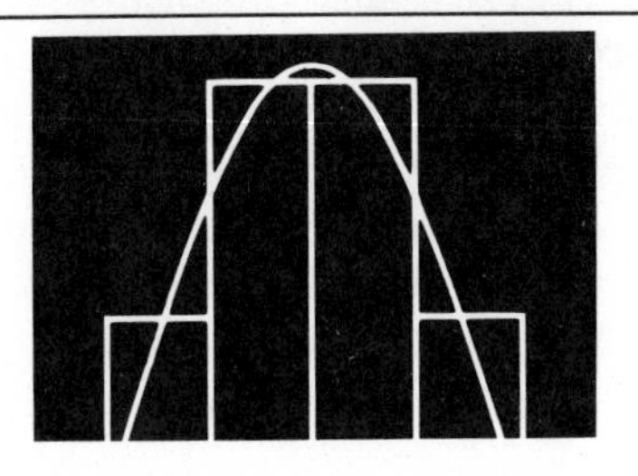

This chapter develops methods that are useful in computing the probabilities of events involving numerical attributes of the outcomes of a random experiment. The cumulative distribution function is introduced. The probability of all events that are intervals of the real line or unions of such intervals can be expressed in terms of the cumulative distribution function. The probability density function is also introduced. The probability of events can be expressed as integrals of the probability density function. The notion of the expected value of a random variable is shown to correspond to our intuitive notion of average. These concepts provide us with the tools that enable us to evaluate the probabilities and averages that are of interest in the design of systems that involve randomness.

## 3.1

## THE NOTION OF A RANDOM VARIABLE

The outcome of a random experiment need not be a number. However, we are usually interested not in the outcome itself, but rather in some measurement or numerical attribute of the outcome. For example, in $n$ tosses of a coin, we may be interested in the total number of heads and not in the specific order in which heads and tails occur. In a randomly selected computer job, we may be interested only in the execution time of the job. In the selection of a student's name from an urn, we may be interested in the weight of the student. In each of these examples, *a measurement assigns a numerical value to the outcome of the random experiment.* Since the outcomes are random, the results of the measurements will also be random. Hence it makes sense to talk about the probabilities of the resulting numerical values. The concept of a random variable formalizes this notion.

A **random variable** $X$ is a function that assigns a real number, $X(\zeta)$, to each outcome $\zeta$ in the sample space of a random experiment. Recall that a function is simply a rule for assigning a numerical value to each element of a set as shown pictorially in Fig. 3.1. The specification of a measurement

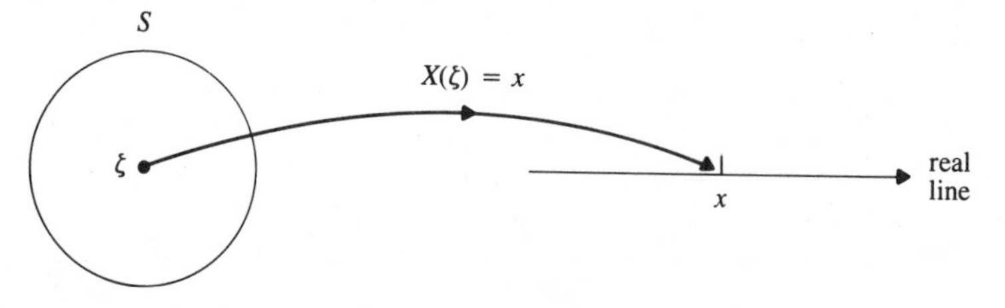


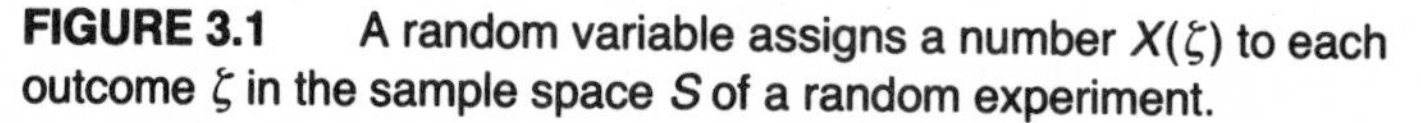
**FIGURE 3.1** A random variable assigns a number $X(\zeta)$ to each outcome $\zeta$ in the sample space $S$ of a random experiment.

on the outcome of a random experiment defines a function on the sample space, and hence a random variable.

**■■ Example 3.1**

Suppose that a coin is tossed three times and the sequence of heads and tails is noted. The sample space for this experiment is $S = \{\text{hhh, hht, hth, htt, thh, tht, tth, ttt}\}$. Now let $X$ be the number of heads in three coin tosses. $X$ assigns each outcome $\zeta$ in $S$ a number from the set $S_X = \{0, 1, 2, 3\}$. The table below lists the eight outcomes of $S$ and the corresponding values of $X$.

| $\zeta$: | hhh | hht | hth | thh | htt | tht | tth | ttt |
|---|---|---|---|---|---|---|---|---|
| $X(\zeta)$: | 3 | 2 | 2 | 2 | 1 | 1 | 1 | 0 |

$X$ is then a random variable taking on values in the set $S_X = \{0, 1, 2, 3\}$. ■■

**■■ Example 3.2**

If the outcome $\zeta$ of some experiment is already a numerical value, we can immediately reinterpret the outcome as a random variable defined by the identity function, $X(\zeta) = \zeta$. Thus many of the examples considered in the previous chapter can now be viewed as random variables. ■■

The function or rule that assigns values to each outcome is fixed and deterministic, as for example, in the rule "count the number of heads in three tosses of a coin." The randomness in the observed values is due to the underlying randomness of the arguments of the function $X$, namely the experiment outcomes $\zeta$. In other words, the randomness in the observed values of $X$ are *induced* by the underlying random experiment, and we should therefore be able to compute the probabilities of the observed values in terms of the probabilities of the underlying outcomes.

**■■ Example 3.3**

The event $\{X = k\} = \{k \text{ heads in three coin tosses}\}$ occurs when the outcome of the coin tossing experiment contains $k$ heads. The probability of the event $\{X = k\}$ is therefore given by the sum of the probabilities of the corresponding outcomes or elementary events. In Example 2.34, we found the probabilities of the elementary events of the coin tossing experiment. Thus we have

$$\begin{aligned}
p_0 &= P[X = 0] = P[\{\text{ttt}\}] = (1 - p)^3, \\
p_1 &= P[X = 1] = P[\{\text{htt}\}] + P[\{\text{tht}\}] + P[\{\text{tth}\}] = 3(1 - p)^2 p, \\
p_2 &= P[X = 2] = P[\{\text{hht}\}] + P[\{\text{hth}\}] + P[\{\text{thh}\}] = 3(1 - p)p^2,
\end{aligned}$$

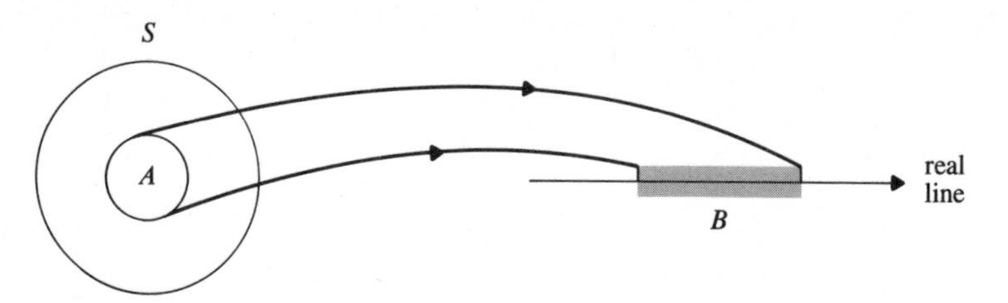


**FIGURE 3.2** $P[X \text{ in } B] = P[\zeta \text{ in } A]$ since $X$ is in $B$ only when $\zeta$ is in $A$, where $A = \{\zeta: X(\zeta) \text{ in } B\}$.

and

$$p_3 = P[X = 3] = P[\{\text{hhh}\}] = p^3.$$

The $p_k$'s can be used to obtain the probabilities of all events that involve $X$. As long as we are concerned only with the values of $X$, we can ignore the underlying experiment with sample space $S$, and proceed as if the experiment consisted of sample space $S_X$ with probabilities $p_k$. ■■

Example 3.3 illustrates the following general technique for finding the probabilities of events involving the random variable $X$. Let $S_X$ be the set of values that can be taken on by $X$, and let $B$ be some subset of $S_X$. $S_X$ can be viewed as a new sample space, and $B$ as an event in the sample space. Let $A$ be the set of outcomes $\zeta$ in $S$ that lead to values $X(\zeta)$ in $B$, as shown in Fig. 3.2, that is,

$$A = \{\zeta: X(\zeta) \text{ in } B\},$$

then the event $B$ in $S_X$ occurs whenever the event $A$ in $S$ occurs. Thus the probability of event $B$ is given by

$$P[B] = P[A] = P[\{\zeta: X(\zeta) \text{ in } B\}].$$

We refer to events $A$ and $B$ as **equivalent events.**

All numerical events of practical interest involve events of the form $\{X = x\}$, where $x$ is a number, or $\{X \text{ in } I\}$, where $I$ is some interval or union of intervals. In the next section we show that the probability of all such events can be expressed in terms of $P[\{X \le x\}]$, where $x$ is a real number. We can therefore compute the probability of events in $S_X$ if we know the probability of the underlying event $\{\zeta: X(\zeta) \le x\}$.

## 3.2

## THE CUMULATIVE DISTRIBUTION FUNCTION

The **cumulative distribution function** (cdf) of a random variable $X$ is defined as the probability of the event $\{X \le x\}$:

$$F_X(x) = P[X \le x] \qquad \text{for } -\infty < x < +\infty, \tag{3.1}$$

that is, it is the probability that the random variable $X$ takes on a value in the set $(-\infty, x]$. In terms of the underlying sample space, the cdf is the probability of the event $\{\zeta : X(\zeta) \le x\}$. The event $\{X \le x\}$ and its probability vary as $x$ is varied; in other words $F_X(x)$ is a function of the variable $x$.

The cdf is simply a convenient way of specifying the probability of all semi-infinite intervals of the real line of the form $(-\infty, x]$. The events of interest when dealing with numbers are intervals of the real line, and their complements, unions, and intersections. We show below that the probabilities of all of these events can be expressed in terms of the cdf.

The cdf has the following interpretation in terms of relative frequency. Suppose that the experiment that yields the outcome $\zeta$, and hence $X(\zeta)$, is performed a large number of times. $F_X(b)$ is then the long-term proportion of times in which $X(\zeta) \le b$.

The axioms of probability and their corollaries imply that the cdf has the following properties:

i. $0 \le F_X(x) \le 1$.

ii. $\lim\limits_{x\to\infty} F_X(x) = 1$.

iii. $\lim\limits_{x\to-\infty} F_X(x) = 0$.

iv. $F_X(x)$ is a nondecreasing function of $x$, that is, if $a < b$, then $F_X(a) \le F_X(b)$.

v. $F_X(x)$ is continuous from the right, that is, for $h > 0$ $F_X(b) = \lim\limits_{h\to 0} F_X(b + h) = F_X(b^+)$.

The first property follows from the fact that the cdf is a probability and hence must satisfy Axiom I and Corollary 2. The second property follows from the fact that the event $\{X < \infty\}$ consists of all the real numbers and is thus the entire sample space (Axiom II). The third property follows from the fact that all real numbers are greater than $-\infty$, so the event $\{X \le -\infty\}$ is the empty set (Corollary 3). To obtain the fourth property, note that the event $\{X \le a\}$ is a subset of $\{X \le b\}$ so it must have smaller or equal probability (Corollary 7). We will see how the fifth property comes about in Example 3.4.[1]

The probability of events that correspond to intervals of the form $\{a < X \le b\}$ can be expressed in terms of the cdf:

vi. $P[a < X \le b] = F_X(b) - F_X(a)$. (3.2)

1. The proof that the cdf is continuous on the right is beyond the level intended here. The proof can be found in Davenport (1970, 116–121).

To show Eq. (3.2) we note that since

$$\{X \le a\} \cup \{a < X \le b\} = \{X \le b\},$$

and since the two events on the left-hand side are mutually exclusive, we have by Axiom III and by Eq. (3.1) that

$$F_X(a) + P[a < X \le b] = F_X(b).$$

Equation (3.2) then follows.

Equation (3.2) allows us to compute the probability of the event $\{X = b\}$. Let $a = b - \varepsilon$ in Eq. (3.2), $\varepsilon > 0$, then

$$P[b - \varepsilon < X \le b] = F_X(b) - F_X(b - \varepsilon).$$

As $\varepsilon \to 0$, the left side of the above equation approaches $P[X = b]$, thus

$$\text{vii.} \qquad P[X = b] = F_X(b) - F_X(b^-). \tag{3.3}$$

Thus the probability that an arbitrary random variable $X$ takes on a specific value, say $b$, is given by the magnitude of the jump of the cdf at the point $b$. It follows that if the cdf is continuous at a point $b$, then the event $\{X = b\}$ has probability zero.

Equation (3.3) can be combined with Eq. (3.2) to compute the probabilities of other types of intervals. For example, since

$$\{a \le X \le b\} = \{X = a\} \cup \{a < X \le b\},$$

then

$$\begin{aligned} P[a \le X \le b] &= P[X = a] + P[a < X \le b] \\ &= F_X(a) - F_X(a^-) + F_X(b) - F_X(a) \\ &= F_X(b) - F_X(a^-). \end{aligned} \tag{3.4}$$

Note that if the cdf is continuous at the endpoints of an interval, then the endpoints have zero probability, and therefore they can be included in, or excluded from, the interval without affecting the probability. In other words, if the cdf is continuous at the points $x = a$ and $x = b$, then the following probabilities are equal:

$$P[a < X < b], \quad P[a \le X < b], \quad P[a < X \le b], \text{ and } P[a \le X \le b].$$

## ■■ Example 3.4

Figure 3.3(a) shows the cdf of the random variable $X$, which is defined as the number of heads in three tosses of a fair coin. From Example 3.1 we know that $X$ takes on only the values 0, 1, 2, and 3 with probabilities 1/8, 3/8, 3/8, and 1/8, respectively, so $F_X(x)$ is simply the sum of the probabilities of the outcomes from $\{0, 1, 2, 3\}$ that are less than or equal to $x$. The resulting cdf is seen to have discontinuities at the points $0, 1, 2, 3$.

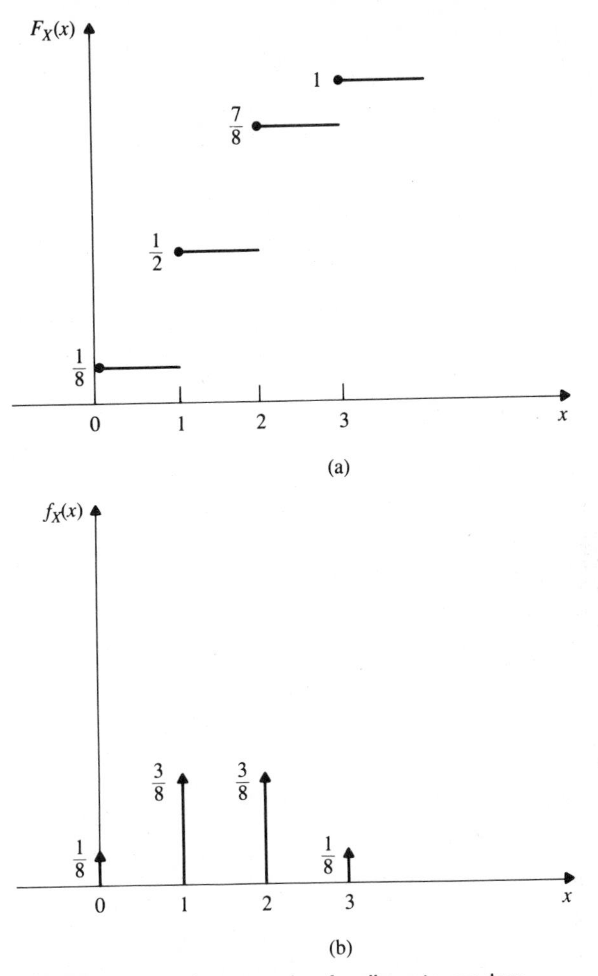


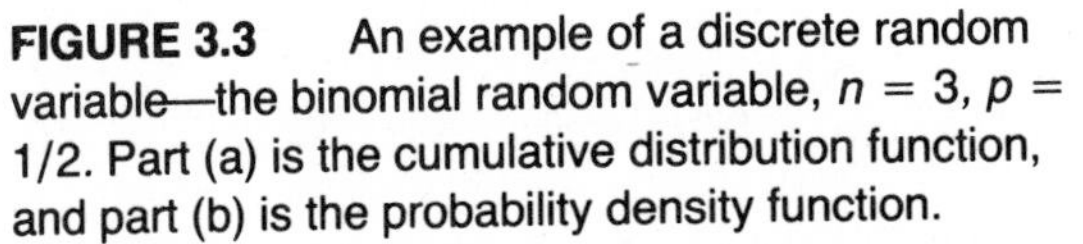

**FIGURE 3.3** An example of a discrete random variable—the binomial random variable, $n = 3$, $p = 1/2$. Part (a) is the cumulative distribution function, and part (b) is the probability density function.

Let us take a closer look at one of these discontinuities. Consider the cdf in the vicinity of the point $x = 1$. For $\delta$ a small positive number, we have

$$F_X(1 - \delta) = P[X \leq 1 - \delta] = P\{0 \text{ heads}\} = \frac{1}{8},$$

so the limit of the cdf as $x$ approaches 1 from the left is 1/8. However,

$$F_X(1) = P[X \leq 1] = P[0 \text{ or } 1 \text{ heads}] = \frac{1}{8} + \frac{3}{8} = \frac{1}{2},$$

and also

$$F_X(1+\delta) = P[X \le 1+\delta] = P[0 \text{ or } 1 \text{ heads}] = \frac{1}{2}.$$

Thus the cdf is continuous from the right and equal to 1/2 at the point $x = 1$. Indeed we note the magnitude of the jump at the point $x = 1$ is equal to $P[X = 1] = 1/2 - 1/8 = 3/8$.

The cdf can be written compactly in terms of the unit step function:

$$u(x) = \begin{cases} 0 & x < 0 \\ 1 & x \ge 0, \end{cases}$$

then

$$F_X(x) = \frac{1}{8}u(x) + \frac{3}{8}u(x-1) + \frac{3}{8}u(x-2) + \frac{1}{8}u(x-3).$$ ■■

---

### ■■ Example 3.5
Exponential Random Variable

The transmission time $X$ of messages in a communication system obey the exponential probability law with parameter $\lambda$, that is,

$$P[X > x] = e^{-\lambda x} \qquad x > 0.$$

Find the cdf of $X$. Find $P[T < X \le 2T]$, where $T = 1/\lambda$.

The cdf of $X$ is $F_X(x) = P[X \le x] = 1 - P[X > x]$:

$$F_X(x) = \begin{cases} 0 & x < 0 \\ 1 - e^{-\lambda x} & x \ge 0. \end{cases}$$

The cdf is shown in Fig. 3.4(a). From property (vi) we have

$$P[T < X \le 2T] = 1 - e^{-2} - (1 - e^{-1}) = e^{-1} - e^{-2} \simeq .233.$$

Note that $F_X(x)$ is continuous for all $x$. Note also that its derivative exists everywhere except at $x = 0$:

$$F'_X(x) = \begin{cases} 0 & x < 0 \\ \lambda e^{-\lambda x} & x > 0. \end{cases}$$

$F'(x)$ is shown in Fig. 3.4(b). ■■

---

### ■■ Example 3.6

The waiting time $W$ of a customer in a queueing system is zero if he finds the system idle, and an exponentially distributed random length of time if he finds the system busy. The probabilities that he finds the system idle or busy are $p$ and $1 - p$, respectively. Find the cdf of $W$.

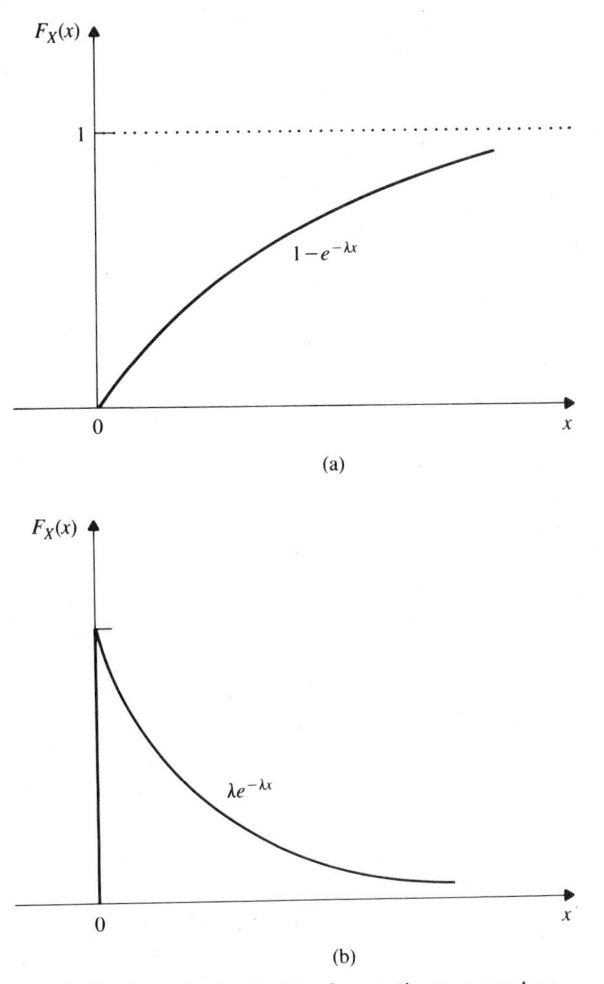


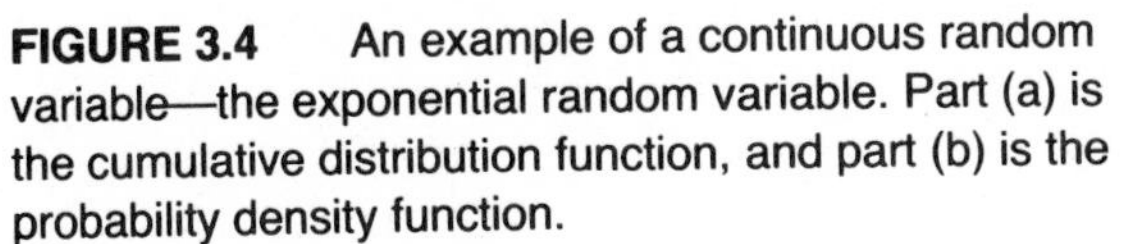

**FIGURE 3.4** An example of a continuous random variable—the exponential random variable. Part (a) is the cumulative distribution function, and part (b) is the probability density function.

The cdf of $W$ is found as follows:

$$\begin{aligned} F_X(x) &= P[W \leq x] \\ &= P[W \leq x \mid \text{idle}]p + P[W \leq x \mid \text{busy}](1-p), \end{aligned}$$

where the last equality used the theorem of total probability, Eq. (2.26). Noting that $P[W \leq x \mid \text{idle}] = 1$ when $x \geq 0$ and 0 otherwise, we have

$$F_X(x) = \begin{cases} 0 & x < 0 \\ p + (1-p)(1-e^{-\lambda x}) & x \geq 0. \end{cases}$$

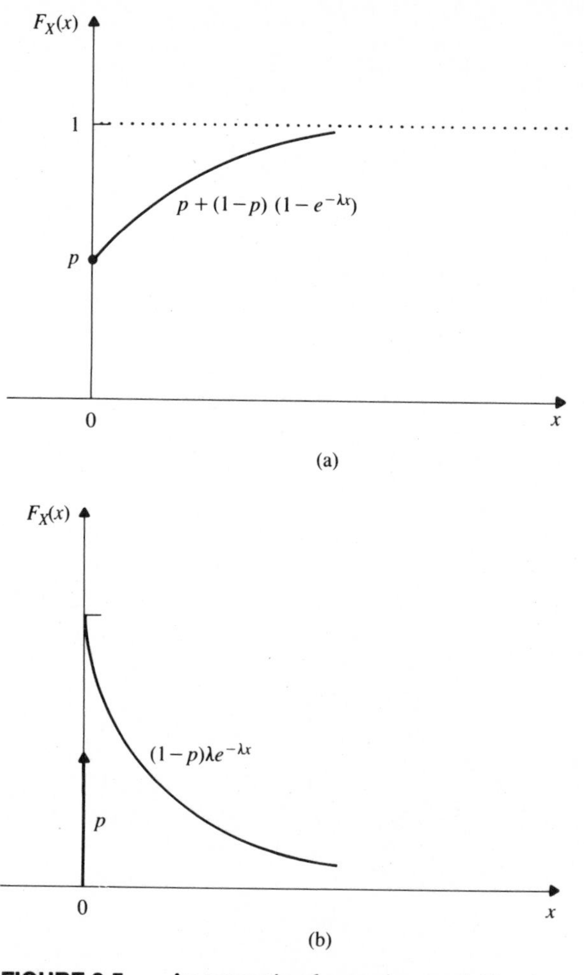


**FIGURE 3.5** An example of a random variable of mixed type. Part (a) is the cumulative distributive function, and part (b) is the probability density function.

The cdf is shown in Fig. 3.5(a). Note that $F_X(x)$ can be expressed as the sum of a step function with amplitude $p$ and a continuous function of $x$. ■■

---

### The Three Types of Random Variables

The random variables in Examples 3.4, 3.5, and 3.6 are typical of the three basic types of random variable that we will be interested in.

A **discrete random variable** is defined as a random variable whose cdf is a right-continuous, staircase function of $x$, with jumps at a countable set of points $x_0, x_1, x_2, \ldots$. The random variable in Example 3.4 is an example of a discrete random variable. Discrete random variables take on values from a finite or at most a countably infinite set $S_X = \{x_0, x_1, \ldots\}$. They arise mostly in applications that involve counting, so we usually have $S_X = \{0, 1, 2, \ldots\}$.

The cdf of a discrete random variable can be written as the weighted sum of unit step functions as in Example 3.4:

$$F_X(x) = \sum_k p_X(x_k) u(x - x_k), \tag{3.5}$$

where $p_X(x_k) = P[X = x_k]$ gives the magnitude of the jumps in the cdf. The set of probabilities $p_X(x_k)$ is called the **probability mass function (pmf) of *X*.**

A **continuous random variable** is defined as a random variable whose cdf $F_X(x)$ is continuous everywhere, and which, in addition, is sufficiently smooth that it can be written as an integral of some nonnegative function $f(x)$:

$$F_X(x) = \int_{-\infty}^{x} f(t)\, dt. \tag{3.6}$$

For continuous random variables, the cdf is continuous everywhere, so property (vii) implies that $P[X = x] = 0$ for all $x$. The cdf of the random variable discussed in Example 3.5 is a continuous random variable since its cdf is continuous everywhere, and since Eq. (3.6) is satisfied if we let $f(x) = F_X'(x)$ as given in the example.

A **random variable of mixed type** is a random variable with a cdf that has jumps on a countable set of points $x_0, x_1, x_2, \ldots$ but that also increases continuously over at least one interval of values of $x$. The cdf for these random variables has the form

$$F_X(x) = pF_1(x) + (1 - p)F_2(x),$$

where $0 < p < 1$, and $F_1(x)$ is the cdf of a discrete random variable and $F_2(x)$ is the cdf of a continuous random variable. The random variable in Example 3.6 is of mixed type.

Random variables of mixed type can be viewed as being produced by a two-step process: A coin is tossed; if the outcome of the toss is heads, a discrete random variable is generated according to $F_1(x)$, otherwise, a continuous random variable is generated according to $F_2(x)$.

## 3.3

## THE PROBABILITY DENSITY FUNCTION

The **probability density function of $X$** (pdf), if it exists, is defined as the derivative of $F_X(x)$:

$$f_X(x) = \frac{dF_X(x)}{dx}. \tag{3.7}$$

In this section we show that the pdf is an alternative, and more useful, way of specifying the information contained in the cumulative distribution function.

The pdf represents the "density" of probability at the point $x$ in the following sense: The probability that $X$ is in a small interval in the vicinity of $x$, that is $\{x < X \leq x + h\}$, is

$$\begin{aligned} P[x < X \leq x + h] &= F_X(x + h) - F_X(x) \\ &= \frac{F_X(x + h) - F_X(x)}{h} h. \end{aligned} \tag{3.8}$$

If the cdf has a derivative at $x$, then as $h$ becomes very small

$$P[x < X \leq x + h] \simeq f_X(x)h. \tag{3.9}$$

Thus $f_X(x)$ represents the "density" of probability at the point $x$ in the sense that the probability that $X$ is in a small interval in the vicinity of $x$ is approximately $f_X(x)h$. The derivative of the cdf, when it exists, is positive since the cdf is a nondecreasing function of $x$, thus

i. $$f_X(x) \geq 0. \tag{3.10}$$

Equations (3.9) and (3.10) provide us with an alternative approach to specifying the probabilities involving the random variable $X$. We can begin by stating a nonnegative function $f_X(x)$, called the probability density function, which specifies the probabilities of events of the form "$X$ falls in a small interval of width $dx$ about the point $x$," as shown in Fig. 3.6(a). The probabilities of events involving $X$ are then expressed in terms of the pdf by adding the probabilities of intervals of width $dx$. As the widths of the intervals approach zero, we obtain an integral in terms of the pdf. For example, the probability of an interval $[a, b]$, is

ii. $$P[a \leq X \leq b] = \int_a^b f_X(x)\,dx. \tag{3.11}$$

*The probability of an interval is therefore the area under $f_X(x)$ in that interval* as shown in Fig. 3.6(b). The probability of any event that consists of the union of disjoint intervals can thus be found by adding the integrals of the pdf over each of the intervals.

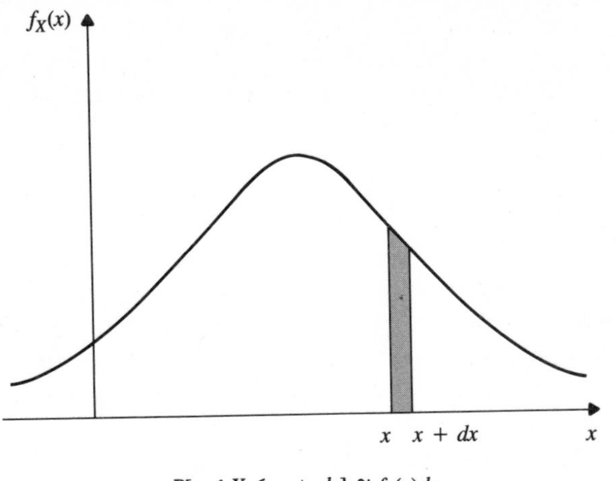


$P[x < X \leqslant x + dx] \cong f_X(x)dx$

(a)

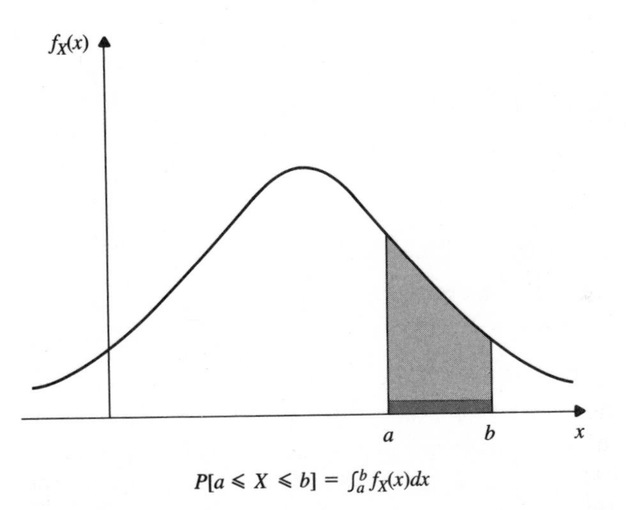


$P[a \leqslant X \leqslant b] = \int_a^b f_X(x)dx$

(b)

**FIGURE 3.6** (a) The probability density function specifies the probability of intervals of infinitesimal width. (b) The probability of an interval $[a, b]$ is the area under the pdf in that interval.

The cdf of $X$ can be obtained by integrating the pdf:

iii. $$F_X(x) = \int_{-\infty}^{x} f_X(t)\,dt. \qquad (3.12)$$

In Section 3.2, we defined a *continuous random variable* as a random variable $X$ whose cdf was given by Eq. (3.12). Since the probabilities of all events involving $X$ can be written in terms of the cdf, it then follows that these probabilities can be written in terms of the pdf. Thus *the pdf completely specifies the behavior of continuous random variables.*

By letting $x$ tend to infinity in Eq. (3.12), we obtain a *normalization* condition for pdf's:

$$\text{iv.} \quad 1 = \int_{-\infty}^{+\infty} f_X(t)\,dt. \tag{3.13}$$

The pdf reinforces the intuitive notion of probability as having attributes similar to "physical mass." Thus Eq. (3.11) states that the probability "mass" in an interval is the integral of the "density of probability mass" over the interval. Equation (3.13) states that the total mass available is one unit.

*A valid pdf can be formed from any nonnegative, piecewise continuous function $g(x)$ that has a finite integral*:

$$\int_{-\infty}^{\infty} g(x)\,dx = c < \infty. \tag{3.14}$$

By letting $f_X(x) = g(x)/c$, we obtain a function that satisfies the normalization condition. Note that the pdf must be defined for all real values of $x$; if $X$ does not take on values from some region of the real line, we simply set $f_X(x) = 0$ in the region.

### ■■ Example 3.7
Uniform Random Variable

The pdf of the uniform random variable is given by:

$$f_X(x) = \begin{cases} \dfrac{1}{b-a} & a \le x \le b \\ 0 & x < a \quad \text{and} \quad x > b \end{cases} \tag{3.15a}$$

and is shown in Fig. 3.7(a). The cdf is found from Eq. (3.12):

$$F_X(x) = \begin{cases} 0 & x < a \\ \dfrac{x-a}{b-a} & a \le x \le b \\ 1 & x > b \end{cases} \tag{3.15b}$$

The cdf is shown in Fig. 3.7(b) ■■

### ■■ Example 3.8

The pdf of the samples of speech waveforms is found to decay exponentially at a rate $\alpha$, so the following pdf is proposed:

$$f_X(x) = ce^{-\alpha|x|} \qquad -\infty < x < \infty. \tag{3.16}$$

Find the constant $c$, and then find the probability $P[|X| < v]$.

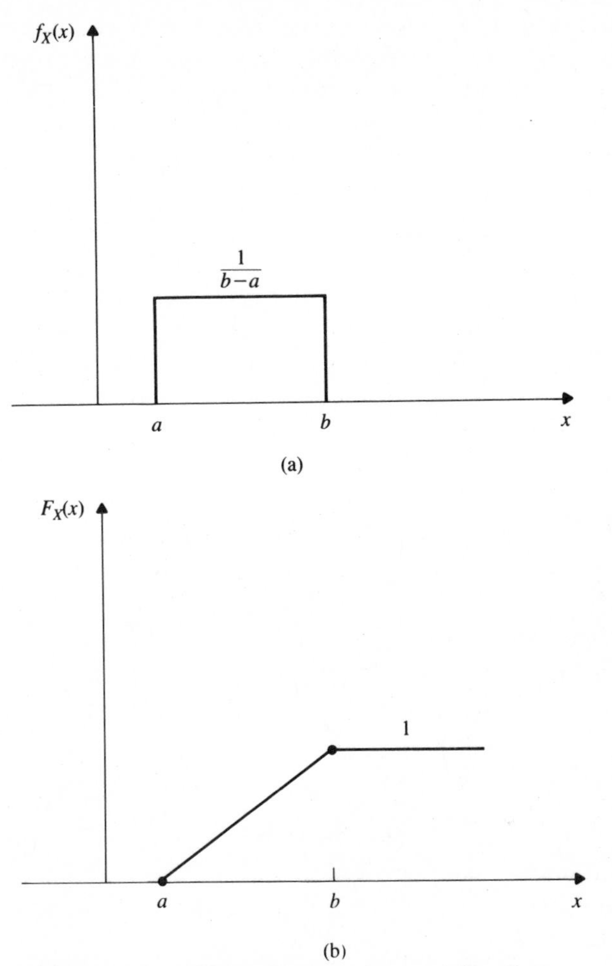


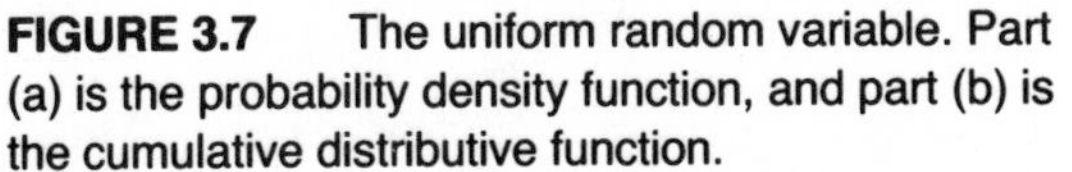


**FIGURE 3.7** The uniform random variable. Part (a) is the probability density function, and part (b) is the cumulative distributive function.

We use the normalization condition in (iv) to find $c$:

$$1 = \int_{-\infty}^{\infty} ce^{-\alpha|x|}\,dx = 2\int_{0}^{\infty} ce^{-\alpha x}\,dx = \frac{2c}{\alpha}.$$

Therefore $c = \alpha/2$. The probability $P[|X| < v]$ is found by integrating the pdf:

$$P[|X| < v] = \frac{\alpha}{2}\int_{-v}^{v} e^{-\alpha|x|}\,dx = 2\left(\frac{\alpha}{2}\right)\int_{0}^{v} e^{-\alpha x}\,dx$$
$$= 1 - e^{-\alpha v}$$

The pdf in Eq. (3.16) is called the **Laplacian pdf.** ■■

The derivative of the cdf does not exist at points where the cdf is not continuous. Thus the notion of pdf as defined by Eq. (3.7) does not apply to discrete random variables at the points where the cdf is discontinuous. We can generalize the definition of the probability density function by noting the relation between the unit step function and the delta function. The **unit step function** is defined as

$$u(x) = \begin{cases} 0 & x < 0 \\ 1 & x \geq 0. \end{cases} \tag{3.17}$$

The **delta function** $\delta(t)$ is defined in terms of the unit step function by the following equation:

$$u(x) = \int_{-\infty}^{x} \delta(t)\, dt. \tag{3.18}$$

Recall from Section 3.2, that the cdf of a discrete random can be written as a weighted sum of unit step functions:

$$F_X(x) = \sum_k p_X(x_k) u(x - x_k), \tag{3.19}$$

where the probability mass function is $p_X(x_k) = P[X = x_k]$. We would like to generalize the definition of the pdf $f_X(x)$ so that Eq. (3.12) holds for discrete random variables:

$$\text{iii.} \quad F_X(x) = \int_{-\infty}^{x} f_X(t)\, dt. \tag{3.20}$$

The integral of a delta function located at $x = b$, that is, $\delta(x - b)$, will yield a step function that begins at $x = b$, that is, $u(x - b)$. This suggests that we define the **pdf for a discrete random variable** by

$$f_X(x) = \sum_k p_X(x_k) \delta(x - x_k). \tag{3.21}$$

Substitution of Eq. (3.21) into Eq. (3.20) then yields Eq. (3.19) as required. Thus the generalized definition of pdf places a delta function of weight $P[X = x_k]$ at the points $x_k$ where the cdf is discontinuous.

The pdf for the discrete random variable discussed in Example 3.4 is shown in Fig. 3.3(b). The pdf of a random variable of mixed type will also contain delta functions at the points where its cdf is not continuous. The pdf for the random variable discussed in Example 3.6 is shown in Fig. 3.5(b).

### Conditional cdf's and pdf's

Conditional cdf's can be defined in a straightforward manner by replacing the probability in Eq. (3.1) by a conditional probability. For example, if some event $A$ concerning $X$ is given, then the **conditional cdf of $X$ given $A$** is defined by

$$F_X(x \mid A) = \frac{P[\{X \le x\} \cap A]}{P[A]} \qquad \text{if } P[A] > 0. \tag{3.22}$$

The **conditional pdf of $X$ given $A$** is then defined by

$$f_X(x \mid A) = \frac{d}{dx} F_X(x \mid A). \tag{3.23}$$

### ■■ Example 3.9

The lifetime $X$ of a machine has a continuous cdf $F_X(x)$. Find the conditional cdf and pdf given the event $A = \{X > t\}$ (i.e., "the machine is still working at time $t$").

The conditional cdf is

$$\begin{aligned} F_X(x \mid X > t) &= P[X \le x \mid X > t] \\ &= \frac{P[\{X \le x\} \cap \{X > t\}}{P[X > t]}. \end{aligned}$$

The intersection of the two events in the numerator is equal to the empty set when $x < t$ and to $\{t < X \le x\}$ when $x \ge t$. Thus

$$F_X(x \mid X > t) = \begin{cases} 0 & x < t \\ \dfrac{F_X(x) - F_X(t)}{1 - F_X(t)} & x \ge t. \end{cases}$$

The conditional pdf is found by differentiating with respect to $x$:

$$f_X(x \mid X > t) = \frac{f_X(x)}{1 - F_X(t)} \qquad x \ge t.$$

■■

## 3.4 SOME IMPORTANT RANDOM VARIABLES

There are a number of random variables that arise in many diverse, unrelated applications. The pervasiveness of these random variables is due to the fact that they model fundamental mechanisms that underlie random behavior. In this section we introduce the cdf and pdf of several of these random variables and discuss how they arise and how they are interrelated. Tables 3.1 and 3.2 list the basic properties of these and other random variables. In later sections, other random variables are derived as functions of the random variables discussed here.

**TABLE 3.1** Discrete Random Variables

**Bernoulli Random Variable**

$S_X = \{0, 1\}$

$p_0 = q = 1 - p \qquad p_1 = p \qquad 0 \le p \le 1$

$E[X] = p \qquad \text{VAR}[X] = p(1 - p)$

$G_X(z) = (q + pz)$

*Remarks*: The Bernoulli random variable is the value of the indicator function $I_A$ for some event $A$; $X = 1$ if $A$ occurs and 0 otherwise.

**Binomial Random Variable**

$S_x = \{0, 1, \ldots, n\}$

$p_k = \binom{n}{k} p^k (1 - p)^{n-k} \qquad k = 0, 1, \ldots, n$

$E[X] = np \qquad \text{VAR}[X] = np(1 - p)$

$G_X(z) = (q + pz)^n$

*Remarks*: $X$ is the number of successes in $n$ Bernoulli trials and hence the sum of $n$ independent, identically distributed Bernoulli random variables.

**Geometric Random Variable**

*First Version*: $S_x = \{0, 1, 2, \ldots\}$

$p_k = p(1 - p)^k \qquad k = 0, 1, \ldots$

$E[X] = \dfrac{1 - p}{p} \qquad \text{VAR}[X] = \dfrac{1 - p}{p^2}$

$G_X(z) = \dfrac{p}{1 - qz}$

*Remarks*: $X$ is the number of failures before the first success in a sequence of independent Bernoulli trials.

The geometric random variable is the only discrete random variable with the memoryless property.

*Second Version*: $S_{X'} = \{1, 2, \ldots\}$

$p_k = p(1 - p)^{k-1} \qquad k = 1, 2, \ldots$

$E[X'] = \dfrac{1}{p} \qquad \text{VAR}[X'] = \dfrac{1 - p}{p^2}$

$G_{X'}(z) = \dfrac{pz}{1 - qz}$

*Remarks*: $X' = X + 1$ is the number of trials until the first success in a sequence of independent Bernoulli trials.

**TABLE 3.1** (continued)

**Negative Binomial Random Variable**

$S_X = \{r, r+1, \ldots\}$ where $r$ is a positive integer

$$p_k = \binom{k-1}{r-1} p^r (1-p)^{k-r} \qquad k = r, r+1, \ldots$$

$$E[X] = \frac{r}{p} \qquad \text{VAR}[X] = \frac{r(1-p)}{p^2}$$

$$G_X(z) = \left(\frac{pz}{1-qz}\right)^r$$

*Remarks*: $X$ is the number of trials until the $r$th success in a sequence of independent Bernoulli trials.

**Poisson Random Variable**

$S_X = \{0, 1, 2, \ldots\}$

$$p_k = \frac{\alpha^k}{k!} e^{-\alpha} \qquad k = 0, 1, \ldots \quad \text{and} \quad \alpha > 0$$

$$E[X] = \alpha \qquad \text{VAR}[X] = \alpha$$

$$G_X(z) = e^{\alpha(z-1)}$$

*Remarks*: $X$ is the number of events that occur in one time unit when the time between events is exponentially distributed with mean $1/\alpha$.

## Discrete Random Variables

Discrete random variables arise mostly in applications where counting is involved. We begin with the Bernoulli random variable as a model for a single coin toss. By counting the outcomes of multiple coin tosses we obtain the binomial, geometric, and Poisson random variables.

**The Bernoulli Random Variable.** Let $A$ be an event related to the outcomes of some random experiment. The **indicator function for $A$** is defined by

$$I_A(\zeta) = \begin{cases} 0 & \text{if } \zeta \text{ not in } A \\ 1 & \text{if } \zeta \text{ in } A, \end{cases} \tag{3.24}$$

that is, $I_A(\zeta)$ equals one if the event $A$ occurs, and zero otherwise. $I_A$ is a random variable since it assigns a number to each outcome of $S$. It is a discrete random variable that takes on values from the set $\{0, 1\}$ and its pmf is

$$p_I(0) = 1 - p \qquad \text{and} \qquad p_I(1) = p, \tag{3.25}$$

where $P[A] = p$. $I_A$ is called the **Bernoulli random variable** since it describes the outcome of a Bernoulli trial if we identify $I_A = 1$ with a "success."

**TABLE 3.2** Continuous Random Variables

---

**Uniform Random Variable**

$S_X = [a, b]$

$$f_X(x) = \frac{1}{b - a} \qquad a \leq x \leq b$$

$$E[X] = \frac{a + b}{2} \qquad \text{VAR}[X] = \frac{(b - a)^2}{12}$$

$$\Phi_X(\omega) = \frac{e^{j\omega b} - e^{j\omega a}}{j\omega(b - a)}$$

---

**Exponential Random Variable**

$S_X = [0, \infty)$

$f_X(x) = \lambda e^{-\lambda x} \qquad x \geq 0 \qquad \text{and } \lambda > 0$

$$E[X] = \frac{1}{\lambda} \qquad \text{VAR}[X] = \frac{1}{\lambda^2}$$

$$\Phi_X(\omega) = \frac{\lambda}{\lambda - j\omega}$$

*Remarks*: The exponential random variable is the only random variable with the memoryless property.

---

**Gaussian (Normal) Random Variable**

$S_X = (-\infty, +\infty)$

$$f_X(x) = \frac{e^{-(x-m)^2/2\sigma^2}}{\sqrt{2\pi}\,\sigma} \qquad -\infty < x < +\infty \quad \text{and} \quad \sigma > 0$$

$E[X] = m \qquad \text{VAR}[X] = \sigma^2$

$\Phi_X(\omega) = e^{jm\omega - \sigma^2\omega^2/2}$

*Remarks*: Under a wide range of conditions $X$ can be used to approximate the sum of a large number of independent random variables.

---

**Gamma Random Variable**

$S_X = (0, +\infty)$

$$f_X(x) = \frac{\lambda(\lambda x)^{\alpha-1}e^{-\lambda x}}{\Gamma(\alpha)} \qquad x > 0 \quad \text{and} \quad \alpha > 0, \lambda > 0$$

where $\Gamma(z)$ is the gamma function (Eq. 3.46).

$E[X] = \alpha/\lambda \qquad \text{VAR}[X] = \alpha/\lambda^2$

$$\Phi_x(\omega) = \frac{1}{(1 - j\omega/\lambda)^\alpha}$$

**TABLE 3.2** (continued)

*Special Cases of Gamma Random Variable*

m-Erlang Random Variable: $\alpha = m$, a positive integer

$$f_X(x) = \frac{\lambda e^{-\lambda x}(\lambda x)^{m-1}}{(m-1)!} \qquad x > 0$$

$$\Phi_X(\omega) = \left(\frac{\lambda}{\lambda - j\omega}\right)^m$$

*Remarks*: An $m$-Erlang random variable is obtained by adding $m$ independent exponentially distributed random variables with parameter $\lambda$.

*Chi-Square Random Variable with k degrees of freedom*: $\alpha = k/2$, $k$ a positive integer and $\lambda = \frac{1}{2}$

$$f_X(x) = \frac{x^{(k-2)/2}e^{-x/2}}{2^{k/2}\Gamma(k/2)} \qquad x > 0$$

$$\Phi_X(\omega) = \left(\frac{1}{1 - j2\omega}\right)^{k/2}$$

*Remarks*: The sum of $k$ mutually independent, squared zero-mean unit-variance Gaussian random variables is a chi-squared random variable with $k$ degrees of freedom.

**Rayleigh Random Variable**

$$S_X = [0, \infty)$$

$$f_X(x) = \frac{x}{\alpha^2}e^{-x^2/2\alpha^2} \qquad x \geq 0 \qquad \alpha > 0$$

$$E[X] = \alpha\sqrt{\pi/2} \qquad \text{VAR}[X] = (2 - \pi/2)\alpha^2$$

**Cauchy Random Variable**

$$S_X = (-\infty, \infty)$$

$$f_X(x) = \frac{\alpha/\pi}{x^2 + \alpha^2} \qquad -\infty < x < \infty \qquad \alpha > 0$$

Mean and Variance do not exist.

$$\Phi_X(\omega) = e^{-\alpha|\omega|}$$

**Laplacian Random Variable**

$$S_X = (-\infty, \infty)$$

$$f_X(x) = \frac{\alpha}{2}e^{-\alpha|x|} \qquad -\infty < x < \infty \qquad \alpha > 0$$

$$E[X] = 0 \qquad \text{VAR}[X] = 2/\alpha^2$$

$$\Phi_X(\omega) = \frac{\alpha^2}{\omega^2 + \alpha^2}$$

Every Bernoulli trial, regardless of the definition of $A$, is equivalent to the tossing of a biased coin. In this sense, coin tossing can be viewed as representative of a fundamental mechanism for generating randomness, and the Bernoulli random variable is the model associated with it.

**The Binomial Random Variable.** Suppose that a random experiment is repeated $n$ independent times. Let $X$ be the number of times a certain event $A$ occurs in these $n$ trials. $X$ is then a random variable taking on values from the set $\{0, 1, \ldots, n\}$. For example, $X$ could be the number of heads in $n$ tosses of a coin. If we let $I_j$ be the indicator function for the event $A$ in the $j$th trial, then

$$X = I_1 + I_2 + \cdots + I_n,$$

that is, $X$ is the sum of the Bernoulli random variables associated with each of the $n$ independent trials.

In Section 2.6, we found that $X$ has the following pmf:

$$P[X = k] = \binom{n}{k} p^k (1 - p)^{n-k} \qquad \text{for } k = 0, \ldots, n. \tag{3.26}$$

$X$ is called the **binomial random variable.** Figure 3.8 shows the pdf of $X$ for $n = 24$ and $p = .2$ and $p = .5$. Note that $P[X = k]$ is maximum at $k_{\max} = [(n + 1)p]$, where $[x]$ denotes the largest integer that is smaller than $x$. When $(n + 1)p$ is an integer, then the maximum is achieved at $k_{\max}$ and $k_{\max} - 1$.

The binomial random variable arises in applications where there are two types of objects (i.e., heads/tails, correct/erroneous bits, good/defective items, active/silent speakers), and we are interested in the number of type 1 objects in a randomly selected batch of size $n$, where the type of each object is independent of the types of the other objects in the batch. Examples involving the binomial random variable were given in Section 2.6.

**The Geometric Random Variable.** The binomial random variable is obtained by fixing the number of Bernoulli trials and counting the number of successes. Suppose that instead we count the number $M$ of independent Bernoulli trials until the first occurrence of a success. $M$ is called the **geometric random variable** and it takes on values from the set $\{1, 2, \ldots\}$. In Section 2.6, we found that the pmf of $M$ is given by

$$P[M = k] = (1 - p)^{k-1} p \qquad k = 1, 2, \ldots, \tag{3.27}$$

where $p = P[A]$ is the probability of "success" in each Bernoulli trial. Figure 3.9 shows the geometric pdf for several values of $p$. Note that $P[M = k]$ decays geometrically with $k$.

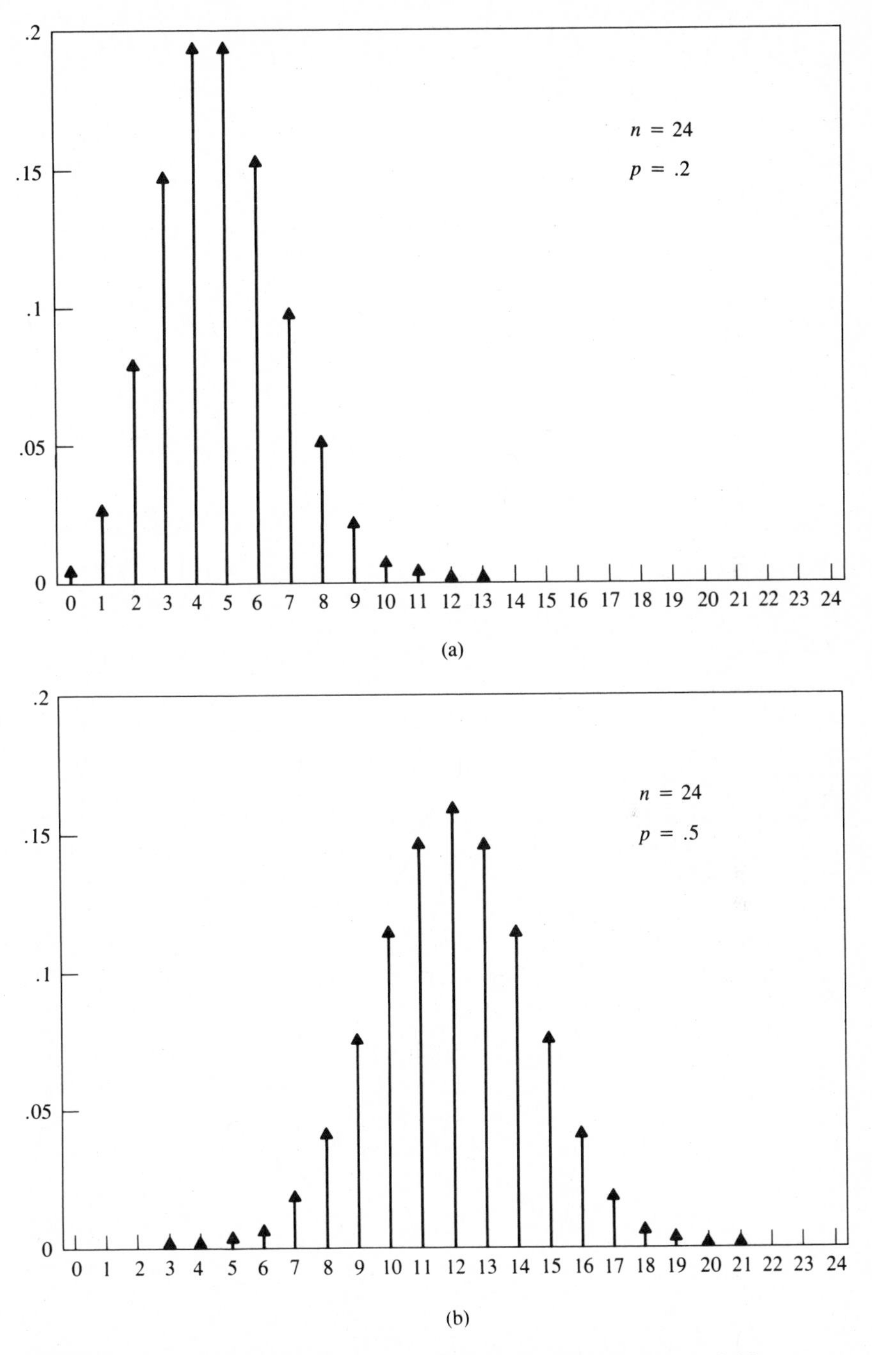


**FIGURE 3.8** Probability density function of binomial random variable. (a) $p = .2$; (b) $p = .5$.

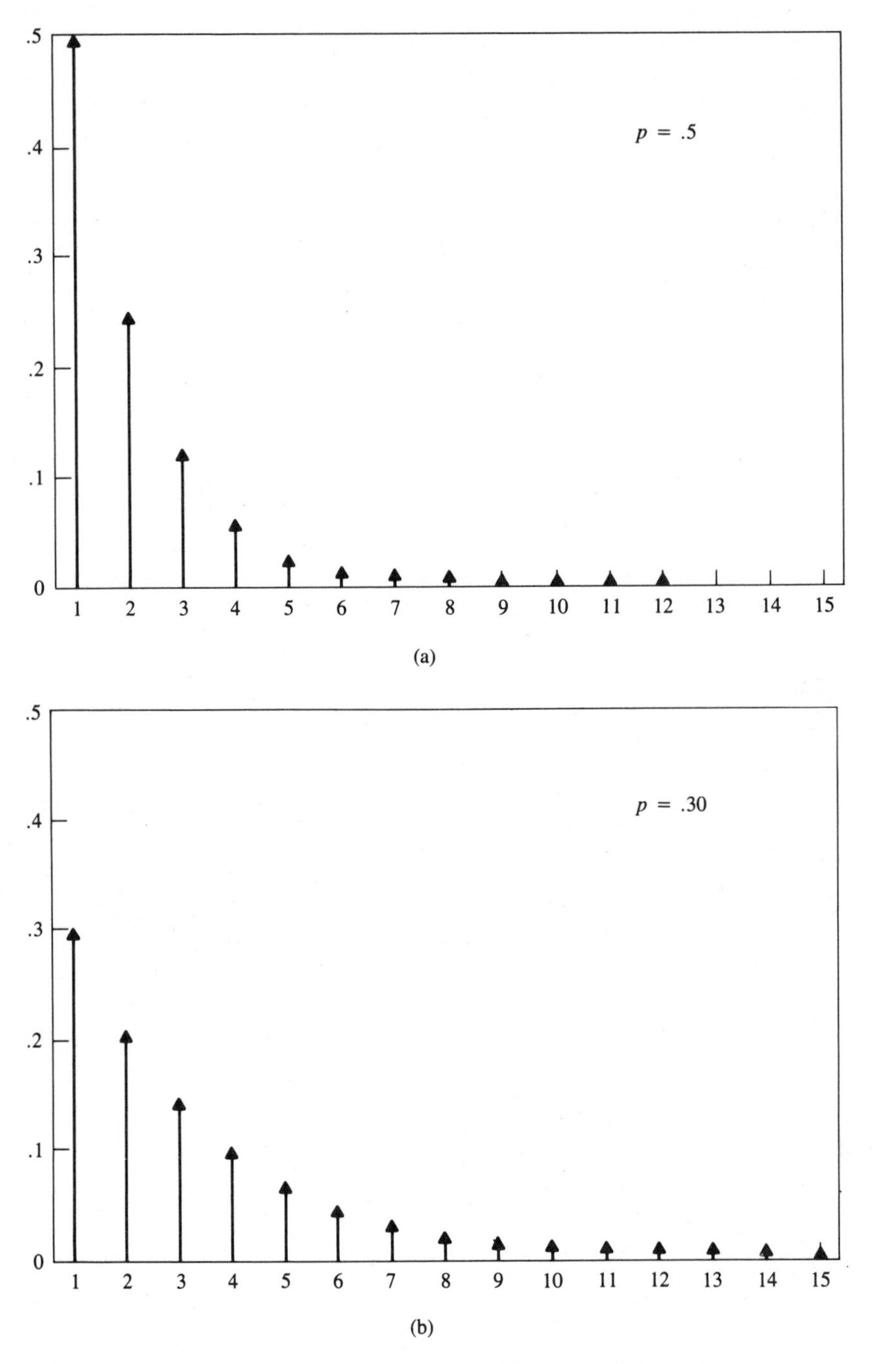


**FIGURE 3.9** Probability density function of the geometric random variable. (a) $p = .5$; (b) $p = .3$.

The cdf of $M$ can be written in closed form:

$$P[M \leq k] = \sum_{j=1}^{k} pq^{j-1} = p \sum_{j'=0}^{k-1} q^{j'} = p \frac{1 - q^k}{1 - q} = 1 - q^k, \tag{3.28}$$

where $q = 1 - p$.

Sometimes we are interested in $M' = M - 1$, the number of *failures* before a success occurs:

$$\begin{aligned} P[M' = k] &= P[M = k + 1] \\ &= (1 - p)^k p \qquad k = 0, 1, 2, \ldots. \end{aligned} \tag{3.29}$$

We also refer to $M'$ as the geometric random variable.

The geometric random variable arises in applications where one is interested in the time (i.e., number of trials) that elapses between the occurrence of events in a sequence of independent experiments, as in Examples 2.8 and 2.40. The modified geometric random variable $M'$ arises as the pmf for the number of customers in many queueing system models.

**The Poisson Random Variable.** In many applications, we are interested in counting the number of occurrences of an event in a certain time period or in a certain region in space. The Poisson random variable arises in situations where the events occur "completely at random" in time or space. For example, the Poisson random variable arises in counts of emissions from radioactive substances, in counts of demands for telephone connections, and in counts of defects in a semiconductor chip.

The pmf for the **Poisson random variable** is given by

$$P[N = k] = \frac{\alpha^k}{k!} e^{-\alpha} \qquad k = 0, 1, 2, \ldots, \tag{3.30}$$

where $\alpha$ is the average number of event occurrences in a specified time interval or region in space. Figure 3.10 shows the Poisson pdf for several values of $\alpha$. For $\alpha < 1$, $P[N = k]$ is maximum at $k = 0$; for $\alpha > 1$, $P[N = k]$ is maximum at $[\alpha]$; if $\alpha$ is a positive integer, the $P[N = k]$ is maximum at $k = \alpha$ and at $k = \alpha - 1$.

### ■■ Example 3.10

Verify that the pmf of the Poisson random variable sums to one.

$$\sum_{k=0}^{\infty} \frac{\alpha^k}{k!} e^{-\alpha} = e^{-\alpha} \sum_{k=0}^{\infty} \frac{\alpha^k}{k!} = e^{-\alpha} e^{\alpha} = 1,$$

where we used the fact that the second summation is the infinite series expansion for $e^{\alpha}$. ■■

---

One of the applications of the Poisson probabilities in Eq. (3.30) is to approximate the binomial probabilities. We will show that *if n is large and*

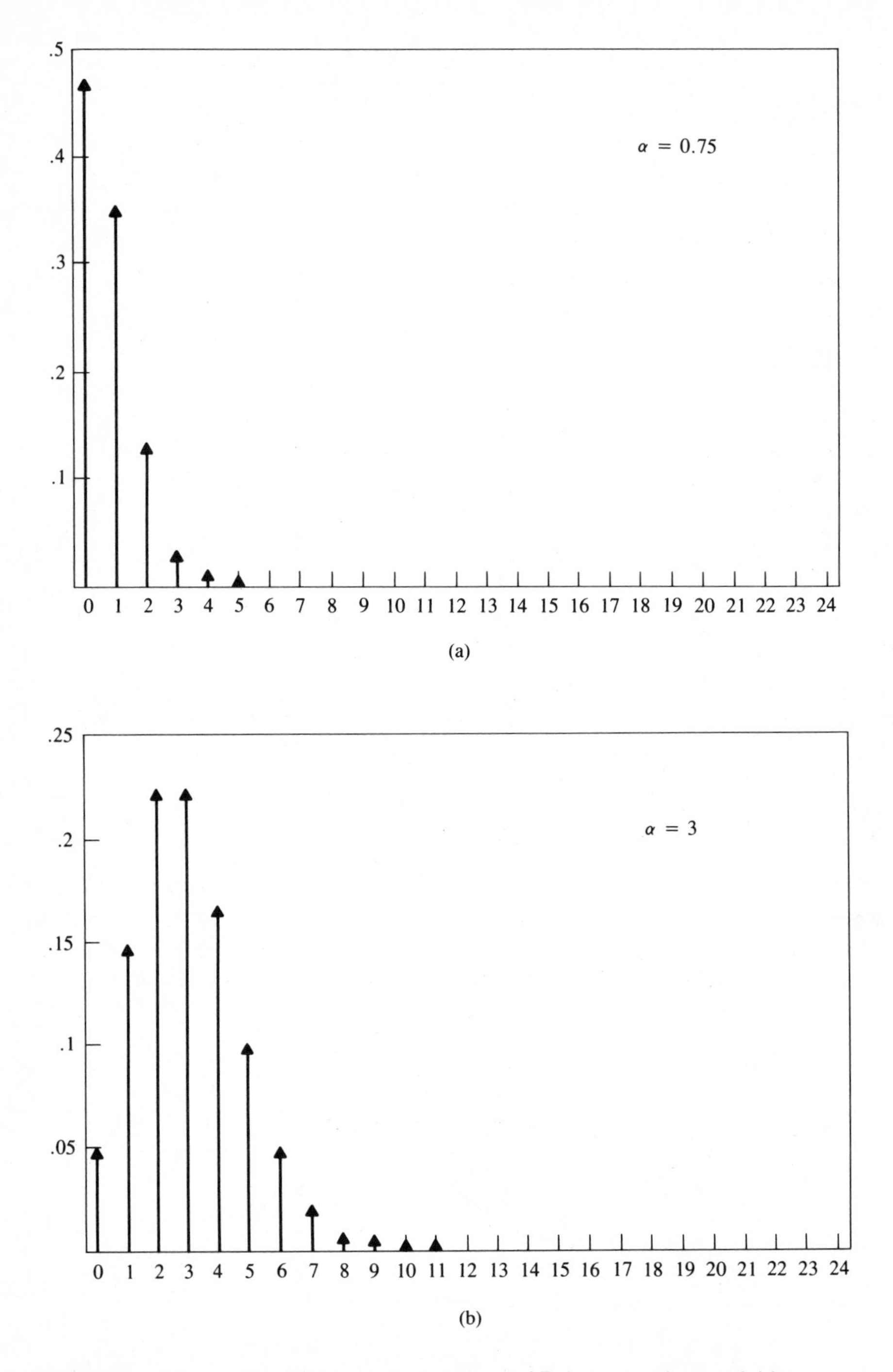


**FIGURE 3.10** Probability density functions of Poisson random variable. (a) $\alpha = 0.75$; (b) $\alpha = 3$; (c) $\alpha = 9$.

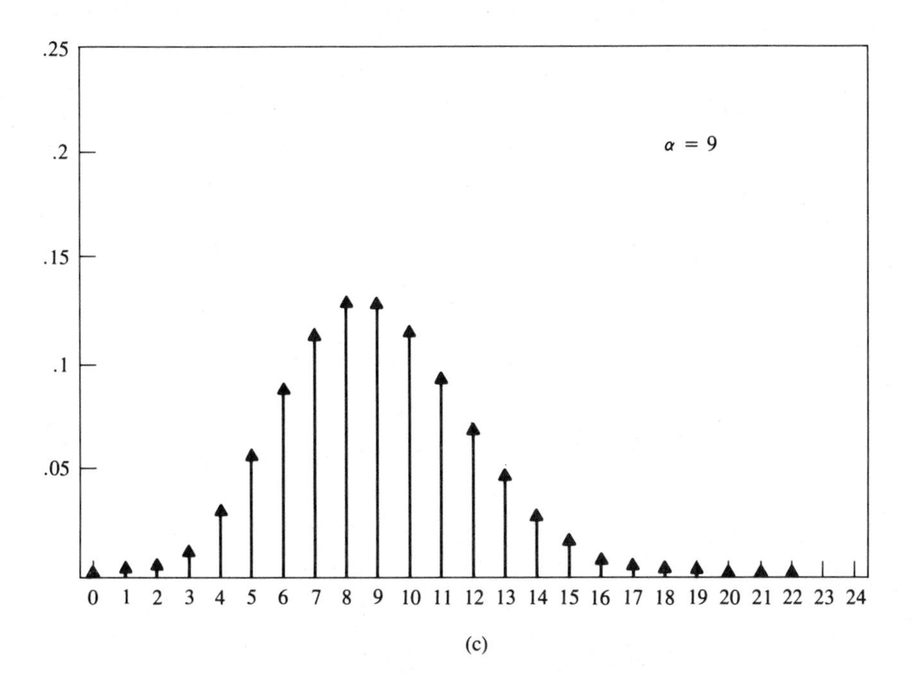


(c)

*p is small, then for $\alpha = np$*

$$p_k = \binom{n}{k} p^k (1-p)^{n-k} \simeq \frac{\alpha^k}{k!} e^{-\alpha} \qquad k = 0, 1, \ldots. \tag{3.31}$$

The approximation in Eq. (3.31) is obtained by taking the limit $n \to \infty$ in the expression for $p_k$, while keeping $\alpha = np$ fixed. First, consider the probability that no events occur in $n$ trials:

$$p_0 = (1-p)^n = \left(1 - \frac{\alpha}{n}\right)^n \to e^{-\alpha} \qquad \text{as } n \to \infty, \tag{3.32}$$

where the limit in the last expression is a well-known result from calculus. The rest of the probabilities are found by noting that

$$\begin{aligned} \frac{p_{k+1}}{p_k} &= \frac{\binom{n}{k+1} p^{k+1} q^{n-k-1}}{\binom{n}{k} p^k q^{n-k}} = \frac{k!\,(n-k)!\,p}{(k+1)!\,(n-k-1)!\,q} \\ &= \frac{(n-k)p}{(k+1)q} = \frac{(1-k/n)\alpha}{(k+1)(1-\alpha/n)} \\ &\to \frac{\alpha}{k+1} \qquad \text{as } n \to \infty. \end{aligned}$$

Thus the limiting probabilities satisfy

$$p_{k+1} = \frac{\alpha}{k+1} p_k \qquad \text{for } k = 0, 1, 2, \ldots \tag{3.33a}$$

and

$$p_0 = e^{-\alpha}. \tag{3.33b}$$

Equations (3.33a) and (3.33b) for $k = 0$ and 1 imply that

$$p_1 = \frac{\alpha}{1} p_0 = \frac{\alpha}{1} e^{-\alpha}$$

$$p_2 = \frac{\alpha}{2} p_1 = \frac{\alpha^2}{2(1)} e^{-\alpha}.$$

A simple induction argument then shows that

$$p_k = \frac{\alpha^k}{k!} e^{-\alpha} \qquad \text{for } k = 0, 1, 2, \ldots.$$

Thus the Poisson pmf is the limiting form of the binomial pmf when the number of Bernoulli trials $n$ is made very large and the probability of success is kept small, so that $\alpha = np$.

### ■■ Example 3.11

The probability of a bit error in a communication line is $10^{-3}$. Find the probability that a block of 1000 bits has five or more errors.

Each bit transmission corresponds to a Bernoulli trial with a "success" corresponding to a bit error in transmission. The probability of $k$ errors in 1000 transmissions is then given by the binomial probability with $n = 1000$ and $p = 10^{-3}$. The Poisson approximation to the binomial probabilities uses the parameter $\alpha = np = 1000(10^{-3}) = 1$. Thus

$$\begin{aligned} P[N \geq 5] &= 1 - P[N < 5] = 1 - \sum_{k=0}^{4} \frac{\alpha^k}{k!} e^{-\alpha} \\ &= 1 - e^{-1}\left\{1 + \frac{1}{1!} + \frac{1}{2!} + \frac{1}{3!} + \frac{1}{4!}\right\} \\ &= .00366. \end{aligned}$$

■■

The Poisson random variable appears naturally in many physical situations.

For example, the Poisson pmf gives an accurate prediction for the relative frequencies of the number of particles emitted by a radioactive mass during a fixed time period. This correspondence can be explained as follows. A radioactive mass is composed of a large number of atoms, say $n$. In a fixed time interval each atom has a very small probability $p$ of disintegrating and emitting a radioactive particle. If atoms disintegrate independently of other atoms, then the number of emissions in a time interval can be viewed as the number of successes in $n$ trials. For example,

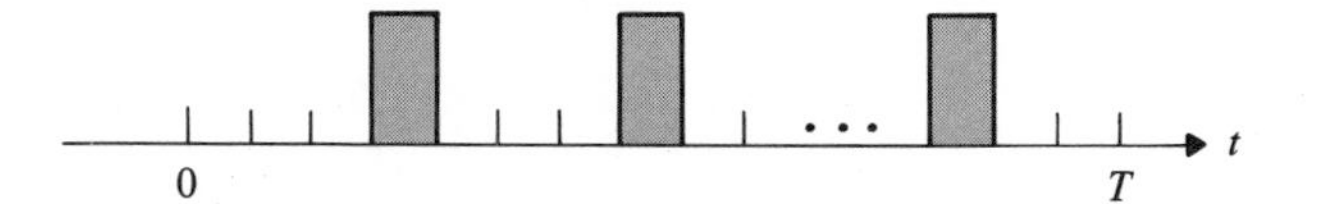


**FIGURE 3.11(a)** The interval [0, $T$] has been divided into $n$ subintervals. A pulse in a subinterval indicates the occurrence of an event. The number $N$ of events is a binomial random variable with parameters $n$ and $p = \alpha/n$. As $n \to \infty$, $N$ becomes a Poisson random variable with parameter $\alpha/T$.

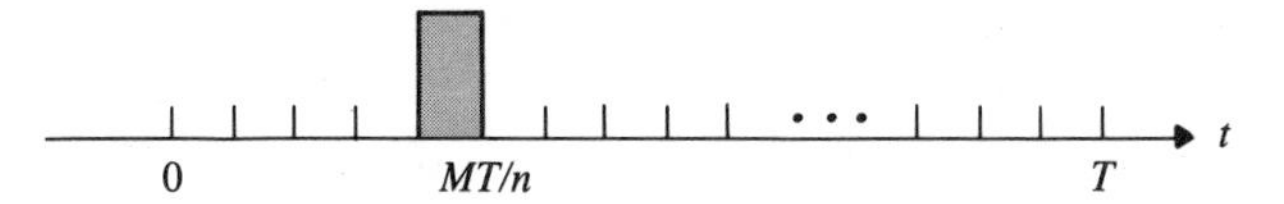


**FIGURE 3.11(b)** The number of trials $M$ until the occurrence of an event is a geometric random variable with parameter $p = \alpha/n$. As $n \to \infty$, the time $Y = MT/n$ until the occurrence of the first event approaches an exponential random variable with parameter $\lambda = \alpha/T$.

one microgram of radium contains about $n = 10^{16}$ atoms, and the probability that a single atom will disintegrate during a one millisecond time interval is $p = 10^{-15}$ (Rozanov 1969, 58). Thus it is an understatement to say that the conditions for the approximation in Eq. (3.31) hold: $n$ is so large and $p$ so small that one could argue that the limit $n \to \infty$ has been carried out and that the number of emissions is *exactly* a Poisson random variable.

The Poisson random variable also comes up in situations where we can imagine a sequence of Bernoulli trials taking place in time or space. Suppose the number of event occurrences in a $T$-second time interval is being counted. Divide the time interval into a very large number, $n$, of subintervals as shown in Fig. 3.11(a). Each subinterval can be viewed as a Bernoulli trial if the following conditions hold: (1) At most one event can occur in a subinterval, that is, the probability of more than one event occurrence is negligible; (2) the outcomes in different subintervals are independent; and (3) the probability of an event occurrence in a subinterval is $p = \alpha/n$ where $\alpha$ is the average number of events observed in a $T$-second interval. If $\alpha$ is finite, then the conditions leading to the limit in Eq. (3.31) will hold as $n$ becomes large. In Chapter 6 we develop this result when we discuss the Poisson random process.

### Continuous Random Variables

We are always limited to measurements of finite precision, so in effect, every random variable found in practice is a discrete random variable.

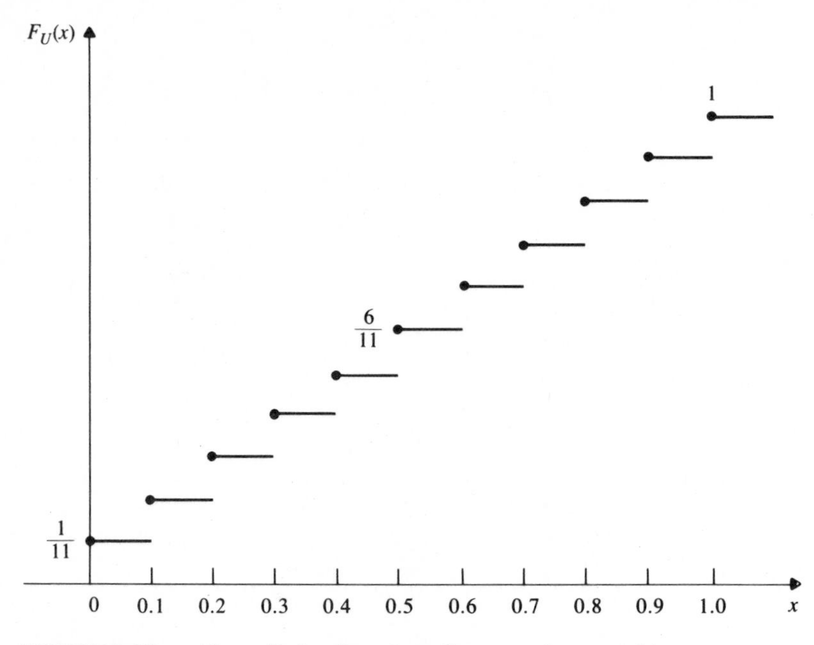


**FIGURE 3.12** The cdf of a discrete uniform random variable.

Nevertheless, there are several compelling reasons for using continuous random variable models. First, in general, continuous random variables are easier to handle analytically. Second, the limiting form of many discrete random variables yield continuous random variables. Finally, there are a number of "families" of continuous random variables that can be used to model a wide variety of situations by adjusting a few parameters.

## ■■ Example 3.12

A random number generator produces a random variable $X$ that assumes values from the set $\{0, 1, \ldots, n-1\}$ with equal probability, $1/n$. Define the random variable $U$ by $U = X/n$. $U$ takes on values from the set $S = \{0, 1/n, \ldots, 1 - 1/n\}$. The cdf of $U$ for $n = 11$ is the staircase function shown in Fig. 3.12. The probability that $U$ falls in a certain interval $I$ is

$$P[U \text{ in } I] = \frac{\text{number of elements in } S \text{ that fall in } I}{n}.$$

For example, if $n = 11$ then $P[0 \leq U \leq 0.5] = 6/11$.

In practice, the value $n$ will be very large. For example, the random number generator discussed in Section 2.7 has $n = 2{,}147{,}483{,}647$.

Consider what happens as $n$ increases: (1) The set of points in $S$ become increasingly dense in the unit interval $[0, 1]$; and (2) the cdf of $X$ approaches that of a continuous random variable uniformly distributed in $[0, 1]$. Thus for very large values of $n$, the continuous cdf can be used to obtain the probability of intervals involving $U$ with great accuracy and much less effort. Thus it becomes extremely convenient to assume that $U$ is a continuous random variable. ■■

---

**The Exponential Random Variable.** The exponential random variable arises in the modelling of the time between occurrence of events (e.g., the time between customer demands for call connections), and in the modelling of the lifetime of devices and systems. The **exponential random variable** $X$ with parameter $\lambda$ has pdf

$$f_X(x) = \begin{cases} 0 & x < 0 \\ \lambda e^{-\lambda x} & x \geq 0 \end{cases} \tag{3.34}$$

and cdf

$$F_X(x) = \begin{cases} 0 & x < 0 \\ 1 - e^{-\lambda x} & x \geq 0. \end{cases} \tag{3.35}$$

The cdf and pdf of $X$ were shown in Fig. 3.4.

The exponential random variable can be obtained as a limiting form of the geometric random variable. Consider the limiting procedure that was used to derive the Poisson random variable. An interval of duration $T$ is divided into subintervals of length $T/n$ as shown in Fig. 3.11(b). The sequence of subintervals corresponds to a sequence of independent Bernoulli trials with probability of success $p = \alpha/n$, where $\alpha$ is the average number of events per $T$ seconds. The number of subintervals until the occurrence of an event is a geometric random variable $M$. Thus the time until the occurrence of the first event is $X = M(T/n)$, and the probability that this time exceeds $t$ seconds is

$$\begin{aligned} P[X > t] &= P\left[M > n\frac{t}{T}\right] \\ &= (1 - p)^{nt/T} \\ &= \left\{\left(1 - \frac{\alpha}{n}\right)^n\right\}^{t/T} \\ &\to e^{-\alpha t/T} \qquad \text{as } n \to \infty. \end{aligned}$$

Thus the exponential random variable is obtained as a limiting form of the geometric random variable. Note that this result implies that for a Poisson random variable, the time between events is an exponentially distributed random variable with parameter $\lambda = \alpha/T$ customers per second.

The exponential random variable satisfies the **memoryless property**:

$$P[X > t + h \mid X > t] = P[X > h]. \tag{3.36}$$

The expression on the left side is the probability of having to wait $h$ additional seconds given that one has already been waiting $t$ seconds. The expression on the right side is the probability of waiting $h$ seconds when one first begins to wait. Thus the probability of waiting an additional $h$ seconds is the same regardless of how long one has already been waiting! We see in Chapters 8 and 9 that the memoryless property of the exponential random variable makes it the cornerstone for the theory of Markov chains, which is used extensively in evaluating the performance of computer systems and communications networks.

We now prove the memoryless property:

$$\begin{aligned} P[X > t + h \mid X > t] &= \frac{P[\{X > t + h\} \cap \{X > t\}]}{P[X > t]} \qquad \text{for } h > 0 \\ &= \frac{P[X > t + h]}{P[X > t]} = \frac{e^{-\lambda(t+h)}}{e^{-\lambda t}} \\ &= e^{-\lambda h} = P[X > h]. \end{aligned}$$

It can be shown that the exponential random variable is the only continuous random variable that satisfies the memoryless property.

Examples 2.10, 2.25, and 2.27 dealt with the exponential random variable.

**The Gaussian (Normal) Random Variable.** There are many situations in man-made and in natural phenomena where one deals with a random variable $X$ that consists of the sum of a large number of "small" random variables. The exact description of the pdf of $X$ in terms of the component random variables can become quite complex and unwieldy. However one finds that under very general conditions, as the number of components becomes large, the cdf of $X$ approaches that of the **Gaussian (normal) random variable.**[2] This random variable appears so often in problems involving randomness that it came to be known as the "normal" random variable.

The pdf for the Gaussian random variable $X$ is given by

$$f_X(x) = \frac{1}{\sqrt{2\pi}\,\sigma} e^{-(x-m)^2/2\sigma^2} \qquad -\infty < x < \infty, \tag{3.37}$$

where $m$ and $\sigma > 0$ are real numbers, which will later be shown to be the mean and standard deviation of $X$. Figure 3.13 shows that the Gaussian

2. This result, called the central limit theorem, will be discussed in Chapter 5.

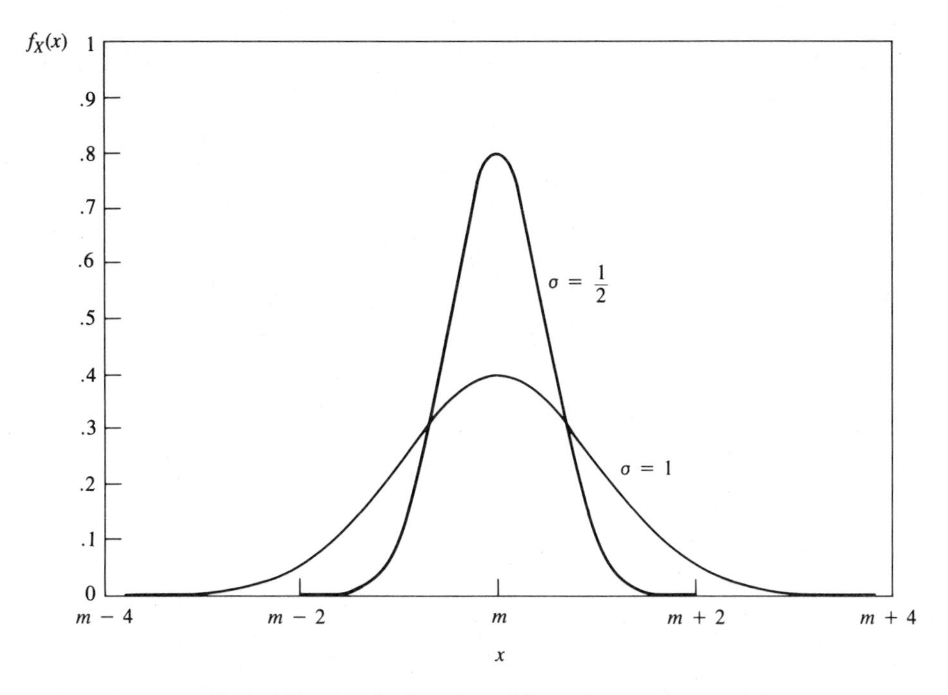


**FIGURE 3.13** Probability density function of Gaussian random variable.

pdf is a "bell-shaped" curve centered and symmetric about $m$ and whose "width" increases with $\sigma$.

The cdf of the Gaussian random variable is given by

$$P[X \leq x] = \frac{1}{\sqrt{2\pi}\,\sigma} \int_{-\infty}^{x} e^{-(x'-m)^2/2\sigma^2}\, dx'. \tag{3.38}$$

The change of variable $t = (x' - m)/\sigma$ results in

$$\begin{aligned} F_X(x) &= \frac{1}{\sqrt{2\pi}} \int_{-\infty}^{(x-m)/\sigma} e^{-t^2/2}\, dt \\ &= \Phi\left(\frac{x - m}{\sigma}\right) \end{aligned} \tag{3.39}$$

where $\Phi(x)$ is the cdf of a Gaussian random variable with $m = 0$ and $\sigma = 1$:

$$\Phi(x) = \frac{1}{\sqrt{2\pi}} \int_{-\infty}^{x} e^{-t^2/2}\, dt. \tag{3.40}$$

Therefore any probability involving an arbitrary Gaussian random variable can be expressed in terms of $\Phi(x)$.

## ■■ Example 3.13

Show that the Gaussian pdf integrates to one. Consider the square of the integral of the pdf:

$$\left[\frac{1}{\sqrt{2\pi}}\int_{-\infty}^{\infty} e^{-x^2/2}\,dx\right]^2 = \frac{1}{2\pi}\int_{-\infty}^{\infty} e^{-x^2/2}\,dx \int_{-\infty}^{\infty} e^{-y^2/2}\,dy$$
$$= \frac{1}{2\pi}\int_{-\infty}^{\infty}\int_{-\infty}^{\infty} e^{-(x^2+y^2)/2}\,dx\,dy.$$

Let $x = r\cos\theta$ and $y = r\sin\theta$ and carry out the change from Cartesian to Polar coordinates, then we obtain:

$$\frac{1}{2\pi}\int_0^{\infty}\int_0^{2\pi} e^{-r^2/2} r\,dr\,d\theta = \int_0^{\infty} re^{-r^2/2}\,dr$$
$$= [-e^{-r^2/2}]_0^{\infty}$$
$$= 1.$$

■■

In electrical engineering it is customary to work with the $Q$-function, which is defined by

$$Q(x) = 1 - \Phi(x) \tag{3.41}$$
$$= \frac{1}{\sqrt{2\pi}}\int_x^{\infty} e^{-t^2/2}\,dt. \tag{3.42}$$

$Q(x)$ is simply the probability of the "tail" of the pdf. The symmetry of the pdf implies that

$$Q(0) = 1/2 \qquad \text{and} \qquad Q(-x) = 1 - Q(x). \tag{3.43}$$

The integral in Eq. (3.40) does not have a closed-form expression. Traditionally the integrals have been evaluated by looking up tables that list $Q(x)$ or by using approximations that require numerical evaluation [8]. Recently, the following expression has been found to give good accuracy for $Q(x)$ over the entire range $0 < x < \infty$:

$$Q(x) \simeq \left[\frac{1}{(1-a)x + a\sqrt{x^2+b}}\right]\frac{1}{\sqrt{2\pi}}e^{-x^2/2}, \tag{3.44}$$

where $a = 1/\pi$ and $b = 2\pi$, [11]. Table 3.3 shows $Q(x)$ and the value given by the above approximation. In some problems, we are interested in finding the value of $x$ for which $Q(x) = 10^{-k}$. Table 3.4 gives these values for $k = 1, \ldots, 10$.

The Gaussian random variable plays a very important role in communication systems, where transmission signals are corrupted by noise voltages that result from the thermal motion of electrons. It can be shown from physical principles that these voltages will have a Gaussian pdf.

**TABLE 3.3** Comparison of $Q(x)$ and Approximation Given by Eq. (3.44)

| $x$ | $Q(x)$ | Approximation | $x$ | $Q(x)$ | Approximation |
|---|---|---|---|---|---|
| 0 | 5.00E-01 | 5.00E-01 | 2.7 | 3.47E-03 | 3.46E-03 |
| 0.1 | 4.60E-01 | 4.58E-01 | 2.8 | 2.56E-03 | 2.55E-03 |
| 0.2 | 4.21E-01 | 4.17E-01 | 2.9 | 1.87E-03 | 1.86E-03 |
| 0.3 | 3.82E-01 | 3.78E-01 | 3.0 | 1.35E-03 | 1.35E-03 |
| 0.4 | 3.45E-01 | 3.41E-01 | 3.1 | 9.68E-04 | 9.66E-04 |
| 0.5 | 3.09E-01 | 3.05E-01 | 3.2 | 6.87E-04 | 6.86E-04 |
| 0.6 | 2.74E-01 | 2.71E-01 | 3.3 | 4.83E-04 | 4.83E-04 |
| 0.7 | 2.42E-01 | 2.39E-01 | 3.4 | 3.37E-04 | 3.36E-04 |
| 0.8 | 2.12E-01 | 2.09E-01 | 3.5 | 2.33E-04 | 2.32E-04 |
| 0.9 | 1.84E-01 | 1.82E-01 | 3.6 | 1.59E-04 | 1.59E-04 |
| 1.0 | 1.59E-01 | 1.57E-01 | 3.7 | 1.08E-04 | 1.08E-04 |
| 1.1 | 1.36E-01 | 1.34E-01 | 3.8 | 7.24E-05 | 7.23E-05 |
| 1.2 | 1.15E-01 | 1.14E-01 | 3.9 | 4.81E-05 | 4.81E-05 |
| 1.3 | 9.68E-02 | 9.60E-02 | 4.0 | 3.17E-05 | 3.16E-05 |
| 1.4 | 8.08E-02 | 8.01E-02 | 4.5 | 3.40E-06 | 3.40E-06 |
| 1.5 | 6.68E-02 | 6.63E-02 | 5.0 | 2.87E-07 | 2.87E-07 |
| 1.6 | 5.48E-02 | 5.44E-02 | 5.5 | 1.90E-08 | 1.90E-08 |
| 1.7 | 4.46E-02 | 4.43E-02 | 6.0 | 9.87E-10 | 9.86E-10 |
| 1.8 | 3.59E-02 | 3.57E-02 | 6.5 | 4.02E-11 | 4.02E-11 |
| 1.9 | 2.87E-02 | 2.86E-02 | 7.0 | 1.28E-12 | 1.28E-12 |
| 2.0 | 2.28E-02 | 2.26E-02 | 7.5 | 3.19E-14 | 3.19E-14 |
| 2.1 | 1.79E-02 | 1.78E-02 | 8.0 | 6.22E-16 | 6.22E-16 |
| 2.2 | 1.39E-02 | 1.39E-02 | 8.5 | 9.48E-18 | 9.48E-18 |
| 2.3 | 1.07E-02 | 1.07E-02 | 9.0 | 1.13E-19 | 1.13E-19 |
| 2.4 | 8.20E-03 | 8.17E-03 | 9.5 | 1.05E-21 | 1.05E-21 |
| 2.5 | 6.21E-03 | 6.19E-03 | 10.0 | 7.62E-24 | 7.62E-24 |
| 2.6 | 4.66E-03 | 4.65E-03 | | | |

**TABLE 3.4** $Q(x) = 10^{-k}$

| $k$ | $x = Q^{-1}(10^{-k})$ |
|---|---|
| 1 | 1.2815 |
| 2 | 2.3263 |
| 3 | 3.0902 |
| 4 | 3.7190 |
| 5 | 4.2649 |
| 6 | 4.7535 |
| 7 | 5.1993 |
| 8 | 5.6120 |
| 9 | 5.9978 |
| 10 | 6.3613 |

### ■■ Example 3.14

A communication system accepts a positive voltage $V$ as input and outputs a voltage $Y = \alpha V + N$, where $\alpha = 10^{-2}$ and $N$ is a Gaussian random variable with parameters $m = 0$ and $\sigma = 2$. Find the value of $V$ that gives $P[Y < 0] = 10^{-6}$.

The probability $P[Y < 0]$ is written in terms of $N$ as follows:

$$P[Y < 0] = P[\alpha V + N < 0]$$
$$= P[N < -\alpha V] = \Phi\left(\frac{-\alpha V}{\sigma}\right) = Q\left(\frac{\alpha V}{\sigma}\right) = 10^{-6}.$$

From Table 3.4 we see that the argument of the $Q$-function should be $\alpha V/\sigma = 4.753$. Thus $V = (4.753)\sigma/\alpha = 950.6$. ■■

---

**The Gamma Random Variable.** The gamma random variable is a versatile random variable that appears in many applications. For example, it is used to model the time required to service customers in queueing systems, the lifetime of devices and systems in reliability studies, and the defect clustering behavior in VLSI chips [12].

The pdf of the **gamma random variable** has two parameters $\alpha > 0$, $\lambda > 0$ and is given by

$$f_X(x) = \frac{\lambda(\lambda x)^{\alpha-1}e^{-\lambda x}}{\Gamma(\alpha)} \qquad 0 < x < \infty, \tag{3.45}$$

where $\Gamma(z)$ is the gamma function, which is defined by the integral

$$\Gamma(z) = \int_0^\infty x^{z-1}e^{-x}\,dx \qquad z > 0. \tag{3.46}$$

The gamma function has the following properties:

$$\Gamma\left(\frac{1}{2}\right) = \sqrt{\pi},$$
$$\Gamma(z+1) = z\Gamma(z) \qquad \text{for } z > 0, \text{ and}$$
$$\Gamma(m+1) = m! \qquad \text{for } m \text{ a nonnegative integer.}$$

The versatility of the gamma random variable is due to the richness of the gamma function $\Gamma(z)$. The pdf of the gamma random variable can assume a variety of shapes as shown in Fig. 3.14. By varying the parameters $\alpha$ and $\lambda$ it is possible to fit the gamma pdf to many types of experimental data. In addition, many random variables are special cases of the gamma random variable. The exponential random variable is obtained by letting $\alpha = 1$. By letting $\lambda = 1/2$ and $\alpha = k/2$, where $k$ is a positive integer, we obtain the **chi-square random variable**, which appears in certain statistical problems. The **$m$-Erlang random variable**

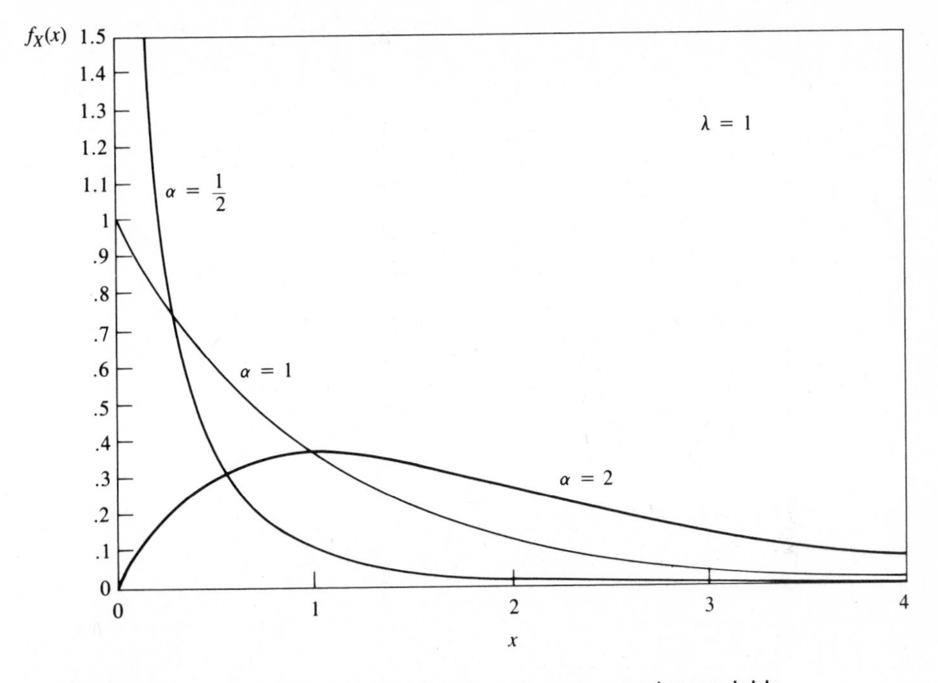


**FIGURE 3.14** Probability density function of gamma random variable.

is obtained when $\alpha = m$, a positive integer. The $m$-Erlang random variable is used in the system reliability models and in queueing systems models. Both of these random variables are discussed in later examples.

### ■■ Example 3.15

Show that the pdf of a gamma random variable integrates to one.

The integral of the pdf is

$$\int_0^{\infty} f_X(x)\,dx = \int_0^{\infty} \frac{\lambda(\lambda x)^{\alpha-1}e^{-\lambda x}}{\Gamma(\alpha)}\,dx$$

$$= \frac{\lambda^{\alpha}}{\Gamma(\alpha)} \int_0^{\infty} x^{\alpha-1}e^{-\lambda x}\,dx.$$

Let $y = \lambda x$, then $dx = dy/\lambda$ and the integral becomes

$$\frac{\lambda^{\alpha}}{\Gamma(\alpha)\lambda^{\alpha}} \int_0^{\infty} y^{\alpha-1}e^{-y}\,dy = 1,$$

where we used the fact that the integral equals $\Gamma(\alpha)$. ■■

In general, the cdf of the gamma random variable does not have a closed-form expression. The next example shows that the $m$-Erlang random variable does have a closed-form expression for its cdf. Later we see that the $m$-Erlang random variable results from the sum of $m$ independent, exponentially distributed times with parameter $\lambda$.

### ■■ Example 3.16

Find the cdf $F_X(x)$ of the $m$-Erlang random variable. The cdf is given by the integral

$$F_X(x) = \frac{\lambda^{m-1}}{(m-1)!} \int_0^x y^{m-1} \lambda e^{-\lambda y}\, dy, \tag{3.47}$$

since $\Gamma(m) = (m-1)!$. Integrate by parts using $u = y^{m-1}$ and $dv = \lambda e^{-\lambda y}\, dy$ so that $du = (m-1)y^{m-2}\, dy$ and $v = -e^{-\lambda y}$:

$$\begin{aligned} F_X(x) &= \frac{\lambda^{m-1}}{(m-1)!} \left\{ -y^{m-1} e^{-\lambda y} \Big|_0^x - \int_0^x -(m-1) y^{m-2} e^{-\lambda y}\, dy \right\} \\ &= -\frac{(\lambda x)^{m-1}}{(m-1)!} e^{-\lambda x} + \frac{\lambda^{m-2}}{(m-2)!} \int_0^x y^{m-2} \lambda e^{-\lambda y}\, dy. \end{aligned}$$

The integral on the right-hand side is identical to that in Eq. (3.47) with $m-1$ replaced by $m-2$. We can therefore repeatedly perform integration by parts to obtain the cdf of $X$:

$$\begin{aligned} F_X(x) &= -\frac{(\lambda x)^{m-1}}{(m-1)!} e^{-\lambda x} - \frac{(\lambda x)^{m-2}}{(m-2)!} e^{-\lambda x} - \cdots + \int_0^x e^{-\lambda y}\, dy \\ &= -\frac{(\lambda x)^{m-1}}{(m-1)!} e^{-\lambda x} - \frac{(\lambda x)^{m-2}}{(m-2)!} e^{-\lambda x} - \cdots - e^{-\lambda x} + 1 \\ &= 1 - \sum_{k=0}^{m-1} \frac{(\lambda x)^k}{k!} e^{-\lambda x}. \end{aligned} \tag{3.48}$$

The sum on the right-hand side involves the first $m$ terms of a Poisson pmf. We explain the relationship between the $m$-Erlang random and the Poisson random variables in Section 6.4. ■■

### ■■ Example 3.17

A factory has two spares of a critical system component that has an average lifetime of $1/\lambda = 1$ month. Find the probability that the three components (the operating one and the two spares) will last more than 6 months. Assume the component lifetimes are exponential random variables.

The remaining lifetime of the component in service is an exponential random variable with rate $\lambda$ by the memoryless property. Thus, the total lifetime $X$ of the three components is the sum of three exponential random variables with parameter $\lambda = 1$. Thus $X$ has a 3-Erlang distribution with

$\lambda = 1$. From Eq. (3.48) the probability that $X$ is greater than 6 is

$$P[X > 6] = 1 - P[X \leq 6]$$
$$= \sum_{k=0}^{2} \frac{6^k}{k!} e^{-6} = .06197$$

■■

## 3.5

## FUNCTIONS OF A RANDOM VARIABLE

Let $X$ be a random variable and let $g(x)$ be a real-valued function defined on the real line. Define $Y = g(X)$, that is, $Y$ is determined by evaluating the function $g(x)$ at the value assumed by the random variable $X$. Then $Y$ is also a random variable. The probabilities with which $Y$ takes on various values depend on the function $g(x)$ as well as the cumulative distribution function of $X$. In this section we consider the problem of finding the cdf and pdf of $Y$.

### ■■ Example 3.18

Let the function $h(x) = (x)^+$ be defined as follows:

$$(x)^+ = \begin{cases} 0 & \text{if } x < 0 \\ x & \text{if } x \geq 0. \end{cases}$$

For example, let $X$ be the number of active speakers in a group of $N$ speakers, and let $Y$ be the number of active speakers in excess of $M$, then $Y = (X - M)^+$. In another example, let $X$ be a voltage input to a half-wave rectifier then $Y = (X)^+$ is the output. ■■

### ■■ Example 3.19

Let the function $q(x)$ be defined as shown in Fig. 3.15, where the set of points on the real line are mapped into the nearest representation point from the set $S_Y = \{-3.5d, -2.5d, -1.5d, -0.5d, 0.5d, 1.5d, 2.5d, 3.5d\}$. Thus, for example, all the points in the interval $(0, d)$ are mapped into the point $d/2$. The function $q(x)$ represents an eight-level uniform quantizer. ■■

### ■■ Example 3.20

Consider the linear function $c(x) = ax + b$, where $a$ and $b$ are constants. This function arises in many situations. For example, $c(x)$ could be the cost associated with the quantity $x$, with the constant $a$ being the cost per unit of $x$, and $b$ being a fixed cost component. In a signal processing

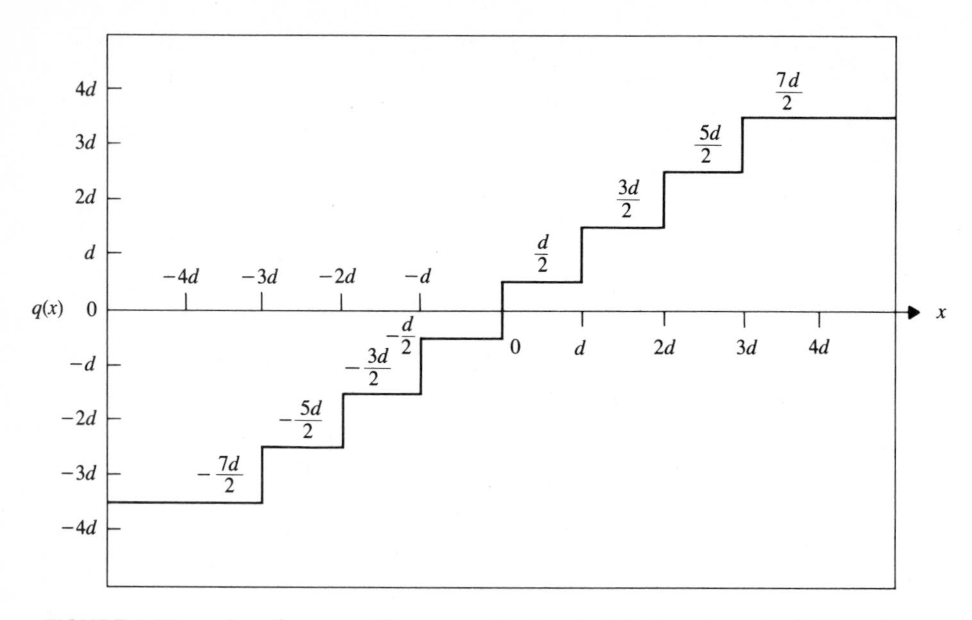


**FIGURE 3.15** A uniform quantizer maps the input $x$ into the closest point from the set $\{\pm d/2, \pm 3d/2, \pm 5d/2, \pm 7d/2\}$.

context, $c(x) = ax$ could be the amplified version (if $a > 1$) or attenuated version (if $a < 1$) of the voltage $x$. ■■

The probability of an event $C$ involving $Y$ is equal to the probability of the equivalent event $B$ of values of $X$ such that $g(X)$ is in $C$:

$$P[Y \text{ in } C] = P[g(X) \text{ in } C] = P[X \text{ in } B].$$

Three types of equivalent events are useful in determining the cdf and pdf of $Y = g(X)$: (1) The event $\{g(X) = y_k\}$ is used to determine the magnitude of the jump at a point $y_k$ where the cdf of $Y$ is known to have a discontinuity; (2) the event $\{g(X) \leq y\}$ is used to find the cdf of $Y$ directly; and (3) the event $\{y < g(X) \leq y + h\}$ is useful in determining the pdf of $Y$. We will demonstrate the use of these three methods in a series of examples.

The next two examples demonstrate how the pmf is computed in cases where $Y = g(X)$ is discrete. In the first example, $X$ is discrete. In the second example, $X$ is continuous.

## ■■ Example 3.21

Let $X$ be the number of active speakers in a group of $N$ independent speakers. Let $p$ be the probability that a speaker is active. In Example

2.36 it was shown that $X$ has a binomial distribution with parameters $N$ and $p$. Suppose that a voice transmission system can transmit up to $M$ voice signals at a time, and that when $X$ exceeds $M$, $X - M$ randomly selected signals are discarded. Let $Y$ be the number of signals discarded, then

$$Y = (X - M)^+.$$

$Y$ takes on values from the set $S_Y = \{0, 1, \ldots, N - M\}$. $Y$ will equal zero whenever $X$ is less than or equal to $M$, and $Y$ will equal $k > 0$ when $X$ is equal to $M + k$. Therefore

$$P[Y = 0] = P[X \text{ in } \{0, 1, \ldots, M\}] = \sum_{j=0}^{M} p_j$$

and

$$P[Y = k] = P[X = M + k] = p_{M+k} \qquad 0 < k \leq N - M,$$

where $p_j$ is the pmf of $X$. ■■

---

## ■■ Example 3.22

Let $X$ be a sample voltage of a speech waveform, and suppose that $X$ has a uniform distribution in the interval $[-4d, 4d]$. Let $Y = q(X)$, where the quantizer input-output characteristic is as shown in Fig. 3.15. Find the pmf for $Y$.

The event $\{Y = q\}$ for $q$ in $S_Y$, is equivalent to the event $\{X \text{ in } I_q\}$, where $I_q$ is an interval of points mapped into the representation point $q$. The pmf of $Y$ is therefore found by evaluating

$$P[Y = q] = \int_{I_q} f_X(t)\, dt.$$

It is easy to see that the representation point has an interval of length $d$ mapped into it. Thus the eight possible outputs are equiprobable, that is, $P[Y = q] = 1/8$, for $q$ in $S_Y$. ■■

---

In Example 3.22, each constant section of the function $q(X)$ produces a delta function in the pdf of $Y$. In general, if the function $g(X)$ is constant during certain intervals and if the pdf of $X$ is nonzero in these intervals, then the pdf of $Y$ will contain delta functions. $Y$ will then be either discrete or of mixed type.

The cdf of $Y$ is defined as the probability of the event $\{Y \leq y\}$. In principle, it can always be obtained by finding the probability of the equivalent event $\{g(X) \leq y\}$ as shown in the next examples.

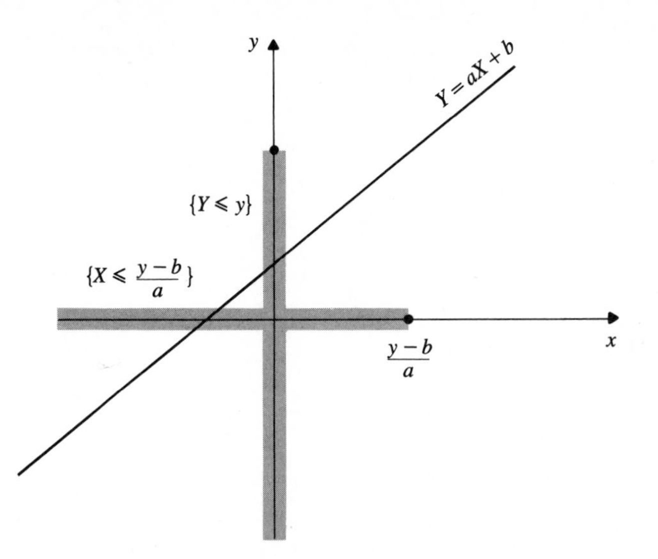


**FIGURE 3.16** The equivalent event for $\{Y \le y\}$ is the event $\{X \le (y - b)/a\}$, if $a > 0$.

## ■■ Example 3.23
### A Linear Function

Let the random variable $Y$ be defined by

$$Y = aX + b,$$

where $a$ is a nonzero constant. Suppose that $X$ has cdf $F_X(x)$, find $F_Y(y)$.

The event $\{Y \le y\}$ occurs when $A = \{aX + b \le y\}$ occurs. If $a > 0$, then $A = \{X \le (y - b)/a\}$ (see Fig. 3.16), and thus

$$F_Y(y) = P\left[X \le \frac{y - b}{a}\right] = F_X\left(\frac{y - b}{a}\right) \qquad a > 0.$$

On the other hand, if $a < 0$ then $A = \{X \ge (y - b)/a\}$, and

$$F_Y(y) = P\left[X \ge \frac{y - b}{a}\right] = 1 - F_X\left(\frac{y - b}{a}\right) \qquad a < 0.$$

We can obtain the pdf of $Y$ by differentiating with respect to $y$. To do this we need to use the chain rule for derivatives:

$$\frac{dF}{dy} = \frac{dF}{du}\frac{du}{dy},$$

where $u$ is the argument of $F$. In this case, $u = (y - b)/a$, and we then obtain

$$f_Y(y) = \frac{1}{a} f_X\left(\frac{y-b}{a}\right) \qquad a > 0$$

and

$$f_Y(y) = \frac{1}{-a} f_X\left(\frac{y-b}{a}\right) \qquad a < 0.$$

The above two results can be written compactly as

$$f_Y(y) = \frac{1}{|a|} f_X\left(\frac{y-b}{a}\right). \tag{3.49}$$

■■

---

### ■■ Example 3.24
A Linear Function of a Gaussian Random Variable

Let $X$ be a random variable with a Gaussian pdf with mean $m$ and standard deviation $\sigma$:

$$f_X(x) = \frac{1}{\sqrt{2\pi}\,\sigma} e^{-(x-m)^2/2\sigma^2} \qquad -\infty < x < \infty. \tag{3.50}$$

Let $Y = aX + b$, find the pdf of $Y$.

Substitution of Eq. (3.50) into Eq. (3.49) yields

$$f_Y(y) = \frac{1}{\sqrt{2\pi}\,|a\sigma|} e^{-(y-b-am)^2/2(a\sigma)^2}.$$

Note that $Y$ also has a Gaussian distribution with mean $b + am$ and standard deviation $|a|\,\sigma$. Therefore *a linear function of a Gaussian random variable is also a Gaussian random variable.* ■■

---

### ■■ Example 3.25

Let the random variable $Y$ be defined by

$$Y = X^2,$$

where $X$ is a continuous random variable. Find the cdf and pdf of $Y$.

The event $\{Y \le y\}$ occurs when $\{X^2 \le y\}$ or equivalently when $\{-\sqrt{y} \le X \le \sqrt{y}\}$ for $y$ nonnegative, see Fig. 3.17. The event is null when $y$ is negative. Thus

$$F_Y(y) = \begin{cases} 0 & y < 0 \\ F_X(\sqrt{y}) - F_X(-\sqrt{y}) & y > 0 \end{cases}$$

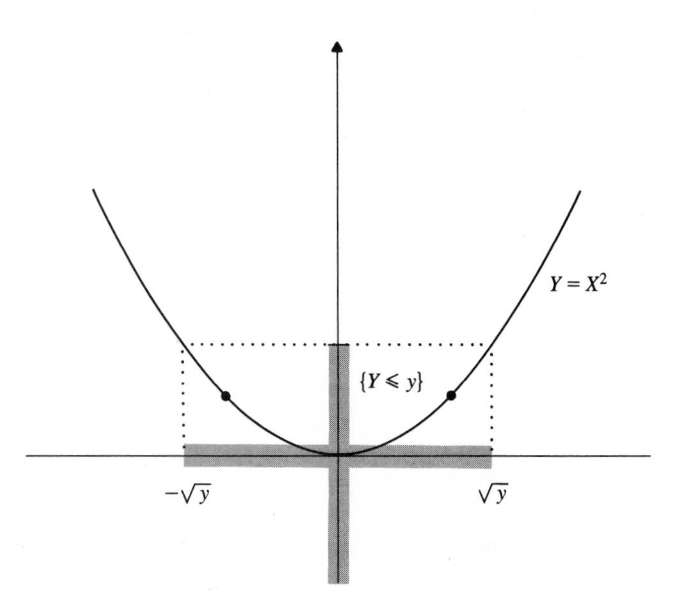


**FIGURE 3.17** The equivalent event for $\{Y \leq y\}$ is the event $\{-\sqrt{y} \leq X \leq \sqrt{y}\}$, if $y \geq 0$.

and

$$f_Y(y) = \frac{f_X(\sqrt{y})}{2\sqrt{y}} - \frac{f_X(-\sqrt{y})}{-2\sqrt{y}} \qquad y > 0$$

$$= \frac{f_X(\sqrt{y})}{2\sqrt{y}} + \frac{f_X(-\sqrt{y})}{2\sqrt{y}}. \tag{3.51}$$

■■

## ■■ Example 3.26
A Chi-Square Random Variable

Let $X$ be a Gaussian random variable with mean $m = 0$ and standard deviation $\sigma = 1$. $X$ is then said to be a standard normal random variable. Let $Y = X^2$. Find the pdf of $Y$.

Substitution of Eq. (3.50) into Eq. (3.51) yields

$$f_Y(y) = \frac{e^{-y/2}}{\sqrt{2y\pi}} \qquad y \geq 0. \tag{3.52}$$

From Table 3.2 we see that $f_Y(y)$ is the pdf of a *chi-square random variable with one degree of freedom.* ■■

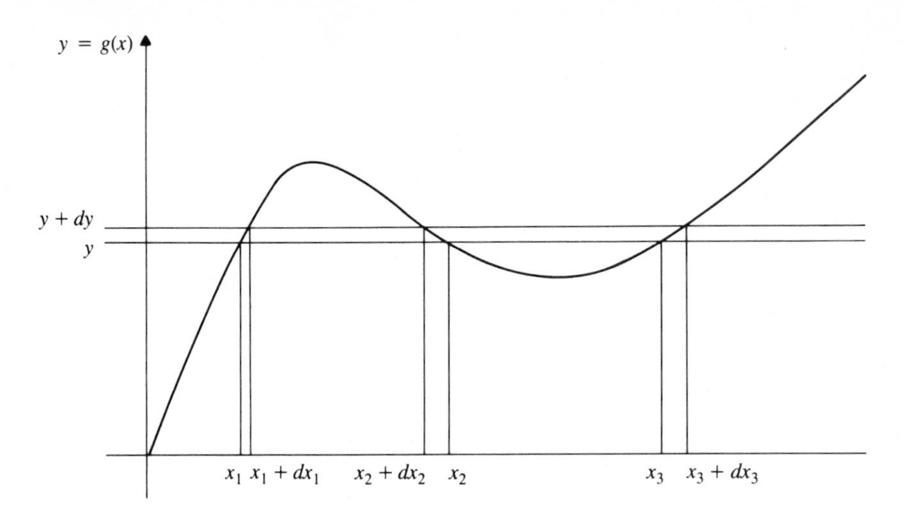


**FIGURE 3.18** The equivalent event of $\{y < Y \leq y + dy\}$ is $\{x_1 < X \leq x_1 + dx_1\} \cup \{x_2 + dx_2 < X \leq x_2\} \cup \{x_3 < X \leq x_3 + dx_3\}$.

The result in Example 3.25 suggests that if the equation $y_0 = g(x)$ has $n$ solutions, $x_0, x_1, \ldots, x_n$, then $f_Y(y_0)$ will be equal to $n$ terms of the type on the right-hand side of Eq. (3.51). We now show that this is generally true by using a method for directly obtaining the pdf of $Y$ in terms of the pdf of $X$.

Consider a nonlinear function $Y = g(X)$ such as the one shown in Fig. 3.18. Consider the event $C_y = \{y < Y < y + dy\}$ and let $B_y$ be its equivalent event. For $y$ indicated in the figure, the equation $g(x) = y$ has three solutions $x_1$, $x_2$, and $x_3$, and the equivalent event $B_y$ has a segment corresponding to each solution:

$$B_y = \{x_1 < X < x_1 + dx_1\} \cup \{x_2 + dx_2 < X < x_2\} \cup \{x_3 < X < x_3 + dx_3\}.$$

The probability of the event $C_y$ is approximately

$$P[C_y] = f_Y(y)\,|dy|, \tag{3.53}$$

where $|dy|$ is the length of the interval $y < Y \leq y + dy$. Similarly, the probability of the event $B_y$ is approximately

$$P[B_y] = f_X(x_1)\,|dx_1| + f_X(x_2)\,|dx_2| + f_X(x_3)\,|dx_3|. \tag{3.54}$$

Since $C_y$ and $B_y$ are equivalent events, their probabilities must be equal.

By equating Eqs. (3.53) and (3.54) we obtain

$$f_Y(y) = \sum_k \frac{f_X(x)}{|dy/dx|}\bigg|_{x=x_k} \tag{3.55}$$

$$= \sum_k f_X(x) \left|\frac{dx}{dy}\right|\bigg|_{x=x_k} \tag{3.56}$$

It is clear that if the equation $g(x) = y$ has $n$ solutions, the expression for the pdf of $Y$ at that point is given by Eqs. (3.55) and (3.56), and contains $n$ terms.

### ■■ Example 3.27

Let $Y = X^2$ as in Example 3.26. For $y \geq 0$, the equation $y = x^2$ has two solutions, $x_0 = \sqrt{y}$ and $x_1 = -\sqrt{y}$, so Eq. (3.55) has two terms. Since $dy/dx = 2x$, Eq. (3.55) yields

$$f_Y(y) = \frac{f_X(\sqrt{y})}{2\sqrt{y}} + \frac{f_X(-\sqrt{y})}{2\sqrt{y}}.$$

This result is in agreement with Eq. (3.51). To use Eq. (3.56), we note that

$$\frac{dx}{dy} = \frac{d}{dy} \pm \sqrt{y} = \pm \frac{1}{2\sqrt{y}},$$

which when substituted into Eq. (3.56) then yields Eq. (3.51) again. ■■

---

### ■■ Example 3.28
Amplitude Samples of a Sinusoidal Waveform

Let $Y = \cos(X)$, where $X$ is uniformly distributed in the interval $(0, 2\pi]$. $Y$ can be viewed as the sample of a sinusoidal waveform at a random instant of time that is uniformly distributed over the period of the sinusoid. Find the pdf of $Y$.

It can be seen in Fig. 3.19, that for $-1 < y < 1$ the equation $y = \cos(x)$ has two solutions in the interval of interest, $x_0 = \cos^{-1}(y)$ and $x_1 = 2\pi - x_0$. Since (see an introductory calculus textbook)

$$\frac{dy}{dx}\bigg|_{x_0} = -\sin(x_0) = -\sin(\cos^{-1}(y)) = -\sqrt{1 - y^2},$$

and since $f_X(x) = 1/2\pi$ in the interval of interest, Eq. (3.55) yields

$$f_Y(y) = \frac{1}{2\pi\sqrt{1 - y^2}} + \frac{1}{2\pi\sqrt{1 - y^2}}$$

$$= \frac{1}{\pi\sqrt{1 - y^2}} \qquad \text{for } -1 < y < 1.$$

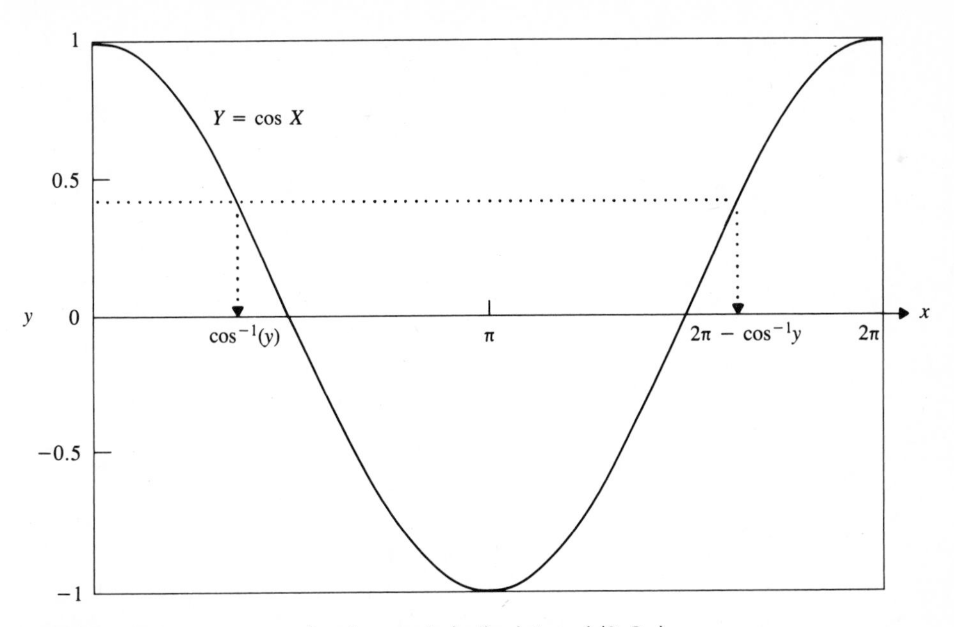


**FIGURE 3.19** $y = \cos x$ has two roots in the interval $(0, 2\pi)$.

The cdf of $Y$ is found by integrating the above:

$$F_Y(y) = \begin{cases} 0 & y < -1 \\ \dfrac{1}{2} + \dfrac{\sin^{-1} y}{\pi} & -1 \leqslant y \leqslant 1 \\ 1 & y > 1. \end{cases}$$

$Y$ is said to have the **arcsine distribution**. ■■

## 3.6

## THE EXPECTED VALUE OF RANDOM VARIABLES

In order to completely describe the behavior of a random variable, an entire function, namely the cdf or pdf, must be given. In some situations we are interested in a few parameters that summarize the information provided by these functions. For example, Fig. 3.20 shows the results of many repetitions of an experiment that produces two random variables. The random variable $X$ varies about the value 0 whereas the random variable $Y$ varies about the value 5. It is also clear that $Y$ is more spread out than $X$. In this section we introduce parameters that quantify these properties.

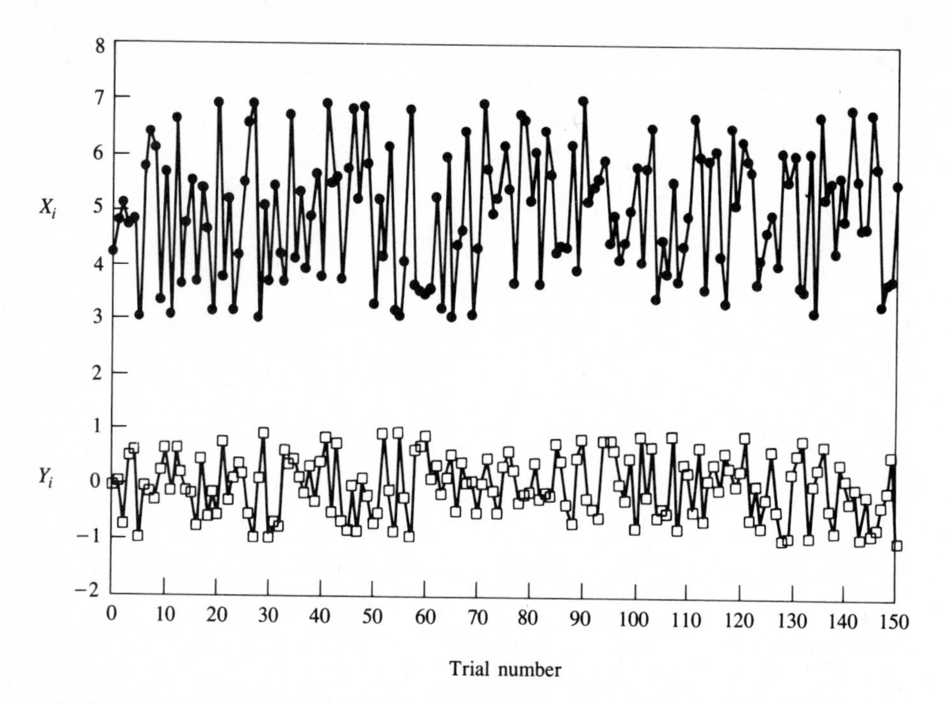


**FIGURE 3.20** The graphs show 150 repetitions of the experiments yielding the random variable $X$ and the random variable $Y$. It is clear that $X$ takes on values centered about the value 5 while $Y$ takes on values centered about 0. It is also clear that $X$ is more spread out than $Y$.

### The Expected Value of $X$

The **expected value** or **mean** of a random variable $X$ is defined by

$$E[X] = \int_{-\infty}^{+\infty} t f_X(t)\, dt. \tag{3.57}$$

If $X$ is a discrete random variable, substitution of Eq. (3.21) into Eq. (3.57) yields

$$E[X] = \sum_k x_k p_X(x_k). \tag{3.58}$$

The expected value $E[X]$ is defined if the above integral or sum converge absolutely, that is,

$$E[X] = \int_{-\infty}^{+\infty} |t|\, f_X(t)\, dt \qquad < \infty$$

or

$$E[X] = \sum_k |x_k|\, p_X(x_k) \qquad < \infty.$$

There are random variables for which the above expressions do not converge. See Problems 59 and 60 for examples of such random variables.

If we view $f_X(x)$ as the distribution of mass on the real line, then $E[X]$ represents the center-of-mass of this distribution.

Equation (3.58) appeared in Chapter 1, Eq. (1.9), where it was pointed out that the arithmetic average of a large number of independent observations of a random variable $X$ will tend to converge to $E[X]$. In this sense, the expected value of a random variable corresponds to our intuitive notion of the "average of $X$."

### Example 3.29
Mean of a Uniform Random Variable

The mean for a uniform random variable is given by

$$E[X] = (b - a)^{-1}\int_a^b t\,dt = \frac{a + b}{2},$$

which is exactly the midpoint of the interval $[a, b]$. The results shown in Fig. 3.20 were obtained by repeating experiments in which outcomes were random variables $X$ and $Y$ that had uniform cdf's in the intervals $[-1, 1]$ and $[3, 7]$, respectively. The respective expected values, 0 and 5, correspond to the values about which $X$ and $Y$ tend to vary. ■■

The result in Example 3.29 could have been found immediately by noting that $E[X] = m$ when the pdf is symmetric about a point $m$. That is, if

$$f_X(m - x) = f_X(m + x) \qquad \text{for all } x,$$

then, assuming that the mean exists,

$$0 = \int_{-\infty}^{+\infty} (m - t)f_X(t)\,dt = m - \int_{-\infty}^{+\infty} tf_X(t)\,dt.$$

The first equality above follows from the symmetry of $f_X(t)$ about $t = m$ and the odd symmetry of $(m - t)$ about the same point. We then have that $E[X] = m$.

### Example 3.30
Mean of a Gaussian Random Variable

The pdf of a Gaussian random variable is symmetric about the point $x = m$. Therefore $E[X] = m$. ■■

The following expressions are useful when $X$ is a nonnegative random variable:

$$E[X] = \int_0^{\infty} (1 - F_X(t))\,dt \qquad \text{if } X \text{ continuous and nonnegative} \tag{3.59}$$

and

$$E[X] = \sum_{k=0}^{\infty} P[X > k] \qquad \text{if } X \text{ nonnegative, integer-valued.} \tag{3.60}$$

The derivation of these formulas is discussed in the problems.

### ■■ Example 3.31

The time $X$ between customer arrivals at a service station has an exponential pdf with parameter $\lambda$. Find the mean interarrival time.

Substituting Eq. (3.34) into Eq. (3.57) we obtain

$$E[X] = \int_0^{\infty} t\lambda e^{-\lambda t}\, dt.$$

We evaluate the integral using integration by parts ($\int u\,dv = uv - \int v\,du$), with $u = t$ and $dv = \lambda e^{-\lambda t} dt$:

$$\begin{aligned} E[X] &= \lambda t e^{-\lambda t}\Big|_0^{\infty} - \int_0^{\infty} e^{-\lambda t}\, dt \\ &= \lim_{t\to\infty} te^{-\lambda t} - 0 - \left\{\frac{-e^{-\lambda t}}{\lambda}\right\}_0^{\infty} \\ &= \lim_{t\to\infty} \frac{e^{-\lambda t}}{\lambda} + \frac{1}{\lambda} = \frac{1}{\lambda}, \end{aligned}$$

where we have used the fact that $e^{-\lambda t}$ and $te^{-\lambda t}$ go to zero as $t$ approaches infinity.

For this example, Eq. (3.59) is much easier to evaluate:

$$E[X] = \int_0^{\infty} e^{-\lambda t}\, dt = \frac{1}{\lambda}.$$

■■

---

### ■■ Example 3.32

Let $N$ be the number of times a computer polls a terminal until the terminal has a message ready for transmission. If we suppose that the terminal produces messages according to a sequence of independent Bernoulli trials, then $N$ has a geometric distribution. Find the mean of $N$.

The expected value of the geometric random variable using Eq. (3.58) is

$$E[N] = \sum_{k=1}^{\infty} kpq^{k-1}.$$

This expression is readily evaluated by differentiating the series

$$\frac{1}{1-x} = \sum_{k=0}^{\infty} x^k$$

to obtain

$$\frac{1}{(1-x)^2} = \sum_{k=0}^{\infty} kx^{k-1}.$$

Letting $x = q$, we obtain

$$E[N] = p\frac{1}{(1-q)^2} = \frac{1}{p}.$$

The direct evaluation of Eq. (3.60) is easy in this case:

$$E[N] = \sum_{k=0}^{\infty} P[N > k] = \sum_{k=0}^{\infty} q^k = \frac{1}{1-q},$$

where we have used the fact that $P[N > k] = q^k$ for $k = 0, 1, 2, \ldots$. ■■

Suppose that $q = .6$ in the above example. Then $E[N] = 2.5$, which is not a value that can be assumed by $N$. Thus the statement "$N$ equals 2.5 on the average" does not make sense. (This reminds us of reading in the newspaper that the "average household has 3.5 persons.") What does make sense is the statement that the arithmetic average of a large number of repetitions of an experiment will be close to 2.5.

Tables 3.1 and 3.2 list the expected values of other important random variables.

### The Expected Value of $Y = g(X)$

Suppose that we are interested in finding the expected value of $Y = g(X)$. The direct approach involves first finding the pdf of $Y$, and then the evaluation of $E[Y]$ using Eq. (3.57). We now show that $E[Y]$ can be found directly in terms of the pdf of $X$:

$$E[Y] = \int_{-\infty}^{\infty} g(x) f_X(x)\, dx. \tag{3.61}$$

To see how Eq. (3.61) comes about, suppose that we divide the $y$-axis into intervals of length $h$, we index the intervals with the index $k$ and we let $y_k$ be the value in the center of the $k$th interval. The expected value of $Y$ is approximated by the following sum:

$$E[Y] \simeq \sum_k y_k f_Y(y_k) h.$$

Suppose that $g(x)$ is strictly increasing, then the $k$th interval in the $y$-axis has a unique corresponding equivalent event of width $h_k$ in the $x$-axis as shown in Fig. 3.21. Let $x_k$ be the value in the $k$th interval such that $g(x_k) = y_k$, then

$$E[Y] \simeq \sum_k g(x_k) f_X(x_k) h_k.$$

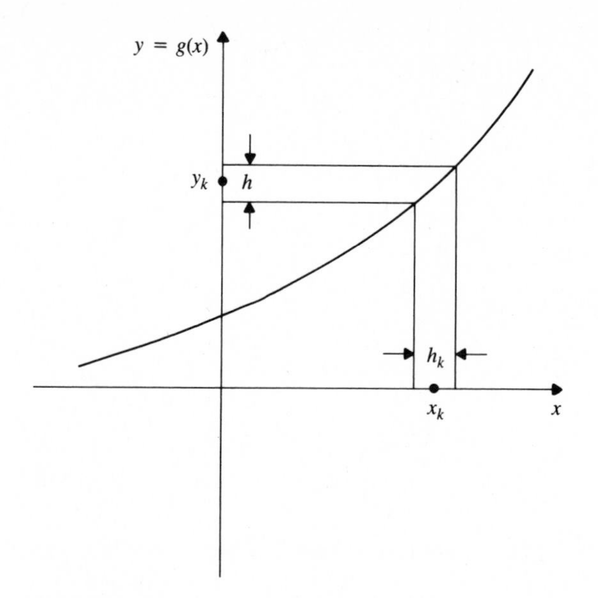


**FIGURE 3.21** Two infinitesimal equivalent events.

By letting $h$ approach zero, we obtain Eq. (3.61). This equation is valid even if $g(x)$ is not strictly increasing. The derivation for the general case involves taking into account the fact that each $y$-interval has an equivalent event as in Figure 3.18.

### ■■ Example 3.33

Let $Y = g(X) = A\cos(X)$, where $A$ is a constant and $X$ is uniformly distributed in the interval $(0, 2\pi]$. The expected value of $Y$ is then

$$E[Y] = \int_0^{2\pi} A\cos(x)\frac{1}{2\pi}\,dx = -A\sin(x)\Big|_0^{2\pi} = 0.$$ ■■

---

### ■■ Example 3.34
Expected Values of the Indicator Function

Let $g(X) = I_C(X)$ be the indicator function for the event $\{X \text{ in } C\}$, where $C$ is some interval or union of intervals in the real line:

$$g(X) = \begin{cases} 0 & X \text{ not in } C \\ 1 & X \text{ in } C, \end{cases}$$

then

$$E[Y] = \int_{-\infty}^{+\infty} g(X)f_X(x)\,dx = \int_C f_X(x)\,dx = P[X \text{ in } C].$$

Thus the expected value of the indicator of an event is equal to the probability of the event. ■■

The following two simple, but useful, properties follow immediately from Eq. (3.61). Let $c$ be some constant, then

$$E[c] = \int_{-\infty}^{\infty} cf_X(x)\,dx = c\int_{-\infty}^{\infty} f_X(x)\,dx = c \tag{3.62}$$

and

$$E[cX] = \int_{-\infty}^{\infty} cxf_X(x)\,dx = c\int_{-\infty}^{\infty} xf_X(x)\,dx = cE[X]. \tag{3.63}$$

The expected value of a sum of functions of a random variable is equal to the sum of the expected values of the individual functions:

$$\begin{aligned} E[Y] &= E\left[\sum_{k=1}^{n} g_k(X)\right] \\ &= \int_{-\infty}^{\infty} \sum_{k=1}^{n} g_k(x)f_X(x)\,dx = \sum_{k=1}^{n} \int_{-\infty}^{\infty} g_k(x)f_X(x)\,dx \\ &= \sum_{k=1}^{n} E[g_k(X)]. \end{aligned} \tag{3.64}$$

**■■ Example 3.35**

Let $Y = g(X) = a_0 + a_1X + a_2X^2 + \cdots + a_nX^n$, where $a_k$ are constants, then

$$\begin{aligned} E[Y] &= E[a_0] + E[a_1X] + \cdots + E[a_nX^n] \\ &= a_0 + a_1E[X] + a_2E[X^2] + \cdots + a_nE[X^n], \end{aligned}$$

where we have used Eq. (3.64), and Eqs. (3.62) and (3.63). A special case of this result is that

$$E[X + c] = E[X] + c,$$

that is, *we can shift the mean of a random variable by adding a constant to it.* ■■

### Variance of $X$

The expected value $E[X]$, by itself, provides us with very limited information about $X$. For example, if we know that $E[X] = 0$, then it could be that $X$ is zero all the time. However it could be as well that $X$ is equally likely to take on extremely large positive and negative values. We are therefore interested not only in the mean of a random variable, but also in the extent of the random variable's variation about its mean. Let the deviation of $X$ about its mean be $D = X - E[X]$, then $D$ can take on positive and negative values. Since we are interested in the magnitude of the variations only, it is convenient to work with $D^2$, which is always positive. The **variance of the random variable** $X$ is defined as the mean squared variation $E[D^2]$:

$$\text{VAR}[X] = E[(X - E[X])^2]. \tag{3.65}$$

By taking the square root of the variance we obtain a quantity with the same units as $X$. The **standard deviation of the random variable** $X$ is defined by

$$\text{STD}[X] = \text{VAR}[X]^{1/2}. \tag{3.66}$$

The STD[$X$] is used as a measure of the "width" or "spread" of a distribution.

The expression in Eq. (3.65) can be simplified as follows:

$$\begin{aligned}\text{VAR}[X] &= E[(X^2 - 2E[X]X + E[X]^2] \\ &= E[X^2] - 2E[X]E[X] + E[X]^2 \\ &= E[X^2] - E[X]^2,\end{aligned} \tag{3.67}$$

where the fact that $E[X]$ is a constant and Eqs. (3.62) and (3.63) have been used.

## ■■ Example 3.36
Variance of Uniform Random Variable

Find the variance the random variable $X$ that is uniformly distributed in the interval $[a, b]$.

Since the mean of $X$ is $(a + b)/2$,

$$\text{VAR}[X] = \frac{1}{b-a}\int_a^b \left(x - \frac{a+b}{2}\right)^2 dx.$$

Let $y = (x - (a + b)/2)$,

$$\text{VAR}[X] = \frac{1}{b-a}\int_{(b-a)/2}^{(b-a)/2} y^2\, dy = \frac{(b-a)^2}{12}.$$

The random variables in Fig. 3.20 were uniformly distributed in the interval $[-1, 1]$ and $[3, 7]$, respectively. Their variances are then 1/3 and 4/3. The corresponding standard deviations are 0.577 and 1.155. ■■

**Example 3.37**
Variance of Geometric Random Variable

Find the variance of the geometric random variable.

Differentiate the term $(1 - x)^{-2}$ in Example 3.32 to obtain

$$\frac{2}{(1-x)^3} = \sum_{k=0}^{\infty} k(k-1)x^{k-2}.$$

Letting $x = q$ and multiplying both sides by $pq$, we obtain

$$\frac{2q}{(1-q)^2} = E[N^2] - E[N].$$

Recalling that $E[N] = 1/p$, we find that $E[N^2] = (1 + q)/p^2$. The variance is then obtained using Eq. (3.67).

$$\text{VAR}[N] = E[N^2] - E[N]^2 = \frac{q}{p^2}.$$

---

**Example 3.38**
Variance of Gaussian Random Variable

Find the variance of a Gaussian random variable.

First multiply the integral of the pdf of $X$ by $\sqrt{2\pi}\,\sigma$ to obtain

$$\int_{-\infty}^{\infty} e^{-(x-m)^2/2\sigma^2}\,dx = \sqrt{2\pi}\,\sigma.$$

Differentiate both sides with respect to $\sigma$

$$\int_{-\infty}^{\infty} \left(\frac{(x-m)^2}{\sigma^3}\right) e^{-(x-m)^2/2\sigma^2}\,dx = \sqrt{2\pi}$$

By rearranging the above equation, we obtain

$$\text{VAR}[X] = \frac{1}{\sqrt{2\pi}\,\sigma}\int_{-\infty}^{\infty} (x-m)^2 e^{-(x-m)^2/2\sigma^2}\,dx = \sigma^2.$$

This result can also be obtained by direct integration. See Problem 57. Figure 3.13 shows the Gaussian pdf for several values of $\sigma$; it is evident that the "width" of the pdf increases with $\sigma$.

---

The following properties are easy to show (see Problem 61).

$$\text{VAR}[c] = 0 \tag{3.68}$$

$$\text{VAR}[X + c] = \text{VAR}[X] \tag{3.69}$$

$$\text{VAR}[cX] = c^2\,\text{VAR}[X], \tag{3.70}$$

where $c$ is a constant.

The mean and variance are the two most important parameters used in summarizing the pdf of a random variable. Other parameters are occasionally used. For example, the skewness defined by $E[(X - E[X])^3]/\text{STD}[X]^3$ measures the degree of asymmetry about the mean. It is easy to show that if a pdf is symmetric about its mean, then its skewness is zero. The point to note with these parameters of the pdf is that each involves the expected value of a higher power of $X$. Indeed we show in a later section that, under certain conditions, a pdf is completely specified if the expected value of all the powers of $X$ are known. These expected values are called the moments of $X$.

The $n$th **moment of the random variable** $X$ is defined by

$$E[X^n] = \int_{-\infty}^{\infty} x^n f_X(x)\, dx. \tag{3.71}$$

The mean and variance can be seen to be defined in terms of the first two moments, $E[X]$ and $E[X^2]$.

## ■■ *Example 3.39
### Analog-to-Digital Conversion: A Detailed Example

The purpose of quantizer is to map a random voltage $X$ into the nearest point $q(X)$ from a set of $2^R$ representation values (see Example 3.19). The value $X$ is then approximated by $q(X)$, which is identified by an $R$ bit binary number. In this manner, an "analog" voltage $X$, that can assume a continuum of values, is converted into an $R$ bit number.

The quantizer introduces an error $Z = X - q(X)$ as shown in Fig. 3.22. Note that $Z$ is a function of $X$ and that it ranges in value between $-d/2$ and $d/2$, where $d$ is the quantizer step size. Suppose that $X$ has a uniform distribution in the interval $[-x_{\max}, x_{\max}]$, that the quantizer has $2^R$ levels, and that $2x_{\max} = 2^R d$. It is easy to show that $Z$ is uniformly distributed in the interval $[-d/2, d/2]$ (see Problem 50).

Therefore from Example 3.29,

$$E[Z] = \frac{d/2 - d/2}{2} = 0.$$

The error $Z$ thus has mean zero.

By Example 3.36

$$\text{VAR}[Z] = \frac{(d/2 - (-d/2))^2}{12} = \frac{d^2}{12}.$$

This result is approximately correct for any pdf that is approximately flat over each quantizer interval. This is the case when $2^R$ is large.

The approximation $q(x)$ can be viewed as a "noisy" version of $X$ since

$$Q(X) = X - Z,$$

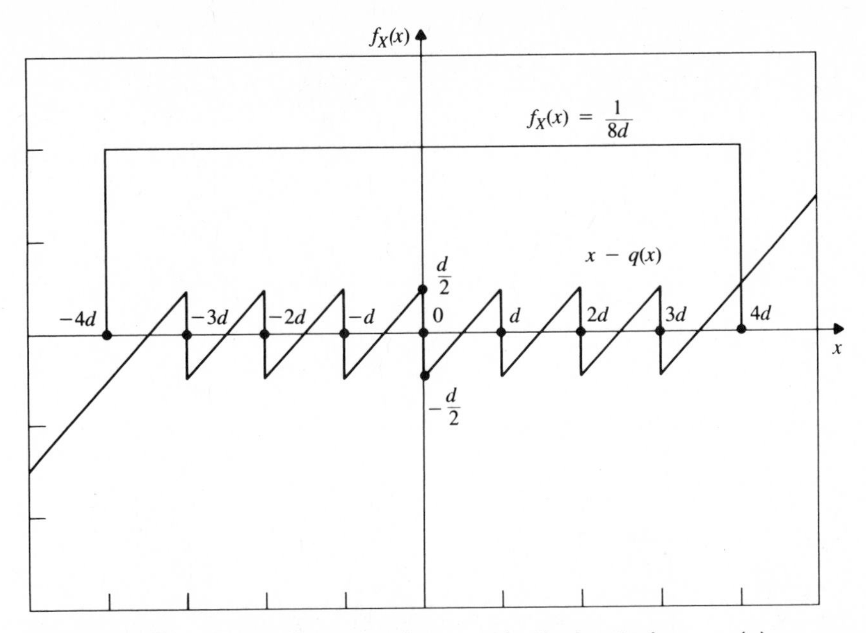


**FIGURE 3.22** The uniform quantizer error for the input $x$ is $x - q(x)$.

where $Z$ is the quantization error $Z$. The measure of goodness of a quantizer is specified by the SNR ratio, which is defined as the ratio of the variance of the "signal" $X$ to the variance of the distortion or "noise" $Z$:

$$\text{SNR} = \frac{\text{VAR}[X]}{\text{VAR}[Z]} = \frac{\text{VAR}[X]}{d^2/12}$$

$$= \frac{\text{VAR}[X]}{x_{\max}^2/3} 2^{2R},$$

where we have used the fact that $d = 2x_{\max}/2^R$. When $X$ is nonuniform, the value $x_{\max}$ is selected so that $P[|X| > x_{\max}]$ is small. A typical choice is $x_{\max} = 4\text{STD}[X]$. The SNR is then

$$\text{SNR} = \frac{3}{16} 2^{2R}.$$

This important formula is often quoted in decibels:

$$\text{SNR dB} = 10 \log_{10} \text{SNR} = 6R - 7.3 \text{ dB}.$$

The SNR increases by a factor of 4 (6 dB) with each additional bit used to represent $X$. This makes sense since each additional bit doubles the number of quantizer levels, which in turn reduces the step size by a factor of 2. The variance of the error should then be reduced by the square of this, namely $2^2 = 4$. ■■

## 3.7

## THE MARKOV AND CHEBYSHEV INEQUALITIES

In general, the mean and variance of a random variable do not provide enough information to determine the cdf/pdf. However, the mean and variance of a random variable $X$ do allow us to obtain bounds for probabilities of the form $P[|X| \geq t]$. Suppose first that $X$ is a nonnegative random variable with mean $E[X]$, **Markov's inequality** then states that

$$P[X \geq a] \leq \frac{E[X]}{a} \quad \text{for } X \text{ nonnegative.} \tag{3.72}$$

We obtain Eq. (3.72) as follows:

$$E[X] = \int_0^a t f_X(t)\,dt + \int_a^\infty t f_X(t)\,dt \geq \int_a^\infty t f_X(t)\,dt$$
$$\geq \int_a^\infty a f_X(t)\,dt = aP[X \geq a].$$

The first inequality results from discarding the integral from zero to $a$; the second inequality results from replacing $t$ by the smaller number $a$.

### ■■ Example 3.40

The mean height of children in a kindergarden class is 3 feet, 6 inches. Find the bound on the probability that a kid in the class is taller than 9 feet. Markov's inequality gives $P[H \geq 9] \leq 42/108 = .389$. ■■

The bound in the above example appears to be ridiculous. However, a bound, by its very nature, must take the worst case into consideration. One can easily construct a random variable for which the bound given by the Markov inequality is exact. The reason we know that the bound in the above example is ridiculous is that we have knowledge about the variability of the children's height about their mean.

Now suppose that the mean $E[X] = m$ and the variance $\text{VAR}[X] = \sigma$ of a random variable are known, and that we are interested in bounding $P[|X - m| \geq a]$. The **Chebyshev inequality** states that

$$P[|X - m| \geq a] \leq \frac{\sigma^2}{a^2}. \tag{3.73}$$

The Chebyshev inequality is a consequence of the Markov inequality. Let $D^2 = (X - m)^2$ be the squared deviation from the mean. Then Markov's inequality applied to $D^2$ gives

$$P[D^2 \geq a^2] \leq \frac{E[(X - m)^2]}{a^2} = \frac{\sigma^2}{a^2}.$$

Equation (3.73) follows when we note that $\{D^2 \geq a^2\}$ and $\{|X - m| \geq a\}$ are equivalent events.

Suppose that a random variable $X$ has zero variance, then the Chebyshev inequality implies that

$$P[X = m] = 1, \tag{3.74}$$

that is, the random variable is equal to its mean with probability one. In other words, $X$ is equal to the constant $m$ in almost all experiments.

**■■ Example 3.41**

The mean response time and the standard deviation in a multi-user computer system are known to be 15 seconds and 3 seconds, respectively. Estimate the probability that the response time is more than 5 seconds from the mean.

The Chebyshev inequality with $m = 15$ seconds, $\sigma = 3$ seconds, and $a = 5$ seconds gives

$$P[|X - 15| \geq 5] \leq \frac{9}{25} = .36.$$

■■

---

**■■ Example 3.42**

If $X$ has mean $m$ and variance $\sigma^2$, then the Chebyshev inequality for $a = k\sigma$ gives

$$P[|X - m| \geq k\sigma] \leq \frac{1}{k^2}.$$

Now suppose that we know that $X$ is a Gaussian random variable, then for $k = 2$, $P[|X - m| \geq 2\sigma] = .0228$, whereas the Chebyshev inequality gives the upper bound .25.

■■

---

We see from Example 3.42 that for certain random variables, the Chebyshev inequality can give rather loose bounds. Nevertheless, the inequality is useful in situations in which we have no knowledge about the distribution of a given random variable other than its mean and variance.

## 3.8 TESTING THE FIT OF A DISTRIBUTION TO DATA

How well does the model fit the data? Suppose you have postulated a probability model for some random experiment, and you are now interested in determining how well the model fits your experimental data. What do you do? In this section we present the chi-square test, which is

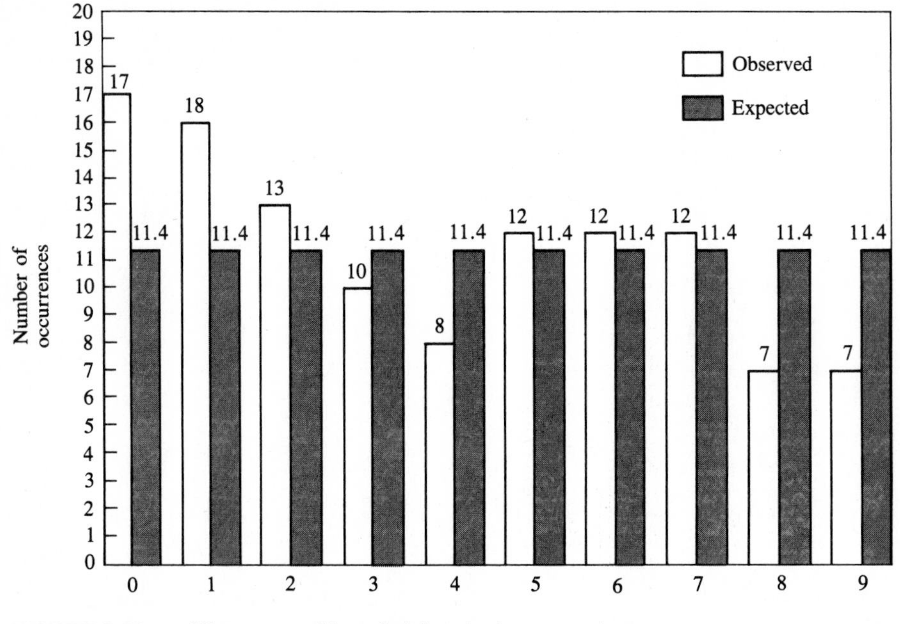


**FIGURE 3.23** Histogram of last digit in telephone numbers.

widely used to determine the goodness-of-fit of a distribution to a set of experimental data.[3]

The natural first test to carry out is to do an "eyeball" comparison of the postulated pmf, pdf, or cdf and an experimentally determined counterpart. If the outcome of the experiment, $X$, is discrete, we can compare the relative frequency of outcomes with the probability specified by the pmf as shown in Fig. 3.23. If $X$ is continuous, we can partition the real axis into $k$ mutually exclusive intervals and determine the relative frequency with which outcomes fall into each interval. These numbers would be compared to the probability of $X$ falling in the interval as shown in Fig. 3.24. If the relative frequencies and corresponding probabilities are in good agreement, then we have established that a good fit exists.

The chi-square test is a more formal way of carrying out the above comparison. Before considering the general case, we demonstrate the basic ideas of the method using the simplest case: Bernoulli trials.

## ■■ Example 3.43

Suppose we flip a coin 100 times and obtain 64 heads. Is it reasonable to assume that the coin is unbiased (i.e., $P$ [heads] $= 1/2$)?

3. Another useful goodness-of-fit test is the Kolmogorov-Smirnov test (Allen 1978, 311–317).

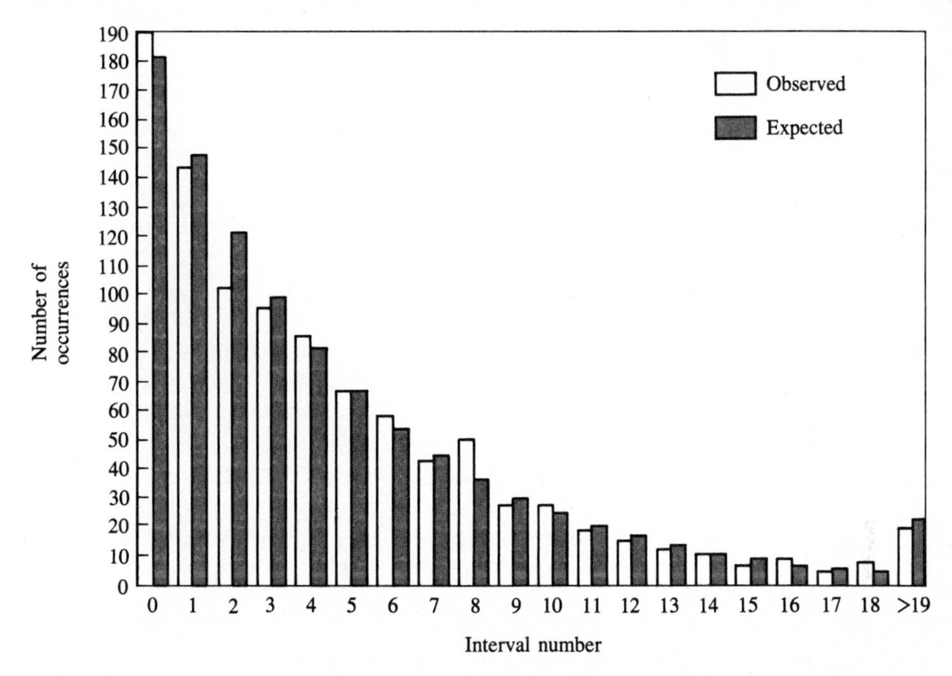


**FIGURE 3.24** Histogram of computer simulation of exponential random variables.

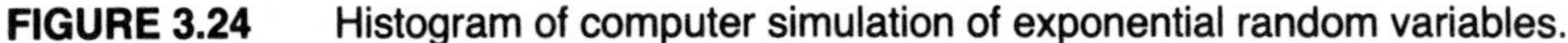

We reason as follows: If the hypothesis $H_0$ is true, that is the coin is unbiased, then the expected number of heads in 100 tosses is 50. Therefore we must accept the hypothesis if $|N - 50|$ is small, and reject it if $|N - 50|$ is "too large," where $N$ is the number of heads observed in the experiment.

There is a standard procedure in statistics for determining what is "too large." We compute the probability of obtaining a result as extreme or more extreme than that observed in the experiment under the assumption that the hypothesis is true. In the example under consideration this is

$$P[|N - 50| \geq 14 | H_0] = 1 - \sum_{k=37}^{63} \binom{100}{k} \left(\frac{1}{2}\right)^{100} \simeq .0093.$$

Therefore there is less than a 1% chance of obtaining a result as extreme as 64 if we are really tossing a fair coin. Thus we would consider this to be a significant reason to reject the hypothesis that the coin is fair. The above probability is called the "observed significance level" since it is a function of the observation.

The problem is usually posed in a slightly different way. The investigator specifies a probability, or "significance level," $\alpha$. This probability defines, for the given application, what constitutes a significant enough deviation from expected behavior to justify rejection of the hypothesis. He then finds a threshold value $t_\alpha$ such that

$$P[|N - 50| \geq t_\alpha \mid H_0] = \alpha.$$

If the difference between the observation and 50 exceeds the threshold, then the observed significance level will be less than $\alpha$, and the hypothesis will be rejected. Thus the decision on whether to reject the hypothesis is made on the basis of the comparison of $|N - 50|$ to the threshold.

The significance level of a test is usually set at 1% or 5%. ■■

---

There are two basic elements in the method used in Example 3.43. First, a measure is defined between the experimentally observed value and the value that would be expected if the postulated pmf/pdf were correct. Second, this measure is compared to a threshold to determine if the difference between the observations and expected results is too large. This threshold is determined by the significance level of the test, which is selected by the investigator.

The **chi-square test** contains the above two elements and it works as follows:

1. Partition the sample space $S_X$ into the union of $K$ disjoint intervals.
2. Compute the probability $b_k$ that an outcome falls in the $k$th interval under the assumption that $X$ has the postulated cdf. Then $m_k = nb_k$ is the expected number of outcomes that fall in the $k$th interval in $n$ repetitions of the experiment. (To see this, imagine performing Bernoulli trials in which a "success" corresponds to an outcome in the $k$th interval.)
3. The chi-square statistic is defined as the weighted difference between the observed number of outcomes, $N_k$, that fall in the $k$th interval, and the expected number $m_k$:

$$D^2 = \sum_{k=1}^{K} \frac{(N_k - m_k)^2}{m_k}. \tag{3.75}$$

4. If the fit is good, then $D^2$ will be small. Therefore the hypothesis is rejected if $D^2$ is too large, that is, if $D^2 \geq t_\alpha$, where $t_\alpha$ is a threshold determined by the significance level of the test.

The chi-square test is based on the fact that for large $n$, the random variable $D^2$ has a pdf that is approximately a chi-square pdf with $K - 1$ degrees of freedom. Thus the threshold $t_\alpha$ can be computed by finding the point at which

$$P[X \geq t_\alpha] = \alpha,$$

where $X$ is a chi-square random variable with $K - 1$ degrees of freedom (see Fig. 3.25). The thresholds for 1% and 5% significance levels and various degrees of freedom are given in Table 3.5.

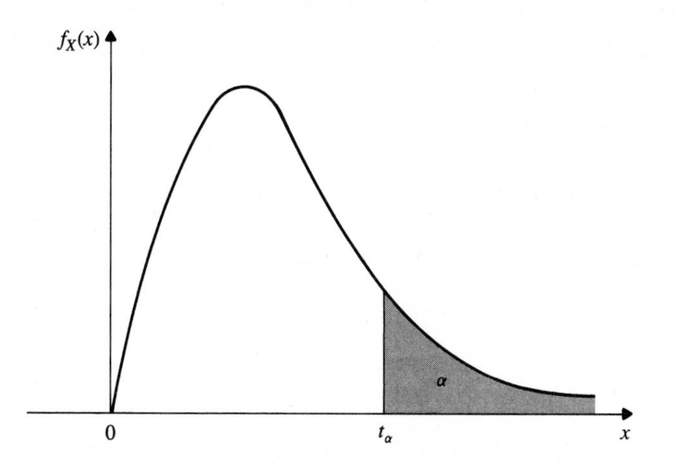


**FIGURE 3.25** Threshold in chi-square test is selected so that $P[D^2 > t_\alpha] = \alpha$.

**TABLE 3.5** Threshold Values for Chi-Square Test

| $k$ | 5% | 1% |
|---|---|---|
| 1 | 3.84 | 6.63 |
| 2 | 5.99 | 9.21 |
| 3 | 7.81 | 11.35 |
| 4 | 9.49 | 13.28 |
| 5 | 11.07 | 15.09 |
| 6 | 12.59 | 16.81 |
| 7 | 14.07 | 18.48 |
| 8 | 15.51 | 20.09 |
| 9 | 16.92 | 21.67 |
| 10 | 18.31 | 23.21 |
| 11 | 19.68 | 24.76 |
| 12 | 21.03 | 26.22 |
| 13 | 22.36 | 27.69 |
| 14 | 23.69 | 29.14 |
| 15 | 25.00 | 30.58 |
| 16 | 26.30 | 32.00 |
| 17 | 27.59 | 33.41 |
| 18 | 28.87 | 34.81 |
| 19 | 30.14 | 36.19 |
| 20 | 31.41 | 37.57 |
| 25 | 37.65 | 44.31 |
| 30 | 43.77 | 50.89 |

### ■■ Example 3.44

The histogram over the set $\{0, 1, 2, \ldots, 9\}$ in Fig. 3.23 was obtained by taking the last digit of the 114 telephone numbers in one column in a telephone directory. Are these observations consistent with the assumption that they have a discrete uniform pmf?

If the outcomes are uniformly distributed, then each has probability 1/10. The expected number of occurrences of each outcome in 114 trials is $114/10 = 11.4$. The chi-square statistic is then

$$D^2 = \frac{(17 - 11.4)^2}{10} + \frac{(16 - 11.4)^2}{10} + \cdots + \frac{(7 - 11.4)^2}{10}$$
$$= 9.51.$$

The number of degrees of freedom is $K - 1 = 10 - 1 = 9$, so from Table 3.5 the threshold for a 1% significance level is 21.7. $D^2$ does not exceed the threshold, so we conclude that the data is consistent with that of a uniformly distributed random variable. ■■

### ■■ Example 3.45

The histogram in Fig. 3.24 was obtained by generating 1000 samples from a program designed to generate exponentially distributed random variables with parameter 1. The histogram was obtained by dividing the positive real line into 20 intervals of equal length 0.2. The exact numbers are given in Table 3.6. A second histogram was also taken using 20 intervals of equal probability. The numbers for this histogram are given in Table 3.7.

From Table 3.5 we find that the threshold for a 5% significance level is 30.1. The chi-square values for the two histograms are 14.2 and 11.6, respectively. Both histograms pass the goodness-of-fit test in this case, but it is apparent that the method of selecting the intervals can significantly affect the value of the chi-square measure. ■■

Example 3.45 shows that there are many ways of selecting the intervals in the partition, and that these can yield different results. The following rules of thumb are recommended. First, to the extent possible the intervals should be selected so that they are equiprobable. Second, the intervals should be selected so that the expected number of outcomes in each interval is five or more. This improves the accuracy of approximating the cdf of $D^2$ by a chi-square cdf.

The discussion so far has assumed that the postulated distribution is completely specified. In the typical case, however, one or two parameters of the distribution, namely the mean and variance, are estimated from the data. It is often recommended that if $r$ of the parameters of a cdf are estimated from the data, then $D^2$ is better approximated by a chi-square

**TABLE 3.6** Chi-Square Test for Exponential Random Variable, Equal-Length Intervals

| Interval | Observed | Expected | $(O - E)^2/E$ |
|---|---|---|---|
| 0 | 190 | 181.3 | 0.417484 |
| 1 | 144 | 148.4 | 0.130458 |
| 2 | 102 | 121.5 | 3.129629 |
| 3 | 96 | 99.5 | 0.123115 |
| 4 | 86 | 81.44 | 0.255324 |
| 5 | 67 | 66.7 | 0.001349 |
| 6 | 59 | 54.6 | 0.354578 |
| 7 | 43 | 44.7 | 0.064653 |
| 8 | 51 | 36.6 | 5.665573 |
| 9 | 28 | 30 | 0.133333 |
| 10 | 28 | 24.5 | 0.5 |
| 11 | 19 | 20.1 | 0.060199 |
| 12 | 15 | 16.4 | 0.119512 |
| 13 | 12 | 13.5 | 0.166666 |
| 14 | 11 | 11 | 0 |
| 15 | 7 | 9 | 0.444444 |
| 16 | 9 | 7.4 | 0.345945 |
| 17 | 5 | 6 | 0.166666 |
| 18 | 8 | 5 | 1.8 |
| >19 | 20 | 22.4 | 0.257142 |
| | Chi-Squared Value = | | 14.13607 |

distribution with $K - r - 1$ degrees of freedom.[4] In effect, each estimated parameter decreases the degrees of freedom by 1.

## ■■ Example 3.46

The histogram in Table 3.8 was reported by Rutherford, Chadwick, and Ellis in a famous paper published in 1920. The number of particles emitted by a radioactive mass in a time period of 7.5 seconds was counted. A total number of 2608 periods were observed. It is postulated that the number of particles emitted in a time period is a random variable with a Poisson distribution. Perform the chi-square goodness-of-fit test.

In this case, the mean of the Poisson distribution is unknown so it is estimated from the data to be 3.870. $D^2$ for $12 - 1 - 1 = 10$ degrees of freedom is then 12.94. The threshold at a 1% significance level is 23.2. $D^2$ does not exceed this, so we conclude that the data is in good agreement with the Poisson distribution. ■■

4. See (Allen 1978, 308).

**TABLE 3.7** Chi-Square Test for Exponential Random Variable, Equiprobable Intervals

| Interval | Observed | Expected | $(O-E)^2/E$ |
|---|---|---|---|
| 0 | 49 | 50 | 0.02 |
| 1 | 61 | 50 | 2.42 |
| 2 | 50 | 50 | 0 |
| 3 | 50 | 50 | 0 |
| 4 | 40 | 50 | 2 |
| 5 | 52 | 50 | 0.08 |
| 6 | 48 | 50 | 0.08 |
| 7 | 40 | 50 | 2 |
| 8 | 45 | 50 | 0.5 |
| 9 | 46 | 50 | 0.32 |
| 10 | 50 | 50 | 0 |
| 11 | 51 | 50 | 0.02 |
| 12 | 55 | 50 | 0.5 |
| 13 | 49 | 50 | 0.02 |
| 14 | 54 | 50 | 0.32 |
| 15 | 52 | 50 | 0.08 |
| 16 | 62 | 50 | 2.88 |
| 17 | 46 | 50 | 0.32 |
| 18 | 49 | 50 | 0.02 |
| 19 | 51 | 50 | 0.02 |
| | | Chi-Squared Value = | 11.6 |

**TABLE 3.8** Chi-Square Test for Poisson Random Variable

| Count | Observed | Expected | $(O-E)^2/E$ |
|---|---|---|---|
| 0 | 57.00 | 54.40 | 0.12 |
| 1 | 203.00 | 210.50 | 0.27 |
| 2 | 383.00 | 407.40 | 1.46 |
| 3 | 525.00 | 525.50 | .00 |
| 4 | 532.00 | 508.40 | 1.10 |
| 5 | 408.00 | 393.50 | 0.53 |
| 6 | 273.00 | 253.80 | 1.45 |
| 7 | 139.00 | 140.30 | 0.01 |
| 8 | 45.00 | 67.80 | 7.67 |
| 9 | 27.00 | 29.20 | 0.17 |
| 10 | 10.00 | 11.30 | 0.15 |
| >11 | 6.00 | 5.80 | 0.01 |
| | | | 12.94 |

Based on H. Cramer, *Mathematical Methods of Statistics*, Princeton University Press, Princeton, N. J., 1946, p. 436.

# 3.9

## TRANSFORM METHODS

In the old days, before calculators and computers, it was very handy to have logarithm tables around if your work involved performing a large number of multiplications. If you wanted to multiply the numbers $x$ and $y$, you looked up $\log(x)$ and $\log(y)$, *added* $\log(x)$ and $\log(y)$, and then looked up the inverse logarithm of the result. You probably remember from grade school that long-hand multiplication is more tedious and error-prone than addition. Thus logarithms were very useful as a computational aid.

Transform methods are extremely useful computational aids in the solution of equations that involve derivatives and integrals of functions. In many of these problems, the solution is given by the convolution of two functions: $f_1(x) * f_2(x)$. We will define the convolution operation later. For now, all you need to know is that finding the convolution of two functions can be more tedious and error-prone than long-hand multiplication! In this section we introduce transforms that map the function $f_k(x)$ into another function $\mathcal{F}_k(\omega)$, and that satisfy the property that $\mathcal{F}[f_1(x) * f_2(x)] = \mathcal{F}_1(\omega)\mathcal{F}_2(\omega)$. In other words, the transform of the convolution is equal to the product of the individual transforms. Therefore transforms allow us to replace the convolution operation by the much simpler multiplication operation. You might say, "So what! I'll just have the computer do the convolutions for me." It turns out however that in many problems, the computer solution using transform methods is more computationally efficient than direct convolution.

The transform expressions introduced in this section will prove very useful when we consider sums of random variables in Chapter 5.

### The Characteristic Function

The **characteristic function** of a random variable $X$ is defined by

$$\Phi_X(\omega) = E[e^{j\omega X}] \tag{3.76a}$$

$$= \int_{-\infty}^{\infty} f_X(x)e^{j\omega x}\,dx, \tag{3.76b}$$

where $j = \sqrt{-1}$ is the imaginary unit number. The two expressions on the right-hand side motivate two interpretations of the characteristic function. In the first expression, $\Phi_X(\omega)$ can be viewed as the expected value of a function of $X$, $e^{j\omega X}$, in which the parameter $\omega$ is left unspecified. In the second expression, $\Phi_X(\omega)$ is simply the Fourier transform of the pdf $f_X(x)$ (with a reversal in the sign of the exponent). Both of these interpretations prove useful in different contexts.

If we view $\Phi_X(\omega)$ as a Fourier transform, then we have from the

Fourier transform inversion formula that the pdf of $X$ is given by

$$f_X(x) = \frac{1}{2\pi}\int_{-\infty}^{\infty} \Phi_X(\omega)e^{-j\omega x}\,d\omega. \tag{3.77}$$

It then follows that every pdf and its characteristic function form a unique Fourier transform pair. Table 3.2 gives the characteristic function of some continuous random variables.

### ■■ Example 3.47

The characteristic function for an exponentially distributed random variable with parameter $\lambda$ is given by

$$\Phi_X(\omega) = \int_0^{\infty} \lambda e^{-\lambda x}e^{j\omega x}\,dx = \int_0^{\infty} \lambda e^{-(\lambda - j\omega)x}\,dx$$

$$= \frac{\lambda}{\lambda - j\omega}.$$ ■■

---

If $X$ is a discrete random variable, substitution of Eq. (3.21) into the definition of $\Phi_X(\omega)$ gives

$$\Phi_X(\omega) = \sum_k p_X(x_k)e^{j\omega x_k} \qquad \text{discrete random variables.}$$

Most of the time we deal with discrete random variables that are integer-valued. The characteristic function is then

$$\Phi_X(\omega) = \sum_{k=-\infty}^{\infty} p_X(k)e^{j\omega k} \qquad \text{integer-valued random variables.} \tag{3.78}$$

Equation (3.78) is the **Fourier transform of the sequence** $p_X(k)$. Note that the Fourier transform in Eq. (3.78) is a periodic function of $\omega$ with period $2\pi$, since $e^{j(\omega+2\pi)k} = e^{j\omega k}e^{jk2\pi}$ and $e^{jk2\pi} = 1$. Therefore the characteristic function of integer-valued random variables is a periodic function of $\omega$. The following inversion formula allows us to recover the probabilities $p_X(k)$ from $\Phi_X(\omega)$:

$$p_X(k) = \frac{1}{2\pi}\int_0^{2\pi} \Phi_X(\omega)e^{-j\omega k}\,d\omega \qquad k = 0, \pm 1, \pm 2, \ldots. \tag{3.79}$$

Indeed a comparison of Eqs. (3.78) and (3.79) shows that the $p_X(k)$ are simply the coefficients of the Fourier series of the periodic function $\Phi_X(\omega)$.

### ■■ Example 3.48

The characteristic function for a geometric random variable is given by

$$\Phi_X(\omega) = \sum_{k=0}^{\infty} pq^k e^{j\omega k} = p \sum_{k=0}^{\infty} (qe^{j\omega})^k$$

$$= \frac{p}{1 - qe^{j\omega}}$$

■■

---

Since $f_X(x)$ and $\Phi_X(\omega)$ form a transform pair, we would expect to be able to obtain the moments of $X$ from $\Phi_X(\omega)$. The **moment theorem** states that the moments of $X$ are given by

$$E[X^n] = \frac{1}{j^n} \frac{d^n}{d\omega^n} \Phi_X(\omega) \bigg|_{\omega=0}. \tag{3.80}$$

To show this, first expand $e^{j\omega x}$ in a power series in the definition of $\Phi_X(\omega)$:

$$\Phi_X(\omega) = \int_{-\infty}^{\infty} f_X(x) \left\{1 + j\omega X + \frac{(j\omega X)^2}{2!} + \cdots \right\} dx.$$

Assuming that all the moments of $X$ are finite and that the series can be integrated term-by-term, we obtain

$$\Phi_X(\omega) = 1 + j\omega E[X] + \frac{(j\omega)^2 E[X^2]}{2!} + \cdots + \frac{(j\omega)^n E[X^n]}{n!} + \cdots.$$

If we differentiate the above expression once and evaluate the result at $\omega = 0$ we obtain

$$\frac{d}{d\omega} \Phi_X(\omega) \bigg|_{\omega=0} = jE[X].$$

If we differentiate $n$ times and evaluate at $\omega = 0$, we finally obtain

$$\frac{d^n}{d\omega^n} \Phi_X(\omega) \bigg|_{\omega=0} = j^n E[X^n],$$

which yields Eq. (3.80).

### ■■ Example 3.49

To find the mean of an exponentially distributed random variable, we differentiate $\Phi_X(\omega) = \lambda(\lambda - j\omega)^{-1}$ once, and obtain

$$\Phi'_X(\omega) = \frac{\lambda j}{(\lambda - j\omega)^2}.$$

The moment theorem then implies that $E[X] = \Phi'_X(0)/j = 1/\lambda$.

If we take two derivatives, we obtain

$$\Phi''_X(\omega) = \frac{-2\lambda}{(\lambda - j\omega)^3},$$

so the second moment is then $E[X^2] = \Phi''_X(0)/j^2 = 2/\lambda^2$. The variance of $X$ is then given by

$$\text{VAR}[X] = E[X^2] - E[X]^2 = \frac{2}{\lambda^2} - \frac{1}{\lambda^2} = \frac{1}{\lambda^2}.$$ ■■

---

### The Probability Generating Function

In problems where random variables are nonnegative, it is usually more convenient to use the $z$-transform or the Laplace transform. The **probability generating function** $G_N(z)$ of a nonnegative integer-valued random variable $N$ is defined by

$$G_N(z) = E[z^N] \tag{3.81a}$$

$$= \sum_{k=0}^{\infty} p_N(k) z^k. \tag{3.81b}$$

The second expression is the $z$-transform of the pmf (with a sign change in the exponent). Table 3.1 shows the probability generating function for some discrete random variables. Note that the characteristic function of $N$ is given by $\Phi_N(\omega) = G_N(e^{j\omega})$.

Using a derivation similar to that used in the moment theorem, it is easy to show that the pmf of $N$ is given by

$$p_N(k) = \frac{1}{k!} \frac{d^k}{dz^k} G_N(z) \bigg|_{z=0}. \tag{3.82}$$

This is why $G_N(z)$ is called the probability generating function. By taking the first two derivatives of $G_N(z)$ and evaluating the result at $z = 1$, it is also easy to show that

$$E[N] = G'_N(1) \tag{3.83}$$

and

$$\text{VAR}[N] = G''_N(1) + G'_N(1) - (G'_N(1))^2. \tag{3.84}$$

### ■■ Example 3.50

The probability generating function for the Poisson random variable with parameter $\alpha$ is given by

$$G_N(z) = \sum_{k=0}^{\infty} \frac{\alpha^k}{k!} e^{-\alpha} z^k = e^{-\alpha} \sum_{k=0}^{\infty} \frac{(\alpha z)^k}{k!}$$

$$= e^{-\alpha} e^{\alpha z} = e^{\alpha(z-1)}.$$

The first two derivatives of $G_N(z)$ are given by

$$G'_N(z) = \alpha e^{\alpha(z-1)}$$

and

$$G''_N(z) = \alpha^2 e^{\alpha(z-1)}.$$

Therefore the mean and variance of the Poisson are

$$E[N] = \alpha$$

$$\text{VAR}[N] = \alpha^2 + \alpha - \alpha^2 = \alpha.$$

■■

### The Laplace Transform of the pdf

In queueing theory one deals with service times, waiting times, and delays. All of these are nonnegative continuous random variables. It is therefore customary to work with the **Laplace transform** of the pdf

$$X^*(s) = \int_0^\infty f_X(x)e^{-sx}\,dx = E[e^{-sX}]. \tag{3.85}$$

Note that $X^*(s)$ can be interpreted as a Laplace transform of the pdf or as an expected value of a function of $X$, $e^{-sX}$.

The moment theorem also holds for $X^*(s)$:

$$E[X^n] = (-1)^n \frac{d^n}{ds^n} X^*(s)\bigg|_{s=0}. \tag{3.86}$$

■■ **Example 3.51**

The Laplace transform of the exponential pdf is given by

$$X^*(s) = \int_0^\infty \lambda e^{-\lambda x} e^{-sx}\,dx = \frac{1}{\lambda + s}.$$

■■

## *3.10

## BASIC RELIABILITY CALCULATIONS

In this section we apply some of the tools developed so far to the calculation of measures that are of interest in assessing the reliability of systems. We also show how the reliability of a system can be determined in terms of the reliability of its components.

### The Failure Rate Function

Let $T$ be the lifetime of a component, a subsystem, or a system. The **reliability** at time $t$ is defined as the probability that the component, subsystem, or system is still functioning at time $t$:

$$R(t) = P[T > t]. \tag{3.87}$$

The reliability can be expressed in terms of the cdf of $T$:

$$R(t) = 1 - P[T \le t] = 1 - F_T(t). \tag{3.88}$$

Note that the derivative of $R(t)$ gives the negative of the pdf of $T$:

$$R'(t) = -f_T(t). \tag{3.89}$$

Suppose that we know a system is still functioning at time $t$, what is its future behavior? In Example 3.9, we found that the conditional cdf of $T$ given that $T > t$ is given by

$$F_T(x \mid T > t) = P[T \le x \mid T > t] = \begin{cases} 0 & x < t \\ \dfrac{F_T(x) - F_T(t)}{1 - F_T(t)} & x \ge t. \end{cases} \tag{3.90}$$

The pdf associated with $F_T(x \mid T > t)$ is

$$f_T(x \mid T > t) = \frac{f_T(x)}{1 - F_T(t)} \qquad x \ge t. \tag{3.91}$$

Note that the denominator of Eq. (3.91) is equal to $R(t)$.

The **failure rate function** $r(t)$ is defined as $f_T(x \mid T > t)$ evaluated at $x = t$:

$$r(t) = f_T(t \mid T > t) = \frac{-R'(t)}{R(t)}, \tag{3.92}$$

since by Eq. (3.89), $R'(t) = -f_T(t)$. The failure rate function has the following meaning:

$$P[t < T \le t + dt \mid T > t] = f_T(t \mid T > t)\, dt = r(t)\, dt. \tag{3.93}$$

In words, $r(t)\,dt$ is the probability that a component that has functioned up to time $t$ will fail in the next $dt$ seconds.

### ■■ Example 3.52
Exponential Failure Law

Suppose a component has a constant failure rate function, say $r(t) = \lambda$. Find the pdf for its lifetime $T$.

Equation (3.92) implies that

$$\frac{R'(t)}{R(t)} = -\lambda \tag{3.94}$$

Equation (3.94) is a first-order differential equation with initial condition $R(0) = 1$. If we integrate both sides of Eq. (3.94) from 0 to $t$, we obtain

$$-\int_0^t \lambda\, dt' + k = \int_0^t \frac{R'(t')}{R(t')} dt' = \ln R(t),$$

which implies that

$$R(t) = Ke^{-\lambda t}, \qquad \text{where } K = e^k.$$

The initial condition $R(0) = 1$ implies that $K = 1$. Thus

$$R(t) = e^{-\lambda t} \qquad t > 0 \tag{3.95}$$

and

$$f_T(t) = \lambda e^{-\lambda t} \qquad t > 0.$$

Thus if $T$ has a constant failure rate function, then $T$ is an exponential random variable. This is not surprising, since the exponential random variable satisfies the memoryless property. ■■

The derivation that was used in Example 3.52 can be used to show that, in general, the failure rate function and the reliability are related by

$$R(t) = \exp\left\{-\int_0^t r(t')\, dt'\right\} \tag{3.96}$$

and

$$f_T(t) = r(t) \exp\left\{-\int_0^t r(t')\, dt'\right\}. \tag{3.97}$$

### ■■ Example 3.53
Weibull Failure Law

The Weibull failure law has failure rate function given by

$$r(t) = \alpha\beta t^{\beta-1}, \tag{3.98}$$

where $\alpha$ and $\beta$ are positive constants. Equation (3.97) then implies that the pdf for $T$ is

$$f_T(t) = \alpha\beta t^{\beta-1} e^{-\alpha t^\beta} \qquad t > 0. \tag{3.99}$$

Figure 3.26 shows $f_T(t)$ for $\alpha = 1$ and several values of $\beta$. Note that $\beta = 1$ yields the exponential failure law, which has a constant failure rate. For $\beta > 1$, Eq. (3.98) gives a failure rate function that increases with time. For $\beta < 1$, Eq. (3.98) gives a failure rate function that decreases with time. Further properties of the Weibull random variable are developed in the problems. ■■

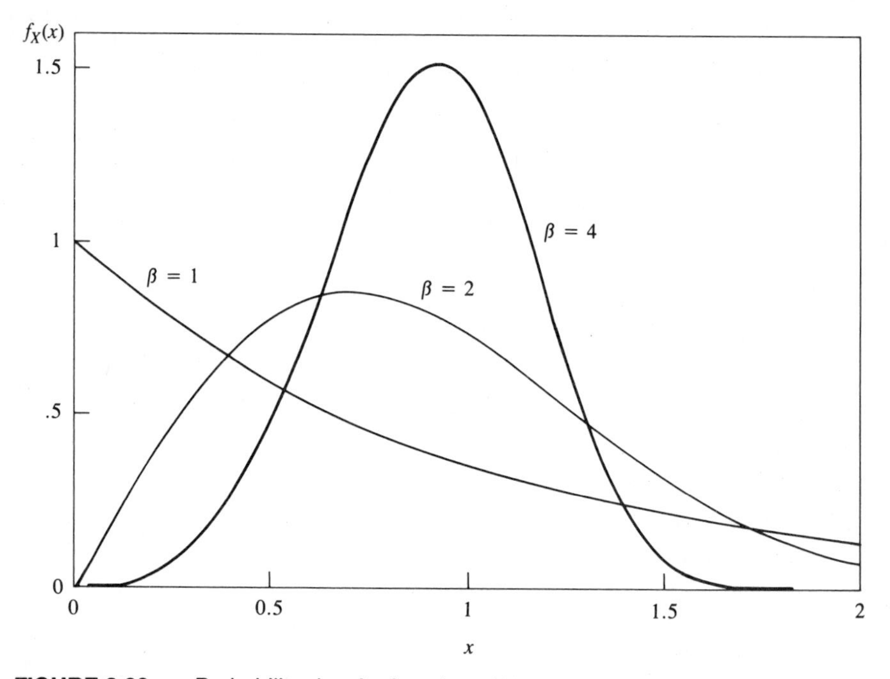


**FIGURE 3.26** Probability density function of Weibull random variable, $\alpha = 1$ and $\beta = 1, 2, 4$.

### Reliability of Systems

Suppose that a system consists of several components or subsystems. We now show how the reliability of a system can be computed in terms of the reliability of its subsystems if the components are assumed to fail independently of each other.

Consider first a system that consists of the series arrangement of $n$ components as shown in Fig. 3.27(a). This system is considered to be functioning only if all the components are functioning. Let $A_s$ be the event "system functioning at time $t$," and let $A_j$ be the event "$j$th component is functioning at time $t$," then the probability that the system is functioning at time $t$ is

$$\begin{aligned} R(t) &= P[A_s] \\ &= P[A_1 \cap A_2 \cap \cdots \cap A_n] = P[A_1]P[A_2]\ldots P[A_n] \\ &= R_1(t)R_2(t)\ldots R_n(t), \end{aligned} \tag{3.100}$$

since $P[A_j] = R_j(t)$, the reliability function of the $j$th component. Since probabilities are numbers that are less than or equal to one, we see that $R(t)$ can be no more reliable than the least reliable of the components, that is, $R(t) \leq \min_j R_j(t)$.

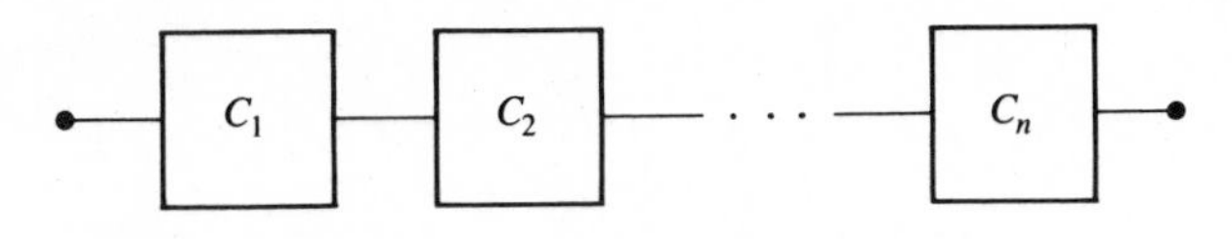


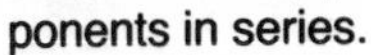
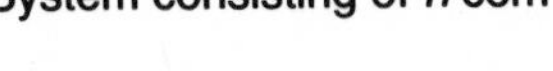

**FIGURE 3.27(a)** System consisting of $n$ components in series.

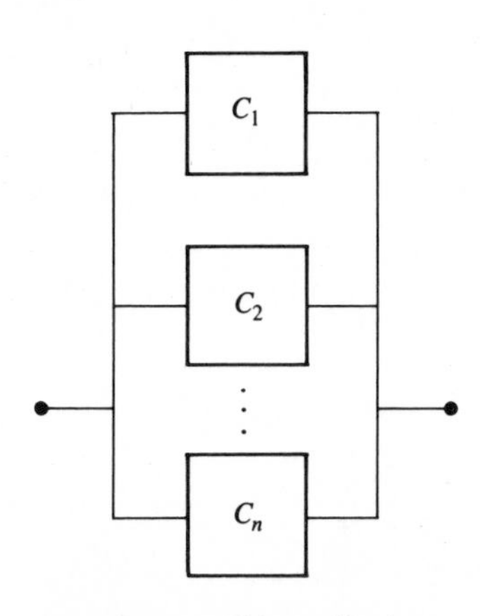


**FIGURE 3.27(b)** System consisting of $n$ components in parallel.

### ■■ Example 3.54

Suppose that a system consists of $n$ components in series and that the component lifetimes are exponential random variables with rates $\lambda_1, \lambda_2, \ldots, \lambda_n$. Find the system reliability.

From Eqs. (3.95) and (3.100), we have

$$\begin{aligned} R(t) &= e^{-\lambda_1 t} e^{-\lambda_2 t} \ldots e^{-\lambda_n t} \\ &= e^{-(\lambda_1 + \cdots + \lambda_n)t}. \end{aligned}$$

Thus the system reliability is exponentially distributed with rate $\lambda_1 + \lambda_2 + \cdots + \lambda_n$. ■■

Now suppose that a system consists of $n$ components in parallel, Fig. 3.27(b). This system is considered to be functioning as long as at least one of the components is functioning. The system will *not* be functioning if and only if all the components have failed, that is,

$$P[A_s^c] = P[A_1^c]P[A_2^c] \ldots P[A_n^c].$$

Thus

$$1 - R(t) = (1 - R_1(t))(1 - R_2(t)) \ldots (1 - R_n(t)),$$

and finally,

$$R(t) = 1 - (1 - R_1(t))(1 - R_2(t)) \ldots (1 - R_n(t)). \tag{3.101}$$

### ■■ Example 3.55

Compare the reliability of a single-unit system against that of a system that operates two units in parallel. Assume all units have exponentially distributed lifetimes with rate 1.

The reliability of the single unit system is

$$R_s(t) = e^{-t}.$$

The reliability of the two-unit system is

$$\begin{aligned} R_p(t) &= 1 - (1 - e^{-t})(1 - e^{-t}) \\ &= e^{-t}(2 - e^{-t}). \end{aligned}$$

The parallel system is more reliable by a factor of

$$(2 - e^{-t}) > 1.$$ ■■

More complex configurations can be obtained by combining subsystems consisting of series and parallel components. The reliability of such systems can then be computed in terms of the subsystem reliabilities. See Example 2.32 for an example of such a calculation.

## *3.11 COMPUTER METHODS FOR GENERATING RANDOM VARIABLES

The computer simulation of any random phenomenon involves the generation of random variables with prescribed distributions. For example, the simulation of a queueing system involves generating the time between customer arrivals as well as the service times of each customer. Once the cdfs that model these random quantities have been selected, an algorithm for generating random variables with these cdf's must be found. In this section we present a number of methods for generating random variables. All of these methods are based on the availability of random numbers that are uniformly distributed between zero and one. Methods for generating these numbers were discussed in Section 2.7.

### The Transformation Method

Suppose that $U$ is uniformly distributed in the interval $[0, 1]$. Let $F_X(x)$ be the cdf of the random variable we are interested in generating. Define the random variable, $Z = F_X^{-1}(U)$, that is, first $U$ is selected and then $Z$ is

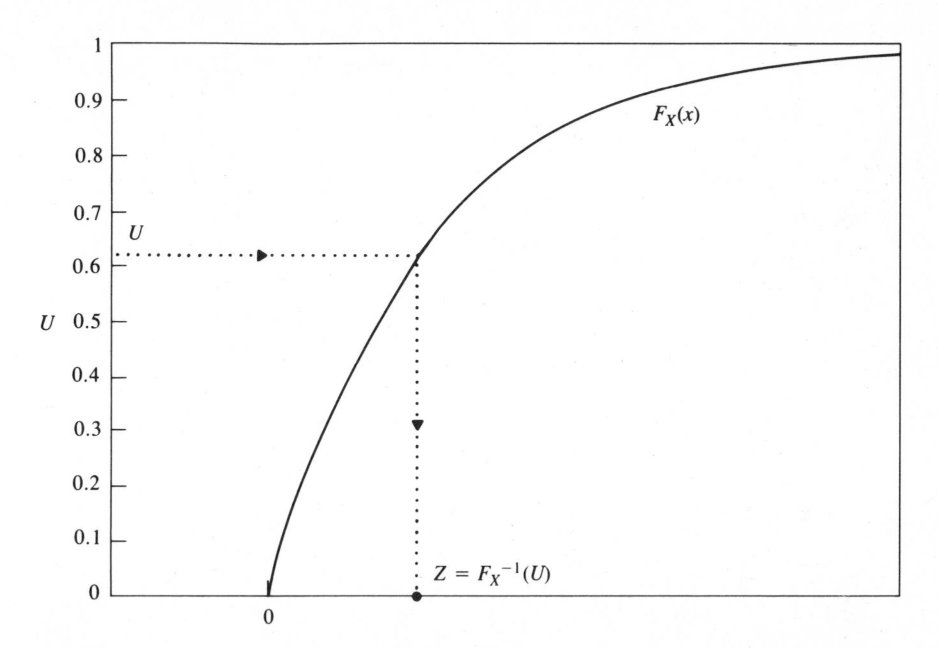


**FIGURE 3.28** Transformation method for generating a random variable with cdf $F_X(x)$.

found as indicated in the Fig. 3.28. The cdf of $Z$ is

$$P[Z \leq x] = P[F_X^{-1}(U) \leq x] = P[U \leq F_X(x)].$$

But if $U$ is uniformly distributed in $[0, 1]$ and $0 \leq h \leq 1$, then $P[U \leq h] = h$ (see Example 3.7). Thus

$$P[Z \leq x] = F_X(x),$$

and $Z = F_X^{-1}(U)$ has the desired cdf.

**Transformation Method for Generating $X$:**

1. Generate $U$ uniformly distributed in $[0, 1]$.
2. Let $Z = F_X^{-1}(U)$.

### ■■ Example 3.56
Exponential Random Variable

To generate an exponentially distributed random variable $X$ with parameter $\lambda$, we need to invert the expression $u = F_X(x) = 1 - e^{-\lambda x}$. We obtain

$$X = -\frac{1}{\lambda}\ln(1 - U).$$

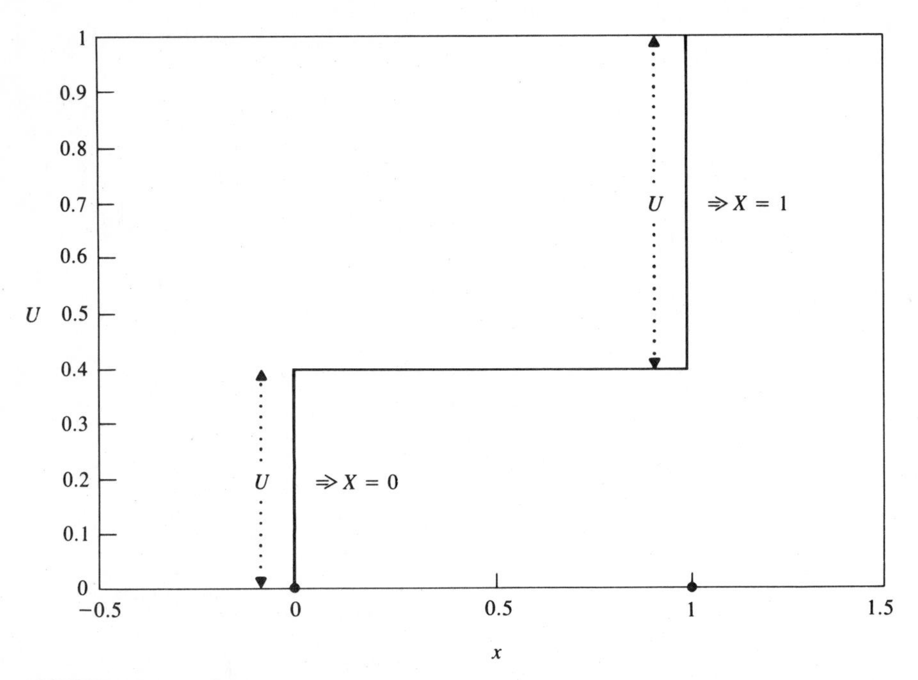


**FIGURE 3.29** Generating a Bernoulli random variable.

Note that we can use the simpler expression $X = -\ln(U)/\lambda$, since $1 - U$ is also uniformly distributed in $[0, 1]$. The random numbers tested in Example 3.45 were generated using this method. ■■

### ■■ Example 3.57
Bernoulli Random Variable

To generate a Bernoulli random variable with probability of success $p$, we note from Fig. 3.29 that

$$X = \begin{cases} 0 & \text{if } U \le p \\ 1 & \text{if } U > p. \end{cases}$$

In other words, we have partitioned the interval $[0, 1]$ into two segments of length $p$ and $1 - p$, respectively. The outcome $X$ is determined by the interval in which $U$ falls. ■■

### ■■ Example 3.58
Binomial Random Variable

To generate a binomial random variable $Y$ with parameters $n = 5$ and $p = 1/2$, we could simply generate five Bernoulli random variables and take $Y$ to be the total number of successes. We can also use the

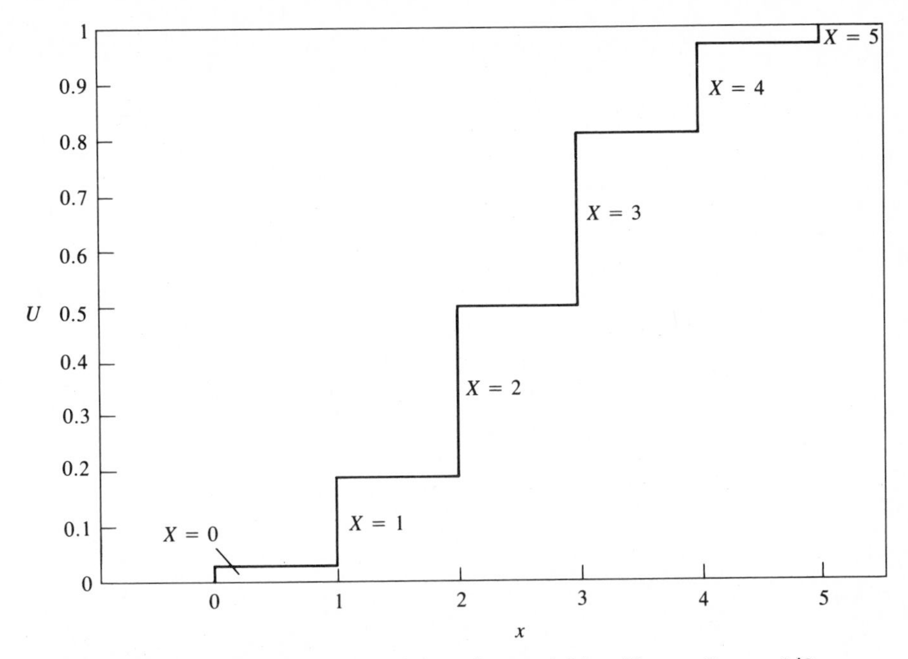


**FIGURE 3.30** Generating a binomial random variable with $n = 5$, $p = 1/2$.

transformation method directly as shown in Fig. 3.30. The unit interval is now partitioned into six segments. The efficiency of the algorithm depends on the order in which the segments are searched to determine into which one $U$ falls. For example, if we search the segments in order from 0 to 5, an average number of 3.5 comparisons are required. If we search the segments in order of decreasing probability the average number of comparisons drops to 2.38 (see Problem 3.93). ■■

The next method is based on the pdf rather than the cdf of $Z$.

### The Rejection Method

We first consider the simple version of this algorithm and explain why it works; then we present it in its general form. Suppose that we are interested in generating a random variable $Z$ with pdf $f_X(x)$ as shown in Fig. 3.31. In particular, we assume that: (1) the pdf is nonzero only in the interval $[0, a]$, and (2) the pdf takes on values in the range $[0, b]$. The **rejection method** in this case works as follows:

1. Generate $X_1$ uniform in the interval $[0, a]$.
2. Generate $Y$ uniform in the interval $[0, b]$.
3. If $Y \leq f_X(X_1)$, then output $Z = X_1$; else, reject $X_1$ and return to step 1.

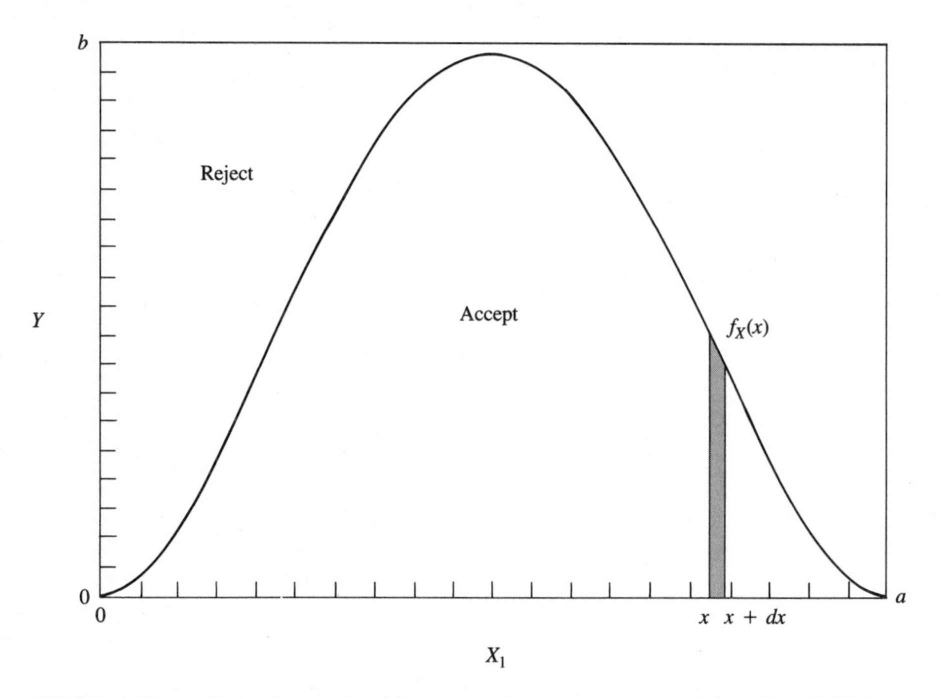


**FIGURE 3.31** Rejection method for generating a random variable with pdf $f_X(x)$.

Note that this algorithm will perform a random number of steps before it produces the output $Z$.

We now show that the output $Z$ has the desired pdf. Steps 1 and 2 select a point at random in a rectangle of width $a$ and height $b$. The probability of selecting a point in any region is simply the area of the region divided by the total area of the rectangle, $ab$. Thus the probability of accepting $X_1$ is the probability of the region below $f_X(x)$ divided by $ab$. But the area under any pdf is 1, so we conclude that the probability of success (i.e., acceptance) is $1/ab$. Consider now the following probability:

$$
\begin{aligned}
P[x \le X_1 < x + dx \mid X_1 \text{ is accepted}] \\
&= \frac{P[\{x \le X_1 < x + dx\} \cap \{X_1 \text{ accepted}\}]}{P[X_1 \text{ accepted}]} \\
&= \frac{\text{shaded area}/ab}{1/ab} = \frac{f_X(x)\,dx/ab}{1/ab} \\
&= f_X(x)\,dx.
\end{aligned}
$$

Therefore $X_1$ when accepted has the desired pdf. Thus $Z$ has the desired pdf.

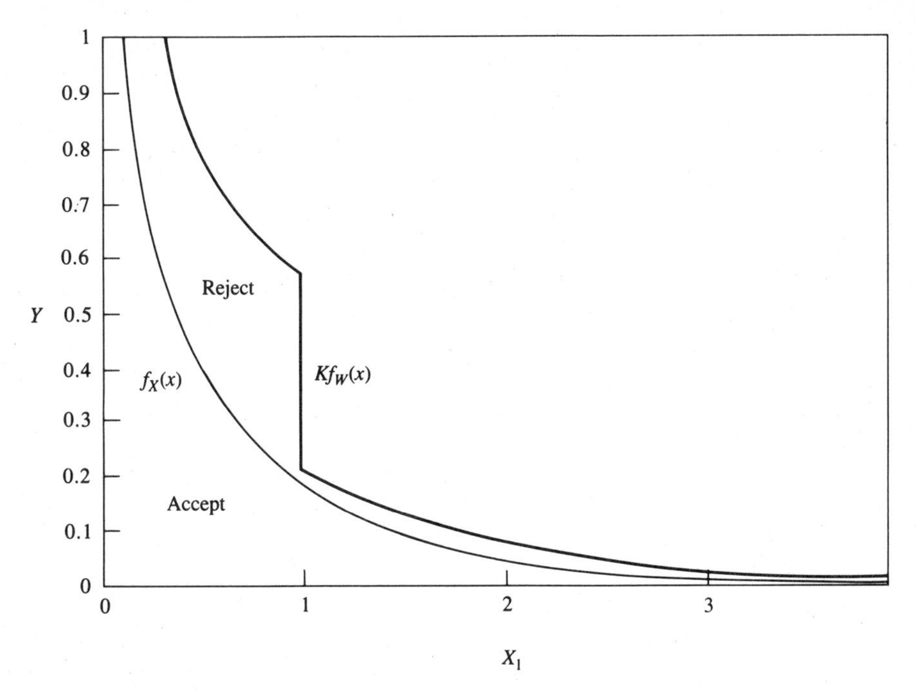


**FIGURE 3.32** Rejection method for generating a random variable with gamma pdf and with $0 < \alpha < 1$.

The algorithm as stated above can have two problems. First, if the rectangle does not fit snugly around $f_X(x)$, the number of $X_1$'s that need to be generated before acceptance may be excessive. Second, the above method cannot be used if $f_X(x)$ is unbounded or if its range is not finite. The general version of this algorithm overcomes both problems. Suppose we want to generate $Z$ with pdf $f_X(x)$. Let $W$ be a random variable with pdf $f_W(x)$ that is *easy* to generate and such that for some constant $K > 1$,

$$Kf_W(x) \geq f_X(x) \qquad \text{for all } x,$$

that is, the region under $Kf_W(x)$ contains $f_X(x)$ as shown in Fig. 3.32.

### Rejection Method for Generating $X$:

1. Generate $X_1$ with pdf $f_W(x)$. Define $B(X_1) = Kf_W(X_1)$.
2. Generate $Y$ uniform in $[0, B(X_1)]$.
3. If $Y \leq f_X(X_1)$, then output $Z = X_1$; else reject $X_1$ and return to step 1.

See Problem 98 for a proof that $Z$ has the desired pdf.

### Example 3.59
Gamma Random Variable

We now show how the rejection method can be used to generate $X$ with gamma pdf and parameters $0 < \alpha < 1$ and $\lambda = 1$. A function $Kf_W(x)$ that "covers" $f_X(x)$ is easily obtained (see Fig. 3.32):

$$f_X(x) = \frac{x^{\alpha-1}e^{-x}}{\Gamma(\alpha)} \leq Kf_W(x) = \begin{cases} \dfrac{x^{\alpha-1}}{\Gamma(\alpha)} & 0 \leq x \leq 1 \\ \dfrac{e^{-x}}{\Gamma(\alpha)} & x > 1. \end{cases}$$

The pdf $f_W(x)$ that corresponds to the function on the right-hand side is

$$f_W(x) = \begin{cases} \dfrac{\alpha e x^{\alpha-1}}{\alpha + e} & 0 \leq x \leq 1 \\ \alpha e \dfrac{e^{-x}}{\alpha + e} & x \geq 1. \end{cases}$$

The cdf of $W$ is

$$F_W(x) = \begin{cases} \dfrac{e x^{\alpha}}{\alpha + e} & 0 \leq x \leq 1 \\ 1 - \alpha e \dfrac{e^{-x}}{\alpha + e} & x > 1. \end{cases}$$

$W$ is easy to generate using the transformation method, with

$$F_W^{-1}(u) = \begin{cases} \left[\dfrac{(\alpha + e)u}{e}\right]^{1/\alpha} & u \leq e/(\alpha + e) \\ -\ln\left[(\alpha + e)\dfrac{(1-u)}{\alpha e}\right] & u > e/(\alpha + e). \end{cases}$$

We can therefore use the transformation method to generate this $f_W(x)$, and then the rejection method to generate any gamma random variable $X$ with parameters $0 < \alpha < 1$ and $\lambda = 1$. Finally we note that if we let $W = \lambda X$, then $W$ will be gamma with parameters $\alpha$ and $\lambda$. The generation of gamma random variables with $\alpha > 1$ is discussed in Problem 97.

## Generation of Functions of a Random Variable

Once we have a simple method of generating a random variable $X$, we can easily generate any random variable that is defined by $Y = g(X)$ or even $Z = h(X_1, X_2, \ldots, X_n)$, where $X_1, \ldots, X_n$ are $n$ outputs of the random variable generator.

■■ **Example 3.60**
Gaussian Random Variable

In Section 4.9 we show that if $U_1$ and $U_2$ are independent random variables uniformly distributed in the unit interval, then

$$X = (-2\ln(U_1))^{1/2}\cos(2\pi U_2),$$

and

$$Y = (-2\ln(U_1))^{1/2}\sin(2\pi U_2)$$

are independent Gaussian random variables with mean 0 and variance 1. This result can therefore be used to produce two Gaussian random variables from two uniformly distributed random variables. ■■

■■ **Example 3.61**
*m*-Erlang Random Variable

Let $X_1, X_2, \ldots$ be independent, exponentially distributed random variables with parameter $\lambda$. In Chapter 5 we show that the random variable

$$Y = X_1 + X_2 + \cdots + X_m$$

has an $m$-Erlang pdf with parameter $\lambda$. We can therefore generate an $m$-Erlang random variable by first generating $m$ exponentially distributed random variables using the transformation method, and then taking the sum. Since the $m$-Erlang random variable is a special case of the gamma random variable, for large $m$ it may be preferable to use the rejection method described in Problem 97. ■■

### Generating Mixtures of Random Variables

We have seen in previous sections that sometimes a random variable consists of a mixture of several random variables. In other words, the generation of the random variable can be viewed as first selecting a random variable type according to some pmf, and then generating a random variable from the selected pdf type. This procedure can be simulated easily.

■■ **Example 3.62**
Hyperexponential Random Variable

A two-stage hyperexponential random variable has pdf

$$f_X(x) = pae^{-ax} + (1-p)be^{-bx}.$$

It is clear from the above expression that $X$ consists of a mixture of two exponential random variables with parameters $a$ and $b$, respectively. $X$ can be generated by first performing a Bernoulli trial with probability of success $p$. If the outcome is a success, we then use the transformation method to generate an exponential random variable with parameter $a$. If the outcome is failure, we generate an exponential random variable with parameter $b$ instead. ■■

## SUMMARY

- A random variable is a function that assigns a real number to each outcome of a random experiment. A random variable is defined if the outcome of a random experiment is a number, or if a numerical attribute of an outcome is of interest.
- The notion of an equivalent event enables us to derive the probabilities of events involving a random variable in terms of the probabilities of events involving the underlying outcomes.
- The cumulative distribution function $F_X(x)$ is the probability that $X$ falls in the interval $(-\infty, x]$. The probability of any event consisting of the union of intervals can be expressed in terms of the cdf.
- A random variable is discrete if it assumes values from some countable set. A random variable is continuous if its cdf can be written as the integral of a nonnegative function. A random variable is mixed if it is a mixture of a discrete and a continuous random variable.
- The probability of events involving a continuous random variable $X$ can be expressed as integrals of the probability density function $f_X(x)$.
- If $X$ is a random variable, then $Y = g(X)$ is also a random variable. The notion of equivalent events allows us to derive expressions for the cdf and pdf of $Y$ in terms of the cdf and pdf of $X$.
- The cdf and pdf of the random variable $X$ are sufficient to compute all probabilities involving $X$ alone. The mean, variance, and moments of a random variable summarize some of the information about the random variable $X$. These parameters are useful in practice because they are easier to measure and estimate than the cdf and pdf.
- The Markov and Chebyshev inequalities allow us to bound probabilities involving $X$ in terms of its first two moments only.
- The chi-square test measures the degree of agreement of a set of data with a postulated pmf or pdf. It is used in the fitting of probability models to experimental data.
- Transforms provide an alternative but equivalent representation of the pmf and pdf. In certain types of problems it is preferable to work with the transforms rather than the pmf or pdf. The moments of a random variable can be obtained from the corresponding transform.

■ The reliability of a system is the probability that it is still functioning after $t$ hours of operation. The reliability of a system can be obtained from the reliability of its subsystems.

■ There are a number of methods for generating random variables with prescribed pmf's or pdf's in terms of a random variable that is uniformly distributed in the unit interval. These methods include the transformation and the rejection methods as well as methods that simulate random experiments (e.g., functions of random variables) and mixtures of random variables.

## CHECKLIST OF IMPORTANT TERMS

Characteristic function
Chebyshev inequality
Chi-square test
Continuous random variable
Cumulative distribution function
Discrete random variable
Equivalent event
Expected value of $X$
Failure rate function
Function of a random variable
Laplace transform of a pdf
Markov inequality
Moment theorem
$n$th moment of $X$
Probability density function
Probability generating function
Probability mass function
Random variable
Random variable of mixed type
Rejection method
Significance level
Standard deviation of $X$
Transformation method
Variance of $X$

## ANNOTATED REFERENCES

Reference [1] is the standard reference for electrical engineers for the material on random variables. References [2] and [3] discuss some of the finer points regarding the concept of a random variable at a level accessible to students of this course. Reference [4] gives a more complete introduction to statistical methods than that given in this book. Reference [5] presents detailed discussions of the various methods for generating random numbers with specified distributions. Reference [6] also discusses the generation of random variables.

1. A. Papoulis, *Probability, Random Variables, and Stochastic Processes,* McGraw-Hill, New York, 1965.
2. K. L. Chung, *Elementary Probability Theory,* Springer-Verlag, New York, 1974.
3. W. B. Davenport, *Probability and Random Processes: An Introduction for Applied Scientists and Engineers,* McGraw-Hill, New York, 1970.

4. A. O. Allen, *Probability, Statistics, and Queueing Theory*, Academic Press, New York, 1978.
5. A. M. Law and W. D. Kelton, *Simulation Modeling and Analysis*, McGraw-Hill, New York, 1982.
6. S. M. Ross, *Introduction to Probability Models*, Academic Press, New York, 1985.
7. H. Cramer, *Mathematical Methods of Statistics*, Princeton University Press, Princeton, N. J., 1946.
8. M. Abramowitz and I. Stegun, *Handbook of Mathematical Functions*, National Bureau of Standards, Washington, D.C., 1964.
9. R. C. Cheng, "The Generation of Gamma Variables with Nonintegral Shape Parameter," *Appl. Statist.*, 26: 71–75, 1977.
10. Y. A. Rozanov, *Probability Theory: A Concise Course*, Dover Publications, New York, 1969.
11. P. O. Börjesson and C. E. W. Sundberg, "Simple Approximations of the Error Function $Q(x)$ for Communications Applications," *IEEE Trans. on Communications*, March 1979, 639–643.
12. C. H. Stapper, F. M. Armstrong, and K. Saji, "Integrated Circuit Yield Statistics," *Proc. IEEE*, April 1983, 453–470.

## PROBLEMS

### Section 3.1
### The Notion of a Random Variable

1. An urn contains 90 \$1 bills, 9 \$5 bills, and 1 \$50 bill. Let the random variable $X$ be the denomination of a bill that is selected at random from the bag.
   a. Describe the underlying space $S$ of this random experiment and specify the probabilities of its elementary events.
   b. Describe the sample space of $X$, $S_X$, and find the probabilities for the various values of $X$.
2. An information source produces symbols at random from a five-letter alphabet: $S = \{a, b, c, d, e\}$. The probabilities of the symbols are

$$p(\mathrm{a}) = \frac{1}{2}, \quad p(\mathrm{b}) = \frac{1}{4}, \quad p(\mathrm{c}) = \frac{1}{8}, \text{ and } \quad p(\mathrm{d}) = p(\mathrm{e}) = \frac{1}{16}.$$

A data compression system encodes the letters into binary strings as follows:

| | |
|---|---|
| a | 1 |
| b | 01 |
| c | 001 |
| d | 0001 |
| e | 0000 |

Let the random variable $Y$ be equal to the length of the binary string output by the system. Specify the sample space of $Y$, $S_Y$, and the probabilities of its values.

3. A dart is thrown onto a square $b$ units wide. Assume that the dart is equally likely to fall anywhere in the square. Let the random variable $Z$ be given by the sum of the two coordinates of the point where the dart lands.
   a. Describe the sample space of $Z$, $S_Z$.
   b. Find the region in the square corresponding to the event $\{Z \leq z\}$ for $-\infty < z < +\infty$.
   c. Find $P[Z \leq z]$.
4. A coin is tossed $n$ times. Let the random variable $Y$ be the difference between the number of heads and the number of tails.
   a. Describe the sample space of $Y$, $S_Y$.
   b. Find the equivalent event for the event $\{Y = 0\}$.
   c. Find the equivalent event for the event $\{Y \leq k\}$ for $k$ a positive integer.

## Section 3.2
## The Cumulative Distribution Function

5. Plot the cdf for the random variable $X$ defined in Problem 1. Specify the type of $X$.
6. Plot the cdf for the random variable $Y$ defined in Problem 2. Specify the type of $Y$.
7. Plot the cdf for the random variable $Z$ defined in Problem 3. Specify the type of $Z$.
8. Plot the cdf for the random variable $Y$ defined in Problem 4 for $n = 4$ and $n = 5$ assuming that a fair coin is tossed.
9. Find the pmf and plot the cdf for the binomial random variable with $n = 8$ and $p = 1/10$, $p = 1/2$, and $p = 9/10$.
10. Let $U$ be a uniform random variable in the interval $[-1, 1]$. Find the following probabilities:

$P[U > 0]$ $\quad P[|U| < 1/3]$ $\quad P[|U| \geq 3/4]$

$P[U < 5]$ $\quad P[1/3 < U < 1/2]$

11. The cdf of a random variable $X$ is shown in Fig. P3.1.
    a. What type of random variable is $X$?
    b. Find the following probabilities in terms of the cdf of $U$:

| | | |
|---|---|---|
| $P[X < -1/2]$ | $P[X < 0]$ | $P[X \leq 0]$ |
| $P[1/4 \leq x < 1]$ | $P[1/4 \leq X \leq 1]$ | $P[X > 1/2]$ |
| $P[X \geq 5]$ | $P[X < 5]$ | |

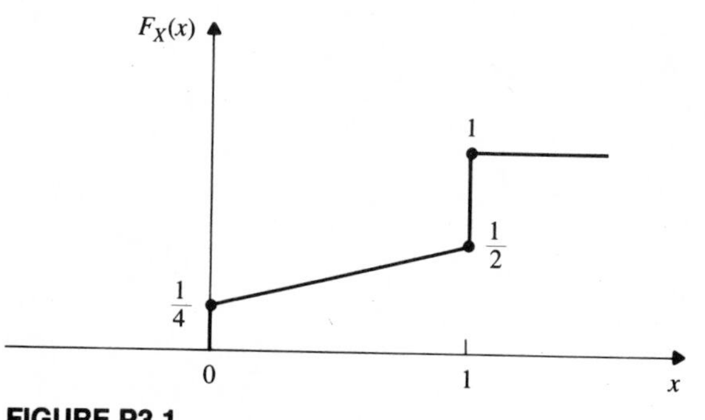

**FIGURE P3.1**

12. A random variable $Y$ has the cdf

$$F_Y(y) = \begin{cases} 0 & y < 1 \\ 1 - y^{-n} & y \geq 1, \end{cases}$$

where $n$ is a positive integer.
    a. Plot the cdf of $Y$.
    b. Find the probability $P[k < Y \leq k + 1]$ for a positive integer $k$.

13. A continuous random variable $X$ has cdf

$$F_X(x) = \begin{cases} 0 & x \leq -\pi/2 \\ c(1 + \sin(x)) & -\pi/2 < x \leq \pi/2 \\ 1 & \pi/2 \leq x \end{cases}$$

    a. Find $c$.
    b. Plot $F_X(x)$.

14. The Rayleigh random variable has cdf

$$F_R(r) = \begin{cases} 0 & r < 0 \\ 1 - e^{-r^2/2\sigma^2} & r \geq 0 \end{cases}$$

Find $P[\sigma \leq R \leq 2\sigma]$ and $P[R > 3\sigma]$.

15. Let $X$ be an exponential random variable with parameter $\lambda$. For $d > 0$ and $k$ a positive integer, find the following probabilities:

$P[X \leq d]$, $P[kd \leq X \leq (k + 1)d]$, and $P[X > kd]$.

16. Let $X$ be an exponential random variable with parameter $\lambda$. Segment the positive real line into five equiprobable disjoint intervals.

### Section 3.3
### The Probability Density Function

17. A random variable $X$ has pdf

$$f_X(x) = \begin{cases} cx(1-x) & 0 \le x \le 1 \\ 0 & \text{elsewhere.} \end{cases}$$

   a. Find $c$.
   b. Find $P[1/2 \le X \le 3/4]$.
   c. Find $F_X(x)$.

18. A random variable $X$ has pdf

$$f_X(x) = \begin{cases} c(1-x^4) & -1 \le x \le 1 \\ 0 & \text{elsewhere.} \end{cases}$$

   a. Find $c$.
   b. Find the cdf of $X$.

19. A random variable $X$ has the pdf shown in Fig. P3.2.

   a. Find $f_X(x)$.
   b. Find the cdf of $X$.

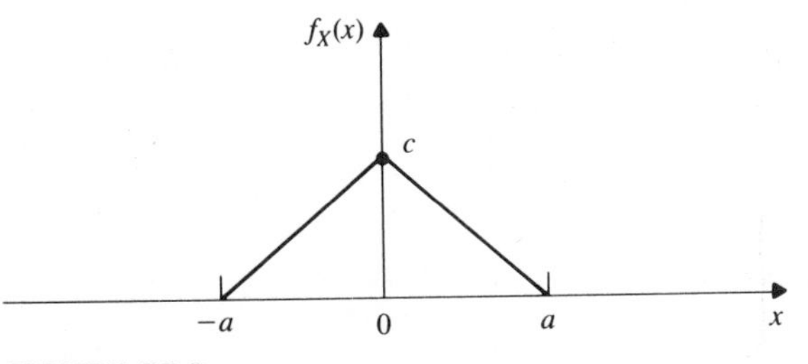


FIGURE P3.2

20. Find the pdf for the random variable defined in Problem 3.
21. Find the pdf for the random variable defined in Problem 11.
22. Find the pdf for the random variable defined in Problem 12.
23. Find the pdf for the random variable defined in Problem 13.

### Section 3.4
### Some Important Random Variables

24. Let $X$ be a binomial random variable that results from the performance of $n$ Bernoulli trials with probability of success $p$.

a. Suppose that $X = 1$, find the probability that the single event occurred in the $k$th Bernoulli trial.
b. Suppose that $X = 2$, find the probability that the two events occurred in the $j$th and $k$th Bernoulli trials, where $j < k$.
c. In light of your answers to Parts a and b, in what sense are successes distributed "completely at random" over the $n$ Bernoulli trials?

25. Let $X$ be the binomial random variable.

a. Show that

$$\frac{p_k}{p_{k+1}} = \frac{(n - k + 1)p}{kq} = 1 + \frac{(n + 1)p - k}{kq}.$$

b. Show that Part a implies that: (1) $P[X = k]$ is maximum at $k_{\max} = [(n + 1)p]$, where $[x]$ denotes the largest integer that is smaller than $x$; and (2) when $(n + 1)p$ is an integer, then the maximum is achieved at $k_{\max}$ and $k_{\max} - 1$.

26. Let $N$ be a geometric random variable with $S_X = \{0, 1, \ldots\}$.

a. Find $P[N > k]$
b. Find the cdf of $N$.
c. Find $P[N \text{ is an even number}]$.

27. Prove the memoryless property of the geometric random variable:

$$P[M \geq k + j \mid M > j] = P[M \geq k] \qquad \text{for all } j, k > 0.$$

In what sense is $M$ memoryless?

28. Let $M$ be a geometric random variable. Find $P[M = k \mid M \leq m]$.

29. Messages arrive at a computer at an average rate of 15 messages per second. The number of messages that arrive in 1 second is known to be a Poisson random variable.

a. Find the probability that no messages arrive in 1 second.
b. Find the probability that more than 10 messages arrive in a 1 second period.

*Hint*: Use Equation (3.33a), $p_{k+1} = (\lambda/(k + 1))p_k$, to compute the probabilities.

30. Compare the Poisson approximation to the binomial probabilities for $k = 0, 1, 2, 3$ and

$n = 10$ and $p = .1$, $\quad n = 20$ and $p = .05$,
$n = 100$ and $p = .01$.

31. One-tenth of 1% of a certain type of RAM chip are defective. A student needs 50 chips for a certain board. How many should he buy in order for there to be a 99% chance or greater of having at least 50 working chips?

32. For the Poisson random variable, show that for $\alpha < 1$, $P[N = k]$ is maximum at $k = 0$; for $\alpha > 1$, $P[N = k]$ is maximum at $[\alpha]$; and if $\alpha$ is a positive integer, the $P[N = k]$ is maximum at $k = \alpha$ and at $k = \alpha - 1$. *Hint*: Use the approach of Problem 25.

33. The $r$th percentile, $\pi(r)$, of a random variable $X$ is defined by $P[X \leq \pi(r)] = r/100$.
    a. Find the 90%, 95%, and 99% percentiles of the exponential random variable with parameter $\lambda$.
    b. Repeat Part a for the Gaussian random variable with parameters $m = 0$ and $\sigma$.

34. Show that the $Q$-function for the Gaussian random variable satisfies

$$Q(-x) = 1 - Q(x).$$

35. Let $X$ be a Gaussian random variable with mean $m$ and variance $\sigma^2$. Find the following probabilities:

$$P[X < m] \text{ and } P[|X - m| > k\sigma] \qquad \text{for } k = 1, 2, 3, 4, 5, \text{ and}$$

$$P[X > m + k\sigma] \qquad \text{for } k = 1.28, 3.09, 4.26, 5.20.$$

36. A communication channel accepts an arbitrary voltage input $v$ and outputs a voltage $Y = v + N$, where $N$ is a Gaussian random variable with mean 0 and variance $\sigma^2 = 1$. Suppose that the channel is used to transmit binary information as follows:

    to transmit 0     input $-1$
    to transmit 1     input $+1$

    The receiver decides a 0 was sent if the voltage is negative and a 1 otherwise. Find the probability of the receiver making an error if a 0 was sent; if a 1 was sent.

37. Two chips are being considered for use in a certain system. The lifetime of chip 1 is modeled by a Gaussian random variable with mean 20,000 hours and standard deviation 4000 hours. (The probability of negative lifetime is negligible.) The lifetime of chip 2 is also a Gaussian random variable but with mean 22,000 hours and standard deviation 1000 hours. Which chip is preferred if the target lifetime of the system is 20,000 hours? 24,000 hours?

38. Messages arrive at a center at a rate of one message per second. Let $X$ be the time for the arrival of five messages. Find the probability that $X < 6$; $X > 8$. Assume that message interarrival times are exponential random variables.

39. Plot the $m$-Erlang pdf for $m = 1, 2, 3$ and $\lambda = 1$.

40. Plot the chi-square pdf for $k = 1, 2, 3$.

41. Equation (3.48) states that the cdf of an $m$-Erlang random variable

is equal to the probability that a Poisson random variable with parameter $\lambda x$ is greater than $m$. Explain why this is true by using the fact that the time between occurrences in a Poisson random variable are exponentially distributed. (See Fig. 3.11.)

42. Let $X$ be an $m$-Erlang random variable with parameter $m$, and $\lambda = m$. Plot the cdf and pdf for $m = 5$ and $m = 10$. Explain why $X$ is approaching a random variable that is always equal to one as $m \to \infty$.

## Section 3.5
## Functions of a Random Variable

43. Let the random variable $X$ have a Laplacian pdf

$$f_X(x) = \frac{\alpha e^{-\alpha|x|}}{2} \qquad \text{where } \alpha > 0,\ -\infty < x < \infty.$$

Suppose that $X$ is input into the eight-level uniform quantizer of Example 3.19. Find the pmf of the quantizer output levels. Find the probability that the input $X$ exceeds the range $\pm 4d$ of the quantizer.

44. Let $X$ be a geometric random variable with parameter $p$. Let $Y = (X - M)^+$ where $M$ is a positive integer. Find the pmf of $Y$.

45. Let $Y = |X|$.
    a. Find the cdf of $Y$ by finding the equivalent event of $\{Y \le y\}$. Find the pdf of $Y$ by differentiation of the cdf.
    b. Find the pdf of $Y$ by finding the equivalent event of $\{y < Y \le y + dy\}$. Does the answer agree with Part a?

46. A *limiter* $Y = g(X)$ is shown in Fig. P3.3.
    a. Find the cdf and pdf of $Y$ in terms of the cdf and pdf of $X$.
    b. Find the cdf and pdf of $Y$ if $X$ has a Laplacian pdf.

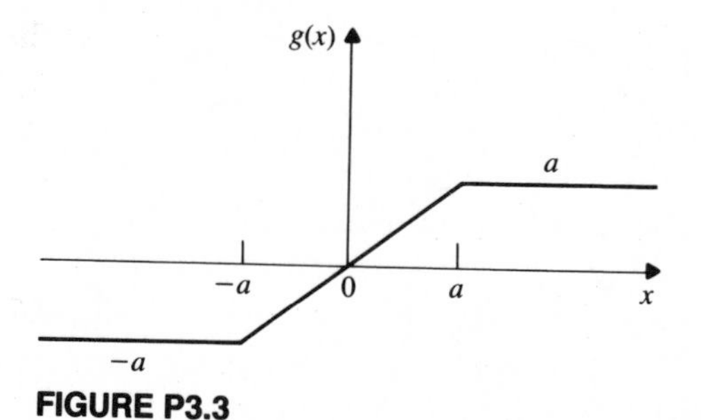


FIGURE P3.3

47. A *center-level clipper* $Y = h(X)$ is shown in Fig. P3.4.
    a. Find the cdf and pdf of $Y$ in terms of the cdf and pdf of $X$.
    b. Find the cdf and pdf of $Y$ if $X$ has a Laplacian pdf.

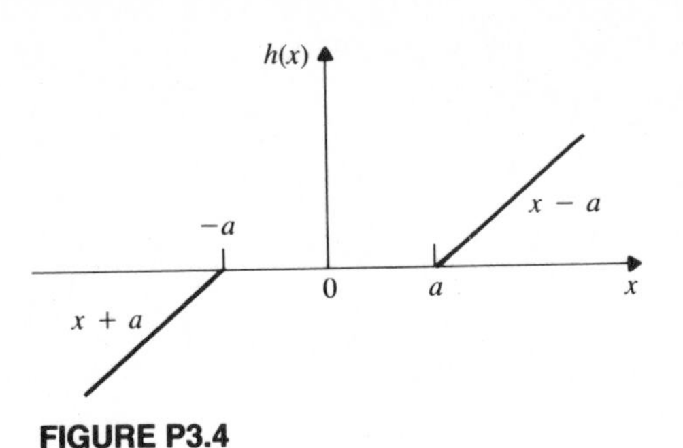


FIGURE P3.4

48. Let $Y = e^x$.
    a. Find the cdf and pdf of $Y$ in terms of the cdf and pdf of $X$.
    b. Find the pdf when $X$ is a Gaussian random variable. In this case $Y$ is said to have a *lognormal pdf*.
49. Let $Y = X^n$, where $n$ is a positive integer. Find the pdf of $Y$ in terms of the pdf of $X$.
50. In Example 3.19, let the input be uniformly distributed in $[-4d, 4d]$. Show that $Z = X - q(X)$ is uniformly distributed in $[-d/2, d/2]$.

## Section 3.6
## The Expected Value of Random Variables

51. Find the mean $m$ of the random variable defined in Problem 1. In what sense is the mean the break-even price of a ticket for the right to draw a single bill from the urn?
52. Find the mean of the random variable defined in Problem 2. How can the value of $m$ be interpreted if a very long sequence of outputs from the information source are to be encoded?
53. Find the mean and variance of the uniform discrete random variable that takes on values for the set $\{1, 2, \ldots, n\}$ with equal probability. You will need the following formulas:

$$\sum_{i=1}^{n} i = \frac{n(n+1)}{2} \quad \text{and} \quad \sum_{i=1}^{n} i^2 = \frac{n(n+1)(2n+1)}{6}.$$

54. Find the mean and variance of the binomial random variable.
55. Find the mean and variance of the Poisson random variable.
56. Find the mean and variance of the gamma random variable.
57. Find the mean and variance of the Gaussian random variable by direct integration in Eq. (3.57) and Eq. (3.65).
58. Prove Eqs. (3.59) and (3.60).

59. Show that $E[X]$ for the random variable with cdf $F_X(x) = 1 - 1/x$, for $x > 1$, does not exist.
60. Show that $E[X]$ for the *Cauchy random variable* $X$ with pdf $f_X(x) = (\pi(1 + x^2))^{-1}$ does not exist.
61. Verify Eqs. (3.68), (3.69), and (3.70).
62. Find the mean and variance of the limiter considered in Problem 46.
63. Find the mean and variance of the center-level clipper considered in Problem 47.
64. Let $X$ be a discrete random variable uniformly distributed in the set $S = \{1, 2, \ldots, n\}$. Let $Y = K + LX$, where $K$ and $L$ are integers. Find the mean and variance of $Y$.
65. Find the $n$th moment of $X$ if $X$ is uniformly distributed in the interval $[0, 1]$. Repeat for an arbitrary interval $[a, b]$.

## Section 3.7
## The Markov and Chebyshev Inequalities

66. Compare the Chebyshev bound and the exact probability for the event $\{|X - m| \geq c\}$ as a function of $c$ for
    a. $X$ a uniform random variable in the interval $[-b, b]$.
    b. $X$ a Laplacian random variable with parameter $a$.
    c. $X$ a zero-mean Gaussian random variable.

## Section 3.8
## Testing the Fit of a Distribution to Data

67. The following histogram was obtained by counting the occurrence of the first digits in telephone numbers in one column of a telephone directory:

| digit | 0 | 1 | 2 | 3 | 4 | 5 | 6 | 7 | 8 | 9 |
|---|---|---|---|---|---|---|---|---|---|---|
| observed | 0 | 0 | 24 | 2 | 25 | 3 | 32 | 15 | 2 | 2 |

    Test the goodness-of-fit of this data to a random variable that is uniformly distributed in the set $\{0, 1, \ldots, 9\}$ at a 1% significance level. Repeat for the set $\{2, 3, \ldots, 9\}$.
68. Test the goodness-of-fit of the interarrival data presented in Problem 9 of Chapter 1 to an exponential random variable at a 5% significance level.
69. A computer simulation program gives pairs of numbers $(X, Y)$ that are supposed to be uniformly distributed in the unit square. How would you use the chi-square to test the goodness-of-fit of the computer output?

## Section 3.9
## Transform Methods

70. Derive the characteristic function of the gamma random variable $X$. Obtain the $n$th moment of $X$ by applying the moment theorem.

71. Find the mean and variance of the Gaussian random variable by applying the moment theorem to the characteristic function given in Table 3.2.

72. Show that the characteristic function for the Cauchy random variable introduced in Problem 60 is $e^{-|\omega|}$.

73. Find the mean and variance of the geometric random variable from its characteristic function.

74. Find the probability generating function for the binomial random variable.

75. Find the probability generating function for the discrete uniform random variable.

76. Find the first two moments of the Poisson random variable from its probability generating function.

77. Find $P[X = r]$ for the negative geometric random variable $X$ from the probability generating function given in Table 3.1. Find the mean of $X$.

78. Find the Laplace transform of the pdf of a gamma random variable. Apply the moment theorem to find the first two moments.

79. Let $X$ be the mixture of two exponential random variables (see Example 3.62). Find the Laplace transform of the pdf of $X$.

80. The Laplace transform of the pdf of a random variable $X$ is given by

$$X^*(s) = \left(\frac{a}{s+a}\right)\left(\frac{b}{s+b}\right).$$

Find the pdf of $X$. *Hint*: Use a partial fraction expansion of $X^*(s)$.

## Section *3.10
## Basic Reliability Calculations

81. The lifetime $T$ of a device has pdf

$$f_T(t) = \begin{cases} \lambda e^{-\lambda(t-T)} & t \geq T \\ 0 & t < T. \end{cases}$$

a. Find the reliability of the device.
b. Find the failure rate function.
c. How many hours of operation can be considered to achieve 99% reliability?

82. The lifetime $T$ of a device has pdf

$$f_T(t) = \begin{cases} \dfrac{1}{T_0} & a < t < a + T_0 \\ 0 & \text{elsewhere.} \end{cases}$$

a. Find the reliability of the device.
b. Find the failure rate function.
c. How many hours of operation can be considered to achieve 99% reliability?

83. The lifetime $T$ of a device is a Rayleigh random variable.

a. Find the reliability of the device.
b. Find the failure rate function.

84. The lifetime of a system is an $m$-Erlang random variable.

a. Find the reliability of the device.
b. Find the failure rate function.

85. Show that the mean and variance of the Weibull random variable are

$$E[T] = \alpha^{-1/\beta}\Gamma\left(1 + \frac{1}{\beta}\right)$$

$$\text{VAR}[T] = \alpha^{-2/\beta}\left\{\Gamma\left(1 + \frac{2}{\beta}\right) - \left[\Gamma\left(1 + \frac{1}{\beta}\right)\right]^2\right\}.$$

86. A system has three identical components and the system is functioning if two or more components are functioning.

a. Find the reliability of the system if the component lifetimes are exponential random variables with mean 1.
b. Find the reliability of the system if one of the components has mean 2.

87. Repeat Problem 86 if the component lifetimes are Rayleigh distributed.

88. A system consists of two processors and three peripheral units. The system is functioning as long as one processor and two peripherals are functioning.

a. Find the system reliability if the processor lifetimes are exponential random variables with mean 5 and the peripheral lifetimes are exponential random variables with mean 10.
b. Find the system reliability if the processor lifetimes are exponential random variables with mean 10 and the peripheral lifetimes are exponential random variables with mean 5.

89. An operation is carried out by a subsystem consisting of three units that operate in a series configuration. The units have exponentially

distributed lifetimes with mean 1. How many subsystems should be operated in parallel to achieve a reliability of 99% in $T$ hours of operation?

### Section *3.11 Generation of Random Variables

90. The random variable $X$ has the triangular pdf shown in Fig. P3.5. Find the transformation needed to generate $X$.

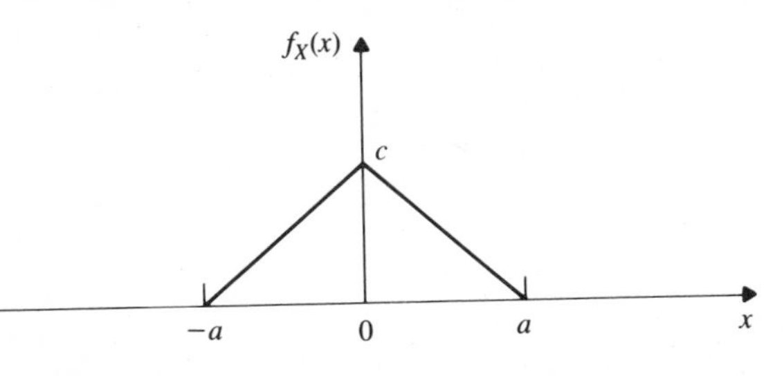


FIGURE P3.5

91. Find the transformation needed to generate the Laplacian random variable.
92. A random variable $Y$ of mixed type has pdf

$$f_Y(x) = p\,\delta(x) + (1 - p)f_X(x),$$

where $X$ is a Laplacian random variable and $p$ is a number between zero and one. Find the transformation required to generate $Y$.
93. Compute the average number of comparisons required in the searches described in Example 3.58.
94. Specify the transformation method needed to generate the geometric random variable with parameter $p = 1/2$. Find the average number of comparisons needed in the search.
95. Specify the transformation method needed to generate the Poisson random variable with small parameter $\alpha$. Compute the average number of comparisons needed in the search.
96. The following rejection method can be used to generate Gaussian random variables:
    1. Generate $U_1$, a uniform random variable in the unit interval.
    2. Let $X_1 = -\ln(U_1)$
    3. Generate $U_2$, a uniform random variable in the unit interval. If $U_2 \le \exp\{-(X_1 - 1)^2/2\}$, accept $X_1$. Otherwise reject $X_1$ and go to step 1.
    4. Generate a random sign (+ or −) with equal probability. Output $X$ equal to $X_1$ with the resulting sign.

a. Show that if $X_1$ is accepted, then its pdf corresponds to the pdf of the absolute value of a Gaussian random variable with mean 0 and variance 1.

b. Show that $X$ is a Gaussian random variable with mean 0 and variance 1.

97. Cheng (1977) has shown that the function $Kf_Z(x)$ bounds the pdf of a gamma random variable with $\alpha > 1$, where

$$f_Z(x) = \frac{\lambda \alpha^{\lambda} x^{\lambda-1}}{(\alpha^{\lambda} + x^{\lambda})^2}$$

and

$$K = (2\alpha - 1)^{1/2}.$$

Find the cdf of $f_Z(x)$ and the corresponding transformation needed to generate $Z$.

98. a. Show that in the modified rejection method, the probability of accepting $X_1$ is $1/K$. *Hint*: Use conditional probability.

b. Show that $Z$ has the desired pdf.

99. Write a computer program to generate the exponential random variable. Generate 500 samples of the random variable and perform a chi-square test for goodness of fit. Order the samples according to increasing value and find the empirical cumulative distribution defined in Problem 1.8.

100. Write a computer program to generate the Gaussian random variable using the method presented in Example 3.60. Generate 500 samples of the random variable and perform a chi-square test for goodness of fit. Order the samples according to increasing value and find the empirical cumulative distribution defined in Problem 1.8.

101. Two methods for generating binomial random variables were presented in Example 3.58. Compare the methods under the following conditions:

a. $p = 1/2$, $n = 5, 25, 50$;

b. $p = 0.1$, $n = 5, 25, 50$.

c. Write computer programs that implement the two methods and verify their relative speeds by generating 1000 binomially distributed samples.

102. The Poisson random variable is the number of event occurrences in a time interval. In Section 3.4, it was found that the time between events for a Poisson random variable is an exponentially distributed random variable. Explain how one can generate Poisson random variables from a sequence of exponentially distributed random variables. How does this method compare with the one presented in

Problem 95? Write computer programs that implement the two methods and compare their relative speeds when $\alpha = 3$, $\alpha = 25$, and $\alpha = 100$.

103. Write a computer program to generate the gamma pdf with $\alpha > 1$ using the rejection method discussed in Problem 97. Use this method to generate $m$-Erlang random variables with $m = 2, 10$ and $\lambda = 1$ and compare the method to the straightforward generation of $m$ exponential random variables as discussed in Example 3.61.

CHAPTER 4

# Multiple Random Variables

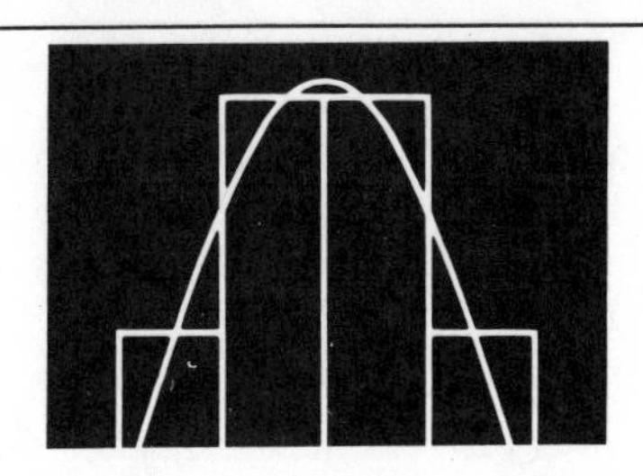

Many random experiments involve dealing with several random variables at a time. In some experiments a number of different quantities are measured. For example, the voltages at several points in a circuit at some specific time may be of interest. Other experiments involve the repeated measurement of a certain quantity. An example of this is the repeated measurement ("sampling") of the amplitude of a voltage waveform that varies with time. In Chapter 3 we developed techniques for calculating the probabilities of events involving a single random variable *in isolation*. In this chapter, we develop techniques for calculating the probabilities of events that involve the *joint* behavior of two or more random variables. We are interested in determining when a set of random variables are independent, as well as in quantifying their degree of "correlation" when they are not independent.

In the next section we present some fundamental notions about multiple random variables. The section is rather sketchy and is intended as a preview of what is to follow. We then discuss the case of two random variables in detail because in this case we can draw on our geometric intuition. Finally we return to the general case of multiple random variables.

## 4.1

## VECTOR RANDOM VARIABLES

The notion of a random variable is easily generalized to the case where several quantities are of interest. A **vector random variable** $\mathbf{X}$ is a function that assigns a vector of real numbers to each outcome $\zeta$ in $S$, the sample space of the random experiment.

### Example 4.1

Let a random experiment consist of selecting a student's name from an urn. Let $\zeta$ denote the outcome of this experiment, and define the following three functions:

$H(\zeta)$ = height of student $\zeta$ in inches,

$W(\zeta)$ = weight of student $\zeta$ in pounds, and

$A(\zeta)$ = age of student $\zeta$.

The vector $(H(\zeta), W(\zeta), A(\zeta))$ is a vector random variable.

### Example 4.2

A random experiment consists of finding the number of defects in a semiconductor chip, and identifying their locations. The outcome of this experiment consists of the vector $\zeta = (n, \mathbf{x}_1, \mathbf{x}_2, \ldots, \mathbf{x}_n)$, where the first component specifies the total number of defects and the remaining components the coordinates of their location. Suppose that the chip consists of $M$ regions. Let $N_1(\zeta), N_2(\zeta), \ldots, N_M(\zeta)$ be the number of defects in each of these regions, that is, $N_k(\zeta)$ is the number of $\mathbf{x}$'s that fall in region $k$. The vector $\mathbf{N}(\zeta) = (N_1, \ldots, N_M)$ is then a vector random variable. ■■

### ■■ Example 4.3

Let the outcome $\zeta$ of some random experiment be a voltage waveform $X(t)$. Let the random variable $X_k = X(kT)$ be the sample of the voltage taken at time $kT$. The vector consisting of the first $n$ samples $(\mathbf{X} = X_1, X_2, \ldots, X_n)$ is then a vector random variable. ■■

## Events and Probabilities

Each event involving an $n$-dimensional random variable $\mathbf{X} = (X_1, X_2, \ldots, X_n)$ has a corresponding region in an $n$-dimensional real space. The methods from set theory introduced in Chapter 2 can be used to find these regions.

### ■■ Example 4.4

Consider the two-dimensional random variable $\mathbf{X} = (X, Y)$. Find the region of the plane corresponding to the events

$$A = \{X + Y \leq 10\},$$

$$B = \{\min(X, Y) \leq 5\}, \text{ and}$$

$$C = \{X^2 + Y^2 \leq 100\}.$$

The regions corresponding to events $A$ and $C$ are straightforward to find and are shown in Fig. 4.1. Event $B$ is found by noting that

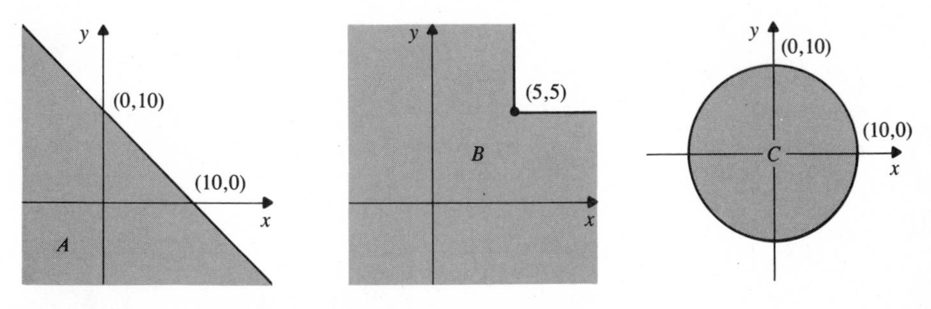


**FIGURE 4.1** Examples of two-dimensional events.

$\{\min(X, Y) \le 5\} = \{X \le 5\} \cup \{Y \le 5\}$, that is, the minimum of $X$ and $Y$ is less than or equal to 5 if either $X$ and/or $Y$ are less than or equal to 5. ■■

---

For the $n$-dimensional random variable $\mathbf{X} = (X_1, \ldots, X_n)$ we are particularly interested in events that have the **product form**

$$A = \{X_1 \text{ in } A_1\} \cap \{X_2 \text{ in } A_2\} \cap \cdots \cap \{X_n \text{ in } A_n\}, \tag{4.1}$$

where $A_k$ is a one-dimensional event (i.e., subset of the real line) that involves $X_k$ only. The event $A$ occurs when all of the events $\{X_k \text{ in } A_k\}$ occur jointly. Figure 4.2 shows some two-dimensional product-form events.

A fundamental problem in modeling a system with a vector random variable $\mathbf{X} = (X_1, \ldots, X_n)$ involves specifying the probability of product-form events:

$$\begin{aligned} P[A] &= P[\{X_1 \text{ in } A_1\} \cap \{X_2 \text{ in } A_2\} \cap \cdots \cap \{X_n \text{ in } A_n\}]. \\ &\triangleq P[X_1 \text{ in } A_1, \ldots, X_n \text{ in } A_n] \end{aligned} \tag{4.2}$$

In principle, the probability in Eq. (4.2) is obtained by finding the probability of the equivalent event in the underlying sample space, that is,

$$P[A] = P[\{\zeta \text{ in } S \text{ such that } \mathbf{X}(\zeta) \text{ in } A\}].$$

Later in the chapter we show that Eq. (4.2) is specified by giving an $n$-dimensional joint cumulative distribution function, joint probability density function, or joint probability mass function.

Many events of interest are not of product form. However, the non–product-form events that we are interested in can be approximated arbitrarily closely by the union of product-form events as shown in Example 4.5.

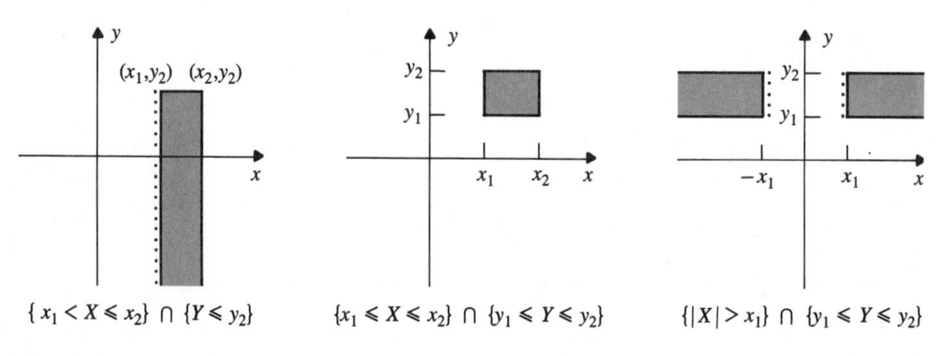

**FIGURE 4.2** Some two-dimensional product-form events.

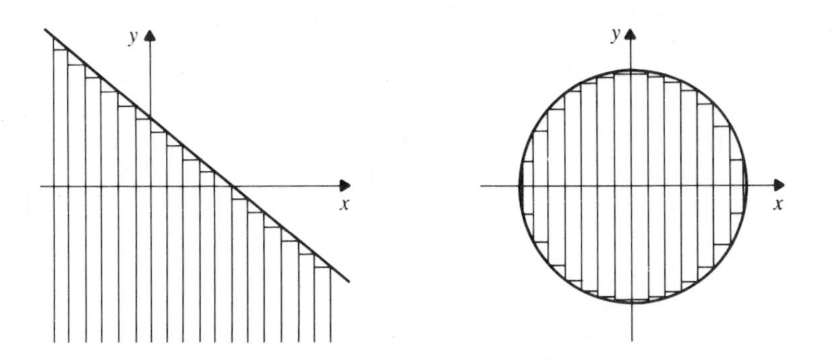


**FIGURE 4.3** Some two-dimensional non–product-form events.

### ■■ Example 4.5

None of the events in Example 4.4 are product form. Event $B$ is the union of two product-form events:

$$B = \{X \leq 5 \text{ and } Y < \infty\} \cup \{X > 5 \text{ and } Y \leq 5\}.$$

Figure 4.3 shows how events $A$ and $C$ are approximated by rectangles of infinitesimal width. This suggests that we will be expressing the probabilities of events as integrals of probability density over the regions that correspond to the events. ■■

---

The probability of a non–product-form event $B$ is found as follows: First, $B$ is approximated by the union of disjoint product-form events, say, $B_1, B_2, \ldots, B_n$; the probability of $B$ is then approximated by

$$P[B] \simeq P\left[\bigcup_k B_k\right] = \sum_k P[B_k].$$

The approximation becomes exact in the limit as the $B_k$'s become arbitrarily fine. We elaborate on this procedure in Section 4.2.

## Independence

Intuitively, we expect that if the one-dimensional random variables $X$ and $Y$ are "independent," then events that involve only $X$ should be independent of events that involve only $Y$. In other words, if $A$ is *any* event that involves $X$ only and $B$ is *any* event that involves $Y$ only, then

$$P[X \text{ in } A, Y \text{ in } B] = P[X \text{ in } A]P[Y \text{ in } B].$$

In the general case of $n$ random variables, we say that the random variables $X_1, X_2, \ldots, X_n$ are **independent** if

$$P[X_1 \text{ in } A_1, \ldots, X_n \text{ in } A_n] = P[X_1 \text{ in } A_1] \ldots P[X_n \text{ in } A_n], \tag{4.3}$$

where the $A_k$ is an event that involves $X_k$ only. Therefore if the random variables are independent, knowledge about the probabilities of the random variables in isolation is sufficient to specify the probabilities of joint events. In Sections 4.3 and 4.5 we see that only a simple set of conditions needs to be checked to verify the independence of a set of random variables.

## 4.2

## PAIRS OF RANDOM VARIABLES

In Chapter 3, we saw that the probabilities of events involving a single random variable $X$ can be expressed in terms of the cumulative distribution function $F_X(x)$. We also saw that these probabilities can be expressed as the sum over the probability mass function if the random variable is discrete, or the integral of the probability density function if the random variable is continuous. We now extend these results to the two-dimensional case.

### Pairs of Discrete Random Variables

Let the vector random variable $\mathbf{X} = (X, Y)$ assume values from some countable set $S = \{(x_j, y_k),\ j = 1, 2, \ldots, k = 1, 2, \ldots\}$. The **joint probability mass function** of $\mathbf{X}$ specifies the probabilities of the product-form event $\{X = x_j\} \cap \{Y = y_k\}$:

$$p(x_j, y_k) = P[\{X = x_j\} \cap \{Y = y_k\}]$$
$$\triangleq P[X = x_j, Y = y_k] \qquad j = 1, 2, \ldots \qquad k = 1, 2, \ldots. \tag{4.4}$$

Thus the joint pmf gives the probability of the occurrence of the pairs $(x_j, y_k)$.

The probability of any event $A$ is the sum of the pmf over the outcomes in $A$:

$$P[A] = \sum_{(x_j, y_k) \text{ in } A} \sum p(x_j, y_k). \tag{4.5}$$

The fact that the probability of the sample space $S$ is 1 gives

$$\sum_{j=1}^{\infty} \sum_{k=1}^{\infty} p(x_j, y_k) = 1. \tag{4.6}$$

We are also interested in the probabilities of events involving each of the random variables in isolation. These can be found in terms of the

**marginal probability mass functions**:

$$\begin{aligned} p_X(x_j) &= P[X = x_j] \\ &= P[X = x_j,\ Y = \text{anything}] \\ &= P[\{X = x_j \text{ and } Y = y_1\} \cup \{X = x_j \text{ and } Y = y_2\} \cup \ldots] \\ &= \sum_{k=1}^{\infty} p(x_j, y_k), \end{aligned} \tag{4.7a}$$

and similarly,

$$\begin{aligned} p_Y(y_k) &= P[Y = y_k] \\ &= \sum_{j=1}^{\infty} p(x_j, y_k). \end{aligned} \tag{4.7b}$$

The marginal pmf's are one-dimensional pmf's and they supply all the information required to compute the probability of events involving the corresponding random variable in isolation.

The probability $p(x_j, y_k)$ can be interpreted as the long-term relative frequency of the joint event $\{X = x_j\} \cap \{Y = y_k\}$ in a sequence of repetitions of the random experiment. Equation (4.7a) corresponds to the fact that relative frequency of the event $\{X = x_j\}$ is found by adding the relative frequencies of all outcome pairs in which $x_j$ appears. In general, it is impossible to deduce the relative frequencies of pairs of values $X$ and $Y$ from the relative frequencies of $X$ and $Y$ in isolation. The same is true for pmf's: In general, knowledge of the marginal pmf's is insufficient to specify the joint pmf.

### ■■ Example 4.6

A random experiment consists of tossing two "loaded" dice and noting the pair of numbers facing up. The joint pmf $p(j, k)$ for $j = 1, \ldots, 6$ and $k = 1, \ldots, 6$ is:

| $j$ \ $k$ | 1 | 2 | 3 | 4 | 5 | 6 |
|---|---|---|---|---|---|---|
| 1 | 2/42 | 1/42 | 1/42 | 1/42 | 1/42 | 1/42 |
| 2 | 1/42 | 2/42 | 1/42 | 1/42 | 1/42 | 1/42 |
| 3 | 1/42 | 1/42 | 2/42 | 1/42 | 1/42 | 1/42 |
| 4 | 1/42 | 1/42 | 1/42 | 2/42 | 1/42 | 1/42 |
| 5 | 1/42 | 1/42 | 1/42 | 1/42 | 2/42 | 1/42 |
| 6 | 1/42 | 1/42 | 1/42 | 1/42 | 1/42 | 2/42 |

Find the marginal pmf's.

In this problem, the marginal probabilities can be viewed as the probability that the outcome falls in a certain row or column. The probability that $X = 1$ is found by summing over the first row:

$$P[X = 1] = \frac{2}{42} + \frac{1}{42} + \cdots + \frac{1}{42} = \frac{1}{6}.$$

Similarly we find that $P[X = j] = 1/6$ for $j = 1, 2, \ldots, 6$. The probability that $Y = k$ is found by summing over the $k$th column. We then find that $P[Y = k] = 1/6$ for $k = 1, 2, \ldots, 6$. Thus each die, in isolation, appears to be fair in the sense that each face is equiprobable. If we knew only these marginal pmf's we would have no idea that the dice are loaded. ■■

### ■■ Example 4.7

The number of bytes $N$ in a message has a geometric distribution with parameter $p$. Suppose that messages are broken into packets of maximum length $M$ bytes. Let $Q$ be the number of full packets in a message and let $R$ be the number of bytes left over. Find the joint pmf and the marginal pmf's of $Q$ and $R$.

If a message has $N$ bytes, then the number of full packets in the message is the quotient $Q$ in the division of $N$ by $M$, and the number of remaining bytes is the remainder $R$ of such division. $Q$ takes on values in the range $0, 1, \ldots,$ and $R$ takes on values in the range $0, 1, \ldots, M - 1$. The probability of the elementary event $\{(q, r)\}$ is given by

$$P[Q = q, R = r] = P[N = qM + r] = (1 - p)p^{qM+r}.$$

The marginal pmf of $Q$ is

$$\begin{aligned} P[Q = q] &= P[N \text{ in } (qM, qM + 1, \ldots, qM + (M - 1)\}] \\ &= \sum_{k=0}^{(M-1)} (1 - p)p^{qM+k} \\ &= (1 - p)p^{qM}\frac{1 - p^M}{1 - p} \qquad q = 0, 1, 2, \ldots \\ &= (1 - p^M)(p^M)^q. \end{aligned}$$

We see that the marginal pmf of $Q$ is geometric with parameter $p^M$. The marginal pmf of $R$ is

$$\begin{aligned} P[R = r] &= P[N \text{ in } \{r, M + r, 2M + r, \ldots\} \\ &= \sum_{q=0}^{\infty} (1 - p)p^{qM+r} \\ &= \frac{(1 - p)}{1 - p^M}p^r \qquad r = 0, 1, \ldots, M - 1. \end{aligned}$$

We see that $R$ has a truncated geometric pmf. As an exercise, you should verify that all of the above pmf's add to one. ■■

### The Joint cdf of *X* and *Y*

In Chapter 3 we saw that semi-infinite intervals of the form $(-\infty, x]$ are a basic building block from which other one-dimensional events can be built. By defining the cdf $F_X(x)$ as the probability of $(-\infty, x]$, we are then able to express the probabilities of other events in terms of the cdf. In this section we repeat the above development for two-dimensional random variables.

A basic building block for events involving two-dimensional random variables is the semi-infinite rectangle defined by $\{(x, y)\colon x \leq x_1$ and $y \leq y_1\}$ as shown in Fig. 4.4. The **joint cumulative distribution function of *X* and *Y*** is defined as the probability of the product-form event $\{X \leq x_1\} \cap \{Y \leq y_1\}$:

$$F_{X,Y}(x_1, y_1) = P[X \leq x_1, Y \leq y_1]. \tag{4.8}$$

In terms of relative frequency, $F_{X,Y}(x_1, y_1)$ represents the long-term proportion of times in which the outcome of the random experiment yields a point $\mathbf{X}$ that falls in the rectangular region shown in Fig. 4.4. In terms of probability "mass," $F_{X,Y}(x_1, y_1)$ represents the amount of mass contained in the rectangular region.

The joint cdf is nondecreasing in the "northeast" direction, that is,

(i) $\quad F_{X,Y}(x_1, y_1) \leq F_{X,Y}(x_2, y_2) \qquad$ if $x_1 \leq x_2$ and $y_1 \leq y_2$,

since the semi-infinite rectangle defined by $(x_1, y_1)$ is contained in that defined by $(x_2, y_2)$.

It is impossible for either $X$ or $Y$ to assume a value less than $-\infty$, therefore

(ii) $\quad F_{X,Y}(-\infty, y_1) = F_{X,Y}(x_1, -\infty) = 0.$

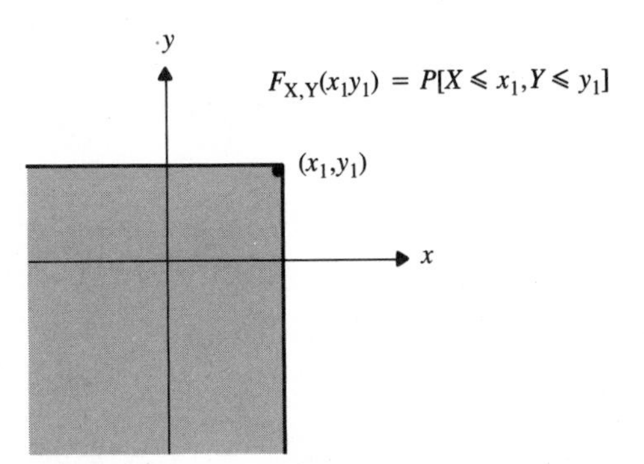


**FIGURE 4.4** The joint cumulative distribution function is defined as the probability of the semi-infinite rectangle defined by the point $(x_1, y_2)$.

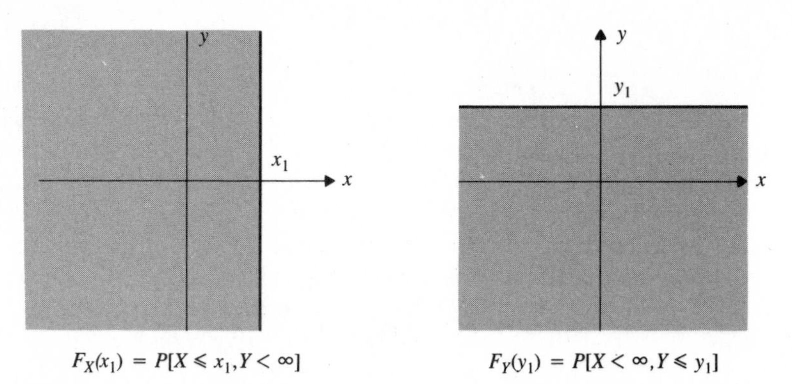


**FIGURE 4.5** The marginal cdf's are the probabilities of these half-planes.

It is certain that $X$ and $Y$ will assume values less than infinity, therefore

(iii) $F_{X,Y}(\infty, \infty) = 1.$

If we let one of the variables approach infinity while keeping the other fixed, we obtain the **marginal cumulative distribution functions**

(iv) $F_X(x) = F_{X,Y}(x, \infty) = P[X \leq x, Y \leq \infty] = P[X \leq x]$

and

$$F_Y(y) = F_{X,Y}(\infty, y) = P[Y \leq y].$$

The marginal cdf's are the probabilities of the regions shown in Fig. 4.5.

Recall that the cdf for a single random variable is continuous from the right. It can be shown that the joint cdf is continuous from the "north" and from the "east," that is,

(v) $\lim_{x \to a^+} F_{X,Y}(x, y) = F_{X,Y}(a, y)$

and

$$\lim_{y \to b^+} F_{X,Y}(x, y) = F_{X,Y}(x, b).$$

### ■■ Example 4.8

The joint cdf for the vector of random variables $\mathbf{X} = (X, Y)$ is given by

$$F_{X,Y}(x, y) = \begin{cases} (1 - e^{-\alpha x})(1 - e^{-\beta y}) & x \geq 0, y \geq 0 \\ 0 & \text{elsewhere.} \end{cases}$$

Find the marginal cdf's.

The marginal cdf's are obtained by letting one of the variables approach infinity:

$$F_X(x) = \lim_{y\to\infty} F_{X,Y}(x, y) = 1 - e^{-\alpha x} \qquad x \geq 0$$

$$F_Y(y) = \lim_{x\to\infty} F_{X,Y}(x, y) = 1 - e^{-\beta y} \qquad y \geq 0.$$

Thus $X$ and $Y$ individually have exponential distributions with parameters $\alpha$ and $\beta$, respectively. ■■

The cdf can be used to find the probability of events that can be expressed as the union and intersection of semi-infinite rectangles. For example consider the strip defined by $\{x_1 < X \leq x_2 \text{ and } Y \leq y_1\}$ denoted by the region $B$ in Fig. 4.6(a). The semi-infinite rectangle defined by $(x_2, y_1)$ is equal to the union of the semi-infinite rectangle defined by $(x_1, y_1)$ and the region $B$. Thus by the third axiom of probability we have that

$$F_{X,Y}(x_2, y_1) = F_{X,Y}(x_1, y_1) + P[x_1 < X \leq x_2, Y \leq y_1].$$

Thus the probability of the semi-infinite strip is

$$P[x_1 < X \leq x_2, Y \leq y_1] = F_{X,Y}(x_2, y_1) - F_{X,Y}(x_1, y_1).$$

Consider next the rectangle $\{x_1 < X \leq x_2, y_1 < Y \leq y_2\}$ denoted by the region $A$ in Fig. 4.6(b). The semi-infinite rectangle defined by $(x_2, y_2)$ is equal to the union of region $A$, $B$, and the semi-infinite rectangle defined by $(x_1, y_2)$. Therefore

$$F_{X,Y}(x_2, y_2) = P[x_1 < X \leq x_2, y_1 < Y \leq y_2] + F_{X,Y}(x_2, y_1) - F_{X,Y}(x_1, y_1) + F_{X,Y}(x_1, y_2).$$

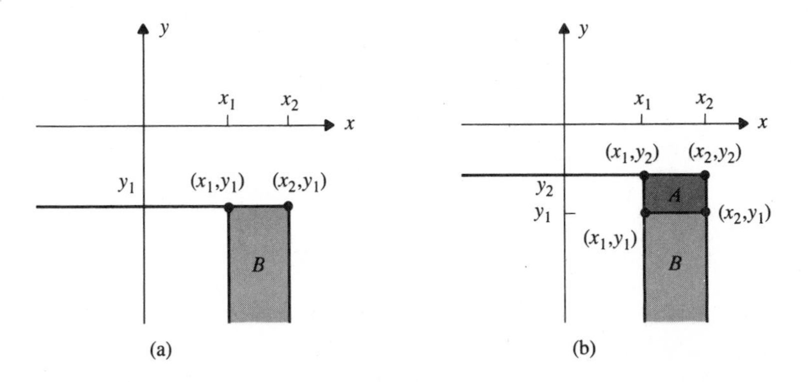


**FIGURE 4.6** The joint cdf can be used to determine the probability of various events.

The probability of the rectangle is thus

$$\text{(vi)} \quad P[x_1 < X \le x_2, y_1 < Y \le y_2] = F_{X,Y}(x_2, y_2) - F_{X,Y}(x_2, y_1) - F_{X,Y}(x_1, y_2) + F_{X,Y}(x_1, y_1).$$

### ■■ Example 4.9

Find the probability of the events $A = \{X \le 1, Y \le 1\}$, $B = \{X > x, Y > y\}$ where $x > 0$ and $y > 0$, and $D = (1 < X \le 2, 2 < Y \le 5\}$ in Example 4.8.

The probability of $A$ is given directly by the cdf:

$$P[A] = P[X \le 1, Y \le 1] = F_{X,Y}(1, 1) = (1 - e^{-\alpha})(1 - e^{-\beta}).$$

The probability of $B$ requires more work. Consider $B^c$:

$$B^c = (\{X > x\} \cap \{Y > y\})^c = \{X \le x\} \cup \{Y \le y\},$$

by DeMorgan's rule. Corollary 5 in Section 2.2 gives the probability of the union of two events:

$$\begin{aligned} P[B^c] &= P[X \le x] + P[Y \le y] - P[X \le x, Y \le y] \\ &= (1 - e^{-\alpha x}) + (1 - e^{-\beta y}) - (1 - e^{-\alpha x})(1 - e^{-\beta y}) \\ &= 1 - e^{-\alpha x}e^{-\beta y}. \end{aligned}$$

Finally we obtain the probability of $B$:

$$P[B] = 1 - P[B^c] = e^{-\alpha x}e^{-\beta y}.$$

You should sketch the region $B$ on the plane and identify the events involved in the calculation of the probability of $B^c$.

The probability of event $D$ is found by applying Property vi of the joint cdf:

$$\begin{aligned} P[1 < X \le 2, 2 < Y \le 5] &= F_{X,Y}(2, 5) - F_{X,Y}(2, 2) - F_{X,Y}(1, 5) + F_{X,Y}(1, 2) \\ &= (1 - e^{-2\alpha})(1 - e^{-5\beta}) - (1 - e^{-2\alpha})(1 - e^{-2\beta}) \\ &\quad - (1 - e^{-\alpha})(1 - e^{-5\beta}) + (1 - e^{-\alpha})(1 - e^{-2\beta}). \end{aligned}$$ ■■

---

### The Joint pdf of Two Jointly Continuous Random Variables

To compute the probability of events corresponding to regions other than rectangles, we note that any reasonable shape (i.e., disk, polygon, or half-plane) can be approximated by the union of rectangles. The probability of such events can therefore be approximated by the sum of the probabilities of the rectangles, as given by Property vi of the joint cdf. If the cdf is sufficiently smooth, then as we increase the fineness of the rectangles, the sum approaches an integral over a probability density function.

We say that the random variables **$X$ and $Y$ are jointly continuous** if the probabilities of events involving $(X, Y)$ can be expressed as an integral of a probability density function. In other words, there is a nonnegative function $f_{X,Y}(x, y)$, called the **joint probability density function**, that is defined on the real plane such that for every event $A$, a subset of the plane,

$$P[\mathbf{X} \text{ in } A] = \int_A \int f_{X,Y}(x', y')\, dx'\, dy', \tag{4.9}$$

as shown in Fig. 4.7. When $A$ is the entire plane, the integral must equal one,

$$1 = \int_{-\infty}^{\infty} \int_{-\infty}^{\infty} f_{X,Y}(x, y)\, dx\, dy. \tag{4.10}$$

Equations (4.9) and (4.10) again suggest that the probability "mass" of an event is found by integrating the density of probability mass over the region corresponding to the event.

The joint cdf can be obtained in terms of the joint pdf of jointly continuous random variables by integrating over the semi-infinite rectangle defined by $(x, y)$:

$$F_{X,Y}(x, y) = \int_{-\infty}^{x} \int_{-\infty}^{y} f_{X,Y}(x', y')\, dx'\, dy'. \tag{4.11}$$

It then follows that *if* $X$ and $Y$ are jointly continuous random variables,

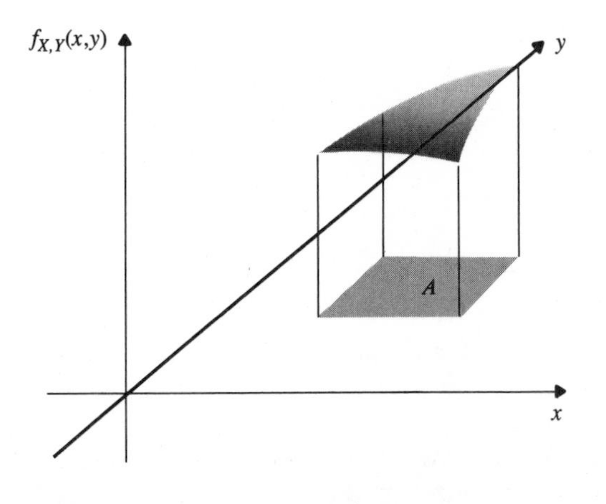


**FIGURE 4.7** The probability of $A$ is the integral of $f_{X,Y}(x, y)$ over the region defined by $A$.

then the pdf can be obtained from the cdf by differentiation:

$$f_{X,Y}(x,y) = \frac{\partial^2 F_{X,Y}(x,y)}{\partial x \, \partial y}. \tag{4.12}$$

Note that if $X$ and $Y$ are not jointly continuous, then it is possible that the above partial derivative does not exist. In particular, if the $F_{X,Y}(x,y)$ is discontinuous or if its partial derivatives are discontinuous, then the joint pdf as defined by Eq. (4.12) will not exist.

The probability of a rectangular region is obtained by letting $A = \{(x,y): a_1 < x \leq b_1 \text{ and } a_2 < y \leq b_2\}$ in Eq. 4.9:

$$P[a_1 < X \leq b_1, a_2 < Y \leq b_2] = \int_{a_1}^{b_1} \int_{a_2}^{b_2} f_{X,Y}(x',y')\, dx'\, dy'. \tag{4.13}$$

It then follows that the probability of an infinitesimal rectangle is the product of the pdf and the area of the rectangle:

$$\begin{aligned} P[x < X \leq x + dx, y < Y \leq y + dy] &= \int_{x}^{x+dx} \int_{y}^{y+dy} f_{X,Y}(x',y')\, dx'\, dy' \\ &\simeq f_{X,Y}(x,y)\, dx\, dy. \end{aligned} \tag{4.14}$$

Equation (4.14) can be interpreted as stating that the joint pdf specifies the probability of the product-form events

$$\{x < X \leq x + dx\} \cap \{y < Y \leq y + dy\}.$$

The **marginal pdf's** $f_X(x)$ and $f_Y(y)$ are obtained by taking the derivative of the corresponding marginal cdf's, $F_X(x) = F_{X,Y}(x, \infty)$ and $F_Y(y) = F_{X,Y}(\infty, y)$. Thus

$$\begin{aligned} f_X(x) &= \frac{d}{dx} \int_{-\infty}^{x} \left\{ \int_{-\infty}^{\infty} f_{X,Y}(x',y')\, dy' \right\} dx' \\ &= \int_{-\infty}^{\infty} f_{X,Y}(x,y')\, dy'. \end{aligned} \tag{4.15a}$$

Similarly,

$$f_Y(y) = \int_{-\infty}^{\infty} f_{X,Y}(x',y)\, dx'. \tag{4.15b}$$

Thus the marginal pdf's are obtained by integrating out the variables that are not of interest.

Note that $f_X(x)\, dx \simeq P[x < X \leq x + dx,\ Y \leq \infty]$ is the probability of the infinitesimal strip shown in Fig. 4.8(a). This reminds us of the interpretation of the marginal pmf's of as the probabilities of columns and rows in the case of discrete random variables. It is not surprising then that the Eqs. (4.15a) and (4.15b) for the marginal pdf's and Eqs. (4.7a) and

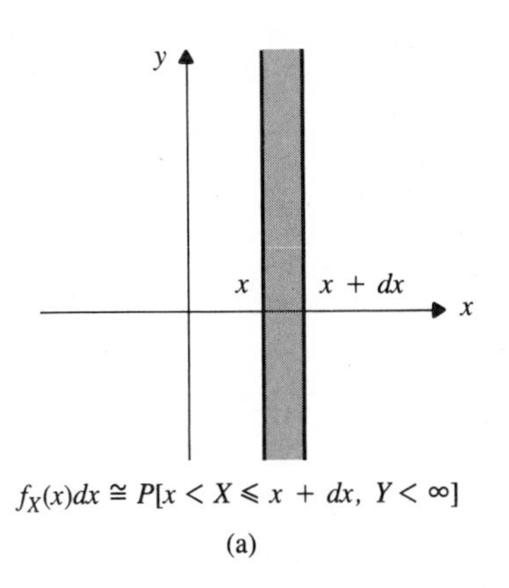


$f_X(x)dx \cong P[x < X \leqslant x + dx,\ Y < \infty]$

(a)

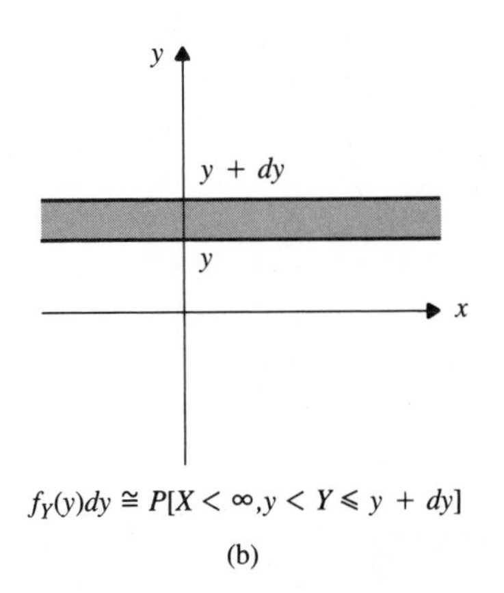


$f_Y(y)dy \cong P[X < \infty, y < Y \leqslant y + dy]$

(b)

**FIGURE 4.8** Interpretation of marginal pdf's.

(4.7b) for the marginal pmf's are identical except for the fact that one contains an integral and the other a summation. As in the case of pmf's, we note that, in general, the joint pdf cannot be obtained from the marginal pdf's.

## ■■ Example 4.10

The uniform pdf on the unit square is given by

$$f_{X,Y}(x, y) = \begin{cases} 1 & 0 \leq x \leq 1 \text{ and } 0 \leq y \leq 1 \\ 0 & \text{elsewhere.} \end{cases}$$

Find the cdf.

The cdf is found by evaluating Eq. (4.11). You must be careful with the limits of the integral: The limits should define the region consisting of the

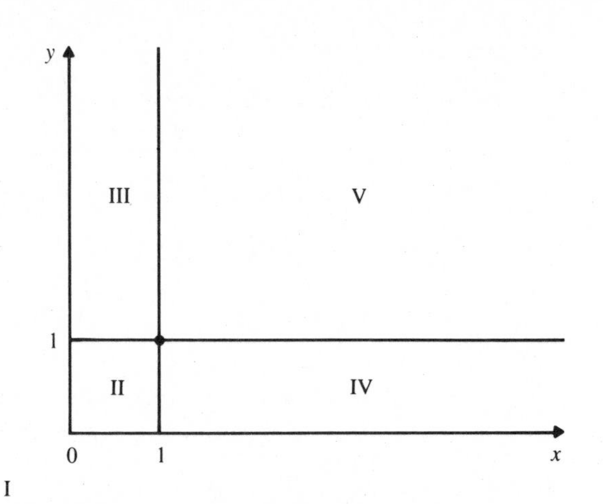


**FIGURE 4.9** Regions that need to be considered separately in computing cdf in Example 4.10.

intersection of the semi-infinite rectangle defined by $(x, y)$ and the region where the pdf is nonzero. There are five cases in this problem corresponding to the five regions shown in Fig. 4.9.

1. If $x < 0$ or $y < 0$, the pdf is zero and Eq. (4.12) implies
   $F_{X,Y}(x, y) = 0.$
2. If $(x, y)$ is inside the unit interval,
   $$F_{X,Y}(x, y) = \int_0^x \int_0^y 1\, dx'\, dy' = xy.$$
3. If $0 \leq x \leq 1$ and $y > 1$,
   $$F_{X,Y}(x, y) = \int_0^x \int_0^1 1\, dx'\, dy' = x.$$
4. Similarly if $x > 1$ and $0 \leq y \leq 1$,
   $F_{X,Y}(x, y) = y.$
5. Finally if $x > 1$ and $y > 1$,
   $$F_{X,Y}(x, y) = \int_0^1 \int_0^1 1\, dx'\, dy' = 1.$$

■■

---

## ■■ Example 4.11

Find the normalization constant $c$ and the marginal pdf's for the following joint pdf:

$$f_{X,Y}(x, y) = \begin{cases} ce^{-x}e^{-y} & 0 \leq y \leq x < \infty \\ 0 & \text{elsewhere.} \end{cases}$$

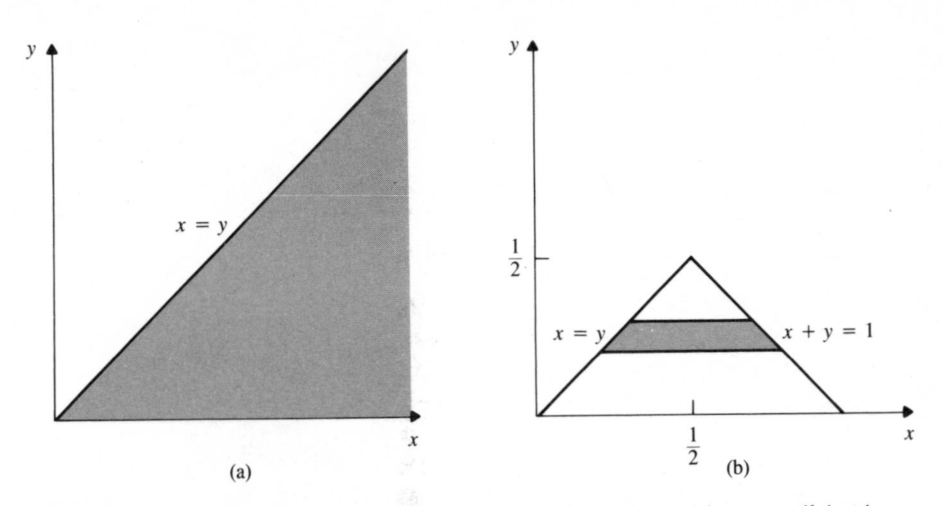


**FIGURE 4.10** The random variables $X$ and $Y$ in Example 4.12 have a pdf that is nonzero only in the shaded region shown in Part (a).

The pdf is nonzero in the shaded region shown in Fig. 4.10(a). The constant $c$ is found from the normalization condition specified by Eq. (4.10):

$$1 = \int_0^\infty \int_0^x ce^{-x}e^{-y}\,dy\,dx = \int_0^\infty ce^{-x}(1 - e^{-x})\,dx = \frac{c}{2}.$$

Therefore $c = 2$. The marginal pdf's are found by evaluating Eqs. (4.15a) and (4.15b):

$$f_X(x) = \int_0^\infty f_{X,Y}(x, y)\,dy = \int_0^x 2e^{-x}e^{-y}\,dy = 2e^{-x}(1 - e^{-x}) \qquad 0 \leq x < \infty$$

and

$$f_Y(y) = \int_0^\infty f_{X,Y}(x, y)\,dx = \int_y^\infty 2e^{-x}e^{-y}\,dx = 2e^{-2y} \qquad 0 \leq y < \infty.$$

You should fill in the steps in the evaluation of the integrals as well as verify that the marginal pdf's integrate to 1. ■■

## ■■ Example 4.12

Find $P[X + Y \leq 1]$ in Example 4.11.

The region corresponding to the intersection of the event $\{X + Y \leq 1\}$ and the region where the pdf is nonzero is shown in Fig. 4.10(b). We obtain the probability of the event by "adding" (actually integrating)

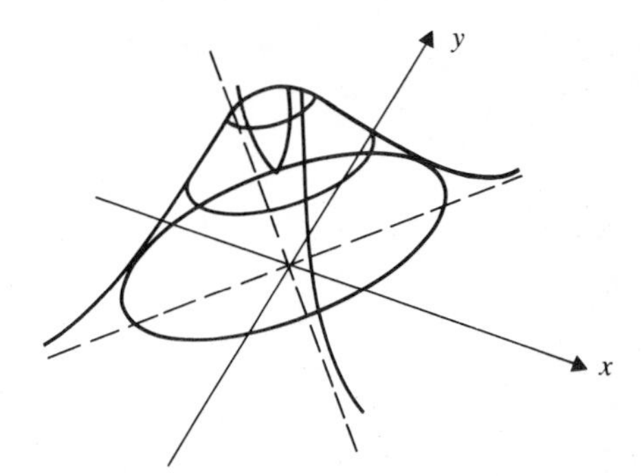


**FIGURE 4.11** Joint pdf of two jointly Gaussian random variables.

infinitesimal rectangles of width $dy$ as indicated in the figure:

$$P[X + Y \leq 1] = \int_0^{.5} \int_y^{1-y} 2e^{-x}e^{-y}\,dx\,dy = \int_0^{.5} 2e^{-y}[e^{-y} - e^{-(1-y)}]\,dy$$
$$= 1 - 2e^{-1}.$$

### ■■ Example 4.13
Jointly Gaussian Random Variables

The joint pdf of $X$ and $Y$, shown in Fig. 4.11, is

$$f_{X,Y}(x,y) = \frac{1}{2\pi\sqrt{1-\rho^2}} e^{-(x^2 - 2\rho xy + y^2)/2(1-\rho^2)} \qquad -\infty < x, y < \infty \tag{4.16}$$

We say that $X$ and $Y$ are jointly Gaussian.[1] Find the marginal pdf's.

The marginal pdf of $X$ is found by integrating $f_{X,Y}(x,y)$ over $y$:

$$f_X(x) = \frac{e^{-x^2/2(1-\rho^2)}}{2\pi\sqrt{1-\rho^2}} \int_{-\infty}^{\infty} e^{-(y^2 - 2\rho xy)/2(1-\rho^2)}\,dy.$$

We complete the square of the argument of the exponent by adding and subtracting $\rho^2x^2$, that is, $y^2 - 2\rho xy + \rho^2x^2 - \rho^2x^2 = (y - \rho x)^2 - \rho^2x^2$. Therefore

$$f_X(x) = \frac{e^{-x^2/2(1-\rho^2)}}{2\pi\sqrt{1-\rho^2}} \int_{-\infty}^{\infty} e^{-[(y-\rho x)^2 - \rho^2x^2]/2(1-\rho^2)}\,dy$$
$$= \frac{e^{-x^2/2}}{\sqrt{2\pi}} \int_{-\infty}^{\infty} \frac{e^{-(y-\rho x)^2/2(1-\rho^2)}}{\sqrt{2\pi(1-\rho^2)}}\,dy$$
$$= \frac{e^{-x^2/2}}{\sqrt{2\pi}},$$

1. This is an important special case of jointly Gaussian random variables. The general case is discussed in Section 4.8.

where we have noted that the last integral equals one since its integrand is a Gaussian pdf with mean $\rho x$ and variance $1 - \rho^2$. The marginal pdf of $X$ is therefore a one-dimensional Gaussian pdf with mean 0 and variance 1. From the symmetry of $f_{X,Y}(x, y)$ in $x$ and $y$, we conclude that the marginal pdf of $Y$ is also a one-dimensional Gaussian pdf with zero mean and unit variance. ■■

### Random Variables that Differ in Type

In some problems it is necessary to work with joint random variables that differ in type, that is, one is discrete and the other is continuous. Usually it is rather clumsy to work with the joint cdf, and it is preferable to work with either $P[X = k, Y \le y]$ or $P[X = k, y_1 < Y \le y_2]$. These probabilities are sufficient to compute the joint cdf should we have to.

**■■ Example 4.14**
A Communication Channel with Discrete Input and Continuous Output

Let $X$ be the input to a communication channel and let $Y$ be the output. The input to the channel is +1 volt or −1 volt with equal probability. The output of the channel is the input plus a noise voltage $N$ that is uniformly distributed in the interval from −2 volts to +2 volts. Find $P[X = +1, Y \le 0]$.

This problem lends itself to the use of conditional probability $P[X = k, Y \le y] = P[Y \le y \mid X = k]P[X = k]$, therefore

$$P[X = +1, Y \le y] = P[Y \le y \mid X = +1]P[X = +1],$$

where $P[X = +1] = 1/2$ and

$$P[Y \le y \mid X = +1] = \frac{y + 1}{4} \qquad \text{for } -1 \le y \le 3,$$

since $Y$ is uniformly distributed in the interval $[-1, 3]$ when the input is +1. Thus

$$P[X = +1, Y \le 0] = P[Y \le 0 \mid X = +1]P[X = +1] = \frac{1}{4}\frac{1}{2} = \frac{1}{8}.$$ ■■

## 4.3 INDEPENDENCE OF TWO RANDOM VARIABLES

Two random variables **$X$ and $Y$ are independent** if for *any* one-dimensional events $A$ and $B$

$$P[X \text{ in } A, Y \text{ in } B] = P[X \text{ in } A]P[Y \text{ in } B]. \qquad (4.17)$$

In this section we present a simple set of conditions for determining when $X$ and $Y$ are independent.

Suppose that $X$ and $Y$ are a pair of discrete random variables, and suppose we are interested in the probability of the event $A = A_1 \cap A_2$, where $A_1$ involves only $X$ and $A_2$ involves only $Y$. In particular, if $X$ and $Y$ are independent, then $A_1$ and $A_2$ are independent events. If we let $A_1 = \{X = x_j\}$ and $A_2 = \{Y = y_k\}$, then the independence of $X$ and $Y$ implies that

$$\begin{aligned} p(x_j, y_k) &= P[X = x_j, Y = y_k] \\ &= P[X = x_j]P[Y = x_k] \\ &= p_X(x_j)p_Y(y_k) \qquad \text{for all } x_j \text{ and } y_k. \end{aligned} \tag{4.18}$$

Therefore, *if $X$ and $Y$ are independent discrete random variables, then the joint pmf is equal to the product of the marginal pmf's.*

Now suppose that we don't know if $X$ and $Y$ are independent, but we do know that the pmf satisfies Eq. (4.18). Let $A = A_1 \cap A_2$ be a product-form event as above, then

$$\begin{aligned} P[A] &= \sum_{x_j \text{ in } A_1} \sum_{y_k \text{ in } A_2} p(x_j, y_k) \\ &= \sum_{x_j \text{ in } A_1} \sum_{y_k \text{ in } A_2} p_X(x_j)p_Y(y_k) \\ &= \sum_{x_j \text{ in } A_1} p_X(x_j) \sum_{y_k \text{ in } A_2} p_Y(y_k) \\ &= P[A_1]P[A_2], \end{aligned} \tag{4.19}$$

which implies that $A_1$ and $A_2$ are independent events. Therefore *if the joint pmf of $X$ and $Y$ equals the product of the marginal pmf's, then $X$ and $Y$ are independent.* We have just proved that the statement "$X$ and $Y$ are independent" is equivalent to the statement "the joint pmf is equal to the product of the marginal pmf's." In mathematical language, we say "*$X$ and $Y$ are independent if and only if the joint pmf is equal to the product of the marginal pmf's for all $x_j$, $y_k$.*"

### ■■ Example 4.15

Is the pmf in Example 4.6 consistent with an experiment that consists of the independent tosses of two fair dice?

The probability of each face in a toss of a fair die is 1/6. If two fair dice are tossed and if the tosses are independent, then the probability of any pair of faces, say $j$ and $k$, is:

$$P[X = j, Y = k] = P[X = j]P[Y = k] = \frac{1}{36}.$$

Thus all possible pairs of outcomes should be equiprobable. This is not the case for the joint pmf given in Example 4.6. Therefore the tosses in Example 4.6 are not independent. ■■

---

### ■■ Example 4.16

Are $Q$ and $R$ in Example 4.7 independent? From Example 4.7 we have

$$
\begin{aligned}
P[Q = q]P[R = r] &= (1 - p^M)(p^M)^q \frac{(1-p)}{1-p^M} p^r \\
&= (1-p)p^{Mq+r} \\
&= P[Q = q, R = r] \qquad \text{for all } q = 0, 1, \ldots \\
&\qquad\qquad\qquad\qquad\qquad\quad r = 0, \ldots, M-1.
\end{aligned}
$$

Therefore $Q$ and $R$ are independent. ■■

---

In general, it can be shown that the random variables $X$ *and* $Y$ *are independent if and only if their joint cdf is equal to the product of its marginal cdf's*:

$$F_{X,Y}(x, y) = F_X(x)F_Y(y) \qquad \text{for all } x \text{ and } y. \tag{4.20}$$

Similarly, if $X$ and $Y$ are jointly continuous, then $X$ *and* $Y$ *are independent if and only if their joint pdf is equal to the product of the marginal pdf's*:

$$f_{X,Y}(x, y) = f_X(x)f_Y(y) \qquad \text{for all } x \text{ and } y. \tag{4.21}$$

Equation (4.21) is obtained from Eq. (4.20) by differentiation. Conversely, Eq. (4.20) is obtained from Eq. (4.21) by integration.

### ■■ Example 4.17

Are the random variables $X$ and $Y$ in Example 4.11 independent?

Note that $f_X(x)$ and $f_Y(y)$ are nonzero for all $x > 0$ and all $y > 0$. Hence $f_X(x)f_Y(y)$ is nonzero in the entire positive quadrant. However $f_{X,Y}(x, y)$ is nonzero only in the region $y < x$ inside the positive quadrant. Hence Eq. (4.21) does not hold for all $x$, $y$ and the random variables are not independent. You should note that in this example the joint pdf appears to factor, but nevertheless it is not the product of the marginal pdf's. ■■

---

### ■■ Example 4.18

Are the random variables $X$ and $Y$ in Example 4.13 independent? The product of the marginal pdf's of $X$ and $Y$ in Example 4.13 is

$$f_X(x)f_Y(y) = \frac{1}{2\pi} e^{-(x^2+y^2)/2} \qquad -\infty < x, y < \infty.$$

By comparing to Eq. (4.16) we see that the product of the marginals is equal to the joint pdf if and only if $\rho = 0$. Therefore the jointly Gaussian random variables $X$ and $Y$ are independent if and only if $\rho = 0$. We see in a later section that $\rho$ is the *correlation coefficient* between $X$ and $Y$. ■■

**■■ Example 4.19**

Are the random variables $X$ and $Y$ independent in Example 4.8? If we multiply the marginal cdf's found in Example 4.8 we find

$$F_X(x)F_Y(y) = (1 - e^{-\alpha x})(1 - e^{-\beta y}) = F_{X,Y}(x, y) \qquad \text{for all } x \text{ and } y.$$

Therefore Eq. (4.20) is satisfied so $X$ and $Y$ are independent. ■■

If $X$ and $Y$ are independent random variables, then the random variables defined by any pair of functions $g(X)$ and $h(Y)$ are also independent. To show this consider the one-dimensional events $A$ and $B$. Let $A'$ be the set of all values of $x$ such that if $x$ is in $A'$ then $g(x)$ is in $A$, and let $B'$ be the set of all values of $y$ such that if $y$ is in $B'$ then $h(y)$ is in $B$. (In Chapter 3 we called $A'$ and $B'$ the equivalent events of $A$ and $B$.) Then

$$\begin{aligned} P[g(X) \text{ in } A, h(Y) \text{ in } B] &= P[X \text{ in } A', Y \text{ in } B'] \\ &= P[X \text{ in } A']P[Y \text{ in } B'] \\ &= P[g(X) \text{ in } A]P[h(Y) \text{ in } B]. \end{aligned}$$

The first and third equalities follow from the fact that $A$ and $A'$ and $B$ and $B'$ are equivalent events. The second equality follows from the independence of $X$ and $Y$. Thus $g(X)$ and $h(Y)$ are independent random variables.

## 4.4

## CONDITIONAL PROBABILITY AND CONDITIONAL EXPECTATION

Many random variables of practical interest are not independent: The output $Y$ of a communication channel must be dependent on the input $X$ in order to convey information; consecutive samples of a waveform that varies slowly are very likely to be close in value and hence are not independent. In this section we are interested in computing the probability of events concerning the random variable $Y$ given that we know that $X = x$. We are also interested in the expected value of $Y$ given $X = x$. We show that the notions of conditional probability and conditional expectation are extremely useful tools in solving problems, even in situations where we are only concerned with one of the random variables.

### Conditional Probability

The definition of conditional probability in Section 2.4 provides the formula for computing the probability that $Y$ is in $A$ given that we know the exact value of $X$, that is, $X = x$:

$$P[Y \text{ in } A \mid X = x] = \frac{P[Y \text{ in } A, X = x]}{P[X = x]}. \tag{4.22}$$

*If $X$ and $Y$ are discrete,* then the **conditional pmf of $Y$ given $X = x_k$** is given by

$$P(y_j \mid x_k) = P[Y = y_j \mid X = x_k] = \frac{P[X = x_k, Y = y_j]}{P[X = x_k]} \tag{4.23}$$

for $x_k$ such that $P[X = x_k] > 0$. We define $p(y_j \mid x_k) = 0$ for $x_k$ such that $P[X = x_k] = 0$. The conditional pmf satisfies all the properties of a pmf. In particular, the probability of any event $A$ given $X = x_k$ is found by summing the pmf over the event:

$$P[Y \text{ in } A \mid X = x_k] = \sum_{y_j \text{ in } A} p(y_j \mid x_k) \tag{4.24}$$

Note that *if $X$ and $Y$ are independent,* then

$$p(y_j \mid x_k) = \frac{P[X = x_k]P[Y = y_j]}{P[X = x_k]} = P[Y = y_j].$$

*If $X$ is discrete and $Y$ is continuous,* then Eq. (4.22) can be used to obtain the **conditional cdf of $Y$ given $X = x_k$:**

$$F_Y(y \mid x_k) = \frac{P[Y \leq y, X = x_k]}{P[X = x_k]}, \qquad \text{for } P[X = x_k] > 0. \tag{4.25}$$

The conditional pdf of $Y$ given $X = x_k$, if the derivative exists, is given by

$$f_Y(y \mid x_k) = \frac{d}{dy} F_Y(y \mid x_k). \tag{4.26}$$

The probability of an event $A$ given $X = x_k$ is obtained by integrating the conditional pdf:

$$P[Y \text{ in } A \mid X = x_k] = \int_{y \text{ in } A} f_Y(y \mid x_k)\, dy \tag{4.27}$$

Note that *if $X$ and $Y$ are independent,* $P[Y \leq y, X = x_k] = P[Y \leq y]$ $P[X = x_k]$ so $F_Y(y \mid x) = F_Y(y)$ and $f_Y(y \mid x) = f_Y(y)$.

### ■■ Example 4.20

Let $X$ be the input and $Y$ the output of the communication channel discussed in Example 4.14. Find the probability that $Y$ is negative given that $X$ is $+1$.

If $X = +1$, then $Y$ is uniformly distributed in the interval $[-1, 3]$, that is,

$$f_Y(y \mid 1) = \begin{cases} \frac{1}{4} & -1 \le y \le 3 \\ 0 & \text{elsewhere.} \end{cases}$$

Thus

$$P[Y < 0 \mid X = +1] = \int_{-1}^{0} \frac{dy}{4} = \frac{1}{4}.$$

■■

*If $X$ is a continuous random variable,* then $P[X = x] = 0$ so Eq. (4.22) is undefined. Suppose that $X$ and $Y$ are jointly continuous random variables with a joint pdf that is continuous and nonzero over some region of the plane. We define the **conditional cdf of $Y$ given $X = x$** by the following limiting procedure:

$$F_Y(y \mid x) = \lim_{h \to 0} F_Y(y \mid x < X \le x + h). \tag{4.28}$$

The conditional cdf on the right side of Eq. (4.28) is:

$$\begin{aligned} F_Y(y \mid x < X \le x + h) &= \frac{P[Y \le y, x < X \le x + h]}{P[x < X \le x + h]} \\ &= \frac{\int_{-\infty}^{y} \int_{x}^{x+h} f_{X,Y}(x', y')\, dx'\, dy'}{\int_{x}^{x+h} f_X(x')\, dx'} \\ &\simeq \frac{\int_{-\infty}^{y} f_{X,Y}(x, y')\, dy'\, h}{f_X(x) h}. \end{aligned} \tag{4.29}$$

As we let $h$ approach zero, Eqs. (4.28) and (4.29) imply that

$$F_Y(y \mid x) = \frac{\int_{-\infty}^{y} f_{X,Y}(x, y')\, dy'}{f_X(x)} \tag{4.30}$$

The **conditional pdf of $Y$ given $X = x$** is obtained by taking the

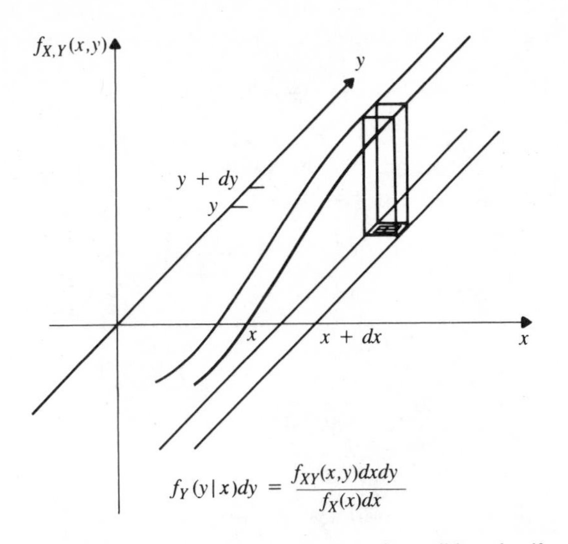

**FIGURE 4.12** Interpretation of conditional pdf.

derivative of $F_Y(y \mid x)$:

$$f_Y(y \mid x) = \frac{f_{X,Y}(x, y)}{f_X(x)}. \tag{4.31}$$

Note that we can interpret $f_Y(y \mid x)\, dy$ as the probability that $Y$ is in the infinitesimal strip defined by $(y, y + dy)$ given that $X$ is in the infinitesimal strip defined by $(x, x + dx)$ as shown in Fig. 4.12. In this sense, Eq. (4.31) can be viewed as a form of Bayes' theorem for continuous random variables.

Note that *if X and Y are independent,* then $f_{X,Y}(x, y) = f_X(x)f_Y(y)$ and $f_Y(y \mid x) = f_Y(y)$ and $F_Y(y \mid x) = F_Y(y)$.

### ■■ Example 4.21

Let $X$ and $Y$ be the random variables introduced in Example 4.11. Find $f_X(x \mid y)$ and $f_Y(y \mid x)$.

Using the marginal pdf's obtained in Example 4.11, we have

$$f_X(x \mid y) = \frac{2e^{-x}e^{-y}}{2e^{-2y}} = e^{-(x-y)} \qquad \text{for } x \geq y$$

and

$$f_Y(y \mid x) = \frac{2e^{-x}e^{-y}}{2e^{-x}(1 - e^{-x})} = \frac{e^{-y}}{1 - e^{-x}} \qquad \text{for } 0 < y < x.$$

The conditional pdf of $X$ is an exponential pdf shifted by $y$ to the right. The conditional pdf of $Y$ is an exponential pdf that has been truncated to the interval $[0, x]$. ■■

---

If we multiply Eq. (4.23) by $P[X = x_k]$, we find that the joint probability can be obtained in terms of the product of a conditional probability and a marginal probability:

$$P[X = x_k, Y = y_j] = P[Y = y_j \mid X = x_k]P[X = x_k]. \tag{4.32}$$

Suppose we are interested in the probability that $Y$ is in $A$:

$$\begin{aligned} P[Y \text{ in } A] &= \sum_{\text{all } x_k} \sum_{y_j \text{ in } A} P[X = x_k, Y = y_j] \\ &= \sum_{\text{all } x_k} \sum_{y_j \text{ in } A} P[Y = y_j \mid X = x_k]P[X = x_k] \\ &= \sum_{\text{all } x_k} P[X = x_k] \sum_{y_j \text{ in } A} P[Y = y_j \mid X = x_k] \end{aligned}$$

$$P[Y \text{ in } A] = \sum_{\text{all } x_k} P[Y \text{ in } A \mid X = x_k]P[X = x_k]. \tag{4.33}$$

Equation (4.33) is simply a restatement of the theorem on total probability discussed in Chapter 2. In words, the above result states that to compute $P[Y \text{ in } A]$ we can first compute $P[Y \text{ in } A \mid X = x_k]$ and then average over $x_k$. It can be shown that Eq. (4.33) also holds when $X$ is discrete and $Y$ is continuous.

If $X$ and $Y$ are continuous, we multiply Eq. (4.31) by $f_X(x)$ to obtain

$$f_{X,Y}(x, y) = f_Y(y \mid x)f_X(x). \tag{4.34}$$

If we replace summations with integrals and pmf's with pdf's, the same arguments that led to Eq. (4.33) in this case lead to

$$P[Y \text{ in } A] = \int_{-\infty}^{\infty} P[Y \text{ in } A \mid X = x]f_X(x)\,dx. \tag{4.35}$$

You can think of Eq. (4.35) as the "continuous" version of the theorem on total probability. The following examples show the usefulness of the above results in calculating the probabilities of complicated events.

### ■■ Example 4.22

The total number of defects $X$ on a chip is a Poisson random variable with mean $\alpha$. Suppose that each defect has a probability $p$ of falling in a specific region $R$ and that the location of each defect is independent of the locations of all other defects. Find the pmf of the number of defects $Y$ that fall in the region $R$.

From Eq. (4.33) we have

$$P[Y = j] = \sum_{k=0}^{\infty} P[Y = j \mid X = k] P[X = k].$$

We can imagine performing a Bernoulli trial each time a defect occurs with a "success" occurring when the defect falls in the region $R$. If the total number of defects is $X = k$, then the number of defects that fall in the region $R$ is a Binomial random variable with parameters $k$ and $p$:

$$P[Y = j \mid X = k] = \binom{k}{j} p^j (1 - p)^{k-j} \qquad 0 \le j \le k.$$

Therefore

$$\begin{aligned} P[Y = j] &= \sum_{k=j}^{\infty} \frac{k!}{j!\,(k - j)!} p^j (1 - p)^{k-j} \frac{\alpha^k}{k!} e^{-\alpha} \\ &= \frac{(\alpha p)^j e^{-\alpha}}{j!} \sum_{k=j}^{\infty} \frac{\{(1 - p)\alpha\}^{k-j}}{(k - j)!} \\ &= \frac{(\alpha p)^j e^{-\alpha}}{j!} e^{(1-p)\alpha} = \frac{(\alpha p)^j}{j!} e^{-\alpha p}. \end{aligned}$$

Thus $Y$ is a Poisson random variable with mean $\alpha p$. ■■

---

### ■■ Example 4.23

The number of customers that arrive at a service station during a time $t$ is a Poisson random variable with parameter $\beta t$. The time required to service each customer is an exponential random variable with parameter $\alpha$. Find the pmf for the number of customers $N$ that arrive during the service time $T$ of a specific customer. Assume that the customer arrivals are independent of the customer service time.

Equation (4.35) holds even if $Y$ is a discrete random variable, thus

$$\begin{aligned} P[N = k] &= \int_0^{\infty} P[N = k \mid T = t] f_T(t)\, dt \\ &= \int_0^{\infty} \frac{(\beta t)^k}{k!} e^{-\beta t} \alpha e^{-\alpha t}\, dt \\ &= \frac{\alpha \beta^k}{k!} \int_0^{\infty} t^k e^{-(\alpha+\beta)t}\, dt \end{aligned}$$

Let $r = (\alpha + \beta)t$, then:

$$= \frac{\alpha\beta^k}{k!\,(\alpha + \beta)^{k+1}} \int_0^\infty r^k e^{-r}\, dr$$

$$= \frac{\alpha\beta^k}{(\alpha + \beta)^{k+1}} = \left(\frac{\alpha}{\alpha + \beta}\right)\left(\frac{\beta}{\alpha + \beta}\right)^k,$$

where we have used the fact that the last integral is a gamma function and is equal to $k!$. Thus $N$ is a geometric random variable with probability of "success" $\beta/(\alpha + \beta)$. ■■

### ■■ Example 4.24

The random variable $X$ is selected at random from the unit interval; the random variable $Y$ is then selected at random from the interval $(0, X)$. Find the cdf of $Y$.

Equation (4.35) yields

$$F_Y(y) = P[Y \le y] = \int_0^1 P[Y \le y \mid X = x] f_X(x)\, dx.$$

When $X = x$, $Y$ is uniformly distributed in $(0, x)$ so the conditional cdf given $X = x$ is

$$P[Y \le y \mid X = x] = \begin{cases} \dfrac{y}{x} & 0 \le y \le x \\ 1 & x < y. \end{cases}$$

Substitution into the above integral yields

$$F_Y(y) = \int_0^y 1\, dx' + \int_y^1 \frac{y}{x'}\, dx' = y - y \ln y.$$

The corresponding pdf is obtained by taking the derivative of the cdf:

$$f_Y(y) = -\ln y \qquad 0 \le y \le 1.$$ ■■

## Conditional Expectation

The **conditional expectation of $Y$ given $X = x$** is defined by

$$E[Y \mid x] = \int_{-\infty}^{\infty} y f_Y(y \mid x)\, dx \qquad \text{if } Y \text{ is continuous} \tag{4.36a}$$

and

$$E[Y \mid x] = \sum_{y_j} y_j P[Y = y_j \mid X = x] \qquad \text{if } Y \text{ is discrete.} \tag{4.36b}$$

Clearly, $E[Y|x]$ is simply the center of mass associated with the conditional pdf or pmf.

The conditional expectation $E[Y|x]$ can be viewed as defining a function of $x$: $g(x) = E[Y|x]$. It therefore makes sense to talk about the random variable $g(X) = E[Y|X]$. Thus we can imagine that a random experiment is performed and a value for $X$ is obtained say $X = x_0$, and finally the value $g(x_0) = E[Y|x_0]$ is produced. We are interested in $E[g(X)] = E[E[Y|X]]$. In particular, we now show that

$$E[Y] = E[E[Y|X]], \tag{4.37}$$

where the right-hand side is

$$E[E[Y|X]] = \int_{-\infty}^{\infty} E[Y|x] f_X(x)\,dx \qquad X \text{ continuous}$$

and

$$E[E[Y|X]] = \sum_{x_k} E[Y|x_k] P[X = x_k] \qquad X \text{ discrete.}$$

We prove Eq. (4.37) for the case where $X$ and $Y$ are jointly continuous random variables, then

$$\begin{aligned} E[E[Y|X]] &= \int_{-\infty}^{\infty} E[Y|x] f_X(x)\,dx \\ &= \int_{-\infty}^{\infty}\int_{-\infty}^{\infty} y f_Y(y|x)\,dy f_X(x)\,dx \\ &= \int_{-\infty}^{\infty} y \int_{-\infty}^{\infty} f_{X,Y}(x, y)\,dx\,dy \\ &= \int_{-\infty}^{\infty} y f_Y(y)\,dy = E[Y]. \end{aligned}$$

The above result also holds for the expected value of a function of $Y$:

$$E[h(Y)] = E[E[h(Y)|X]].$$

In particular, the $k$th moment of $Y$ is given by

$$E[Y^k] = E[E[Y^k|X]].$$

### ■■ Example 4.25

Find the mean of $Y$ in Example 4.22 using conditional expectation.

$$E[Y] = \sum_{k=0}^{\infty} E[Y|X = k]P[X = k] = \sum_{k=0}^{\infty} kpP[X = k] = pE[X] = p\alpha.$$

The second equality uses the fact that $E[Y|X = k] = kp$ since $Y$ is binomial with parameters $k$ and $p$. Note that the second to the last equality holds for *any* pmf of $X$. The fact that $X$ is Poisson with mean $\alpha$ is not used until the last equality. ■■

### ■■ Example 4.26

Find the mean and variance of the number of customer arrivals $N$ during the service time $T$ of a specific customer in Example 4.23.

We will need the first two conditional moments of $N$ given $T = t$:

$$E[N \mid T = t] = \beta t \qquad E[N^2 \mid T = t] = (\beta t) + (\beta t)^2,$$

where we have used the fact that $N$ is a Poisson random variable with parameter $\beta t$ when the condition $T = t$ is given. The first two moments of $N$ are

$$E[N] = \int_0^\infty E[N \mid T = t] f_T(t)\, dt = \int_0^\infty \beta t f_T(t)\, dt = \beta E[T]$$

$$\begin{aligned} E[N^2] &= \int_0^\infty E[N^2 \mid T = t] f_T(t)\, dt = \int_0^\infty \{\beta t + \beta^2 t^2\} f_T(t)\, dt \\ &= \beta E[T] + \beta^2 E[T^2]. \end{aligned}$$

The variance of $N$ is then

$$\begin{aligned} \text{VAR}[N] &= E[N^2] - (E[N])^2 \\ &= \beta^2 E[T^2] + \beta E[T] - \beta^2 (E[T])^2 \\ &= \beta^2 \,\text{VAR}[T] + \beta E[T]. \end{aligned}$$

Note that if $T$ is not random (i.e., $E[T] =$ constant and $\text{VAR}[T] = 0$) then the mean and variance of $N$ are those of a Poisson random variable with parameter $\beta E[T]$. When $T$ is random, the mean of $N$ remains the same but the variance of $N$ increases by the term $\beta^2 \,\text{VAR}[T]$, that is, the variability of $T$ causes greater variability in $N$. Up to this point, we have intentionally avoided using the fact that $T$ has an exponential distribution to emphasize that the above results hold for *any* service time distribution $f_T(t)$. If $T$ is exponential with parameter $\alpha$, then $E[T] = 1/\alpha$ and $\text{VAR}[T] = 1/\alpha^2$, so

$$E[N] = \frac{\beta}{\alpha} \qquad \text{and} \qquad \text{VAR}[N] = \frac{\beta^2}{\alpha^2} + \frac{\beta}{\alpha}.$$

■■

# 4.5

## MULTIPLE RANDOM VARIABLES

Let $X_1, X_2, \ldots, X_n$ be the components of an $n$-dimensional vector random variable. In this section we show how the methods for specifying the probabilities of pairs of random variables are readily extended to the case of $n$ random variables.

### Joint Distributions

The **joint cumulative distribution function** of $X_1, X_2, \ldots, X_n$ is defined as the probability of an $n$-dimensional semi-infinite rectangle associated with the point $(x_1, \ldots, x_n)$:

$$F_{X_1,X_2,\ldots,X_n}(x_1, x_2, \ldots, x_n) = P[X_1 \leq x_1, X_2 \leq x_2, \ldots, X_n \leq x_n]. \tag{4.38}$$

The joint cdf is defined for discrete, continuous, and random variables of mixed type. The probability of all product-form events $\{X_1 \text{ in } A_1\} \cap \{X_2 \text{ in } A_2\} \cap \cdots \cap \{X_n \text{ in } A_n\}$ can be expressed in terms of the joint cdf.

Equation (4.38) generates a family of *marginal cdf's* for subsets of the random variables $X_1, \ldots, X_n$. These marginal cdf's are obtained by setting the appropriate entries to $+\infty$ in the joint cdf in Eq. (4.38). For example, the joint cdf for $X_1, \ldots, X_{n-1}$ is given by $F_{X_1,X_2,\ldots,X_n}(x_1, x_2, \ldots, x_{n-1}, \infty)$, and the joint cdf for $X_1$ and $X_2$ is given by $F_{X_1,X_2,\ldots,X_n}(x_1, x_2, \infty, \ldots, \infty)$

#### Example 4.27

Let the event $A$ be defined as follows:

$$A = \{\max(X_1, X_2, X_3) \leq 5\}.$$

Find the probability of $A$.

The maximum of three numbers is less than 5 if and only if each of the three numbers is less than 5, therefore

$$\begin{aligned} P[A] &= P[\{X_1 \leq 5\} \cap \{X_2 \leq 5\} \cap \{X_3 \leq 5\}] \\ &= F_{X_1,X_2,X_3}(5, 5, 5). \end{aligned}$$

The **joint probability mass function** of $n$ discrete random variables is defined by

$$p(x_1, x_2, \ldots, x_n) = P[X_1 = x_1, X_2 = x_2, \ldots, X_n = x_n]. \tag{4.39}$$

The probability of any $n$-dimensional event $A$ is found by summing the pmf over the points in the event

$$P[(X_1, \ldots, X_n) \text{ in } A] = \sum_{\mathbf{x} \text{ in } A} \cdots \sum p(x_1, \ldots, x_n). \tag{4.40}$$

Equation (4.39) generates a family of *marginal pmf's* that specify the joint probabilities for subsets of the $n$ random variables. For example, the one-dimensional pmf of $X_j$ is found by adding the joint pmf over all variables other than $x_j$:

$$p(x_j) = P[X_j = x_j] = \sum_{x_1} \cdots \sum_{x_{j-1}} \sum_{x_{j+1}} \cdots \sum_{x_n} p(x_1, x_2, \ldots, x_n). \tag{4.41}$$

The two-dimensional joint pmf of any pair $X_j$ and $X_k$ is found by adding the joint pmf over all $n - 2$ other variables, and so on. Thus, the marginal pmf for $X_1, \ldots, X_{n-1}$ is given by

$$p(x_1, x_2, \ldots, x_{n-1}) = \sum_{x_n} p(x_1, x_2, \ldots, x_n). \tag{4.42}$$

A family of *conditional pmf's* is obtained from the joint pmf by conditioning on different subsets of the random variables. For example,

$$p(x_n \mid x_1, \ldots, x_{n-1}) = \frac{p(x_1, \ldots, x_n)}{p(x_1, \ldots, x_{n-1})}, \tag{4.43}$$

if $p(x_1, \ldots, x_{n-1}) > 0$.

## ■■ Example 4.28

A computer system receives messages over three communications lines. Let $X_j$ be the number of messages received on line $j$ in one hour. Suppose that the joint pmf of $X_1$, $X_2$, and $X_3$ is given by

$$p(x_1, x_2, x_3) = (1 - a_1)(1 - a_2)(1 - a_3)a_1^{x_1}a_2^{x_2}a_3^{x_3} \quad x_1 \geq 0, x_2 \geq 0, x_3 \geq 0.$$

Find $p(x_1, x_2)$ and $p(x_1)$ given that $0 < a_i < 1$.

The marginal pmf of $X_1$ and $X_2$ is obtained as follows:

$$\begin{aligned} p(x_1, x_2) &= (1 - a_1)(1 - a_2)(1 - a_3) \sum_{x_3=0}^{\infty} a_1^{x_1}a_2^{x_2}a_3^{x_3} \\ &= (1 - a_1)(1 - a_2)a_1^{x_1}a_2^{x_2}. \end{aligned}$$

The pmf of $X_1$ is then obtained summing over $x_2$:

$$\begin{aligned} p(x_1) &= (1 - a_1)(1 - a_2) \sum_{x_2=0}^{\infty} a_1^{x_1}a_2^{x_2} \\ &= (1 - a_1)a_1^{x_1} \end{aligned}$$

■■

---

We say that the random variables $X_1, X_2, \ldots, X_n$ are jointly continuous random variables if the probability of any $n$-dimensional event $A$ is given by an $n$-dimensional integral of a probability density function:

$$P[(X_1, \ldots, X_n) \text{ in } A] = \int_{\mathbf{x} \text{ in } A} \cdots \int f_{X_1,\ldots,X_n}(x_1, \ldots, x_n)\, dx_1 \ldots dx_n, \tag{4.44}$$

where $f_{x_1,\ldots,x_n}(x_1, \ldots, x_n)$ is the **joint probability density function.**

The joint cdf of $\mathbf{X}$ is obtained from the joint pdf by integration

$$F_{X_1,X_2,\ldots,X_n}(x_1, x_2, \ldots, x_n) = \int_{-\infty}^{x_1} \cdots \int_{-\infty}^{x_n} f_{X_1,\ldots,X_n}(x'_1, \ldots, x'_n)\, dx'_1 \ldots dx'_n. \tag{4.45}$$

The joint pdf (if the derivative exists) is given by

$$f_{X_1,X_2,\ldots,X_n}(x_1, x_2, \ldots, x_n) = \frac{\partial^n}{\partial x_1 \ldots \partial x_n} F_{X_1,\ldots,X_n}(x_1, \ldots, x_n). \tag{4.46}$$

A family of *marginal pdf's* is associated with the joint pdf in Eq. (4.46). The marginal pdf for a subset of the random variables is obtained by integrating the other variables out. For example, the marginal pdf of $X_1$ is

$$f_{X_1}(x_1) = \int_{-\infty}^{\infty} \cdots \int_{-\infty}^{\infty} f_{X_1,X_2,\ldots,X_n}(x_1, \ldots, x_n)\, dx_2 \ldots dx_n. \tag{4.47}$$

As another example, the marginal pdf for $X_1, \ldots, X_{n-1}$ is given by

$$f_{X_1,\ldots,X_{n-1}}(x_1, \ldots, x_{n-1}) = \int_{-\infty}^{\infty} f_{X_1,\ldots,X_n}(x_1, \ldots, x_n)\, dx_n. \tag{4.48}$$

A family of *conditional pdf's* is also associated with the joint pdf. For example, the pdf of $X_n$ given the values of $X_1, \ldots, X_{n-1}$ is given by

$$f_{X_n}(x_n \mid x_1, \ldots, x_{n-1}) = \frac{f_{X_1,\ldots,X_n}(x_1, \ldots, x_n)}{f_{X_1\ldots X_{n-1}}(x_1, \ldots, x_{n-1})} \tag{4.49}$$

if $f_{X_1\ldots X_{n-1}}(x_1, \ldots, x_{n-1}) > 0$.

### Example 4.29

The random variables $X_1$, $X_2$, and $X_3$ have the joint Gaussian pdf

$$f_{X_1,X_2,X_3}(x_1, x_2, x_3) = \frac{e^{-(x_1^2 + x_2^2 - \sqrt{2}x_1x_2 + \frac{1}{2}x_3^2)}}{2\pi\sqrt{\pi}}.$$

Find the marginal pdf of $X_1$ and $X_3$.

The marginal pdf for the pair $X_1$ and $X_3$ is found by integrating the joint pdf over $x_2$:

$$f_{X_1,X_3}(x_1, x_3) = \frac{e^{-x_3^2/2}}{\sqrt{2\pi}} \int_{-\infty}^{\infty} \frac{e^{-(x_1^2+x_2^2-\sqrt{2}x_1x_2)}}{2\pi/\sqrt{2}} dx_2.$$

The above integral was carried out in Example 4.13 with $\rho = 1/\sqrt{2}$. By substituting the result of the integration above, we obtain

$$f_{X_1,X_3}(x_1, x_3) = \frac{e^{-x_3^2/2}}{\sqrt{2\pi}} \frac{e^{-x_1^2/2}}{\sqrt{2\pi}}.$$

Therefore $X_1$ and $X_3$ are independent zero mean, unit variance Gaussian random variables.[2] ■■

### Independence

Recall from Section 4.1 that $X_1, \ldots, X_n$ are independent if

$$P[X_1 \text{ in } A_1, \ldots, X_n \text{ in } A_n] = P[X_1 \text{ in } A_1] \ldots P[X_n \text{ in } A_n]$$

for any one-dimensional events $A_1, \ldots, A_n$. It can be shown that $X_1, \ldots, X_n$ *are independent if and only if*

$$F_{X_1,\ldots,X_n}(x_1, \ldots, x_n) = F_{X_1}(x_1) \ldots F_{X_n}(x_n) \tag{4.50}$$

for all $x_1, \ldots, x_n$. If the random variables are discrete, Eq. (4.50) is equivalent to

$$p(x_1, \ldots, x_n) = p_{X_1}(x_1) \ldots p_{X_n}(x_n) \qquad \text{for all } x_1, \ldots, x_n.$$

If the random variables are jointly continuous, Eq. (4.50) is equivalent to

$$f_{X_1,\ldots,X_n}(x_1, \ldots, x_n) = f_{X_1}(x_1) \ldots f_{X_n}(x_n)$$

for all $x_1, \ldots, x_n$.

### ■■ Example 4.30

The joint pdf of $n$ zero-mean, unit variance random variables is given by

$$f_{X_1,\ldots,X_n}(x_1, \ldots, x_n) = \frac{e^{-(x_1^2+\cdots+x_n^2)/2}}{(2\pi)^{n/2}} \qquad \text{for all } x_1, \ldots, x_n.$$

It is clear that the above is the product of $n$ one-dimensional Gaussian pdf's. Thus $X_1, \ldots, X_n$ are independent Gaussian random variables. ■■

2. Jointly Gaussian random variables are discussed in Section 4.8.

## 4.6

## FUNCTIONS OF SEVERAL RANDOM VARIABLES

Quite often one is interested in one or more functions of the random variables associated with some experiment. For example, if $X_1, X_2, \ldots, X_n$ represent repeated measurements of the same random quantity, we might be interested in the maximum and minimum value in the set, as well as the sample mean and sample variance. In this section we present methods for determining the probabilities of events involving functions of several random variables.

### One Function of Several Random Variables

Let the random variable $Z$ be defined as a function of several random variables:

$$Z = g(X_1, X_2, \ldots, X_n). \tag{4.51}$$

The cdf of $Z$ is found by first finding the equivalent event of $\{Z \leq z\}$, that is, the set $R_z = \{\mathbf{x} = (x_1, \ldots, x_n) \text{ such that } g(\mathbf{x}) \leq z\}$, then

$$\begin{aligned} F_Z(z) &= P[\mathbf{X} \text{ in } R_z] \\ &= \int \cdots \int_{\mathbf{x} \text{ in } R_z} f_{X_1,\ldots,X_n}(x_1, \ldots, x_n)\, dx_1 \ldots dx_n. \end{aligned} \tag{4.52}$$

The pdf of $Z$ is then found by taking the derivative of $F_Z(z)$.

**Example 4.31**
Sum of Two Random Variables

Let $Z = X + Y$. Find $F_Z(z)$ and $f_Z(z)$ in terms of the joint pdf of $X$ and $Y$.

The cdf of $Z$ is found by integrating the joint pdf of $X$ and $Y$ over the region of the plane corresponding to the event $\{Z \leq z\}$ as shown in Fig. 4.13:

$$F_Z(z) = \int_{-\infty}^{\infty} \int_{-\infty}^{z-x'} f_{X,Y}(x', y')\, dy'\, dx'.$$

The pdf of $Z$ is

$$f_Z(z) = \frac{d}{dz} F_Z(z) = \int_{-\infty}^{\infty} f_{X,Y}(x', z - x')\, dx'. \tag{4.53}$$

Thus the pdf for the sum of two random variables is given by a *superposition* integral.

If $X$ and $Y$ are independent random variables, then by Eq. (4.21) the

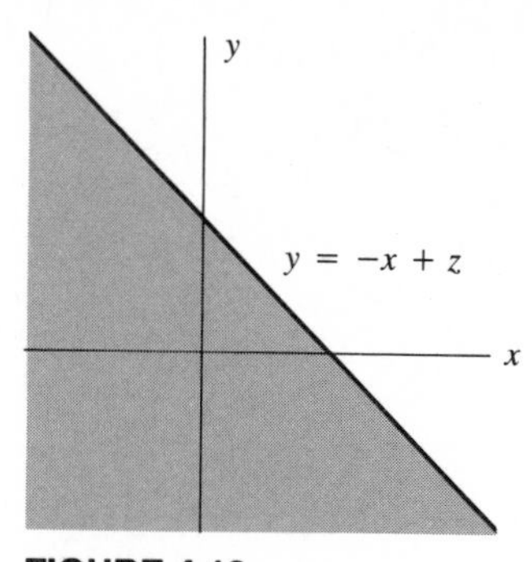


**FIGURE 4.13**
$P[Z \leq z] = P[X + Y \leq z]$.

pdf is given by the *convolution integral* of the marginal pdf's of $X$ and $Y$:

$$f_Z(z) = \int_{-\infty}^{\infty} f_X(x')f_Y(z - x')\,dx'. \tag{4.54}$$

In Section 5.1 we show how transform methods are used to evaluate convolution integrals such as Eq. (4.54). ■■

---

### ■■ Example 4.32
Sum of Nonindependent Gaussian Random Variables

Find the pdf of the sum $Z = X + Y$ of two zero-mean, unit variance Gaussian random variables with correlation coefficient $\rho = -1/2$.

The joint pdf for this pair of random variables was given in Example 4.13. The pdf of $Z$ is obtained by substituting the pdf for the joint Gaussian random variables into the superposition integral found in Example 4.31:

$$\begin{aligned} f_Z(z) &= \int_{-\infty}^{\infty} f_{X,Y}(x', z - x')\,dx' \\ &= \frac{1}{2\pi(1-\rho^2)^{1/2}} \int_{-\infty}^{\infty} e^{-[x'^2 - 2\rho x'(z-x') + (z-x')^2]/2(1-\rho^2)}\,dx' \\ &= \frac{1}{2\pi(3/4)^{1/2}} \int_{-\infty}^{\infty} e^{-(x'^2 - x'z + z^2)/2(3/4)}\,dx'. \end{aligned}$$

After completing the square of the argument in the exponent we obtain

$$f_Z(z) = \frac{e^{-z^2/2}}{\sqrt{2\pi}}.$$

Thus the sum of these two nonindependent Gaussian random variables is also a Gaussian random variable. ■■

---

The conditional pdf can be used to find the pdf of a function of several random variables. Let $Z = g(X, Y)$, and suppose that we are given that $Y = y$, then $Z = g(X, y)$ is a function of one random variable. Therefore we can use the methods developed in Section 3.5 for single random variables to find the pdf of $Z$ *given* $Y = y$: $f_Z(z \mid Y = y)$. The pdf of $Z$ is then found from

$$f_Z(z) = \int_{-\infty}^{\infty} f_Z(z \mid y') f_Y(y')\, dy'.$$

### ■■ Example 4.33

Let $Z = X/Y$. Find the pdf of $Z$ if $X$ and $Y$ are independent and both exponentially distributed with mean one.

Assume $Y = y$, then $Z = X/y$ is simply a scaled version of $X$. Therefore from Example 3.23

$$f_Z(z \mid y) = |y| f_X(yz \mid y).$$

The pdf of $Z$ is therefore

$$f_Z(z) = \int_{-\infty}^{\infty} |y| f_X(yz \mid y) f_Y(y)\, dy = \int_{-\infty}^{\infty} |y| f_{X,Y}(yz, y)\, dy.$$

We now use the fact that $X$ and $Y$ are independent and exponentially distributed with mean one:

$$\begin{aligned} f_Z(z) &= \int_0^{\infty} y f_X(yz) f_Y(y)\, dy \qquad z > 0 \\ &= \int_0^{\infty} y e^{-yz} e^{-y}\, dy \\ &= \frac{1}{(1+z)^2} \qquad z > 0. \end{aligned}$$

■■

### Transformations of Random Vectors

Let $X_1, \ldots, X_n$ be random variables associated with some experiment, and let the random variables $Z_1, \ldots, Z_n$ be defined by $n$ functions of $\mathbf{X} = (X_1, \ldots, X_n)$:

$$Z_1 = g_1(\mathbf{X}) \qquad Z_2 = g_2(\mathbf{X}) \qquad \ldots \qquad Z_n = g_n(\mathbf{X}).$$

We now consider the problem of finding the joint cdf and pdf of $Z_1, \ldots, Z_n$.

The joint cdf of $Z_1, \ldots, Z_n$ at the point $\mathbf{z} = (z_1, \ldots, z_n)$ is equal to the probability of the region of $\mathbf{x}$ where $g_k(\mathbf{x}) \le z_k$ for $k = 1, \ldots, n$:

$$F_{Z_1,\ldots,Z_n}(z_1, \ldots, z_n) = P[g_1(\mathbf{X}) \le z_1, \ldots, g_n(\mathbf{X}) \le z_n]. \tag{4.55a}$$

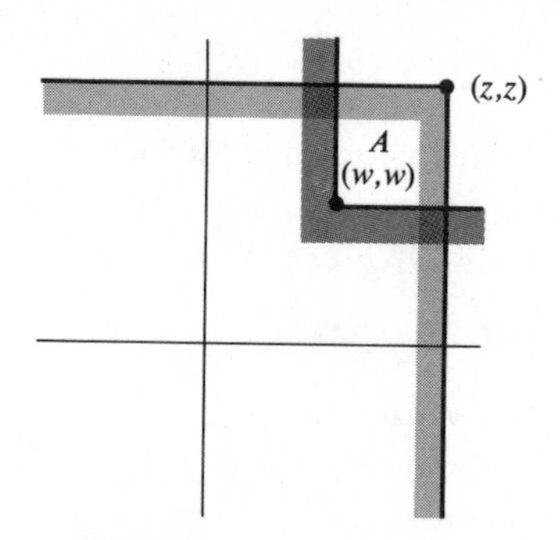


**FIGURE 4.14**
$\{\min(X, Y) \leq w\}$
$= \{X \leq w\} \cup \{Y \leq w\}$
and
$\{\max(X, Y) \leq z\}$
$= \{X \leq z\} \cap \{Y \leq z\}$.

If $X_1, \ldots, X_n$ have a joint pdf, then

$$F_{Z_1,\ldots,Z_n}(z_1, \ldots, z_n) = \int_{\mathbf{x}:g_k(\mathbf{x})\leq z_k} \cdots \int f_{X_1,\ldots,X_n}(x_1, \ldots, x_n)\, dx_1 \ldots dx_n. \tag{4.55b}$$

## Example 4.34

Let the random variables $W$ and $Z$ be defined by

$$W = \min(X, Y) \quad \text{and} \quad Z = \max(X, Y).$$

Find the joint cdf of $W$ and $Z$ in terms of the joint cdf of $X$ and $Y$.

Equation (4.55a) implies that

$$F_{W,Z}(w, z) = P[\{\min(X, Y) \leq w\} \cap \{\max(X, Y) \leq z\}].$$

The region corresponding to this event is shown in Fig. 4.14. From the figure it is clear that if $z > w$, the above probability is the probability of the semi-infinite rectangle defined by the point $(z, z)$ minus the square region denoted by $A$. Thus if $z > w$

$$\begin{aligned} F_{W,Z}(w, z) &= F_{X,Y}(z, z) - P[A] \\ &= F_{X,Y}(z, z) \\ &\quad - \{F_{X,Y}(z, z) - F_{X,Y}(w, z) - F_{X,Y}(z, w) + F_{X,Y}(w, w)\} \\ &= F_{X,Y}(w, z) + F_{X,Y}(z, w) - F_{X,Y}(w, w). \end{aligned}$$

If $z < w$ then

$$F_{W,Z}(w, z) = F_{X,Y}(z, z).$$

## pdf of Linear Transformations

The joint pdf of **Z** can be found directly in terms of the joint pdf of **X** by finding the equivalent events of infinitesimal rectangles. We consider first the linear transformation of two random variables:

$$\begin{aligned} V &= aX + bY \\ W &= cX + dY \end{aligned} \qquad \text{or} \qquad \begin{bmatrix} V \\ W \end{bmatrix} = \begin{bmatrix} a & b \\ c & e \end{bmatrix} \begin{bmatrix} X \\ Y \end{bmatrix}.$$

Denote the above matrix by $A$. We will assume that $A$ has an inverse, that is, it has determinant $|ae - bc| \neq 0$, so each point $(v, w)$ has a unique corresponding point $(x, y)$ obtained from

$$\begin{bmatrix} x \\ y \end{bmatrix} = A^{-1} \begin{bmatrix} v \\ w \end{bmatrix}. \tag{4.56}$$

Consider the infinitesimal rectangle shown in Fig. 4.15. The points in this rectangle are mapped into the parallelogram shown in the figure. The infinitesimal rectangle and the parallelogram are equivalent events, so their probabilities must be equal. Thus

$$f_{X,Y}(x, y)\, dx\, dy \simeq f_{V,W}(v, w)\, dP$$

where $dP$ is the area of the parallelogram. The joint pdf of $V$ and $W$ is thus

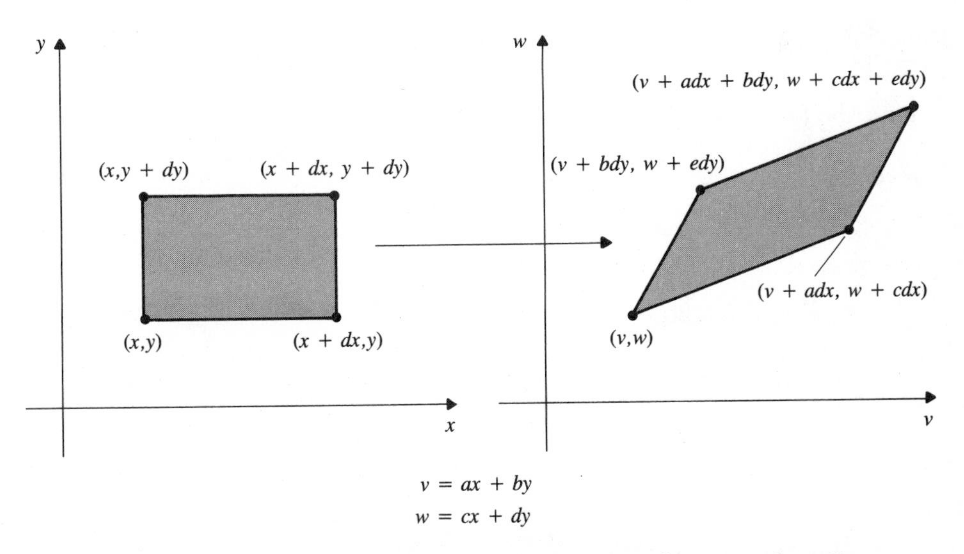


**FIGURE 4.15** Image of an infinitesimal rectangle under a linear transformation.

given by

$$f_{V,W}(v, w) = \frac{f_{X,Y}(x, y)}{\left|\dfrac{dP}{dx\,dy}\right|}, \tag{4.57}$$

where $x$ and $y$ are related to $(v, w)$ by Eq. (4.56). Equation (4.57) states that the joint pdf of $V$ and $W$ at $(v, w)$ is the pdf of $X$ and $Y$ at the corresponding point $(x, y)$, but rescaled by the "stretch factor" $dP/dx\,dy$. It can be shown that $dP = (|ae - bc|)\,dx\,dy$, so the "stretch factor" is

$$\left|\frac{dP}{dx\,dy}\right| = \frac{|ae - bc|\,(dx\,dy)}{(dx\,dy)} = |ae - bc| = |A|,$$

where $|A|$ is the determinant of $A$.

The above result holds for a general linear transformation of $n$ random variables. Let the $n$-dimensional vector $\mathbf{Z}$ be

$$\mathbf{Z} = A\mathbf{X},$$

where $A$ is an $n \times n$ invertible matrix. The joint pdf of $\mathbf{Z}$ is then

$$f_{Z_1,\ldots,Z_n}(z_1, \ldots, z_n) = \frac{f_{X_1,\ldots,X_n}(x_1, \ldots, x_n)}{|A|}\Bigg|_{\mathbf{x}=A^{-1}\mathbf{z}} \tag{4.58}$$

### ■■ Example 4.35

Let $X$ and $Y$ be the jointly Gaussian random variables introduced in Example 4.13. Let $V$ and $W$ be obtained from $(X, Y)$ by

$$\begin{bmatrix} V \\ W \end{bmatrix} = \frac{1}{\sqrt{2}}\begin{bmatrix} 1 & 1 \\ -1 & 1 \end{bmatrix}\begin{bmatrix} X \\ Y \end{bmatrix} = A\begin{bmatrix} X \\ Y \end{bmatrix}.$$

Find the joint pdf of $V$ and $W$.

The determinant of the matrix is $|A| = 1$, and the inverse mapping is given by

$$\begin{bmatrix} X \\ Y \end{bmatrix} = \frac{1}{\sqrt{2}}\begin{bmatrix} 1 & -1 \\ 1 & 1 \end{bmatrix}\begin{bmatrix} V \\ W \end{bmatrix},$$

so $X = (V - W)/\sqrt{2}$ and $Y = (V + W)/\sqrt{2}$. Therefore the pdf of $V$ and $W$ is

$$f_{V,W}(v, w) = f_{X,Y}\left(\frac{v - w}{\sqrt{2}}, \frac{v + w}{\sqrt{2}}\right),$$

where

$$f_{X,Y}(x, y) = \frac{1}{2\pi\sqrt{1 - \rho^2}}e^{-(x^2 - 2\rho xy + y^2)/2(1-\rho^2)}.$$

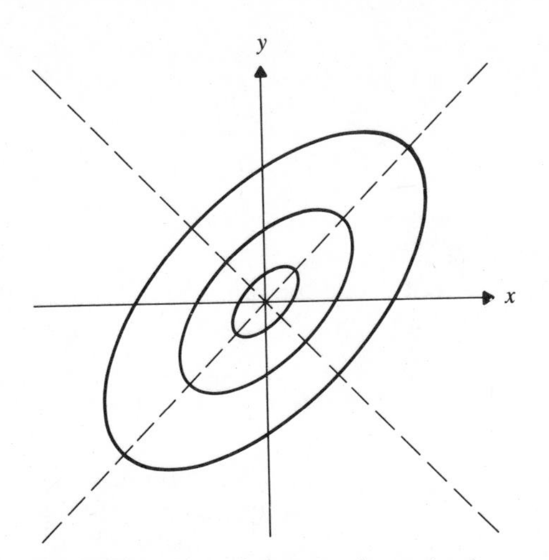


**FIGURE 4.16** Contours of equal value of joint Gaussian pdf discussed in Example 4.35.

By substituting for $x$ and $y$, the argument of the exponent becomes

$$\frac{(v - w)^2/2 - 2\rho(v - w)(v + w)/2 + (v + w)^2/2}{2(1 - \rho^2)}$$

$$= \frac{v^2}{2(1 + \rho)} + \frac{w^2}{2(1 - \rho)}.$$

Thus

$$f_{V,W}(v, w) = \frac{1}{2\pi(1 - \rho^2)^{1/2}} e^{-\{[v^2/2(1+\rho)]+[w/2(1-\rho)]\}}.$$

It can be seen that the transformed variables $V$ and $W$ are independent, zero-mean Gaussian random variables with variance $1 + \rho$ and $1 - \rho$, respectively. Figure 4.16 shows contours of equal value of the joint pdf of $(X, Y)$. It can be seen that the pdf has elliptical symmetry about the origin with principle axes at 45° with respect to the axes of the plane. In Section 4.8 we show that the above linear transformation corresponds to a rotation of the coordinate system so that the axes of the plane are aligned with the axes of the ellipse. ■■

### *pdf of General Transformations

Now let the random variables $V$ and $W$ be defined by two nonlinear functions of $X$ and $Y$:

$$V = g_1(X, Y) \quad \text{and} \quad W = g_2(X, Y). \tag{4.59}$$

Assume that the functions $v(x, y)$ and $w(x, y)$ are invertible in the sense that the equations $v = g_1(x, y)$ and $w = g_2(x, y)$ can be solved for $x$ and $y$, that is,

$$x = h_1(v, w) \qquad \text{and} \qquad y = h_2(v, w).$$

The joint pdf of $V$ and $W$ is found by finding the equivalent event of infinitesimal rectangles. The image of the infinitesimal rectangle is shown in Fig. 4.17(a). The image can be approximated by the parallelogram shown in Fig. 4.17(b) by making the approximation

$$g_k(x + dx, y) \simeq g_k(x, y) + \frac{\partial}{\partial x} g_k(x, y)\, dx \qquad k = 1, 2$$

and similarly for the $y$ variable. The probabilities of the infinitesimal rectangle and the parallelogram are approximately equal, therefore

$$f_{X,Y}(x, y)\, dx\, dy = f_{V,W}(v, w)\, dP$$

and

$$f_{V,W}(v, w) = \frac{f_{X,Y}(h_1(v, w), h_2(v, w))}{\left|\dfrac{dP}{dx\, dy}\right|}, \tag{4.60}$$

where $dP$ is the area of the parallelogram. By analogy with the case of a linear transformation, see Eq. (4.57) above, we can match the derivatives in the above approximations with the coefficients in the linear transforma-

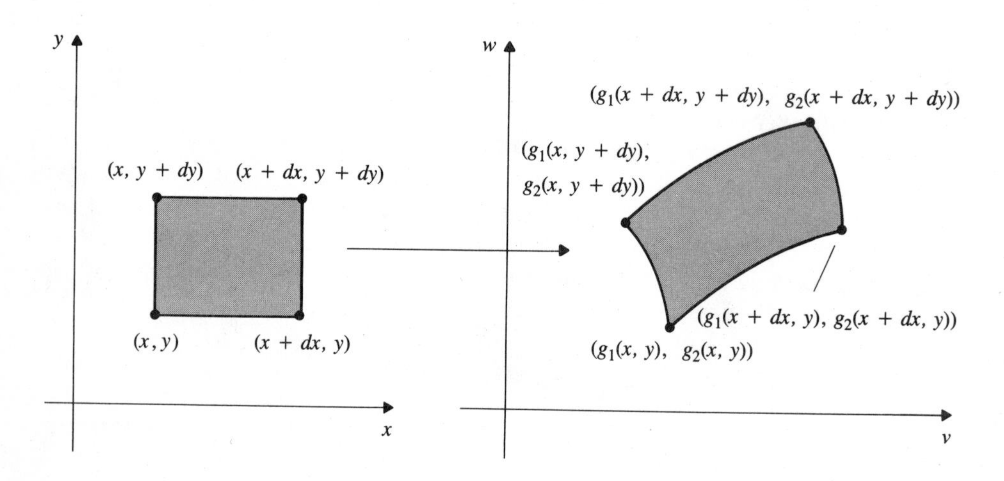

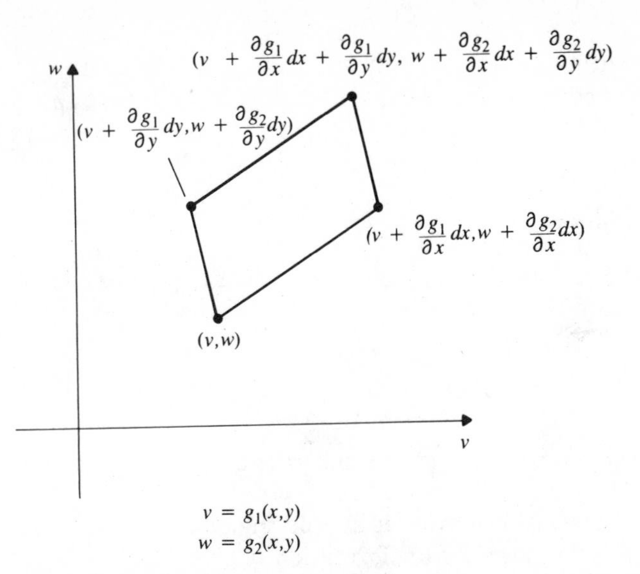


**FIGURE 4.17(b)** Approximation of image by a parallelogram.

tions and conclude that the "stretch factor" at the point $(v, w)$ is given by the determinant of a matrix of partial derivatives:

$$J(x, y) = \det \begin{bmatrix} \dfrac{\partial v}{\partial x} & \dfrac{\partial v}{\partial y} \\ \dfrac{\partial w}{\partial x} & \dfrac{\partial w}{\partial y} \end{bmatrix}.$$

The determinant $J(x, y)$ is called the **Jacobian** of the transformation. The Jacobian of the inverse transformation is given by

$$J(v, w) = \det \begin{bmatrix} \dfrac{\partial x}{\partial v} & \dfrac{\partial x}{\partial w} \\ \dfrac{\partial y}{\partial v} & \dfrac{\partial y}{\partial w} \end{bmatrix}.$$

It can be shown that

$$|J(v, w)| = \frac{1}{|J(x, y)|}.$$

We therefore conclude that the joint pdf of $V$ and $W$ can be found using either of the following expressions:

$$f_{V,W}(v, w) = \frac{f_{X,Y}(h_1(v, w), h_2(v, w))}{|J(x, y)|} \tag{4.61a}$$

$$= f_{X,Y}(h_1(v, w), h_2(v, w))\,|J(v, w)|. \tag{4.61b}$$

It should be noted that Eq. (4.60) is applicable even if Eq. (4.59) has more than one solution; the pdf is then equal to the sum of terms of the form given by Eqs. (4.61a) and (4.61b), with each solution providing one such term.

### ■■ Example 4.36

Let $X$ and $Y$ be zero-mean, unit-variance independent Gaussian random variables. Find the joint pdf of $V$ and $W$ defined by

$$V = (X^2 + Y^2)^{1/2}$$
$$W = \angle(X, Y),$$

where $\angle\theta$ denotes the angle in the range $(0, 2\pi)$ that is defined by the point $(x, y)$.

The above transformation is simply the change from Cartesian to polar coordinates. The inverse transformation is given by

$$x = v \cos w \qquad \text{and} \qquad y = v \sin w.$$

The Jacobian is given by

$$J(v, w) = \begin{vmatrix} \cos w & -v \sin w \\ \sin w & v \cos w \end{vmatrix} = v.$$

Thus

$$f_{V,W}(v, w) = \frac{v}{2\pi} e^{-[v^2\cos^2(w)+v^2\sin^2(w)]/2}$$
$$= \frac{1}{2\pi} v e^{-v^2/2} \qquad v \geq 0, 0 \leq w < 2\pi.$$

The pdf of a **Rayleigh random variable** is given by

$$f_V(v) = ve^{-v^2/2} \qquad v \geq 0.$$

We therefore conclude that the radius $V$ and the angle $W$ are independent random variables, with the radius $V$ having a Rayleigh pdf while the angle $W$ is uniformly distributed in the interval $(0, 2\pi)$. ■■

The method developed above can be used even if we are interested in only one function of a random variable. By defining an "auxiliary" variable, we can use the transformation method to find the joint pdf of both random variables, and then we can find the marginal pdf involving the random variable of interest. The following example demonstrates the method.

### Example 4.37

Let $X$ be a zero-mean, unit-variance Gaussian random variable and let $Y$ be a chi-square random variable with $n$ degrees of freedom. Assume that $X$ and $Y$ are independent. Find the pdf of $V = X/\sqrt{Y/n}$.

Define the auxiliary function $W = Y$. The variables $X$ and $Y$ are then related to $V$ and $W$ by

$$X = V\sqrt{W/n} \quad \text{and} \quad Y = W.$$

The Jacobian of the inverse transformation is

$$|J(v, w)| = \begin{vmatrix} \sqrt{w/n} & (v/2)/\sqrt{wn} \\ 0 & 1 \end{vmatrix} = \sqrt{w/n}.$$

Since $f_{X,Y}(x, y) = f_X(x)f_Y(y)$, the joint pdf of $V$ and $W$ is thus

$$\begin{aligned} f_{V,W}(v, w) &= \frac{e^{-x^2/2}}{\sqrt{2\pi}} \frac{(y/2)^{n/2-1}e^{-y/2}}{2\Gamma(n/2)} |J(v, w)| \Big|_{\substack{x=v\sqrt{w/n} \\ y=w}} \\ &= \frac{(w/2)^{(n-1)/2}e^{-[(w/2)(1+v^2/n)]}}{2\sqrt{n\pi}\,\Gamma(n/2)}. \end{aligned}$$

The pdf of $V$ is found by integrating the joint pdf over $w$:

$$f_V(v) = \frac{1}{2\sqrt{n\pi}\,\Gamma(n/2)} \int_0^\infty (w/2)^{(n-1)/2}e^{-[(w/2)(1+v^2/n)]}\,dw.$$

If we let $w' = (w/2)(v^2/n + 1)$, the integral becomes

$$f_V(v) = \frac{(1 + v^2/n)^{-(n+1)/2}}{\sqrt{n\pi}\,\Gamma(n/2)} \int_0^\infty (w')^{(n-1)/2}e^{-w'}\,dw'.$$

By noting that the above integral is the gamma function evaluated at $(n + 1)/2$, we finally obtain the **Student's *t*-distribution:**

$$f_V(v) = \frac{(1 + v^2/n)^{-(n+1)/2}\Gamma((n + 1)/2)}{\sqrt{n\pi}\,\Gamma(n/2)}$$

This pdf is used extensively in statistical calculations.

---

Finally, consider the problem of finding the joint pdf for $n$ functions of $n$ random variables $\mathbf{X} = (X_1, \ldots, X_n)$:

$$Z_1 = g_1(\mathbf{X}), \quad Z_2 = g_2(\mathbf{X}), \ldots, \quad Z_n = g_n(\mathbf{X}).$$

We assume as before that the set of equations

$$z_1 = g_1(\mathbf{x}), \quad z_2 = g_2(\mathbf{x}), \ldots, \quad z_n = g_n(\mathbf{x}) \tag{4.62}$$

has a unique solution given by

$$x_1 = h_1(\mathbf{z}), \quad x_2 = h_2(\mathbf{z}), \ldots, \quad x_n = h_n(\mathbf{z}).$$

The joint pdf of $\mathbf{Z}$ is then given by

$$f_{Z_1,\ldots,Z_n}(z_1, \ldots, z_n) = \frac{f_{X_1,\ldots,X_n}(h_1(\mathbf{z}), h_2(\mathbf{z}), \ldots, h_n(\mathbf{z}))}{|J(x_1, x_2, \ldots, x_n)|} \tag{4.63a}$$

$$= f_{X_1,\ldots,X_n}(h_1(\mathbf{z}), h_2(\mathbf{z}), \ldots, h_n(\mathbf{z}))\, |J(z_1, z_2, \ldots, z_n)|, \tag{4.63b}$$

where $|J(x_1, \ldots, x_n)|$ and $|J(z_1, \ldots, z_n)|$ are the determinants of the transformation and the inverse transformation, respectively,

$$J(x_1, \ldots, x_n) = \det \begin{bmatrix} \dfrac{\partial g_1}{\partial x_1} & \cdots & \dfrac{\partial g_1}{\partial x_n} \\ \vdots & & \vdots \\ \dfrac{\partial g_n}{\partial x_1} & \cdots & \dfrac{\partial g_n}{\partial x_n} \end{bmatrix}.$$

and

$$J(z_1, \ldots, z_n) = \det \begin{bmatrix} \dfrac{\partial h_1}{\partial z_1} & \cdots & \dfrac{\partial h_1}{\partial z_n} \\ \vdots & & \vdots \\ \dfrac{\partial h_n}{\partial z_1} & \cdots & \dfrac{\partial h_n}{\partial z_n} \end{bmatrix}.$$

## 4.7 EXPECTED VALUE OF FUNCTIONS OF RANDOM VARIABLES

The problem of finding the expected value of a function of two or more random variables is similar to that of finding the expected value of a function of a single random variable. It can be shown that the expected value of $Z = g(X, Y)$ can be found using the following expressions:

$$E[Z] = \begin{cases} \displaystyle\int_{-\infty}^{\infty}\int_{-\infty}^{\infty} g(x, y) f_{X,Y}(x, y)\, dx\, dy & X, Y \text{ jointly continuous} \\ \displaystyle\sum_i \sum_n g(x_i, y_n) p(x_i, y_n) & X, Y \text{ discrete.} \end{cases} \tag{4.64}$$

Equation (4.64) generalizes in the obvious way for a function of $n$ random variables.

■■ **Example 4.38**
Sum of Random Variables

Let $Z = X + Y$. Find $E[Z]$.

$$\begin{aligned} E[Z] &= E[X+Y] \\ &= \int_{-\infty}^{\infty}\int_{-\infty}^{\infty} (x' + y') f_{X,Y}(x', y')\, dx'\, dy' \\ &= \int_{-\infty}^{\infty}\int_{-\infty}^{\infty} x' f_{X,Y}(x', y')\, dy'\, dx' + \int_{-\infty}^{\infty}\int_{-\infty}^{\infty} y'\, f_{X,Y}(x', y')\, dx'\, dy' \\ &= \int_{-\infty}^{\infty} x' f_X(x')\, dx' + \int_{-\infty}^{\infty} y' f_Y(y')\, dy' \\ &= E[X] + E[Y]. \end{aligned} \tag{4.65}$$

Thus the expected value of the sum of two random variables is equal to the sum of the individual expected values. Note that $X$ and $Y$ need not be independent. ■■

The result in Example 4.38 and a simple induction argument show that *the expected value of a sum of n random variables is equal to the sum of the expected values*:

$$E[X_1 + X_2 + \cdots + X_n] = E[X_1] + \cdots + E[X_n]. \tag{4.66}$$

Note the random variables do not have to be independent.

■■ **Example 4.39**
Product of Functions of Independent Random Variables

Suppose that $X$ and $Y$ are independent random variables, and let $g(X, Y) = g_1(X)g_2(Y)$. Find $E[g(X, Y)] = E[g_1(X)g_2(Y)]$.

$$\begin{aligned} E[g_1(X)g_2(Y)] &= \int_{-\infty}^{\infty}\int_{-\infty}^{\infty} g_1(x')g_2(y') f_X(x') f_Y(y')\, dx'\, dy' \\ &= \left\{\int_{-\infty}^{\infty} g_1(x') f_X(x')\, dx'\right\}\left\{\int_{-\infty}^{\infty} g_2(y') f_Y(y')\, dy'\right\} \\ &= E[g_1(X)]E[g_2(Y)]. \end{aligned}$$

■■

In general if $X_1, \ldots, X_n$ are *independent* random variables, then

$$E[g_1(X_1)g_2(X_2)\ldots g_n(X_n)] = E[g_1(X_1)]E[g_2(X_2)]\ldots E[g_n(X_n)]. \tag{4.67}$$

### The Correlation and Covariance of Two Random Variables

The joint moments of two random variables $X$ and $Y$ summarize information about their joint behavior. The $jk$th **joint moment of $X$ and $Y$** is defined by

$$E[X^jY^k] = \begin{cases} \int_{-\infty}^{\infty}\int_{-\infty}^{\infty} x^j y^k f_{X,Y}(x, y)\, dx\, dy & X, Y \text{ jointly continuous} \\ \sum_i \sum_n x_i^j y_n^k p(x_i, y_n) & X, Y \text{ discrete.} \end{cases} \tag{4.68}$$

If $j = 0$, we obtain the moments of $Y$, and if $k = 0$, we obtain the moments of $X$. In electrical engineering, it is customary to call the $j = 1$ $k = 1$ moment, $E[XY]$, the **correlation of $X$ and $Y$.** If $E[XY] = 0$, then we say that **$X$ and $Y$ are orthogonal.**

The $jk$th **central moment of $X$ and $Y$** is defined as the joint moment of the centered random variables, $X - E[X]$ and $Y - E[Y]$:

$$E[(X - E[X])^j(Y - E[Y])^k].$$

Note that $j = 2$ $k = 0$ gives VAR($X$) and $j = 0$ $k = 2$ gives VAR($Y$). The **covariance of $X$ and $Y$** is defined as the $j = k = 1$ central moment:

$$\text{COV}(X, Y) = E[(X - E[X])(Y - E[Y])]. \tag{4.69}$$

The following form for COV($X$, $Y$) is sometimes more convenient to work with:

$$\begin{aligned} \text{COV}(X, Y) &= E[XY - XE[Y] - YE[X] + E[X]E[Y]] \\ &= E[XY] - 2E[X]E[Y] + E[X]E[Y] \\ &= E[XY] - E[X]E[Y]. \end{aligned} \tag{4.70}$$

Note that COV($X$, $Y$) = $E[XY]$ if either of the random variables has mean zero.

### ■■ Example 4.40
Covariance of Independent Random Variables

Let $X$ and $Y$ be independent random variables. Find their covariance.

$$\begin{aligned} \text{COV}(X, Y) &= E[(X - E[X])(Y - E[Y])] \\ &= E[X - E[X]]E[Y - E[Y]] \\ &= 0, \end{aligned}$$

where the second equality follows from the fact that $X$ and $Y$ are independent, and the third equality follows from $E[X - E[X]] = E[X] - E[X] = 0$. Therefore *pairs of independent random variables have covariance zero.* ■■

---

The **correlation coefficient of $X$ and $Y$** is defined by

$$\rho = \frac{\text{COV}(X, Y)}{\sigma_X \sigma_Y} = \frac{E[XY] - E[X]E[Y]}{\sigma_X \sigma_Y}, \tag{4.71}$$

where $\sigma_X = \sqrt{\text{VAR}(X)}$ and $\sigma_Y = \sqrt{\text{VAR}(Y)}$ are the standard deviations of $X$ and $Y$, respectively.

The correlation coefficient is a number that is at most 1 in magnitude:

$$-1 \le \rho \le 1. \tag{4.72}$$

The extreme values of $\rho$ are achieved when $X$ and $Y$ are related linearly, $Y = aX + b$; $\rho = 1$ if $a > 0$ and $\rho = -1$ if $a < 0$. Later in the section we prove Eq. (4.72) and show that $\rho$ can be viewed as a statistical measure of the extent to which $Y$ can be predicted by a linear function of $X$.

$X$ and $Y$ are said to be *uncorrelated* if $\rho = 0$. If $X$ and $Y$ are independent, then $\text{COV}(X, Y) = 0$, so $\rho = 0$. Thus *if $X$ and $Y$ are independent, then $X$ and $Y$ are uncorrelated.* In Example 4.18, we saw that *if $X$ and $Y$ are jointly Gaussian and $\rho = 0$, then $X$ and $Y$ are independent random variables.* The Example 4.41 shows that this is not always true for non-Gaussian random variables: It is possible for $X$ and $Y$ to be uncorrelated but not independent.

### ■■ Example 4.41
Uncorrelated but Dependent Random Variables

Let $\Theta$ be uniformly distributed in the interval $(0, 2\pi)$. Let

$$X = \cos\Theta \qquad \text{and} \qquad Y = \sin\Theta.$$

The point $(X, Y)$ then corresponds to the point on the unit circle specified by the angle $\Theta$ as shown in Fig. 4.18. In Example 3.28, we saw that the

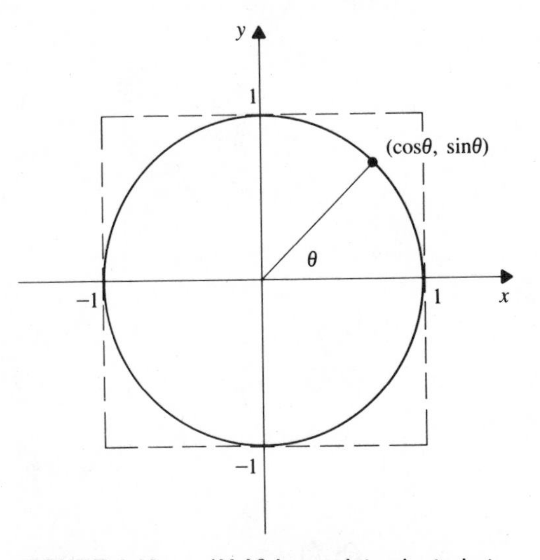


**FIGURE 4.18** $(X, Y)$ is a point selected at random *on* the unit circle. $X$ and $Y$ are uncorrelated but not independent.

marginal pdf's of $X$ and $Y$ are arcsine pdf's, which are nonzero in the interval $(-1, 1)$. The product of the marginals is nonzero in the square defined by $-1 \leq x \leq 1$ and $-1 \leq y \leq 1$, so if $X$ and $Y$ were independent the point $(X, Y)$ would assume all values in this square. This is not the case, so $X$ and $Y$ are dependent.

We now show that $X$ and $Y$ are uncorrelated:

$$E[XY] = E[\sin \Theta \cos \Theta] = \frac{1}{2\pi} \int_0^{2\pi} \sin \phi \cos \phi \, d\phi$$

$$= \frac{1}{4\pi} \int_0^{2\pi} \sin 2\phi \, d\phi = 0.$$

Since $E[X] = E[Y] = 0$, Eq. (4.70) then implies that $X$ and $Y$ are uncorrelated. ■■

### Mean Square Estimation

The correlation coefficient arises in the problem of predicting a random variable $Y$ by a linear function of one or more random variables. Consider first the problem of estimating a random variable $Y$ by a constant $a$ so that the mean square error is minimized:

$$\min_a E[(Y - a)^2] = E[Y^2] - 2aE[Y] + a^2. \tag{4.73}$$

The best $a$ is found by taking the derivative with respect to $a$, setting the result to zero, and solving for $a$. The result is

$$a^* = E[Y], \tag{4.74}$$

and the minimum mean square error is equal to $E[(Y - a^*)^2] = \text{VAR}(Y)$.

Now consider estimating $Y$ by $aX + b$:

$$\min_{a,b} E[(Y - aX - b)^2]. \tag{4.75}$$

Equation (4.75) can be viewed as the approximation of $Y - aX$ by the constant $b$. This is the minimization posed in Eq. (4.73) and the best $b$ is

$$b^* = E[Y - aX] = E[Y] - aE[X].$$

Substitution into Eq. (4.75) implies that the best $a$ is found by

$$\min_a E[\{(Y - E[Y]) - a(X - E[X])\}^2].$$

Note that

$$E[\{(Y - E[Y]) - a(X - E[X])\}^2]$$
$$= \text{VAR}(Y) - 2a\,\text{COV}(X, Y) + a^2\,\text{VAR}(X).$$

We once again differentiate with respect to $a$, set the result to zero, and solve for $a$:

$$0 = -2\,\text{COV}(X, Y) + 2a\,\text{VAR}(X). \tag{4.76}$$

The best coefficient $a$ is found to be

$$a^* = \frac{\text{COV}(X, Y)}{\text{VAR}(X)} = \rho\frac{\sigma_Y}{\sigma_X},$$

where $\sigma_Y = \sqrt{\text{VAR}(Y)}$ and $\sigma_Y = \sqrt{\text{VAR}(Y)}$. Therefore, *the minimum mean square error linear estimator for $Y$ in terms of $X$* is

$$\begin{aligned}\hat{Y} &= a^*X + b^*\\ &= \rho\sigma_Y\left(\frac{X - E[X]}{\sigma_X}\right) + E[Y].\end{aligned} \tag{4.77}$$

The term $(X - E[X])/\sigma_X$ is simply a zero-mean, unit-variance version of $X$. Thus $\sigma_Y(X - E[X])/\sigma_X$ is a rescaled version of $X$ that has the variance of the random variable that is being estimated, namely $\sigma_Y^2$. The term $E[Y]$ simply ensures that the estimator has the correct mean. The key term in the above estimator is the correlation coefficient: *$\rho$ specifies the sign and extent of the estimate of $Y$ relative to $\sigma_Y(X - E[X])/\sigma_X$.* If $X$ and $Y$ are uncorrelated (i.e., $\rho = 0$) then the best estimate for $Y$ is its mean, $E[Y]$. On the other hand, if $\rho = \pm 1$ then the best estimate is equal to $\pm\sigma_Y(X - E[X])/\sigma_X + E[Y]$.

To find the mean square error of the best linear estimator first note that Eq. (4.76) is equivalent to

$$E[\{(Y - E[Y]) - a^*(X - E[X])\}(X - E[X])] = 0. \tag{4.78}$$

This equation is called the **orthogonality condition** because it states that the error of the best linear estimator, the quantity inside the braces, is orthogonal to the observation $X - E[X]$. Thus the mean square error of the best estimator is

$$\begin{aligned}&E[((Y - E[Y]) - a^*(X - E[X]))^2]\\ &\quad = E[((Y - E[Y]) - a^*(X - E[X]))(Y - E[Y])]\\ &\quad\quad - a^*E[((Y - E[Y]) - a^*(X - E[X]))(X - E[X])]\\ &\quad = E[((Y - E[Y]) - a^*(X - E[X]))(Y - E[Y])]\\ &\quad = \text{VAR}(Y) - a^*\,\text{COV}(X, Y)\\ &\quad = \text{VAR}(Y)(1 - \rho^2),\end{aligned} \tag{4.79}$$

where the second equality followed from the orthogonality condition. Note that when $|\rho| = 1$, the mean square error is zero. This implies that $P[|Y - a^*X - b^*| = 0] = P[Y = a^*X + b^*] = 1$ by Eq. (3.74).

Equation (4.79) can be used to prove Eq. (4.72), which stated that $|\rho| \leq 1$. Since the mean square error is a nonnegative number,

$$\text{VAR}(Y)(1 - \rho^2) \geq 0.$$

Since the variance is nonnegative,

$$1 - \rho^2 \geq 0.$$

This implies that $|\rho| \leq 1$.

So far we have only considered estimating $Y$ by a linear function of $X$. In general the estimator for $Y$ that minimizes the mean square error is a *nonlinear* function of $X$. The estimator $g(X)$ that best approximates $Y$ in the sense of minimizing mean square error must satisfy

$$\underset{g(.)}{\text{minimize}}\, E[(Y - g(X))^2]. \tag{4.80}$$

The problem can be solved by using conditional expectation:

$$\begin{aligned} E[(Y - g(X))^2] &= E[E[(Y - g(X))^2 \mid X]] \\ &= \int_{-\infty}^{\infty} E[(Y - g(X))^2 \mid X = x] f_X(x)\, dx. \end{aligned}$$

The integrand above is positive for all $x$; therefore, the integral is minimized by minimizing $E[(Y - g(x))^2 \mid X = x]$ for each $x$. But $g(x)$ is a constant as far as the conditional expectation is concerned, so the problem is equivalent to Eq. (4.73) and the "constant" that minimizes $E[(Y - g(x))^2 \mid X = x]$ is

$$g^*(x) = E[Y \mid X = x]. \tag{4.81}$$

The function $g^*(x) = E[Y \mid X = x]$ is called the **regression curve**. Thus $E[Y \mid X]$ is the estimator for $Y$ in terms of $X$ that yields the smallest possible mean square error. Linear estimators will generally have larger mean square errors.

### ■■ Example 4.42

Let $X$ be uniformly distributed in the interval $(-1, 1)$ and let $Y = X^2$. Find the best linear estimator for $Y$ in terms of $X$. Compare its performance to the best estimator.

The mean of $X$ is zero, and its correlation with $Y$ is

$$E[XY] = E[XX^2] = \int_{-1}^{1} x^3\, dx = 0.$$

Therefore $\text{COV}(X, Y) = 0$ and the best linear estimator for $Y$ is $E[Y]$ by Eq. (4.77). The mean square error of this estimator is the $\text{VAR}(Y)$ by Eq. (4.79).

The best estimator is given by Eq. (4.81)

$$E[Y \mid X = x] = E[X^2 \mid X = x] = x^2.$$

The mean square error of this estimator is

$$E[(Y - g(X))^2] = E[(X^2 - X^2)^2] = 0.$$

Thus in this problem, the best linear estimator performs poorly while the nonlinear estimator gives the smallest possible mean square error, zero. ■■

### *Linear Prediction

Linear prediction methods are used extensively in the processing of random signals. In the remainder of the section we develop the equations that determine the best linear estimator of $Y$ in terms of several random variables.

Let $X_1$, $X_2$, and $X_3$ be zero-mean random variables, and suppose that we wish to approximate $X_3$ by $aX_1 + bX_2$ so as to

$$\underset{a,b}{\text{minimize}}\, E[(X_3 - aX_1 - bX_2)^2].$$

If we take the derivative of the mean square error with respect to $a$ and set the result to zero, we obtain

$$E[-2(X_3 - aX_1 - bX_2)X_1] = 0,$$

or equivalently

$$E[(X_3 - aX_1 - bX_2)X_1] = 0. \tag{4.82a}$$

Similarly, differentiation with respect to $b$ yields

$$E[(X_3 - aX_1 - bX_2)X_2] = 0. \tag{4.82b}$$

Equations (4.82a) and (4.82b) state that the error $X_3 - aX_1 - bX_2$ is orthogonal to each of the observations $X_1$ and $X_2$.

Equations (4.82a) and (4.82b) lead to two equations in two unknowns:

$$\begin{bmatrix} E[X_1^2] & E[X_1X_2] \\ E[X_1X_2] & E[X_2^2] \end{bmatrix} \begin{bmatrix} a \\ b \end{bmatrix} = \begin{bmatrix} E[X_1X_3] \\ E[X_2X_3] \end{bmatrix}. \tag{4.83}$$

The solution to this set of equations is:

$$a = \frac{\text{VAR}(X_2)\,\text{COV}(X_1, X_3) - \text{COV}(X_1, X_2)\,\text{COV}(X_2, X_3)}{\text{VAR}(X_1)\,\text{VAR}(X_2) - \text{COV}(X_1, X_2)^2} \tag{4.84a}$$

$$b = \frac{\text{VAR}(X_1)\,\text{COV}(X_2, X_3) - \text{COV}(X_1, X_2)\,\text{COV}(X_1, X_3)}{\text{VAR}(X_1)\,\text{VAR}(X_2) - \text{COV}(X_1, X_2)^2}. \tag{4.84b}$$

The problem of finding a linear estimator for $X_{n+1}$ in terms of

$X_1, \ldots, X_n$ is solved in the same way. The orthogonality condition provides us with a set of $n$ linear equations that can then be solved for the best set of coefficients. This problem is discussed in detail in Section 7.4.

### ■■ Example 4.43
Second-Order Prediction of Speech

Let $X_1, X_2, \ldots$ be a sequence of samples of a speech voltage waveform, and suppose that the samples are fed into the second-order predictor shown in Fig. 4.19. Find the set of predictors $a$ and $b$ that minimize the mean square value of the predictor error.

We find the best predictor for $X_1$, $X_2$, and $X_3$ and assume that the situation is identical for $X_2$, $X_3$, and $X_4$, and so on. It is common practice to model speech samples as having zero-mean and variance $\sigma^2$, and a covariance that does not depend on the specific index of the samples, but rather on the distance between them:

$$\text{COV}(X_j, X_k) = \rho_{|j-k|}\sigma^2.$$

Equation (4.84) then becomes

$$a = \frac{\rho_2 - \rho_1^2}{1 - \rho_1^2}$$

and

$$b = \frac{\rho_1(1 - \rho_2)}{1 - \rho_1^2}.$$

In Problem 64, you are asked to show that the mean square error using

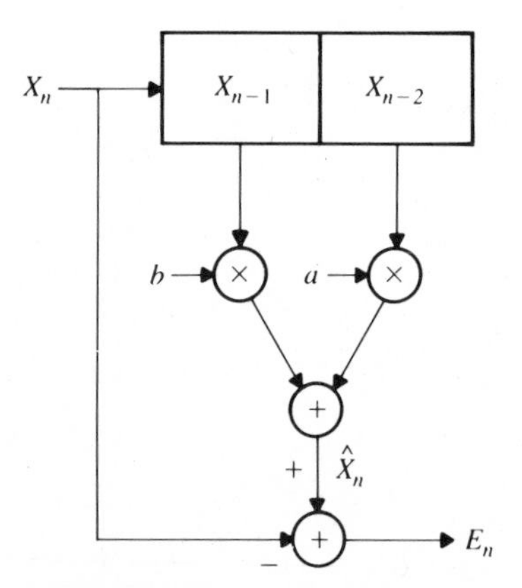


**FIGURE 4.19** A two-tap linear predictor for processing speech.

the above values of $a$ and $b$ is

$$\sigma^2\left\{1 - \rho_1^2 - \frac{(\rho_1^2 - \rho_2)^2}{1 - \rho_1^2}\right\}. \tag{4.85}$$

Jayant and Noll (1984, 263) give the values $\rho_1 = .825$ and $\rho_2 = .562$ for speech signals. The mean square value of the predictor output is thus $.281\ \sigma^2$, which is 5.5 dB lower than variance of the input sequence. It can be shown that the performance of quantizers improves as the variance of its input decreases. For this reason, many speech encoding devices pass the sequence of speech samples through a linear predictor prior to quantization. See reference [4] for more on speech coding. ■■

## 4.8

## JOINTLY GAUSSIAN RANDOM VARIABLES

The random variables $X$ and $Y$ are said to be **jointly Gaussian** if their joint pdf has the form

$$f_{X,Y}(x, y) = \frac{\exp\left\{\frac{-1}{2(1-\rho^2)}\left[\left(\frac{x-m_1}{\sigma_1}\right)^2 - 2\rho\left(\frac{x-m_1}{\sigma_1}\right)\left(\frac{y-m_2}{\sigma_2}\right) + \left(\frac{y-m_2}{\sigma_2}\right)^2\right]\right\}}{2\pi\sigma_1\sigma_2\sqrt{1-\rho^2}} \tag{4.86}$$

for $-\infty < x < \infty$ and $-\infty < y < \infty$.

The pdf is centered at the point $(m_1, m_2)$ and it has a bell shape that depends on the values of $\sigma_1$, $\sigma_2$, and $\rho$ as shown previously in Fig. 4.11. The pdf is constant for values $x$ and $y$ for which the argument of the exponent is constant:

$$\left[\left(\frac{x-m_1}{\sigma_1}\right)^2 - 2\rho\left(\frac{x-m_1}{\sigma_1}\right)\left(\frac{y-m_2}{\sigma_2}\right) + \left(\frac{y-m_2}{\sigma_2}\right)^2\right] = \text{constant}.$$

Figure 4.20 shows contours of constant pdf for various values of $\sigma_1$, $\sigma_2$, and $\rho$. It can be seen that the contours are ellipses that are oriented along the angle

$$\theta = \frac{1}{2}\arctan\left(\frac{2\rho\sigma_1\sigma_2}{\sigma_1^2 - \sigma_2^2}\right). \tag{4.87}$$

Note that the angle is 45° when the variances are equal.

The marginal pdf of $X$ is found by integrating $f_{X,Y}(x, y)$ over all $y$. The integration is carried out by completing the square in the exponent as was

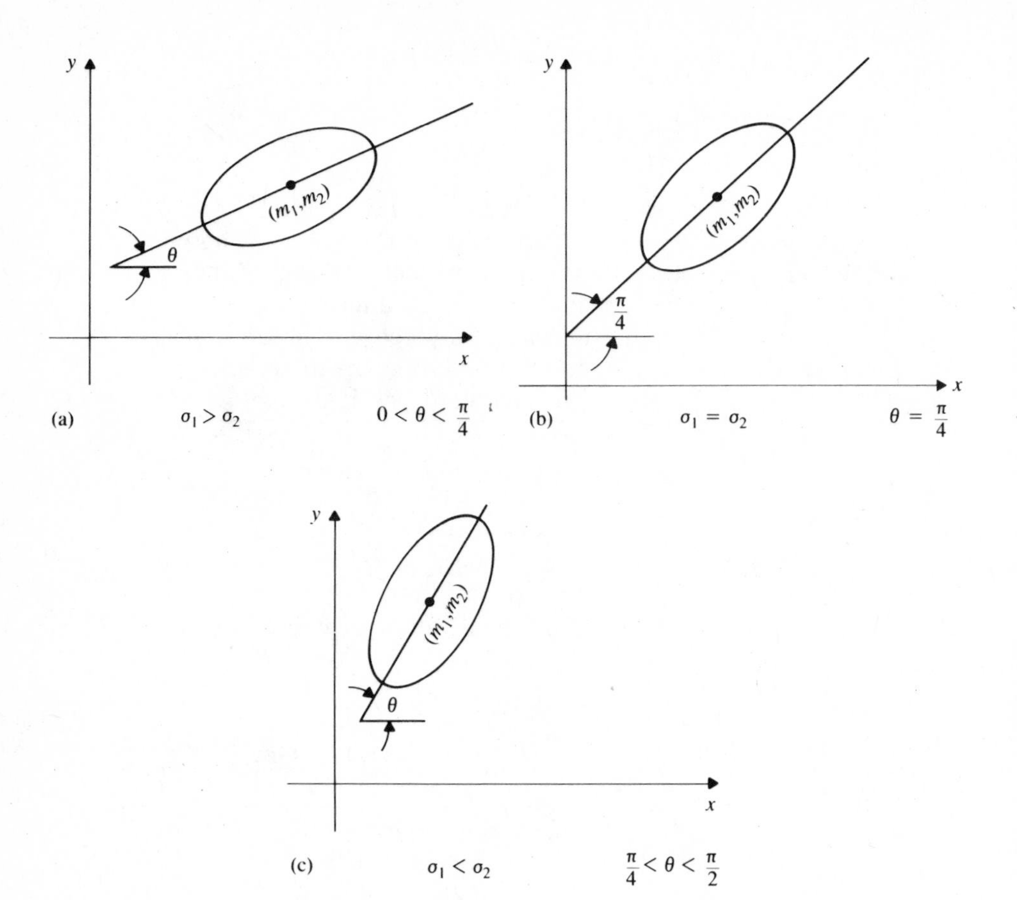


**FIGURE 4.20** Orientation of contours of equal value of joint Gaussian pdf.

done in Example 4.13. The result is that marginal pdf of $X$ is

$$f_X(x) = \frac{e^{-(x-m_1)^2/2\sigma_1^2}}{\sqrt{2\pi}\,\sigma_1}, \tag{4.88}$$

that is, $X$ is a Gaussian random variable with mean $m_1$ and variance $\sigma_1^2$. Similarly, the marginal pdf for $Y$ is found to be Gaussian with pdf mean $m_2$ and variance $\sigma_2^2$.

The conditional pdf's $f_X(x \mid y)$ and $f_Y(y \mid x)$ give us information about the interrelation between $X$ and $Y$. The conditional pdf of $X$ given $Y = y$ is

$$\begin{aligned} f_X(x \mid y) &= \frac{f_{X,Y}(x, y)}{f_Y(y)} \\ &= \frac{\exp\left\{\dfrac{-1}{2(1-\rho^2)\sigma_1^2}\left[x - \rho\dfrac{\sigma_1}{\sigma_2}(y - m_2) - m_1\right]^2\right\}}{\sqrt{2\pi\sigma_1^2(1-\rho^2)}}. \end{aligned} \tag{4.89}$$

Equation (4.89) shows that the conditional pdf of $X$ given $Y = y$ is also Gaussian but with conditional mean $m_1 + \rho(\sigma_1/\sigma_2)(y - m_2)$ and conditional variance $\sigma_1^2(1 - \rho^2)$. Note that when $\rho = 0$, the conditional pdf of $X$ given $Y = y$ equals the marginal pdf of $X$. This is consistent with the fact that $X$ and $Y$ are independent when $\rho = 0$. On the other hand, as $|\rho| \to 1$ the variance of $X$ about the conditional mean approaches zero, so the conditional pdf approaches a delta function at the conditional mean. Thus when $|\rho| = 1$, the conditional variance is zero and $X$ is equal to the conditional mean with probability one.

We now show that the $\rho$ in Eq. (4.86) is indeed the correlation coefficient between $X$ and $Y$. The covariance between $X$ and $Y$ is defined by

$$\begin{aligned} \text{COV}(X, Y) &= E[(X - m_1)(Y - m_2)] \\ &= E[E[(X - m_1)(Y - m_2) \mid Y]]. \end{aligned}$$

Now the conditional expectation of $(X - m_1)(Y - m_2)$ given $Y = y$ is

$$\begin{aligned} E[(X - m_1)(Y - m_2) \mid Y = y] &= (y - m_2)E[X - m_1 \mid Y = y] \\ &= (y - m_2)(E[X \mid Y = y] - m_1) \\ &= (y - m_2)\left(\rho \frac{\sigma_1}{\sigma_2}(y - m_2)\right), \end{aligned}$$

where we have used the fact that the conditional mean of $X$ given $Y = y$ is $m_1 + \rho(\sigma_1/\sigma_2)(y - m_2)$. Therefore

$$E[(X - m_1)(Y - m_2) \mid Y] = \rho \frac{\sigma_1}{\sigma_2}(Y - m_2)^2$$

and

$$\begin{aligned} \text{COV}(X, Y) &= E[E[(X - m_1)(Y - m_2) \mid Y]] = \rho \frac{\sigma_1}{\sigma_2} E[(Y - m_2)^2] \\ &= \rho\sigma_1\sigma_2. \end{aligned}$$

The above equation is consistent with the definition of the correlation coefficient, $\rho = \text{COV}(X, Y)/\sigma_1\sigma_2$. Thus the $\rho$ in Eq. (4.86) is indeed the correlation coefficient between $X$ and $Y$.

### ■■ Example 4.44

The amount of yearly rainfall at city 1 and at city 2 are modeled by a pair of jointly Gaussian random variables, $X$ and $Y$, with pdf given by Eq. (4.86). Find the most likely value of $X$ given that we know $Y = y$.

The most likely value of $X$ given $Y = y$ is the value of $x$ for which $f_X(x \mid y)$ is maximum. The conditional pdf of $X$ given $Y = y$ is given by Eq. (4.89), which is maximum at the conditional mean

$$E[X \mid y] = m_1 + \rho \frac{\sigma_1}{\sigma_2}(y - m_2).$$

Note that this "maximum likelihood" estimate is identical to the minimum mean square error estimate given by Eq. (4.77) in Section 4.7. ■■

### *n* Jointly Gaussian Random Variables

The random variables $X_1, X_2, \ldots, X_n$ are said to be jointly Gaussian if their joint pdf is given by

$$f_{X_1, X_2, \ldots, X_n}(x_1, \ldots, x_n) = \frac{\exp\{-\frac{1}{2}(\mathbf{x} - \mathbf{m})^{\mathrm{T}} K^{-1}(\mathbf{x} - \mathbf{m})\}}{(2\pi)^{n/2} |K|^{1/2}}, \tag{4.90}$$

where $\mathbf{x}$ and $\mathbf{m}$ are column vectors defined by

$$\mathbf{x} = \begin{bmatrix} x_1 \\ x_2 \\ \vdots \\ x_n \end{bmatrix}, \qquad \mathbf{m} = \begin{bmatrix} m_1 \\ m_2 \\ \vdots \\ m_n \end{bmatrix} = \begin{bmatrix} E[X_1] \\ E[X_2] \\ \vdots \\ E[X_n] \end{bmatrix}$$

and $K$ is the **covariance matrix** that is defined by

$$K = \begin{bmatrix} \mathrm{VAR}(X_1) & \mathrm{COV}(X_1, X_2) & \ldots & \mathrm{COV}(X_1, X_n) \\ \mathrm{COV}(X_2, X_1) & \mathrm{VAR}(X_2) & \ldots & \mathrm{COV}(X_2, X_n) \\ \vdots & \vdots & & \vdots \\ \mathrm{COV}(X_n, X_1) & \ldots & & \mathrm{VAR}(X_n) \end{bmatrix}. \tag{4.91}$$

The $(\cdot)^{\mathrm{T}}$ in Eq. (4.90) denotes the transpose of a matrix or vector. Note that the covariance matrix is a symmetric matrix since $\mathrm{COV}(X_i, X_j) = \mathrm{COV}(X_j, X_i)$.

Equation (4.90) shows that *the pdf of jointly Gaussian random variables is completely specified by the individual means and variances, and the pairwise covariances.* It can be shown that all the marginal pdf's associated with Eq. (4.90) are also Gaussian and that these too are completely specified by the same set of means, variances, and covariances.

### ■■ Example 4.45

Verify that the two-dimensional Gaussian pdf given in Eq. (4.86) has the form of Eq. (4.90).

The covariance matrix for the two-dimensional case is given by

$$K = \begin{bmatrix} \sigma_1^2 & \rho\sigma_1\sigma_2 \\ \rho\sigma_1\sigma_2 & \sigma_2^2 \end{bmatrix},$$

where we have used the fact that $\mathrm{COV}(X_1, X_2) = \rho\sigma_1\sigma_2$. The determinant of $K$ is $\sigma_1^2\sigma_2^2(1 - \rho^2)$ so the denominator of the pdf has the correct form.

The inverse of the covariance matrix is

$$K^{-1} = \frac{1}{\sigma_1^2\sigma_2^2(1-\rho^2)}\begin{bmatrix}\sigma_2^2 & -\rho\sigma_1\sigma_2 \\ -\rho\sigma_1\sigma_2 & \sigma_1^2\end{bmatrix}.$$

The term in the exponent is therefore

$$\frac{1}{\sigma_1^2\sigma_2^2(1-\rho^2)}(x-m_1, y-m_2)\begin{bmatrix}\sigma_2^2 & -\rho\sigma_1\sigma_2 \\ -\rho\sigma_1\sigma_2 & \sigma_1^2\end{bmatrix}\begin{bmatrix}x-m_1 \\ y-m_2\end{bmatrix}$$

$$= \frac{1}{\sigma_1^2\sigma_2^2(1-\rho^2)}(x-m_1, y-m_2)\begin{bmatrix}\sigma_2^2(x_1-m_1) - \rho\sigma_1\sigma_2(y-m_2) \\ -\rho\sigma_1\sigma_2(x_1-m_1) + \sigma_1^2(y-m_2)\end{bmatrix}$$

$$= \frac{((x-m_1)/\sigma_1)^2 - 2\rho((x-m_1)/\sigma_1)((y-m_2)/\sigma_2) + ((y-m_2)/\sigma_2)^2}{\{1-\rho^2)}$$

Thus the two-dimensional pdf has the form of Eq. (4.90). ■■

### ■■ Example 4.46

The vector of random variables $(X, Y, Z)$ are jointly Gaussian with zero means and covariance matrix:

$$K = \begin{bmatrix}\text{VAR}(X) & \text{COV}(X,Y) & \text{COV}(X,Z) \\ \text{COV}(Y,X) & \text{VAR}(Y) & \text{COV}(Y,Z) \\ \text{COV}(Z,X) & \text{COV}(Z,Y) & \text{VAR}(Z)\end{bmatrix} = \begin{bmatrix}1.0 & 0.2 & 0.3 \\ 0.2 & 1.0 & 0.4 \\ 0.3 & 0.4 & 1.0\end{bmatrix}.$$

Find the marginal pdf of $X$ and $Z$.

We can solve this problem two ways. The first involves integrating the pdf directly to obtain the marginal pdf. The second involves using the fact that the marginal pdf for $X$ and $Z$ is also Gaussian and has the same set of means, variances, and covariances. We will use the second approach.

The pair $(X, Z)$ have zero mean vector and covariance matrix:

$$K' = \begin{bmatrix}\text{VAR}(X) & \text{COV}(X,Z) \\ \text{COV}(Z,X) & \text{VAR}(Z)\end{bmatrix} = \begin{bmatrix}1.0 & 0.3 \\ 0.3 & 1.0\end{bmatrix}.$$

The joint pdf of $X$ and $Z$ is found by substituting a zero mean vector and this covariance matrix into Eq. (4.90). ■■

### *Linear Transformation of Gaussian Random Variables

A very important property of jointly Gaussian random variables is that *the linear transformation of any n jointly Gaussian random variables results in n random variables that are also jointly Gaussian.* This is easy to show using the matrix notation in Eq. (4.90). Let $\mathbf{X} = (X_1, \ldots, X_n)$ be jointly Gaussian and define $\mathbf{Y} = (Y_1, \ldots, Y_n)$ by

$$\mathbf{Y} = A\mathbf{X},$$

where $A$ is an $n \times n$ invertible matrix. From Eq. (4.58), we know that the pdf of $\mathbf{Y}$ is given by

$$f_{Y_1,\ldots,Y_n}(\mathbf{y}) = \frac{f_{X_1,\ldots,X_n}(A^{-1}\mathbf{y})}{|A|}$$

$$= \frac{\exp\{-\frac{1}{2}(A^{-1}\mathbf{y} - \mathbf{m})^{\mathrm{T}}K^{-1}(A^{-1}\mathbf{y} - \mathbf{m})\}}{(2\pi)^{n/2}\,|A|\,|K|^{1/2}}.$$

From elementary properties of matrices we have that

$$(A^{-1}\mathbf{y} - \mathbf{m}) = A^{-1}(\mathbf{y} - A\mathbf{m})$$

and

$$(A^{-1}\mathbf{y} - \mathbf{m})^{\mathrm{T}} = (\mathbf{y} - A\mathbf{m})^{\mathrm{T}}A^{-1\mathrm{T}}.$$

The argument of the exponential is therefore equal to

$$(\mathbf{y} - A\mathbf{m})^{\mathrm{T}}A^{-1\mathrm{T}}K^{-1}A^{-1}(\mathbf{y} - A\mathbf{m}) = (\mathbf{y} - A\mathbf{m})^{\mathrm{T}}(AKA^{\mathrm{T}})^{-1}(\mathbf{y} - A\mathbf{m})$$

since $A^{-1\mathrm{T}}K^{-1}A^{-1} = (AKA^{\mathrm{T}})^{-1}$. Letting $C = AKA^{\mathrm{T}}$ and $\mathbf{n} = A\mathbf{m}$ and noting that $\det(C) = \det(AKA^{\mathrm{T}}) = \det(A)\det(K)\det(A^{\mathrm{T}}) = \det(A)^2\det(K)$, we finally have that the pdf of $\mathbf{Y}$ is

$$f_{Y_1,Y_2,\ldots,Y_n}(y_1,\ldots,y_n) = \frac{e^{-(1/2)(\mathbf{y}-\mathbf{n})^{\mathrm{T}}C^{-1}(\mathbf{y}-\mathbf{n})}}{(2\pi)^{n/2}\,|C|^{1/2}}.$$

Thus the pdf of $\mathbf{Y}$ has the form of Eq. (4.90) and therefore $Y_1,\ldots,Y_n$ are jointly Gaussian random variables with mean $n$ and covariance matrix $C$.

Since $K$ is a symmetric matrix, it is always possible to find a matrix $A$ such that $AKA^{\mathrm{T}} = D$, where $D$ is a diagonal matrix. (See any book on linear algebra, for example [Anton, 1981, 285–289].) For such a matrix $A$, the pdf of $\mathbf{Y}$ will be

$$f_{Y_1,Y_2,\ldots,Y_n}(y_1,\ldots,y_n) = \frac{e^{-(1/2)(\mathbf{y}-\mathbf{n})^{\mathrm{T}}D^{-1}(\mathbf{y}-\mathbf{n})}}{(2\pi)^{n/2}\,|D|^{1/2}}$$

$$= \frac{\exp\left\{-\frac{1}{2}\sum_{i=1}^{n}(y_i - n_i)^2/d_i\right\}}{[(2\pi d_1)(2\pi d_2)\ldots(2\pi d_n)]^{1/2}},$$

where $d_1 \ldots d_n$ are the diagonal components of $D$. We assume that these values are all nonzero. The above pdf implies that $Y_1,\ldots,Y_n$ are *independent* random variables with means $n_i$ and variance $d_i$. In conclusion, *it is possible to linearly transform a vector of jointly Gaussian random variables into a vector of independent Gaussian random variables.*

It is always possible to select the matrix $A$ that diagonalizes $K$ so that $\det(A) = 1$. The transformation $\mathbf{AX}$ then corresponds to a rotation of the coordinate system so that the principal axes of the ellipsoid corresponding to the pdf are aligned to the axes of the system. Example 4.47 discusses the two-dimensional case.

### ■■ Example 4.47

The ellipse corresponding to an arbitrary two-dimensional Gaussian vector forms an angle

$$\theta = \frac{1}{2}\arctan\left(\frac{2\rho\sigma_1\sigma_2}{\sigma_1^2 - \sigma_2^2}\right)$$

relative to the $x$-axis. Suppose we define a new coordinate system whose axes are aligned with those of the ellipse as shown in Fig. 4.21. This is accomplished by using the following rotation matrix:

$$\begin{bmatrix} V \\ W \end{bmatrix} = \begin{bmatrix} \cos\theta & \sin\theta \\ -\sin\theta & \cos\theta \end{bmatrix} \begin{bmatrix} X \\ Y \end{bmatrix}.$$

To show that the new random variables are independent it suffices to show that they have covariance zero:

$$\begin{aligned}
\text{COV}(V, W) &= E[(V - E[V])(W - E[W])] \\
&= E[\{(X - m_1)\cos\theta + (Y - m_2)\sin\theta\} \\
&\quad \times \{-(X - m_1)\sin\theta + (Y - m_2)\cos\theta\}] \\
&= -\sigma_1^2 \sin\theta\cos\theta + \text{COV}(X, Y)\cos^2\theta \\
&\quad - \text{COV}(X, Y)\sin^2\theta + \sigma_2^2 \sin\theta\cos\theta \\
&= \frac{(\sigma_2^2 - \sigma_1^2)\sin 2\theta + 2\,\text{COV}(X, Y)\cos 2\theta}{2} \\
&= \frac{\cos 2\theta[(\sigma_2^2 - \sigma_1^2)\tan 2\theta + 2\,\text{COV}(X, Y)]}{2}.
\end{aligned}$$

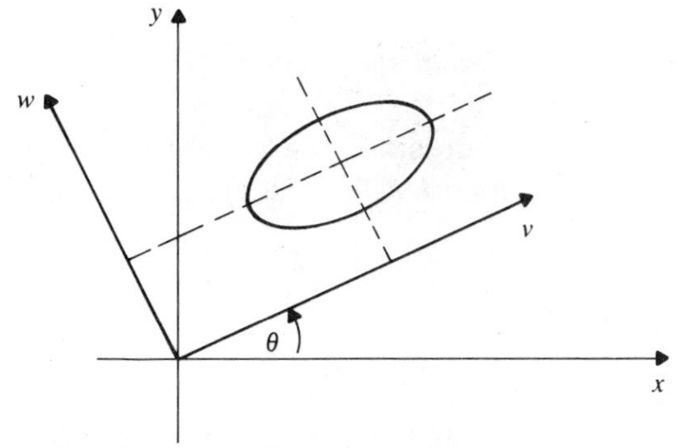

**FIGURE 4.21** A rotation of the coordinate system transforms a pair of dependent Gaussian random variables into a pair of independent Gaussian random variables.

If we let the angle of rotation $\theta$ be such that

$$\tan 2\theta = \frac{2\,\text{COV}(X, Y)}{\sigma_1^2 - \sigma_2^2},$$

then the covariance of $V$ and $W$ is zero as required. ■■

## *4.9

## GENERATING CORRELATED VECTOR RANDOM VARIABLES

Many applications involve vectors or sequences of correlated random variables. Computer simulation models of such applications therefore require methods for generating such random variables. In this section we present methods for generating vectors of random variables with specified covariance matrices. We also discuss the generation of jointly Gaussian vector random variables. In Example 7.13 we discuss the generation of sequences of correlated random variables.

### Generating Vectors of Random Variables with Specified Covariances

Let $\mathbf{X} = (X_1, \ldots, X_n)$ be a vector of $n$ zero-mean, unit-variance, uncorrelated random variables. $\mathbf{X}$ can be generated using the methods discussed in Chapter 2. Suppose that we are interested in generating a zero-mean vector $\mathbf{Y} = (Y_1, Y_2, \ldots, Y_n)$ with some specified covariance matrix $K$.[3] We will show that this can be accomplished by letting $\mathbf{Y} = A\mathbf{X}$ where $A$ is an $n \times n$ matrix.

Let $a_{kj}$ be the element in the $k$th row and $j$th column of $A$. The $k$th element of $\mathbf{Y}$ is then

$$Y_k = \sum_{j=1}^{n} a_{kj} X_j. \tag{4.92}$$

Clearly, the mean of $Y_k$ is zero since the means of the $X_j$'s are zero. The covariance between the elements of $\mathbf{Y}$ is then given by

$$\begin{aligned} E[Y_k Y_{k'}] &= E\left[\sum_{j=1}^{n} a_{kj} X_j \sum_{j'=1}^{n} a_{k'j'} X_{j'}\right] \\ &= \sum_{j=1}^{n} \sum_{j'=1}^{n} a_{kj} a_{k'j'} E[X_j X_{j'}] \end{aligned}$$

Since the $X_j$'s are uncorrelated and have zero mean and unit variance, we

3. $K$ is a valid covariance matrix if it is a nonnegative definite, symmetric matrix ([6] Stark 1986, 158), that is, a real-valued, symmetric matrix with nonnegative eigenvalues.

have that $E[X_j X_{j'}] = 1$ if $j = j'$ and equals zero otherwise. Therefore all terms in the above double summation are zero except when $j = j'$, and

$$E[Y_k Y_{k'}] = \sum_{j=1}^{n} a_{kj} a_{k'j}. \tag{4.93}$$

Equation (4.93) shows that the $k, k'$ element of the covariance matrix of $Y$ is equal to the dot product between the $k$th row of the matrix $A$ and the $k'$th column of the matrix $A^T$, the transpose of $A$. In other words, if $\mathbf{Y} = A\mathbf{X}$ where $\mathbf{X}$ consists of unit-variance, uncorrelated random variables, then the covariance matrix of $\mathbf{Y}$ is

$$K = AA^T. \tag{4.94}$$

Next we need a method for finding a matrix $A$ that satisfies Eq. (4.94) for a given $K$. This problem is solved using elementary methods from linear algebra. Since $K$ is a symmetric matrix, it can be expressed in the form

$$K = PDP^T, \tag{4.95}$$

where $D$ is a diagonal matrix that consists of the eigenvalues of $K$ and $P$ is a matrix whose columns consist of an orthonormal set of eigenvectors of $K$.[4,5] Define $D^{1/2}$ to be the diagonal matrix whose elements are the square root of the elements of $D$. Finally, let $A = PD^{1/2}$, then

$$AA^T = PD^{1/2}D^{1/2}P^T = PDP^T = K.$$

Thus $A$ yields $\mathbf{Y}$ with the desired covariance matrix.

## ■■ Example 4.48

Let $\mathbf{X} = (X_1, X_2)$ consist of two unit-variance, uncorrelated random variables. Find the matrix $A$ such that $\mathbf{Y} = A\mathbf{X}$ has covariance matrix

$$K = \sigma^2 \begin{bmatrix} 1 & \rho \\ \rho & 1 \end{bmatrix}, \tag{4.96}$$

where $|\rho| < 1$.

By computing the eigenvalues and eigenvectors of $K$ we find that

$$P = \frac{1}{\sqrt{2}} \begin{bmatrix} 1 & 1 \\ -1 & 1 \end{bmatrix} \quad \text{and} \quad D^{1/2} = \begin{bmatrix} \sigma(1-\rho)^{1/2} & 0 \\ 0 & \sigma(1-\rho)^{1/2} \end{bmatrix},$$

4. Consult the section on orthogonal diagonalization of symmetric matrices in any introductory text on linear algebra, such as Anton [7] (1981, 277).
5. The Cholesky decomposition can be used to find $AA^T$ efficiently, see reference [5].

and therefore

$$A = \sigma\begin{bmatrix} [(1-\rho)/2]^{1/2} & [(1+\rho)/2]^{1/2} \\ -[(1-\rho)/2]^{1/2} & [(1+\rho)/2]^{1/2} \end{bmatrix}. \tag{4.97}$$

You should verify that $K = AA^{\mathrm{T}}$. ■■

### Generating Vectors of Jointly Gaussian Random Variables

In Section 4.8 we found that if $\mathbf{X}$ is a vector of jointly Gaussian random variables with covariance $K_X$, then $\mathbf{Y} = A\mathbf{X}$ is also jointly Gaussian with covariance matrix $K_Y = AK_XA^{\mathrm{T}}$. If we assume that $\mathbf{X}$ consists of unit-variance, uncorrelated random variables, then $K_X = I$, the identity matrix, and therefore $K_Y = AA^{\mathrm{T}}$.

We can use the method from the first part of this section to find $A$ for any desired covariance matrix $K_Y$. Thus we conclude that we can generate jointly Gaussian random vectors $\mathbf{Y}$ with arbitrary covariance matrix $K_Y$ as follows:

1. Find a matrix $A$ such that $K_Y = AA^{\mathrm{T}}$.
2. Generate $\mathbf{X}$ consisting of $n$ independent, zero-mean, unit-variance Gaussian random variables.
3. Let $\mathbf{Y} = A\mathbf{X}$.

We now present a method for generating unit-variance, uncorrelated (and hence independent) jointly Gaussian random variables. Suppose that $X$ and $Y$ are two independent zero-mean, unit-variance jointly Gaussian random variables; the joint pdf is then

$$f_{X,Y}(x, y) = \frac{1}{2\pi}e^{-(x^2+y^2)/2}.$$

Consider the transformation

$$R^2 = X^2 + Y^2 \qquad \text{and} \qquad \Theta = \angle(X, Y),$$

then the inverse transformation is

$$X = R\cos\Theta \qquad \text{and} \qquad Y = R\sin\Theta. \tag{4.98}$$

It is easy to show that the Jacobian for this transformation is 1/2, therefore the joint pdf of $R^2$ and $\Theta$ is given by

$$f_{R^2,\Theta}(s, t) = \frac{1}{4\pi}e^{-s/2} = f_{R^2}(s)f_\Theta(t),$$

where

$$f_{R^2}(s) = \frac{1}{2}e^{-s/2} \qquad \text{for } s > 0$$

expected value of random variables can be obtained through the use of conditional expectation.

- The covariance between two random variables and its normalized form, the correlation coefficient, are measures of the linear dependence between the random variables. The covariance between random variables is needed in the synthesis of predictors that predict a random variable by a linear combination of other random variables.

- The joint pdf of a vector **X** of jointly Gaussian random variables is determined by the vector of means and by the covariance matrix. All marginal pdf's and conditional pdf's of subsets of **X** have Gaussian pdf's. Any linear function or linear transformation of jointly Gaussian random variables will result in a set of jointly Gaussian random variables.

- A vector of random variables with an arbitrary covariance matrix can be generated by taking a linear transformation of a vector of unit-variance, uncorrelated random variables. A vector of Gaussian random variables with an arbitrary covariance matrix can be generated by taking a linear transformation of a vector of independent, unit-variance jointly Gaussian random variables.

## CHECKLIST OF IMPORTANT TERMS

Conditional cdf
Conditional expectation
Conditional pdf
Conditional pmf
Correlation coefficient
Covariance of $X$ and $Y$
Independent random variable
Jacobian of a transformation
Joint cdf
Joint pdf
Joint pmf
Jointly continuous random variables
Jointly Gaussian random variables
Linear transformation
Marginal cdf
Marginal pdf
Marginal pmf
Mean square estimation
Product-form event
Vector random variable

## ANNOTATED REFERENCES

Papoulis [1] is the standard reference in electrical engineering for random variables. References [2] and [3] present many interesting examples involving multiple random variables. The book by Jayant and Noll [4] gives numerous applications of probability concepts to the digital coding of waveforms.

1. A. Papoulis, *Probability, Random Variables, and Stochastic Processes,* McGraw-Hill, New York, 1965.

and

$$f_{\Theta}(t) = \frac{1}{2\pi} \qquad \text{for } 0 < t < 2\pi.$$

Therefore we can generate $R^2$ by generating an exponential ran variable with parameter 1/2, and we can generate $\theta$ by generatin random variable that is uniformly distributed in the interval $(0, 2\pi)$. If substitute these random variables into Eq. (4.98), we then obtain a pai independent zero-mean, unit-variance Gaussian random variables. ' above discussion thus leads to the following algorithm:

1. Generate $U_1$ and $U_2$, two independent random variables uniforr distributed in the unit interval.
2. Let $R^2 = -2 \log U_1$ and $\Theta = 2\pi\, U_2$
3. Let $X = R \cos \Theta = (-2 \log U_1) \cos 2\pi U_2$ and $Y = R \sin \Theta = (-2 \log U_1) \sin 2\pi U_2$.

Then $X$ and $Y$ are independent, zero-mean, unit-variance Gaussi random variables. By repeating the above procedure we can generate a number of such random variables.

## SUMMARY

- The joint statistical behavior of a vector of random variables **X** is specified by stating the joint cumulative distribution function, the joint probability mass function, or the joint probability density function. The probability of any event involving the joint behavior of these random variables can be computed from these functions.

- The statistical behavior of a subset of random variables from a vector **X** is specified by the marginal cdf, marginal pdf, or marginal pmf that can be obtained from the joint cdf, joint pdf, or joint pmf of **X**.

- A set of random variables is independent if the probability of a product-form event is equal to the product of the probabilities of the component events. Equivalent conditions for the independence of a set of random variables are that the joint cdf, joint pdf, or joint pmf factors into the product of the corresponding marginal functions.

- The statistical behavior of a subset of random variables from a vector **X**, given the exact values of the other random variables in the vector, is specified by the conditional cdf, conditional pmf, or conditional pdf. Many problems naturally lend themselves to a solution that involves conditioning on the values of some of the random variables. In these problems, the

2. L. Breiman, *Probability and Stochastic Processes,* Houghton Mifflin, Boston, 1969.
3. H. J. Larson and B. O. Shubert, *Probabilistic Models in Engineering Sciences,* Volume 1, Wiley, New York, 1979.
4. N. S. Jayant and P. Noll, *Digital Coding of Waveforms,* Prentice-Hall, Englewood Cliffs, N.J., 1984.
5. L. R. Rabiner and R. W. Schafer, *Digital Processing of Speech Signals,* Prentice-Hall, Englewood Cliffs, N.J., 1978.
6. H. Stark and J. W. Woods, *Probability, Random Processes, and Estimation Theory for Engineers,* Prentice-Hall, Englewood Cliffs, N.J., 1986.
7. H. Anton, *Elementary Linear Algebra,* Third Edition, John Wiley and Sons, New York, 1981.


## PROBLEMS

### Section 4.1
### Vector Random Variables

1. For the two-dimensional random variable $\mathbf{X} = (X, Y)$ sketch the region of the plane corresponding to the following events and state whether the events are of product form.

   a. $\{X - Y \leq 2\}$. d. $\{|X - Y| \leq 2\}$. g. $\{X^2 \leq Y\}$.
   b. $\{e^X < 6\}$. e. $\{|X| > |Y|\}$. h. $\{XY \leq 2\}$.
   c. $\{\max(X, Y) < 6\}$. f. $\{X/Y < 1\}$. i. $\{\max(|X|, |Y|) < 3\}$.

2. An urn contains one black ball and three white balls. Four balls are drawn from the urn. Let $I_k = 1$ if the outcome of the $k$th draw is the black ball and let $I_k = 0$ otherwise. Define the following three random variables:

   $X = I_1 + I_2 + I_3 + I_4,$
   $Y = \min\{I_1, I_2, I_3, I_4\},$ and
   $Z = \max\{I_1, I_2, I_3, I_4\}.$

   a. Find the probability law for $(X, Y, Z)$ if each ball is put back into the urn after each draw.
   b. Find the probability law for $(X, Y, Z)$ if each ball is not put back into the urn after each draw.

3. Let the random variables $X$, $Y$, and $Z$ be independent random variables. Find the following probabilities in terms of $F_X(x)$, $F_Y(y)$, and $F_Z(z)$.

a. $P[|X| < 5, Y > 2, Z^2 \geq 2]$.
b. $P[X > 5, Y < 0, Z = 1]$.
c. $P[\min(X, Y, Z) > 2]$.
d. $P[\max(X, Y, Z) < 6]$.

## Section 4.2
## Pairs of Random Variables

4. A die is tossed twice; let $X_1$ and $X_2$ denote the outcome of the first and second toss, respectively.
   a. What is the joint pmf for $(X_1, X_2)$ if the tosses are independent and if the outcomes of each toss are equiprobable.
   b. Let $X = \min(X_1, X_2)$ and $Y = \max(X_1, X_2)$. Find the joint pmf for $(X, Y)$.
   c. Find the marginal pmf's for $X$ and $Y$ in Part b.
5. a. Find the marginal pmf's for the pairs of random variables with the indicated joint pmf.

i.

| $Y$ \ $X$ | $-1$ | $0$ | $1$ |
|---|---|---|---|
| $-1$ | 1/6 | 0 | 1/6 |
| $0$ | 0 | 1/3 | 0 |
| $1$ | 1/6 | 0 | 1/6 |

ii.

| $Y$ \ $X$ | $-1$ | $0$ | $1$ |
|---|---|---|---|
| $-1$ | 1/9 | 1/9 | 1/9 |
| $0$ | 1/9 | 1/9 | 1/9 |
| $1$ | 1/9 | 1/9 | 1/9 |

iii.

| $Y$ \ $X$ | $-1$ | $0$ | $1$ |
|---|---|---|---|
| $-1$ | 0 | 0 | 1/3 |
| $0$ | 0 | 1/3 | 0 |
| $1$ | 1/3 | 0 | 0 |

   b. Find the probability of the events $A = \{X \leq 0\}$, $B = \{X \leq Y\}$, and $C = \{X = -Y\}$.
6. Sketch the three joint cdf's given in Problem 5 Part a and verify that the properties of the joint cdf are satisfied. You may find it helpful to first divide the plane into regions where the cdf is constant.
7. Sketch the joint cdf in Example 4.10 and verify that the properties of the joint cdf are satisfied. Include in your sketch the locus of points where the cdf is equal to 1/10, 1/2, and 9/10.
8. a. Find the joint cdf for the vector random variable introduced in Example 4.11.
   b. Use the result of Part a to find the marginal cdf's.
9. Let $X$ and $Y$ denote the amplitude of noise signals at two antennas. The random vector $(X, Y)$ has the joint pdf

$$f(x, y) = axe^{-ax^2/2}bye^{-by^2/2} \qquad x > 0,\ y > 0,\ a > 0,\ b > 0.$$

   a. Find the joint cdf.
   b. Find $P[X > Y]$.
   c. Find $P[|X - Y| < 1]$.
   d. Find the marginal pdf's.

10. The random vector variable $(X, Y)$ has the joint pdf

$$f(x, y) = k(x + y) \qquad 0 < x < 1,\ 0 < y < 1.$$

a. Find $k$.
b. Find the joint cdf of $(X, Y)$.
c. Find the marginal pdf's of $X$ and of $Y$.

11. The random vector $(X, Y)$ is uniformly distributed (i.e., $f(x, y) = k$) inside the regions shown in Fig. P4.1 and zero elsewhere.
a. Find the value of $k$ in each case.
b. Find the marginal pdf for $X$ and for $Y$ in each case.

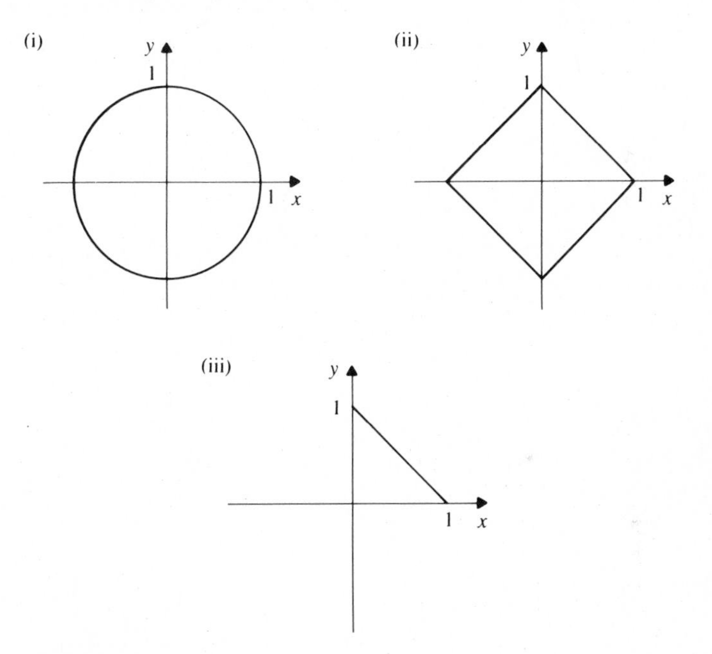


**FIGURE P4.1**

12. The random vector $(X, Y)$ has a joint pdf

$$f_{X,Y}(x, y) = 2e^{-x}e^{-2y} \qquad x > 0,\ y > 0.$$

Find the probability of the following events:
a. $\{X + Y \leq 8)$.
b. $\{X < Y\}$.
c. $\{X - Y \leq 10\}$.
d. $\{X^2 < Y\}$.

13. Let $(X, Y)$ have the joint pdf

$$f_{X,Y}(x, y) = xe^{-x(1+y)} \qquad x > 0,\ y > 0.$$

Find the marginal pdf of $X$ and of $Y$.

14. Let $(X, Y)$ be the jointly Gaussian random variables discussed in Example 4.13. Find $P[X^2 + Y^2 < R^2]$. *Hint*: Use polar coordinates to compute the integral.

15. The general form of the joint pdf for two jointly Gaussian random variables is

$$f_{X,Y}(x, y) = \frac{\exp\left\{\frac{-1}{2(1-\rho^2)}\left[\left(\frac{x-m_1}{\sigma_1}\right)^2 - 2\rho\left(\frac{x-m_1}{\sigma_1}\right)\left(\frac{y-m_2}{\sigma_2}\right) + \left(\frac{y-m_2}{\sigma_2}\right)\right]\right\}}{2\pi\sigma_1\sigma_2\sqrt{1-\rho^2}}$$

for $-\infty < x < \infty$ and $-\infty < y < \infty$.
Show that the marginal pdf's of $X$ and $Y$ are those of Gaussian random variables with means $m_1$ and $m_2$ and variances $\sigma_1^2$ and $\sigma_2^2$, respectively.

16. Find the marginal pdf for the output, $Y$, of the communication channel discussed in Example 4.14.

17. Let $X$ be the input to a communication channel. $X$ takes on the values $\pm 1$ with equal probability. Suppose that the output of the channel is $Y = X + N$, where $N$ is a Laplacian random variable with pdf

$$f_N(z) = \tfrac{1}{2}\alpha e^{-\alpha|z|} \qquad -\infty < z < \infty.$$

a. Find $P[X = i, Y \leq y]$ for $k = \pm 1$.
b. Find the marginal pdf of $Y$.
c. Suppose we are given that $Y > 0$, which is more likely $X = 1$ or $X = -1$?

18. A factory has $n$ machines of a certain type. Let $p$ be the probability that a machine is working on any given day, and let $N$ be the total number of machines working in a certain day. The time $T$ required to manufacture an item is an exponentially distributed random variable with rate $k\alpha$ if $k$ machines are working. Find $P[N = k, T \leq t]$ and $P[T \leq t]$. Find $P[T \leq t]$ as $t \to \infty$ and explain the result.

## Section 4.3
## Independence of Two Random Variables

19. Are $X$ and $Y$ independent random variables in Problem 5?

20. Let $X$ and $Y$ be independent random variables. Find an expression for the probability of the following events in terms of $F_X(x)$ and $F_Y(y)$:

a. $\{a < X \leq b\} \cap \{Y \leq d\}$.
b. $\{a \leq X \leq b\} \cap \{c \leq Y \leq d\}$.
c. $\{|X| > a\} \cap \{c \leq Y \leq d\}$.

21. Let $X$ and $Y$ be independent random variables that are uniformly distributed in $[0, 1]$. Find the probability of the following events:
    a. $P[X^2 < 1/2, |Y - 1| < 1/2]$.
    b. $P[X/2 < 1, Y > 0]$.

22. Let $X$ and $Y$ be the jointly Gaussian random variables introduced in Problem 15.
    a. Show that $X$ and $Y$ are independent random variables if $\rho = 0$.
    b. Suppose $\rho = 0$, find $P[XY > 0]$.

23. Let $X$ and $Y$ be random variables that take on values from the set $\{-1, 0, 1\}$. Find a joint pmf assignment for $X$ and $Y$ such that $X$ and $Y$ are *not* independent, but $X^2$ and $Y^2$ are independent.

## Section 4.4
## Conditional Probability and Conditional Expectation

24. Find the conditional pmf's of $Y$ given $X = -1$ in Problem 5.

25. Find $f_Y(y \mid x)$ in Problem 10.

26. Find $f_Y(y \mid x)$ in Problem 11.

27. Find $f_Y(y \mid x)$ and $f_X(x \mid y)$ in Problem 15.

28. Let $X = \cos\Theta$ and $Y = \sin\Theta$ where $\Theta$ is an angle that is uniformly distributed in the interval $(0, 2\pi)$. Find $f(y \mid x)$ and $E[Y \mid X]$.

29. A customer entering a store is served by clerk $i$ with probability $p_i$, $i = 1, \ldots, n$. The time taken by clerk $i$ to service a customer is an exponentially distributed random variable with parameter $\alpha_i$.
    a. Find the pdf of $T$, the time taken to service a customer.
    b. Find $E[T]$ and VAR$[T]$.

30. A message requires $N$ time units to be transmitted where $N$ is a geometric random variable with pmf $p_j = (1 - a)a^{j-1}, j = 1, 2, \ldots$. A single new message arrives during a time unit with probability $p$, and no messages arrive with probability $1 - p$. Let $K$ be the number of new messages that arrive during the transmission of a single message.
    a. Find the pmf of $K$. *Hint:*

$$(1 - \beta)^{-(k+1)} = \sum_{n=k}^{\infty} \binom{n}{k} \beta^{n-k}.$$

    b. Find $E[K]$ and VAR$[K]$ using conditional expectation.

31. The number of defects in a VLSI chip is a Poisson random variable with rate $r$. However, $r$ is itself a gamma random variable with parameters $\alpha$ and $\lambda$.
    a. Find the pmf for $N$, the number of defects.
    b. Use conditional expectation to find $E[N]$ and VAR$[N]$.

## Section 4.5
## Multiple Random Variables

32. Show that $f_{X,Y,Z}(x, y, z) = f_Z(z \mid x, y) f_Y(y \mid x) f_X(x)$.

33. Let $X_1$ be uniformly distributed in the interval $[0, 1]$, $X_2$ be uniformly distributed in $[0, X_1]$, $X_3$ uniformly distributed $[0, X_2]$, and so on.
    a. Find the joint pdf of $(X_1, \ldots, X_n)$.
    b. Find the marginal pdf of $X_k$ for $k = 1, \ldots, n$.
    c. Find the conditional pdf of $X_3$ given $X_1 = x$.

34. A random experiment has four possible outcomes for which the probabilities are $p_1$, $p_2$, $p_3$, and $1 - p_1 - p_2 - p_3$. Suppose that the experiment is repeated $n$ independent times and let $X_k$ be the number of times outcome $k$ occurs. The joint pmf of $(X_1, X_2, X_3)$ is given by

$$p(k_1, k_2, k_3) = \frac{n! p_1^{k_1} p_2^{k_2} p_3^{k_3} (1 - p_1 - p_2 - p_3)^{n - k_1 - k_2 - k_3}}{k_1! \, k_2! \, k_3! \, (n - k_1 - k_2 - k_3)!},$$

where $k_j \geq 0$ and $k_1 + k_2 + k_3 \leq n$.
    a. Find the marginal pmf of $(X_1, X_2)$. *Hint*: Use the binomial theorem, Eq. (2.36).
    b. Find the marginal pmf of $X_1$.
    c. Find the conditional joint pmf of $(X_2, X_3)$ given $X_1 = m$, where $m$ is a nonnegative integer less than or equal to $n$.

35. The number $N$ of customer arrivals at a service station is a Poisson random variable with mean $\alpha$ customers per second. There are four types of customers. Let $X_k$ be the number of type $k$ arrivals. Suppose

$$P[X_1 = k_1, X_2 = k_2, X_3 = k_3 \mid N = n] = p(k_1, k_2, k_3),$$

the pmf given in the previous problem.
    a. Find the joint pmf of $(N, X_1, X_2, X_3)$.
    b. Find the marginal pmf of $(X_1, X_2, X_3)$.

36. A random experiment has four possible outcomes. Suppose that the experiment is repeated $n$ independent times and let $X_k$ be the number of times outcome $k$ occurs. The joint pmf of $(X_1, X_2, X_3)$ is given by

$$p(k_1, k_2, k_3) = \frac{n! \, 3!}{(n + 3)!} = \binom{n + 3}{3}^{-1},$$

where $k_j \geq 0$ and $k_1 + k_2 + k_3 \leq n$.
    a. Find the marginal pmf of $(X_1, X_2)$.
    b. Find the marginal pmf of $X_1$.
    c. Find the conditional joint pmf of $(X_2, X_3)$ given $X_1 = m$, where $m$ is a nonnegative integer less than or equal to $n$.

37. Let $X$, $Y$, $Z$ have joint pdf

$$f_{X,Y,Z}(x, y, z) = k(x + y + z) \qquad 0 \leq x \leq 1,\ 0 \leq y \leq 1,\ 0 \leq z \leq 1.$$

a. Find $k$.
b. Find $f_Z(z \mid x, y)$.

**Section 4.6**
**Functions of Several Random Variables**

38. The lifetime of a device is a Rayleigh random variable. Let $T$ be the time until the first failure in a batch of $n$ such devices. Find the cdf of $T$. Find the mean of $T$.

39. Let $X$ and $Y$ be independent exponential random variables. Find the pdf of $Z = |X - Y|$.

40. Let $X$ and $Y$ be independent random variables that are uniformly distributed in the interval $[-1, 1]$. Find the pdf of $Z = XY$.

41. Let $X$ and $Y$ be independent Gaussian random variables that are zero-mean and unit-variance. Show that $Z = X/Y$ is a Cauchy random variable.

42. The random variables $X$ and $Y$ have the joint pdf

$$f_{X,Y}(x, y) = 2e^{-(x+y)} \qquad 0 \leq y \leq x < \infty.$$

Find the pdf of $Z = X + Y$. *Note*: $X$ and $Y$ are not independent.

43. Let $X$ and $Y$ be independent Rayleigh random variables. Find the pdf of $Z = X/Y$.

44. a. Find the joint pdf of

$$U = X_1,$$
$$V = X_1 + X_2, \text{ and}$$
$$W = X_1 + X_2 + X_3.$$

b. Evaluate the joint pdf of $(U, V, W)$ if the $X_i$ are independent zero-mean, unit-variance Gaussian random variables.

45. Find the joint pdf of the sample mean and variance

$$M = \frac{X_1 + X_2}{2}$$

$$V = \frac{(X_1 - M)^2 + (X_2 - M)^2}{2}$$

in terms of the joint pdf of $X_1$ and $X_2$. Evaluate the joint pdf if the $X_1$ and $X_2$ are independent exponential random variables with the same parameter.

46. a. Use the auxiliary variable method to find the pdf of

$$Z = \frac{X}{X + Y}$$

b. Find the pdf of $Z$ if $X$ and $Y$ are independent exponential random variables with the same parameter $\alpha$.

47. Let $X$ and $Y$ be zero-mean, unit-variance jointly Gaussian random variables with correlation coefficient $\rho$. Find the joint pdf of $U = X^2$ and $V = Y^4$.

48. Find the joint pdf of $Z = X_1 X_2 X_3$ where the $X_i$ are independent random variables that are uniformly distributed in $[0, 1]$.

49. Let $X$, $Y$, and $Z$ be independent, zero-mean, unit-variance Gaussian random variables. Find the pdf of

$$Z = \sqrt{X^2 + Y^2 + Z^2}.$$

## Section 4.7
## Expected Value of Functions of Random Variables

50. a. Find $E[(X + Y)^2]$.
b. Find the variance of $X + Y$.
c. Under what condition is the variance of the sum equal to the sum of the individual variances?

51. Find $E[|X - Y|]$ if $X$ and $Y$ are independent exponential random variables with parameter $\alpha = 1$.

52. Find $E[X^2 Y]$ where $X$ is a zero-mean, unit-variance Gaussian random variable, and $Y$ is a uniform random variable in the interval $[-1, 3]$, and $X$ and $Y$ are independent.

53. Find $E[M]$ and $E[V]$ in Problem 45.

54. Compute $E[Z]$ in Problem 48 in two ways:
a. by integrating over $f_Z(z)$;
b. by integrating over the joint pdf of $(X_1, X_2, X_3)$.

55. Show that the correlation coefficient equals $\pm 1$ if $Y = aX + b$.

56. Let $Y = X + N$, where $X$ and $N$ are independent, zero-mean Gaussian random variables with different variances. Find the correlation coefficient between the "observed signal" $Y$ and the "desired signal" $X$. Plot the correlation coefficient versus the signal-to-noise ratio $\sigma_X / \sigma_N$.

57. a. Find the correlation coefficient for $X$ and $Y$ in Problem 5.
b. Find the minimum mean square error linear estimator for $Y$.
c. Find the mean square error for a predictor that when observing $X = x$ predicts the value of $y$ that maximizes $p_Y(y \mid x)$.
d. Compare both of the above predictors to the minimum mean square error nonlinear predictor.

58. Repeat Problem 57 for $X$ and $Y$ in Problem 10, except in Part c, the predictor maximizes $f_Y(y \mid x)$.

59. Repeat Problem 58 for $X$ and $Y$ introduced in Problem 56.

60. Find the correlation coefficient for $X$ and $Y$ in Problem 11.

61. Find the correlation coefficient for $X$ and $Y$ in Example 4.11.

62. Find the correlation coefficient for $X$ and $Y$ in Problem 17.

63. (Jayant and Noll, 1984, 264) A second order predictor for samples of an image predicts the sample $E$ in terms of the sample $D$ to its left and the sample $B$ in the previous line, as shown below:

| | | | | | |
|---|---|---|---|---|---|
| line $j$ | .... | $A$ | $B$ | $C$ | .... |
| line $j + 1$ | .... | $D$ | $E$ | | .... |

estimate for $E = aD + bB$.

a. Find $a$ and $b$ if all samples have variance $\sigma^2$ and if the correlation coefficients between $D$ and $E$ is $\rho$, between $B$ and $E$ is $\rho$, and between $D$ and $B$ is $\rho^2$.

b. Find the mean square error of the predictor found in Part a, and determine the reduction in the variance of the signal in going from the input to the output of the predictor.

64. Show that the mean square error of the two-tap linear predictor is given by Eq. (4.85).

## Section 4.8 Jointly Gaussian Random Variables

65. Let $X$ and $Y$ be zero-mean, unit-variance jointly Gaussian random variables whose covariance matrix has determinant zero. Sketch the joint cdf of $X$ and $Y$. Does a joint pdf exist?

66. Find the covariance matrix for the jointly Gaussian random variables discussed in Example 4.29.

67. Let $X$ and $Y$ be zero-mean, unit-variance independent Gaussian random variables.

a. Find the value of $r$ for which the probability that $(X, Y)$ falls inside a circle of radius $r$ is 1/2.

b. Find the conditional pdf of $(X, Y)$ given that $(X, Y)$ is not inside a circle of radius $r$.

68. Let $h(x, y)$ be a joint Gaussian pdf for zero-mean, unit-variance Gaussian random variables with correlation coefficient $\rho_1$. Let $g(x, y)$ be a joint Gaussian pdf for zero-mean, unit-variance Gaussian random variables with correlation coefficient $\rho_2 \neq \rho_1$. Suppose the

random variables $X$ and $Y$ have joint pdf

$$f_{X,Y}(x, y) = \frac{1}{2}[h(x, y) + g(x, y)].$$

a. Find the marginal pdf for $X$ and for $Y$.
b. Explain why $X$ and $Y$ are *not* jointly Gaussian random variables.

69. Use conditional expectation to show that for $X$ and $Y$ jointly Gaussian random variables, $E[X^2Y^2] = E[X^2]E[Y^2] + 2E[XY]^2$.

## Section *4.9 Generating Correlated Vector Random Variables

70. a. Find the transformation required to generate two jointly Gaussian random variables with means $m_1$ and $m_2$ and variances $\sigma_1^2$ and $\sigma_2^2$ and correlation coefficient $\rho$. Assume two independent, zero-mean, unit-variance Gaussian random variables are available. *Hint*: Use Eq. (4.89).
    b. Write a computer program that implements the transformation and generate 1000 pairs of numbers with $m_1 = m_2 = 0$ and $\sigma_1 = \sigma_2$. Display each pair of values generated by the program as a dot on the $x - y$ plane. Is the pattern of dots in agreement with the desired joint pdf?
    c. Re-run the program with nonzero means and nonequal variances. Repeat the above display for the results and verify that the results are as expected.

71. Let $X$ and $Y$ be zero-mean, jointly Gaussian random variables with covariance matrix

$$K = \begin{vmatrix} 2 & 1 \\ 1 & 4 \end{vmatrix}.$$

Find the linear transformation that diagonalizes $K$.

72. Let $X_1, X_2, \ldots, X_n$ be independent zero-mean, unit-variance Gaussian random variables. Let $Y_k = (X_k + X_{k-1})/2$, that is, $Y_k$ is the moving average of pairs of values of $X$. Assume $X_{-1} = 0 = X_{n+1}$.
    a. Find the covariance matrix of the $Y_k$'s.
    b. Write a computer program that generates a very long sequence of sample $Y_1, \ldots, Y_n$. How would you check whether the $Y_k$'s have the correct covariances?

73. Repeat Problem 72 with $Y_k = X_k - X_{k-1}$.

CHAPTER 5

# Sums of Random Variables and Long-Term Averages

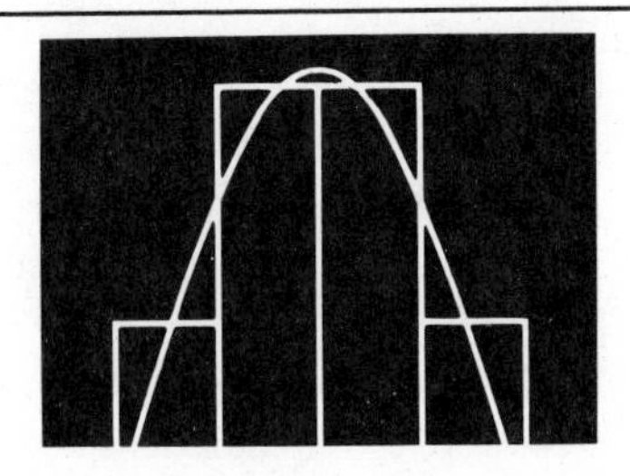

Many problems involve the counting of the number of occurrences of events, the measurement of cumulative effects, or the computation of arithmetic averages in a series of measurements. Usually these problems can be reduced to the problem of finding, exactly or approximately, the distribution of a random variable that consists of the sum of $n$ independent, identically distributed random variables. In this chapter, we investigate sums of random variables and their properties as $n$ becomes large.

In Section 5.1, we show how the characteristic function is used to compute the pdf of the sum of independent random variables. In Section 5.2, we discuss the sample mean estimator for the expected value of a random variable and the relative frequency estimator for the probability of an event. We introduce measures for assessing the goodness of these estimators. We then discuss the laws of large numbers, which are theorems that state that the sample mean and relative frequency estimators converge to the corresponding expected values and probabilities as the number of samples are increased. These theoretical results demonstrate the remarkable consistency between probability theory and observed behavior, and they reinforce the relative frequency interpretation of probability.

In Section 5.3, we present the central limit theorem, which states that, under very general conditions, the cdf of a sum of random variables approaches that of a Gaussian random variable even though the cdf of the individual random variables may be far from Gaussian. This result enables us to approximate the pdf of sums of random variables by the pdf of a Gaussian random variable. The result also explains why the Gaussian random variable appears in so many diverse applications.

In Section 5.4, we discuss the notion of confidence intervals, which plays an important role in the experimental determination of the values of the parameters of a random variable. In Section 5.5 we discuss random experiments in which events occur at random times. In these experiments we are interested in the average rate at which events occur as well as in the rate at which quantities associated with the events grow. Finally, Section 5.6 introduces computer methods based on the discrete Fourier transform that prove very useful in the numerical calculation of pmf's and pdf's from their transforms.

## 5.1 SUMS OF RANDOM VARIABLES

Let $X_1, X_2, \ldots, X_n$ be a sequence of random variables, and let $S_n$ be their sum:

$$S_n = X_1 + X_2 + \cdots + X_n. \tag{5.1}$$

In this section, we find the mean and variance of $S_n$, as well as the pdf of $S_n$ in the important special case where the $X_j$'s are independent random variables.

### Mean and Variance of Sums of Random Variables

In Section 4.7, it was shown that *regardless of statistical dependence, the expected value of a sum of n random variables is equal to the sum of the expected values:*

$$E[X_1 + X_2 + \cdots + X_n] = E[X_1] + \cdots + E[X_n]. \tag{5.2}$$

Thus knowledge of the means of the $X_j$'s suffices to find the mean of $S_n$.

The following example shows that in order to compute the variance of a sum of random variables, we need to know the variances and covariances of the $X_j$'s.

### ■■ Example 5.1

Find the variance of $Z = X + Y$.

From Eq. (5.2), $E[Z] = E[X + Y] = E[X] + E[Y]$. The variance of $Z$ is therefore

$$\begin{aligned}
\text{VAR}(Z) &= E[(Z - E[Z])^2] = E[(X + Y - E[X] - E[Y])^2] \\
&= E[\{(X - E[X]) + (Y - E[Y])\}^2] \\
&= E[(X - E[X])^2 + (Y - E[Y])^2 + (X - E[X])(Y - E[Y]) \\
&\quad + (Y - E[Y])(X - E[X])] \\
&= \text{VAR}[X] + \text{VAR}[Y] + \text{COV}(X, Y) + \text{COV}(Y, X) \\
&= \text{VAR}[X] + \text{VAR}[Y] + 2\,\text{COV}(X, Y)
\end{aligned}$$

In general, the covariance $\text{COV}(X, Y)$ is not equal to zero, so the variance of a sum is not necessarily equal to the sum of the individual variances. ■■

---

The result in Example 5.1 can be generalized to the case of $n$ random variables:

$$\begin{aligned}
\text{VAR}(X_1 + X_2 + \cdots + X_n) &= E\left\{\sum_{j=1}^{n} (X_j - E[X_j]) \sum_{k=1}^{n} (X_k - E[X_k])\right\} \\
&= \sum_{j=1}^{n} \sum_{k=1}^{n} E[(X_j - E[X_j])(X_j - E[X_j])] \\
&= \sum_{k=1}^{n} \text{VAR}(X_k) + \sum_{\substack{j=1 \\ j \neq k}}^{n} \sum_{k=1}^{n} \text{COV}(X_j, X_k).
\end{aligned} \tag{5.3}$$

Thus *in general, the variance of a sum of random variables is not equal to the sum of the individual variances.*

An important special case is when the $X_j$'s are independent random variables. *If $X_1, X_2, \ldots, X_n$ are independent random variables,* then $\text{COV}(X_j, X_k) = 0$ for $j \neq k$ and

$$\text{VAR}(X_1 + X_2 + \cdots + X_n) = \text{VAR}(X_1) + \cdots + \text{VAR}(X_n). \tag{5.4}$$

### Example 5.2
Sum of iid Random Variables

Find the mean and variance of the sum of $n$ **independent, identically distributed** (iid) random variables, each with mean $\mu$ and variance $\sigma^2$.

The mean of $S_n$ is obtained from Eq. (5.2):

$$E[S_n] = E[X_1] + \cdots + E[X_n] = n\mu.$$

The covariance of pairs of independent random variables is zero, so by Eq. (5.4)

$$\text{VAR}[S_n] = n\,\text{VAR}[X_j] = n\sigma^2,$$

since $\text{VAR}[X_j] = \sigma^2$ for $j = 1, \ldots, n$.

### pdf of Sums of Independent Random Variables

Let $X_1, X_2, \ldots, X_n$ be $n$ *independent* random variables. In this section we show how transform methods can be used to find the pdf of $S_n = X_1 + X_2 + \cdots + X_n$.

First, consider the $n = 2$ case, $Z = X + Y$, where $X$ and $Y$ are independent random variables. The characteristic function of $Z$ is given by

$$\begin{aligned}\Phi_Z(\omega) &= E[e^{j\omega Z}] \\ &= E[e^{j\omega(X+Y)}] \\ &= E[e^{j\omega X}e^{j\omega Y}] \\ &= E[e^{j\omega X}]E[e^{j\omega Y}] \\ &= \Phi_X(\omega)\Phi_Y(\omega),\end{aligned} \tag{5.5}$$

where the fourth equality follows from the fact that functions of independent random variables (i.e., $e^{j\omega X}$ and $e^{j\omega Y}$) are also independent random variables, as discussed in Example 4.39. Thus the characteristic function of $Z$ is the product of the individual characteristic functions of $X$ and $Y$.

In Example 4.31, we saw that the pdf of $Z = X + Y$ is given by the convolution of the pdf's of $X$ and $Y$:

$$f_Z(z) = f_X(x) * f_Y(y). \tag{5.6}$$

Recall that $\Phi_Z(\omega)$ can also be viewed as the Fourier transform of the pdf of $Z$:

$$\Phi_Z(\omega) = \mathcal{F}\{f_Z(z)\}.$$

By equating the transform of Eq. (5.6) to Eq. (5.5) we obtain

$$\Phi_Z(\omega) = \mathcal{F}\{f_Z(z)\} = \mathcal{F}\{f_X(x) * f_Y(y)\} = \Phi_X(\omega)\Phi_Y(\omega). \tag{5.7}$$

Equation (5.7) states the well-known result that the Fourier transform of a convolution of two functions is equal to the product of the individual Fourier transforms.

Now consider the sum of $n$ independent random variables:

$$S_n = X_1 + X_2 + \cdots + X_n.$$

The characteristic function of $S_n$ is

$$\begin{aligned}\Phi_{S_n}(\omega) &= E[e^{j\omega S_n}] = E[e^{j\omega(X_1+X_2+\cdots+X_n)}] \\ &= E[e^{j\omega X_1}]\cdots E[e^{j\omega X_n}] \\ &= \Phi_{X_1}(\omega)\cdots\Phi_{X_n}(\omega).\end{aligned} \tag{5.8}$$

Thus the pdf of $S_n$ can then be found by finding the inverse Fourier transform of the product of the individual characteristic functions of the $X_j$'s.

$$f_{S_n}(X) = \mathcal{F}^{-1}\{\Phi_{X_1}(\omega)\cdots\Phi_{X_n}(\omega)\}. \tag{5.9}$$

### Example 5.3
Sum of Independent Gaussian Random Variables

Let $S_n$ be the sum of $n$ independent Gaussian random variables with respective means and variances, $m_1, \ldots, m_n$ and $\sigma_1^2, \ldots, \sigma_n^2$. Find the pdf of $S_n$.

The characteristic function of $X_k$ is

$$\Phi_{X_k}(\omega) = e^{-j\omega m_k + \omega^2\sigma_k^2/2}$$

so by Eq. (5.8),

$$\begin{aligned}\Phi_{S_n}(\omega) &= \prod_{k=1}^{n} e^{-j\omega m_k + \omega^2\sigma_k^2/2} \\ &= e^{-j\omega(m_1+\cdots+m_n)+\omega^2(\sigma_1^2+\cdots+\sigma_n^2)/2}\end{aligned}$$

This is the characteristic function of a Gaussian random variable. Thus $S_n$ is a Gaussian random variable with mean $m_1 + \cdots + m_n$ and variance $\sigma_1^2 + \cdots + \sigma_n^2$.

### Example 5.4
Sum of iid Random Variables

Find the pdf of a sum of $n$ independent, identically distributed random variables with characteristic functions

$$\Phi_{X_k}(\omega) = \Phi_X(\omega) \qquad \text{for } k = 1, \ldots, n.$$

Equation (5.8) immediately implies that the characteristic function of $S_n$ is

$$\Phi_{S_n}(\omega) = \{\Phi_X(\omega)\}^n. \tag{5.10}$$

The pdf of $S_n$ is found by taking the inverse transform of this expression.

---

### Example 5.5
Sum of iid Exponential Random Variables

Find the pdf of a sum of $n$ independent exponentially distributed random variables, all with parameter $\alpha$.

The characteristic function of a single exponential random variable is

$$\Phi_X(\omega) = \frac{\alpha}{\alpha - j\omega}.$$

From the previous example we then have that

$$\Phi_{S_n}(\omega) = \left\{\frac{\alpha}{\alpha - j\omega}\right\}^n.$$

From Table 3.2, we see that $S_n$ is an $m$-Erlang random variable.

---

When dealing with integer-valued random variables it is usually preferable to work with the probability generating function

$$G_N(z) = E[z^N].$$

The generating function for a sum of independent discrete random variables, $N = X_1 + \cdots + X_n$, is

$$\begin{aligned} G_N(z) &= E[z^{X_1 + \cdots + X_n}] = E[z^{X_1}] \cdots E[z^{X_n}] \\ &= G_{X_1}(z) \cdots G_{X_n}(z). \end{aligned} \tag{5.11}$$

### Example 5.6

Find the generating function for a sum of $n$ independent, identically geometrically distributed random variables.

The generating function for a single geometric random variable is

given by

$$G_X(z) = \frac{pz}{1 - qz}.$$

Therefore the generating function for a sum of $n$ such independent random variables is

$$G_N(z) = \left\{\frac{pz}{1 - qz}\right\}^n.$$

From Table 3.1, we see that this is the generating function of a negative binomial random variable with parameters $p$ and $n$. ■■

### *Sum of a Random Number of Random Variables

In some problems we are interested in the sum of a random number $N$ of iid random variables:

$$S_N = \sum_{k=1}^{N} X_k, \tag{5.12}$$

where $N$ is assumed to be a random variable that is independent of the $X_k$'s. For example, $N$ might be the number of computer jobs submitted in an hour and $X_k$ might be the time required to execute the $k$th job.

The mean of $S_N$ is found readily by using conditional expectation:

$$\begin{aligned} E[S_N] &= E[E[S_N \mid N]]. \\ &= E[NE[X]] \\ &= E[N]E[X]. \end{aligned} \tag{5.13}$$

The second equality follows from the fact that

$$E[S_N \mid N = n] = E\left[\sum_{k=1}^{n} X_k\right] = nE[X],$$

so $E[S_N \mid N] = NE[X]$.

The characteristic function of $S_n$ can also be found by using conditional expectation. From Eq. (5.10), we have that

$$E[e^{j\omega S_N} \mid N = n] = E[e^{j\omega(X_1 + \cdots + X_n)}] = \Phi_X(\omega)^n,$$

so

$$E[e^{j\omega S_N} \mid N] = \Phi_X(\omega)^N.$$

Therefore

$$\begin{aligned}\Phi_{S_N}(\omega) &= E[E[e^{j\omega S_N} \mid N]] \\ &= E[\Phi_X(\omega)^N] \\ &= E[z^N]\big|_{z=\Phi_X(\omega)} \\ &= G_N(\Phi_X(\omega)).\end{aligned} \tag{5.14}$$

That is, the characteristic function of $S_n$ is found by evaluating the generating function of $N$ at $z = \Phi_X(\omega)$.

### ■■ Example 5.7

The number of jobs $N$ submitted to a computer in an hour is a geometric random variable with parameter $p$, and the job execution times are independent exponentially distributed random variables with mean $1/\alpha$. Find the pdf for the sum of the execution times of the jobs submitted in an hour.

The generating function for $N$ is

$$G_N(z) = \frac{p}{1 - qz},$$

and the characteristic function for an exponentially distributed random variable is

$$\Phi_X(\omega) = \frac{\alpha}{\alpha - j\omega}.$$

From Eq. (5.14), the characteristic function of $S_n$ is

$$\begin{aligned}\Phi_{S_N}(\omega) &= \frac{p}{1 - q[\alpha/(\alpha - j\omega)]} \\ &= p(\alpha - j\omega)/(p\alpha - j\omega) \\ &= p + (1 - p)\frac{p\alpha}{p\alpha - j\omega}.\end{aligned}$$

The pdf of $S_n$ is found by taking the inverse transform of the above expression:

$$f_{S_N}(x) = p\,\delta(x) + (1 - p)e^{-p\alpha x} \qquad x \geq 0.$$

The pdf has a direct interpretation: With probability $p$ there are no job arrivals and hence the total execution time is zero; with probability $(1 - p)$ there are one or more arrivals, and the total execution time is an exponential random variable with mean $1/p\alpha$. ■■

## 5.2

## THE SAMPLE MEAN AND THE LAWS OF LARGE NUMBERS

Let $X$ be a random variable for which the mean, $E[X] = \mu$, is unknown. Let $X_1, \ldots, X_n$ denote $n$ independent, repeated measurements of $X$, that is, the $X_j$'s are **independent, identically distributed** (iid) random variables with the same pdf as $X$. The **sample mean** of the sequence is used to estimate $E[X]$:

$$M_n = \frac{1}{n}\sum_{j=1}^{n} X_j. \tag{5.15}$$

In this section, we compute the expected value and variance of $M_n$ in order to assess the effectiveness of $M_n$ as an estimator for $E[X]$. We also investigate the behavior of $M_n$ as $n$ becomes large.

The following example shows that relative frequency estimator for the probability of an event is a special case of a sample mean. Thus the results derived below for the sample mean are also applicable to the relative frequency estimator.

### ■■ Example 5.8
Relative Frequency

Consider a sequence of independent repetitions of some random experiment, and let the random variable $I_j$ be the indicator function for the occurrence of event $A$ in the $j$th trial. The total number of occurrences of $A$ in the first $n$ trials is then

$$N_n = I_1 + I_2 + \cdots + I_n.$$

The **relative frequency** of event $A$ in the first $n$ repetitions of the experiment is then

$$f_A(n) = \frac{1}{n}\sum_{j=1}^{n} I_j. \tag{5.16}$$

Thus the relative frequency $f_A(n)$ is simply the sample mean of the random variables $I_j$. ■■

The sample mean is itself a random variable so it will exhibit random variation. A good estimator should have the following two properties: (1) On the average, it should give the correct value of the parameter being estimated, that is, $E[M_n] = \mu$; and (2) It should not vary too much about the correct value of this parameter, that is, $E[(M_n - \mu)^2]$ is small.

The expected value of the sample mean is given by

$$E[M_n] = E\left[\frac{1}{n}\sum_{j=1}^{n} X_j\right] = \frac{1}{n}\sum_{j=1}^{n} E[X_j] = \mu, \tag{5.17}$$

since $E[X_j] = E[X] = \mu$ for all $j$. Thus the sample mean is equal to $E[X] = \mu$, on the average. For this reason, we say that the sample mean is an **unbiased estimator** for $\mu$.

Equation (5.17) implies that the mean square error of the sample mean about $\mu$ is equal to the variance of $M_n$, that is,

$$E[(M_n - \mu)^2] = E[(M_n - E[M_n])^2].$$

Note that $M_n = S_n/n$, where $S_n = X_1 + X_2 + \cdots + X_n$. From Eq. (5.4), $\text{VAR}[S_n] = n\,\text{VAR}[X_j] = n\sigma^2$, since the $X_j$'s are iid random variables. Thus

$$\text{VAR}[M_n] = \frac{1}{n^2}\text{VAR}[S_n] = \frac{n\sigma^2}{n^2} = \frac{\sigma^2}{n}. \tag{5.18}$$

Equation (5.18) states that the variance of the sample mean approaches zero as the number of samples is increased. This implies that the probability that the sample mean is close to the true mean approaches one as $n$ becomes very large. We can formalize this statement by using the Chebyshev inequality, Eq. (3.73):

$$P[|M_n - E[M_n]| \geq \varepsilon] \leq \frac{\text{VAR}[M_n]}{\varepsilon^2}.$$

Substituting for $E[M_n]$ and $\text{VAR}[M_n]$, we obtain

$$P[|M_n - \mu| \geq \varepsilon] \leq \frac{\sigma^2}{n\varepsilon^2}. \tag{5.19}$$

If we consider the complement of the event considered in Eq. (5.19), we obtain

$$P[|M_n - \mu| < \varepsilon] \geq 1 - \frac{\sigma^2}{n\varepsilon^2}. \tag{5.20}$$

Thus for any choice of error $\varepsilon$ and probability $1 - \delta$, we can select the number of samples $n$ so that $M_n$ is within $\varepsilon$ of the true mean with probability $1 - \delta$. The following example illustrates this.

### ■■ Example 5.9

A voltage of constant, but unknown, value is to be measured. Each measurement $X_j$ is actually the sum of the desired voltage $v$ and a noise voltage $N_j$ of zero mean and standard deviation of 1 microvolt:

$$X_j = v + N_j.$$

Assume that the noise voltages are independent random variables. How many measurements are required so that the probability that $M_n$ is within $\varepsilon = 1$ microvolt of the true mean is at least .99?

Each measurement $X_j$ has mean $v$ and variance 1, so from Eq. (5.20) we require that $n$ satisfy

$$1 - \frac{\sigma^2}{n\varepsilon^2} = 1 - \frac{1}{n} = .99.$$

This implies that $n = 100$.

Thus if we were to repeat the measurement 100 times and compute the sample mean, on the average, at least ninety-nine times out of a hundred, the resulting sample mean will be within 1 microvolt of the true mean. ■■

---

Note that if we let $n$ approach infinity in Eq. (5.20) we obtain

$$\lim_{n \to \infty} P[|M_n - \mu| < \varepsilon] = 1.$$

Equation (5.20) requires that the $X_j$'s have finite variance. It can be shown that this limit holds even if the variance of the $X_j$'s does not exist (Gnedenko [1976, 203]). We state this more general result:

**WEAK LAW OF LARGE NUMBERS**

Let $X_1, X_2, \ldots$ be a sequence of iid random variables with finite mean $E[X] = \mu$, then for $\varepsilon > 0$

$$\lim_{n \to \infty} P[|M_n - \mu| < \varepsilon] = 1. \tag{5.21}$$

The weak law of large numbers states that for a large enough *fixed* value of $n$, the sample mean using $n$ samples will be close to the true mean with high probability. The weak law of large numbers does not address the question about what happens to the sample mean as a function of $n$ as we make additional measurements. This question is taken up by the strong law of large numbers, which we discuss next.

Suppose we make a series of independent measurements of the same random variable. Let $X_1, X_2, \ldots,$ be the resulting sequence of iid random variables with mean $\mu$. Now consider the *sequence of sample means* that results from the above measurements: $M_1, M_2, \ldots,$ where $M_j$ is the sample mean computed using $X_1$ through $X_j$. The notion of statistical regularity discussed in Chapter 1 leads us to expect that this sequence of sample means converges to $\mu$, that is, we expect that with high probability, each particular sequence of sample means approaches $\mu$ and stays there as shown in Fig. 5.1. In terms of probabilities, we expect the following:

$$P\left[\lim_{n \to \infty} M_n = \mu\right] = 1,$$

that is, with virtual certainty, every sequence of sample mean calculations

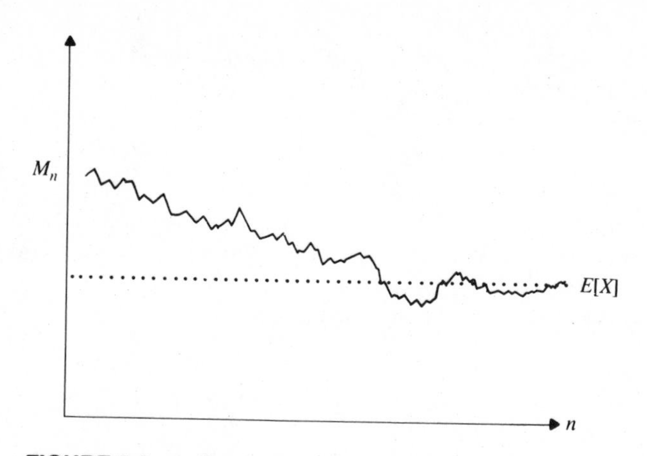


**FIGURE 5.1** Convergence of sequence of sample means to $E[X]$.

converges to the true mean of the quantity. The proof of this result is well beyond the level of this course (see Gnedenko [1976, 216]), but we will have the opportunity in later sections to apply the result in various situations.

**STRONG LAW OF LARGE NUMBERS**

Let $X_1, X_2, \ldots$ be a sequence of iid random variables with finite mean $E[X] = \mu$ and finite variance, then

$$P\left[\lim_{n\to\infty} M_n = \mu\right] = 1. \tag{5.22}$$

Equation (5.22) appears similar to Eq. (5.21), but in fact, it makes a dramatically different statement. It states that *with probability 1, every sequence of sample mean calculations will eventually approach and stay close to* $E[X] = \mu$. This is the type of convergence we expect in physical situations where statistical regularity holds.

With the strong law of large numbers we come around full circle in the modelling process. We began in Chapter 1 by noting that statistical regularity is observed in many physical phenomena, and from this we deduced a number of properties of relative frequency. These properties were used to formulate a set of axioms from which we developed a mathematical theory of probability. We have now come around full circle and shown that, under certain conditions, the *theory* predicts the convergence of sample means to expected values. There are still gaps between the mathematical theory and the real world (i.e., we can never actually carry out an infinite number of measurements and compute an infinite number of sample means). Nevertheless, the strong law of large numbers demonstrates the remarkable consistency between the theory and the observed physical behavior.

We already indicated that relative frequencies are special cases of sample averages. If we apply the weak law of large numbers to the relative frequency of an event $A$, $f_A(n)$, in a sequence of independent repetitions of a random experiment, we obtain

$$\lim_{n\to\infty} P[|f_A(n) - P[A]| < \varepsilon] = 1. \tag{5.23}$$

If we apply the strong law of large numbers, we obtain

$$P\left[\lim_{n\to\infty} f_A(n) = P[A]\right] = 1. \tag{5.24}$$

### ■■ Example 5.10

In order to estimate the probability of an event $A$, a sequence of Bernoulli trials is carried out and the relative frequency of $A$ is observed. How large should $n$ be in order to have a .95 probability that the relative frequency is within 0.01 of $p = P[A]$?

Let $X = I_A$ be the indicator function of $A$. From Table 3.1 we have that the mean of $I_A$ is $\mu = p$ and the variance is $\sigma^2 = p(1 - p)$. Since $p$ is unknown, $\sigma^2$ is also unknown. However it is easy to show that $p(1 - p)$ is at most 1/4 for $0 \le p \le 1$. Therefore by Eq. (5.19)

$$P[|f_A(n) - p| \ge \varepsilon] \le \frac{\sigma^2}{n\varepsilon^2} \le \frac{1}{4n\varepsilon^2}.$$

The desired accuracy is $\varepsilon = 0.01$ and the desired probability is

$$1 - .95 = \frac{1}{4n\varepsilon^2}.$$

We then solve for $n$ and obtain $n = 50{,}000$. It has already been pointed out that the Chebyshev inequality gives very loose bounds, so we expect that this value for $n$ is probably overly conservative. In the next section, we present a better estimate for the required value of $n$. ■■

## 5.3

## THE CENTRAL LIMIT THEOREM

Let $X_1, X_2, \ldots$ be a sequence of iid random variables with finite mean $\mu$ and finite variance $\sigma^2$, and let $S_n$ be the sum of the first $n$ random variables in the sequence:

$$S_n = X_1 + X_2 + \cdots + X_n. \tag{5.25}$$

In Section 5.1, we developed methods for determining the exact pdf of $S_n$. We now present the central limit theorem, which states that, as $n$ becomes large, the cdf of $S_n$ approaches that of a Gaussian random variable. This enables us to approximate the cdf of $S_n$ with that of a Gaussian random variable.

The central limit theorem explains why the Gaussian random variable appears in so many diverse applications. In nature, many macroscopic phenomena result from the addition of numerous independent, microscopic processes; this gives rise to the Gaussian random variable. In many man-made problems, we are interested in averages that often consist of the sum of independent random variables. This again gives rise to the Gaussian random variable.

From Example 5.2, we know that if the $X_j$'s are iid, then $S_n$ has mean $n\mu$ and variance $n\sigma^2$. The central limit theorem states that the cdf of a suitably normalized version of $S_n$ approaches that of a Gaussian random variable.

**CENTRAL LIMIT THEOREM**

Let $S_n$ be the sum of $n$ iid random variables with finite mean $E[X] = \mu$ and finite variance $\sigma^2$, and let $Z_n$ be the zero-mean, unit variance random variable defined by

$$Z_n = \frac{S_n - n\mu}{\sigma\sqrt{n}}, \tag{5.26}$$

then

$$\lim_{n\to\infty} P[Z_n < z] = \frac{1}{\sqrt{2\pi}} \int_{-\infty}^{z} e^{-x^2/2}\,dx. \tag{5.27}$$

The amazing part about the central limit theorem is that the summands $X_j$ can have *any* distribution as long as they have a finite mean and finite variance. This gives the result its wide applicability.

Figures 5.2 through 5.4 compare the exact cdf and the Gaussian approximation for the sums of Bernoulli, uniform, and exponential random variables, respectively. In all three cases, it can be seen that the approximation improves as the number of terms in the sum increases. The proof of the central limit theorem is discussed in the last part of this section.

## ■■ Example 5.11

Suppose that orders at a restaurant are iid random variables with mean $\mu = \$8$ and standard deviation $\sigma = \$2$. Estimate the probability that the first 100 customers spend a total of more than \$840. Estimate the

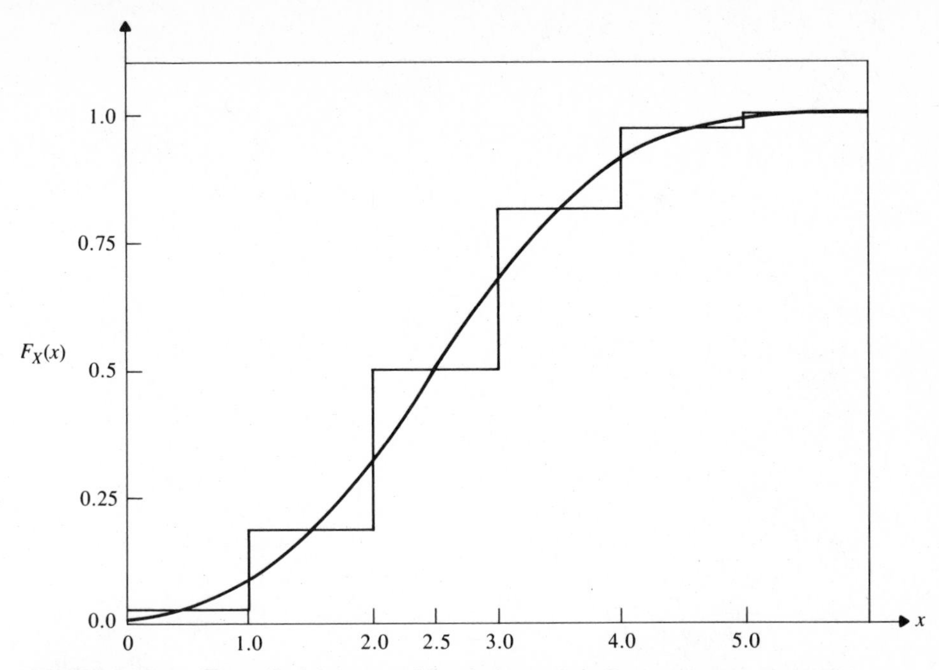


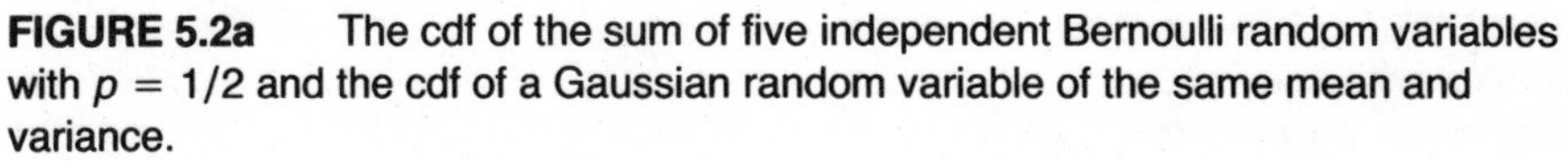

**FIGURE 5.2a** The cdf of the sum of five independent Bernoulli random variables with $p = 1/2$ and the cdf of a Gaussian random variable of the same mean and variance.

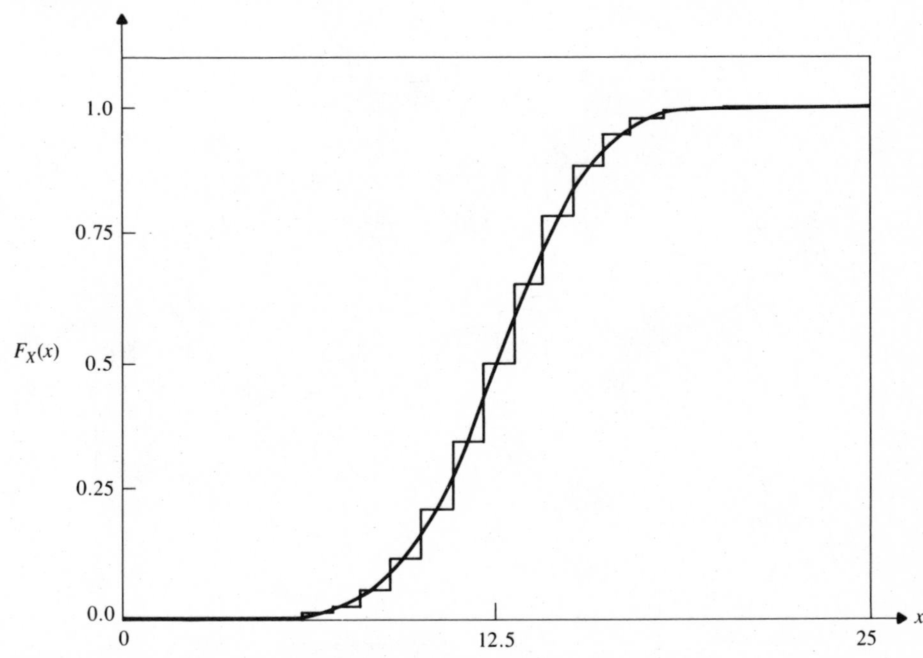


**FIGURE 5.2b** The cdf of the sum of 25 independent Bernoulli random variables with $p = 1/2$ and the cdf of a Gaussian random variable of the same mean and variance.

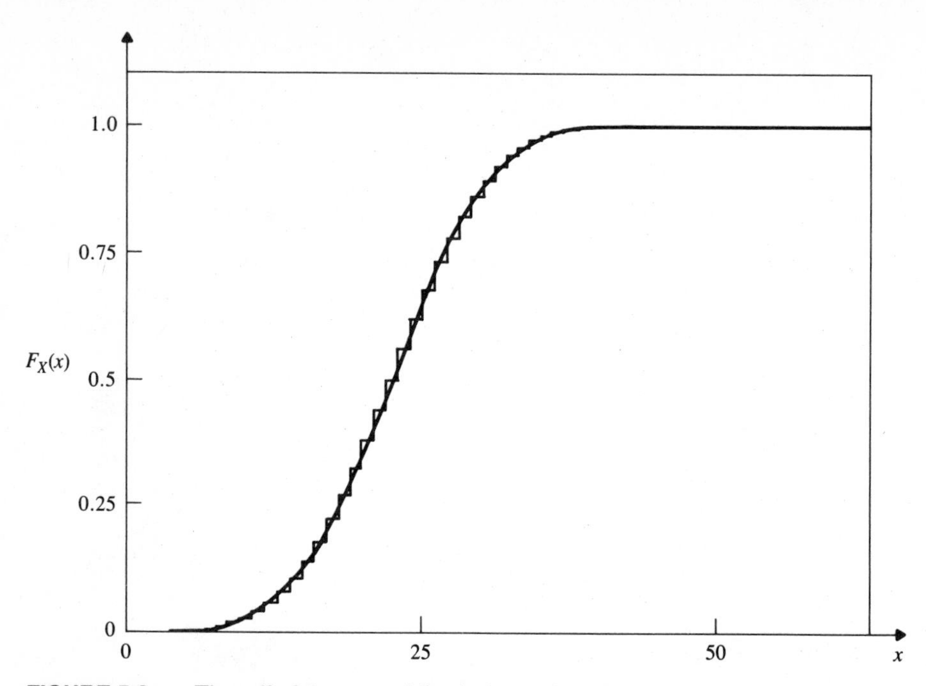


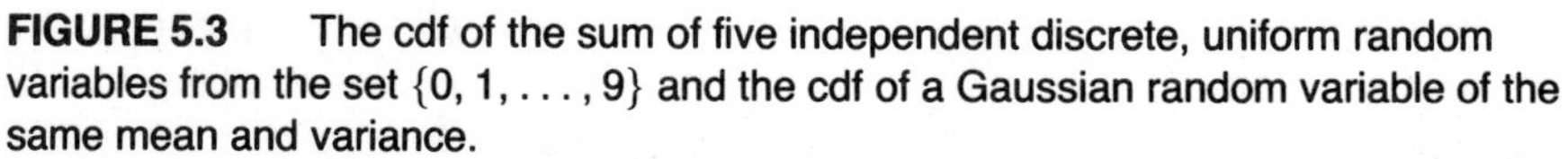

**FIGURE 5.3** The cdf of the sum of five independent discrete, uniform random variables from the set $\{0, 1, \ldots, 9\}$ and the cdf of a Gaussian random variable of the same mean and variance.

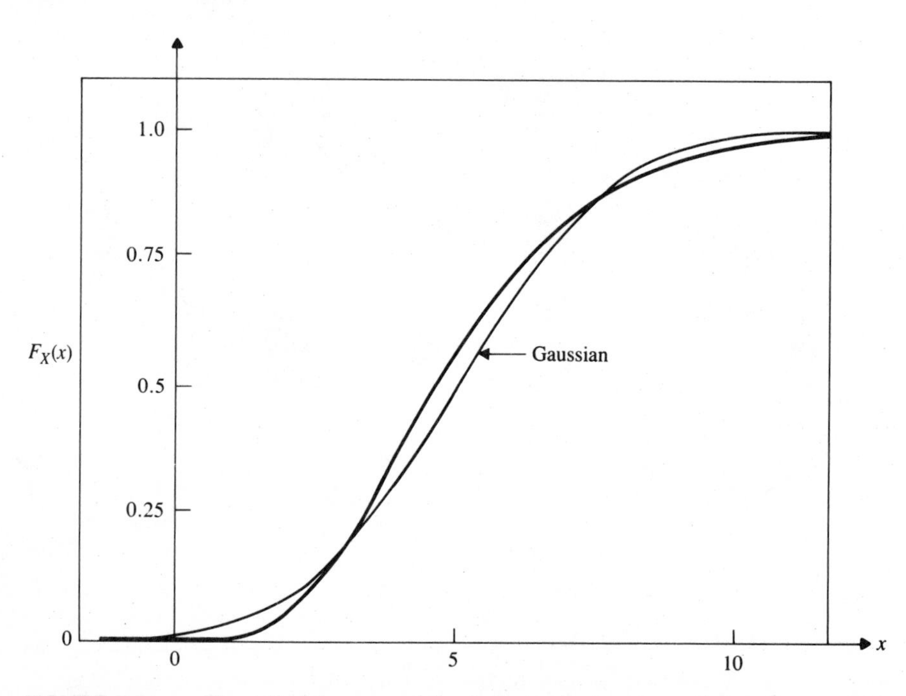


**FIGURE 5.4a** The cdf of the sum of five independent exponential random variables of mean 1 and the cdf of a Gaussian random variable of the same mean and variance.

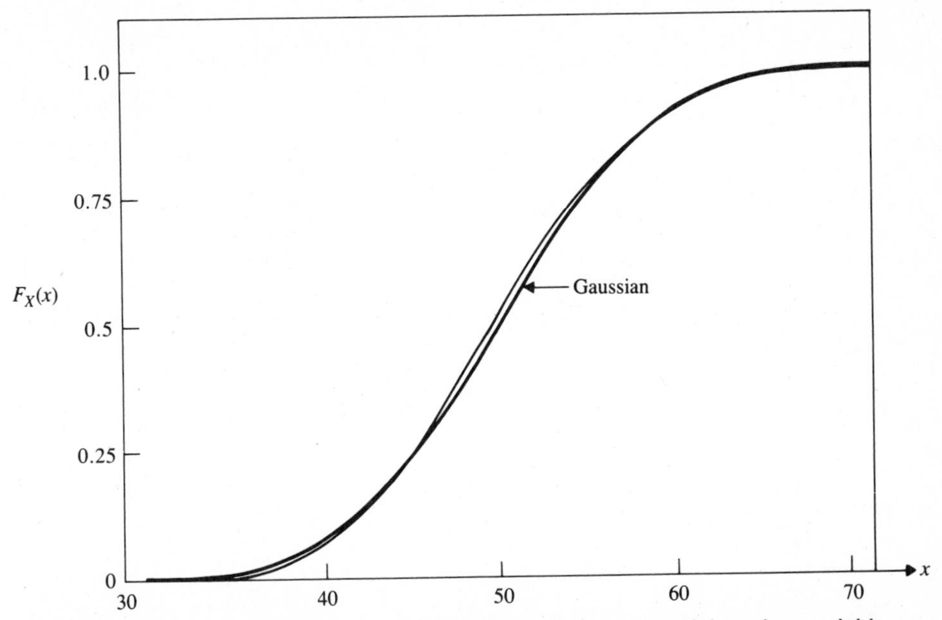


**FIGURE 5.4b** The cdf of the sum of 50 independent exponential random variables of mean 1 and the cdf of a Gaussian random variable of the same mean and variance.

probability that the first 100 customers spend a total of between \$780 and \$820.

Let $X_k$ denote the expenditure of the $k$th customer, then the total spent by the first 100 customers is

$$S_{100} = X_1 + X_2 + \cdots + X_{100}.$$

The mean of $S_{100}$ is $n\mu = 800$ and the variance is $n\sigma^2 = 400$. Figure 5.5 shows the pdf of $S_{100}$ where it can be seen that the pdf is highly concentrated about the mean. The normalized form of $S_{100}$ is

$$Z_{100} = \frac{S_{100} - 800}{20}.$$

Thus

$$P[S_{100} > 840] = P\left[Z_{100} > \frac{840 - 800}{20}\right]$$
$$\simeq Q(2) = 2.28(10^{-2}),$$

where we used Table 3.3 to evaluate $Q(2)$. Similarly,

$$P[780 \le S_{100} \le 820] = P[-1 \le Z_{100} \le 1]$$
$$\simeq 1 - 2Q(1)$$
$$= .682.$$

■■

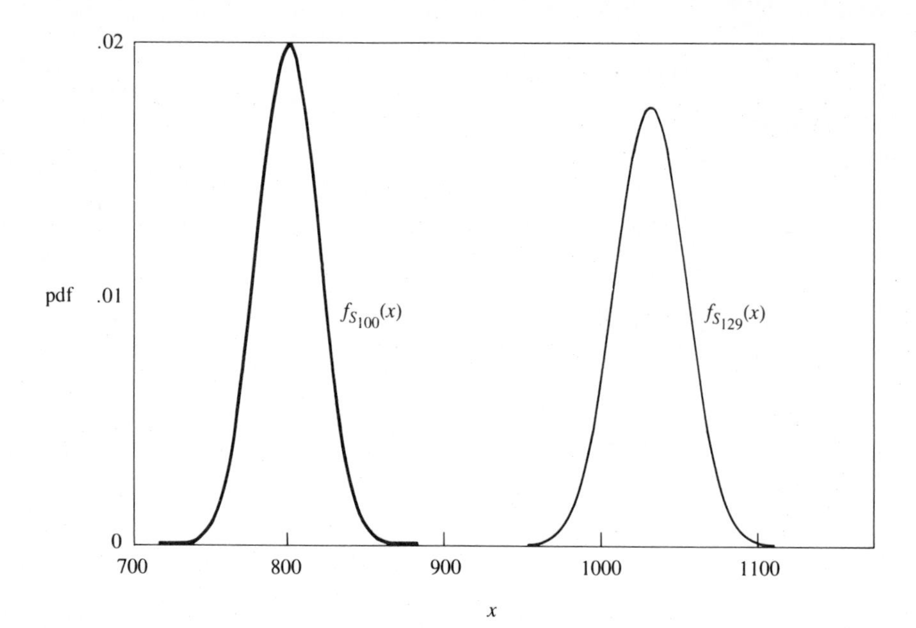


**FIGURE 5.5** Gaussian pdf approximations $S_{100}$ and $S_{129}$ in Examples 5.11 and 5.12.

### ■■ Example 5.12

In Example 5.11, after how many orders can we be 90% sure that the total spent by all customers is more than \$1000?

The problem here is to find the value of $n$ for which

$$P[S_n > 1000] = .90.$$

$S_n$ has mean $8n$ and variance $4n$. Proceeding as in the previous example, we have

$$P[S_n > 1000] = P\left[Z_n > \frac{1000 - 8n}{2\sqrt{n}}\right] = .90.$$

Using the fact that $Q(-x) = 1 - Q(x)$, Table 3.4 implies that $n$ must satisfy

$$\frac{1000 - 8n}{2\sqrt{n}} = -1.2815,$$

which yields the following quadratic equation for $\sqrt{n}$:

$$8n - 1.2815(2)\sqrt{n} - 1000 = 0.$$

The positive root of the equation yields $\sqrt{n} = 11.34$, or $n = 128.6$. Figure 5.5 shows the pdf for $S_{129}$. ■■

### Example 5.13

The time between events in a certain random experiment are iid exponential random variables with mean $m$ seconds. Find the probability that the 1000th event occurs in the time interval $(1000 \pm 50)m$.

Let $X_j$ be the time between events and let $S_n$ be the time of the $n$th event, then $S_n$ is given by Eq. (5.25). From Table 3.2, the mean and variance of $X_j$ is given by $E[X_j] = m$ and $\text{VAR}[X_j] = m^2$. The mean and variance of $S_n$ are then $E[S_n] = nE[X_j] = nm$ and $\text{VAR}[S_n] = n\,\text{VAR}[X_j] = nm^2$. The central limit theorem then gives

$$
\begin{aligned}
P[950m \le S_{1000} \le 1050m] \\
&= P\left[\frac{950m - 1000m}{m\sqrt{1000}} \le Z_n \le \frac{1050m - 1000m}{m\sqrt{1000}}\right] \\
&\simeq Q(1.58) - Q(-1.58) \\
&= 1 - 2Q(1.58) \\
&= 1 - 2(0.0567) = .8866
\end{aligned}
$$

Thus as $n$ becomes large, $S_n$ is very likely to be close to its mean $nm$. We can therefore conjecture that the long-term average rate at which events occur is

$$
\frac{n \text{ events}}{S_n \text{ seconds}} = \frac{n}{nm} = \frac{1}{m} \text{ events/second.}
$$

The calculation of event occurrence rates and related averages is discussed in Section 5.5.

---

### Gaussian Approximation for Binomial Probabilities

We found in Chapter 2 that the binomial random variable becomes difficult to compute directly for large $n$ because of the need to calculate factorial terms. A particularly important application of the central limit theorem is in the approximation of binomial probabilities. Since the binomial random variable is a sum of iid Bernoulli random variables (which have finite mean and variance), its cdf approaches that of a Gaussian random variable. Let $X$ be a binomial random variable with mean $np$ and variance $np(1 - p)$, and let $Y$ be a Gaussian random variable with the same mean and variance, then by the Central limit theorem for $n$ large the probability that $X = k$ is approximately equal to the integral of the Gaussian pdf in an interval of unit length about $k$:

$$
\begin{aligned}
P[X = k] &\simeq P\left[k - \frac{1}{2} < Y < k + \frac{1}{2}\right] \\
&= \frac{1}{\sqrt{2\pi}} \int_{k-1/2}^{k+1/2} e^{-(x-np)^2/2np(1-p)}\, dx. \qquad (5.29)
\end{aligned}
$$

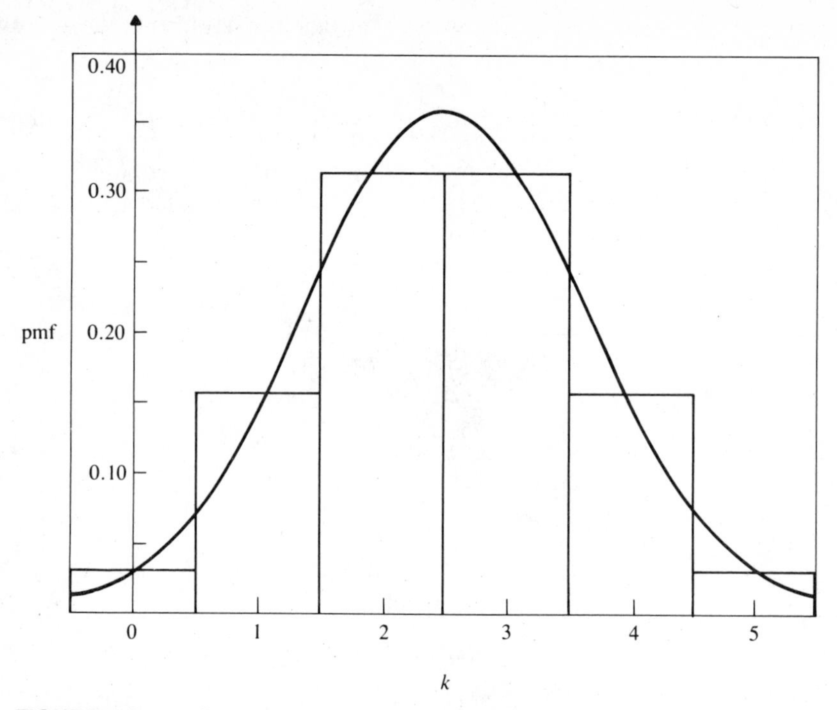


**FIGURE 5.6a** Gaussian approximation for binomial probabilities with $n = 5$ and $p = 1/2$.

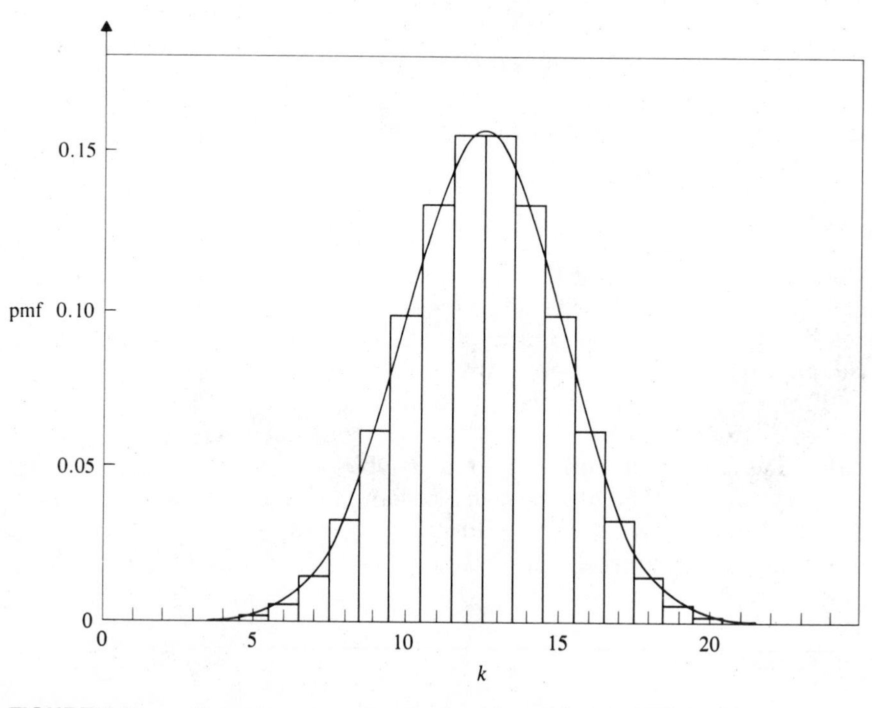


**FIGURE 5.6b** Gaussian approximation for binomial probabilities with $n = 25$ and $p = 1/2$.

The above approximation can be simplified by approximating the integral by the product of the integrand at the center of the interval of integration (that is, $x = k$) and the length of the interval of integration (one):

$$P[X = k] \simeq \frac{1}{\sqrt{2\pi}} e^{-(k-np)^2/2np(1-p)}. \tag{5.30}$$

Figures 5.6(a) and 5.6(b) compare the binomial probabilities and the Gaussian approximation using Eq. (5.30).

### ■■ Example 5.14

In Example 5.10 in Section 5.2, we used the Chebyshev inequality to estimate the number of samples required in order for there to be a .95 probability that the relative frequency estimate for the probability of an event $A$ to be within 0.01 of $P[A]$. We now estimate the required number of samples using the Gaussian approximation for the binomial distribution.

Let $f_A(n)$ be the relative frequency of $A$ in $n$ Bernoulli trials. Since $f_A(n)$ has mean $p$ and variance $p(1-p)/n$, then

$$Z_n = \frac{f_A(n) - p}{\sqrt{p(1-p)/n}}$$

has zero mean and unit variance, and is approximately Gaussian for $n$ sufficiently large. The probability of interest is

$$P[|f_A(n) - p| < \varepsilon] \simeq P\left[|Z_n| < \frac{\varepsilon\sqrt{n}}{\sqrt{p(1-p)}}\right]$$
$$= 1 - 2Q\left(\frac{\varepsilon\sqrt{n}}{\sqrt{p(1-p)}}\right).$$

The above probability cannot be computed because $p$ is unknown. However, it can be easily shown that $p(1-p) \leq 1/4$ for $p$ in the unit interval. It then follows that for such $p$, $\sqrt{p(1-p)} \leq 1/2$, and since $Q(x)$ decreases with increasing argument

$$P[|f_A(n) - p| < \varepsilon] < 1 - 2Q(2\varepsilon\sqrt{n}).$$

We want the above probability to equal .95. This implies that $Q(2\varepsilon\sqrt{n}) = (1 - .95)/2 = .025$. From Table 3.3, we see that the argument of $Q(x)$ should be approximately 1.95, thus

$$2\varepsilon\sqrt{n} = 1.95.$$

Solving for $n$, we obtain

$$n = (.98)^2/\varepsilon^2 = 9506.$$

■■

### *Proof of the Central Limit Theorem

We now sketch a proof of the central limit theorem. First note that

$$Z_n = \frac{S_n - n\mu}{\sigma\sqrt{n}} = \frac{1}{\sigma\sqrt{n}} \sum_{k=1}^{n} (X_k - \mu).$$

The characteristic function of $Z_n$ is given by

$$\begin{aligned}
\Phi_{Z_n}(\omega) &= E[e^{j\omega Z_n}] \\
&= E\left[\exp\left\{\frac{j\omega}{\sigma\sqrt{n}} \sum_{k=1}^{n} (X_k - \mu)\right\}\right] \\
&= E\left[\prod_{k=1}^{n} e^{j\omega(X_k - \mu)/\sigma\sqrt{n}}\right] \\
&= \prod_{k=1}^{n} E\left[e^{j\omega(X_k - \mu)/\sigma\sqrt{n}}\right] \\
&= \{E[e^{j\omega(X - \mu)/\sigma\sqrt{n}}]\}^n.
\end{aligned} \tag{5.31}$$

The third equality follows from the independence of the $X_k$'s and the last equality follows from the fact that the $X_k$'s are identically distributed.

By expanding the exponential in the expression, we obtain an expression in terms of $n$ and the central moments of $X$:

$$\begin{aligned}
&E[e^{j\omega(X - \mu)/\sigma\sqrt{n}}] \\
&= E\left[1 + \frac{j\omega}{\sigma\sqrt{n}}(X - \mu) + \frac{(j\omega)^2}{2!\,n\sigma^2}(X - \mu)^2 + R(\omega)\right] \\
&= 1 + \frac{j\omega}{\sigma\sqrt{n}} E[(X - \mu)] + \frac{(j\omega)^2}{2!\,n\sigma^2} E[(X - \mu)^2\} + E[R(\omega)].
\end{aligned}$$

Noting that $E[(X - \mu)] = 0$ and $E[(X - \mu)^2] = \sigma^2$, we have

$$E[e^{j\omega(X - \mu)/\sigma\sqrt{n}}] = 1 - \frac{\omega^2}{2n} + E[R(\omega)]. \tag{5.32}$$

The term $E[R(\omega)]$ can be neglected relative to $\omega^2/2n$ as $n$ becomes large. If we substitute Eq. (5.32) into Eq. (5.31), we obtain

$$\begin{aligned}
\Phi_{Z_n}(\omega) &= \left\{1 - \frac{\omega^2}{2n}\right\}^n \\
&\to e^{-\omega^2/2} \quad \text{as } n \to \infty.
\end{aligned}$$

The latter expression is the characteristic function of a zero-mean, unit-variance Gaussian random variable. Thus the cdf of $Z_n$ approaches the cdf of a zero-mean, unit-variance Gaussian random variable.

# *5.4 CONFIDENCE INTERVALS

The sample mean estimator $M_n$ provides us with a single numerical value for the estimate of $E[X] = \mu$, namely,

$$M_n = \frac{1}{n}\sum_{j=1}^{n} X_j. \tag{5.33}$$

In order to know how good is the estimate provided by $M_n$, for a particular sample of observations $\mathbf{X} = (X_1, \ldots, X_n)$, we can compute the **sample variance**,[1] which is the average dispersion about $M_n$:

$$V_n^2 = \frac{1}{n-1}\sum_{j=1}^{n} (X_j - M_n)^2. \tag{5.34}$$

If $V_n^2$ is small, then the observations are tightly clustered about $M_n$, and we can be confident that $M_n$ is close to $E[X]$. On the other hand, if $V_n^2$ is large, the samples are widely dispersed about $M_n$ and we cannot be confident that $M_n$ is close to $E[X]$. In this section we introduce the notion of confidence intervals, which approach the question in a different way.

Instead of seeking a single value that we designate to be the "estimate" of the parameter of interest (i.e., $E[X] = \mu$), we can attempt to specify an *interval of values* that is highly likely to contain the true value of the parameter. In particular, we can begin by specifying some high probability, say $1 - \alpha$, and we can then pose the following problem: Find an interval $[l(\mathbf{X}), u(\mathbf{X})]$ such that

$$P[l(\mathbf{X}) \le \mu \le u(\mathbf{X})] = 1 - \alpha, \tag{5.35}$$

that is, the interval contains the true value of the parameter with probability $1 - \alpha$. We say that such an interval is a $(1 - \alpha) \times 100\%$ **confidence interval**.

This approach simultaneously handles the question of the accuracy and confidence of an estimate. The probability $1 - \alpha$ is a measure of the consistency, and hence degree of confidence, with which the interval contains the desired parameter: If we were to compute confidence intervals a large number of times, we would find that approximately $(1 - \alpha) \times 100\%$ of the time, the computed intervals would contain the true value of the parameter. For this reason, $1 - \alpha$ is called the **confidence level**. The width of a confidence interval is a measure of the accuracy with which we can pinpoint the estimate of a parameter. The narrower the confidence interval, the more accurately we can specify the estimate for a parameter.

1. If the sum is divided by $n - 1$ instead of $n$, then $E[V_n^2] = \sigma^2$. See Problem 21.

The probability in Eq. (5.35) clearly depends on the pdf of the $X_j$'s. In the remainder of this section, we will obtain confidence intervals in the cases where the $X_j$'s are Gaussian random variables or can be approximated by Gaussian random variables.

### Case 1: $X_j$'s Gaussian; Unknown Mean and Known Variance

Suppose that the $X_j$'s are iid Gaussian random variables with unknown mean $\mu$ and known variance $\sigma^2$. From Example 5.3 and Eqs. (5.17) and (5.18), $M_n$ is then a Gaussian random variable with mean $\mu$ and variance $\sigma^2/n$, thus

$$1 - 2Q(z) = P\left[-z \le \frac{M_n - \mu}{\sigma/\sqrt{n}} \le z\right]$$

$$= P\left[M_n - \frac{z\sigma}{\sqrt{n}} \le \mu \le M_n + \frac{z\sigma}{\sqrt{n}}\right]. \tag{5.36}$$

This equation states that the interval $(M_n - z\sigma/\sqrt{n}, M_n + z\sigma/\sqrt{n})$ contains $\mu$ with probability $1 - 2Q(z)$. If we let $z_{\alpha/2}$ be such that $\alpha = 2Q(z_{\alpha/2})$, then

$$(M_n - z_{\alpha/2}\sigma/\sqrt{n}, M_n + z_{\alpha/2}\sigma/\sqrt{n}) \tag{5.37}$$

is a $(1 - \alpha) \times 100\%$ confidence interval for $\mu$.

The confidence interval in Eq. (5.37) depends on the sample mean $M_n$, the variance $\sigma^2$ of the $X_j$'s, the number of measurements $n$, and the confidence level $1 - \alpha$. Table 5.1 shows the values of $z_{\alpha/2}$ corresponding to typical values of $\alpha$.

### ■■ Example 5.15

A voltage $Y$ is given by

$$X = v + N,$$

where $v$ is an unknown constant voltage and $N$ is a random noise voltage that has a Gaussian pdf with zero mean, and variance 1 microvolt$^2$. Find the 95% confidence interval for $v$ if the voltage $Y$ is measured 100 independent times and the sample mean is found to be 5.25 microvolts.

**TABLE 5.1** Values of $z_{\alpha/2}$ for Calculating Confidence Intervals in Eq. (5.37)

| $1 - \alpha$ | .90 | .95 | .99 |
|---|---|---|---|
| $z_{\alpha/2}$ | 1.645 | 1.960 | 2.576 |

From Example 3.35, we know that the voltage $X$ is a Gaussian random variable with mean $v$ and variance 1. Thus the 100 measurements $X_1, X_2, \ldots, X_{100}$ are iid Gaussian random variables with mean $v$ and variance 1. The confidence interval is given by Eq. (5.37) with $z_{\alpha/2} = 1.96$:

$$\left(5.25 - \frac{1.96(1)}{10}, 5.25 + \frac{1.96(1)}{10}\right) = (5.05, 5.45)$$

■■

### Case 2: $X_j$'s Gaussian; Mean and Variance Unknown

Suppose that the $X_j$'s are iid Gaussian random variables with unknown mean $\mu$ and unknown variance $\sigma^2$, and that we are interested in finding a confidence interval for the mean $\mu$. Suppose we do the obvious thing in the confidence interval given by Eq. (5.37) by replacing the variance $\sigma^2$ by its estimate, the sample variance $V_n^2$ as given by Eq. (5.34):

$$\left(M_n - \frac{zV_n}{\sqrt{n}}, M_n + \frac{zV_n}{\sqrt{n}}\right). \tag{5.38}$$

The probability for the interval in Eq. (5.38) is

$$P\left[-z \le \frac{M_n - \mu}{V_n/\sqrt{n}} \le z\right] = P\left[M_n - \frac{zV_n}{\sqrt{n}} \le \mu \le M_n + \frac{zV_n}{\sqrt{n}}\right]. \tag{5.39}$$

The random variable involved in Eq. (5.39) is

$$\begin{aligned} W &= \frac{M_n - \mu}{V_n/\sqrt{n}} = \frac{\sqrt{n}\,(M_n - \mu)/\sigma}{V_n/\sigma} \\ &= \frac{(M_n - \mu)/(\sigma/\sqrt{n})}{\{[(n-1)V_n^2/\sigma^2]/(n-1)\}^{1/2}}. \end{aligned} \tag{5.40}$$

The numerator in the above equation is a zero-mean, unit-variance Gaussian random variable since the sample mean is Gaussian with mean $\mu$ and variance $\sigma^2/n$. It can be shown that $(n-1)V_n^2/\sigma^2$ is a chi-square random variable with $n - 1$ degrees of freedom. Furthermore, it can be shown that for independent Gaussian random variables, the sample mean and variance are independent random variables (Ross [1985, 121]), so the numerator and denominator of Eq. (5.39) are independent random variables. In Example 4.37 we showed that $W$, the ratio in Eq. (5.40), has a Student's $t$-distribution[2] with $n - 1$ degrees of freedom:

$$f_{n-1}(y) = \frac{\Gamma(n/2)}{\Gamma((n-1)/2)\sqrt{\pi(n-1)}}\left(1 + \frac{y^2}{n-1}\right)^{-n/2}. \tag{5.41}$$

2. The distribution is named after W. S. Gosset who published under the pseudonym "A. Student."

Let $F_{n-1}(y)$ be the cdf corresponding to $f_{n-1}(y)$, then the probability in Eq. (5.39) is given by

$$P\left[M_n - \frac{zV_n}{\sqrt{n}} \le \mu \le M_n + \frac{zV_n}{\sqrt{n}}\right] = \int_{-z}^{z} f_{n-1}(y)\,dy$$

$$= 1 - 2F_{n-1}(-z), \tag{5.42}$$

where we have used the fact that $f_{n-1}(y)$ is symmetric about $y = 0$. The Student's $t$-distribution for various degrees of freedom can be found in tables, e.g., reference [6]. In order to obtain a confidence interval with confidence level $1 - \alpha$, we need to find the value $z_{\alpha/2,n-1}$ for which $\alpha = 2F_{n-1}(-z_{\alpha/2,n-1})$. The $(1 - \alpha) \times 100\%$ confidence interval for the mean $\mu$ is then given by

$$(M_n - z_{\alpha/2,n-1}V_n/\sqrt{n}, M_n + z_{\alpha/2,n-1}V_n/\sqrt{n}). \tag{5.43}$$

The confidence interval in Eq. (5.43) depends on the sample mean $M_n$ and the sample variance $V_n^2$, the number of measurements $n$, and the confidence level $1 - \alpha$. Table 5.2 shows various values of $z_{\alpha/2,n-1}$ for typical values of $1 - \alpha$ and $n$.

For a given $1 - \alpha$, the confidence intervals given by Eq. (5.43) should be wider than those given by Eq. (5.37) since the former assumes that the

**TABLE 5.2** Values of $z_{\alpha/2,n-1}$ for Calculating Confidence Intervals in Eq. (5.43)

| | $1 - \alpha$ | | |
|---|---|---|---|
| $n - 1$ | .90 | .95 | .99 |
| 1 | 6.314 | 12.706 | 63.657 |
| 2 | 2.920 | 4.303 | 9.925 |
| 3 | 2.353 | 3.182 | 5.841 |
| 4 | 2.132 | 2.776 | 4.604 |
| 5 | 2.015 | 2.571 | 4.032 |
| 6 | 1.943 | 2.447 | 3.707 |
| 7 | 1.895 | 2.365 | 3.499 |
| 8 | 1.860 | 2.306 | 3.355 |
| 9 | 1.833 | 2.262 | 3.250 |
| 10 | 1.812 | 2.228 | 3.169 |
| 15 | 1.753 | 2.131 | 2.947 |
| 20 | 1.725 | 2.086 | 2.845 |
| 30 | 1.697 | 2.042 | 2.750 |
| 40 | 1.684 | 2.021 | 2.704 |
| 60 | 1.671 | 2.000 | 2.660 |
| $\infty$ | 1.645 | 1.960 | 2.576 |

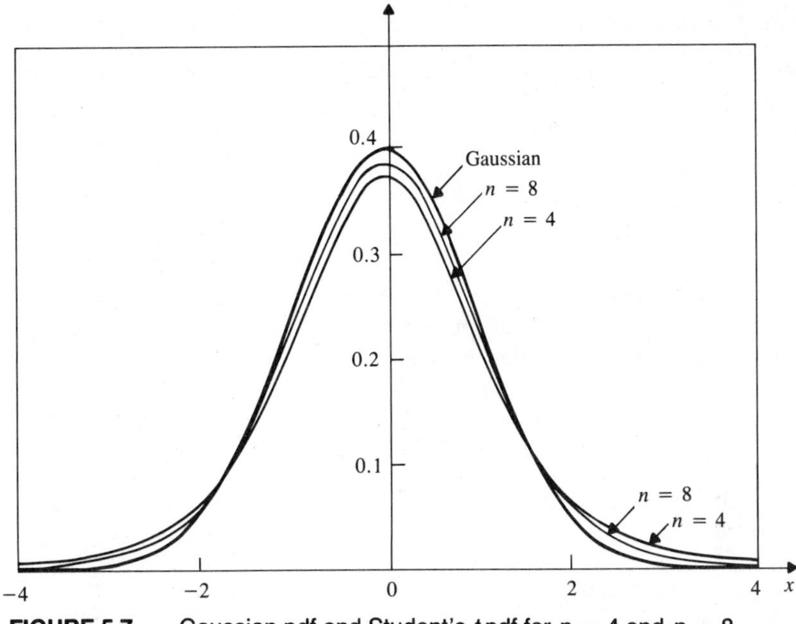


**FIGURE 5.7** Gaussian pdf and Student's $t$ pdf for $n = 4$ and $n = 8$.

variance is unknown. Figure 5.7 compares the Gaussian pdf and the Student's $t$ pdf. It can be seen that the Student's $t$ pdf's are more dispersed than the Gaussian pdf and so they indeed lead to wider confidence intervals. On the other hand, since the accuracy of the sample variance increases with $n$, we can expect that the confidence interval given by Eq. (5.43) should approach that given by Eq. (5.37). It can be seen from Fig. 5.7 that the Student's $t$ pdf's do approach the pdf of a zero-mean, unit-variance Gaussian random variable with increasing $n$. This confirms that Eqs. (5.37) and (5.43) give the same confidence intervals for large $n$. Thus the bottom row ($n = \infty$) of Table 5.2 yields the same confidence intervals as Table 5.1.

## ■■ Example 5.16

The lifetime of a certain device is assumed to have a Gaussian distribution. Eight devices are tested and the sample mean and sample variance for the lifetime obtained are 10 days and 4 days$^2$. Find the 99% confidence interval for the mean lifetime of the device.

For a 99% confidence interval and $n - 1 = 7$ Table 5.2 gives $z_{\alpha/2,7} = 3.499$. Thus the confidence interval is given by

$$\left(10 - \frac{(3.499)(2)}{\sqrt{8}}, 10 + \frac{(3.499)(2)}{\sqrt{8}}\right) = (7.53, 12.47).$$

■■

### Case 3: $X_j$'s Non-Gaussian; Mean and Variance Unknown

Equation (5.43) is used routinely to compute confidence intervals in experimental measurements and in computer simulation studies. The use of the method is justified only if the samples $X_j$ are iid and approximately Gaussian.

If the random variables $X_j$ are not Gaussian, the above method for computing confidence intervals can be modified using the **method of batch means**. This method involves performing a series of $M$ independent experiments in which the sample mean of the random variable is computed. If we assume that in each experiment each sample mean is calculated from a large number $n$ of iid observations, then the central limit theorem implies that the sample mean in each experiment is approximately Gaussian. We can therefore compute a confidence interval from Eq. (5.43) using the set of $M$ sample means as the $X_j$'s.

**■■ Example 5.17**

A computer simulation program generates exponentially distributed random variables of unknown mean. Two hundred samples of these random variables are generated and grouped into 10 batches of 20 samples each. The sample means of the 10 batches are given below:

| | | | | |
|---|---|---|---|---|
| 1.04190 | 0.64064 | 0.80967 | 0.75852 | 1.12439 |
| 1.30220 | 0.98478 | 0.64574 | 1.39064 | 1.26890 |

Find the 90% confidence interval for the mean of the random variable.

The sample mean and the sample variance of the batch sample means are calculated from the above data and found to be

$$M_{10} = 0.99674 \qquad V_{10}^2 = 0.07586.$$

The 90% confidence interval is given by Eq. 5.43 with $z_{\alpha/2,9} = 1.833$ from Table 5.2:

$$(0.83709, 1.15639).$$

This confidence interval suggests that $E[X] \simeq 1$. Indeed the simulation program used to generate the above data was set to produce exponential random variables with mean one. ■■

## *5.5

## LONG-TERM ARRIVAL RATES AND ASSOCIATED AVERAGES

In many problems events of interest occur at random times, and we are interested in the long-term average rate at which the events occur. For example, suppose that a new electronic component is installed at time $t = 0$ and that it fails at time $X_1$; an identical new component is installed immediately, and it fails after $X_2$ seconds, and so on. Let $N(t)$ be the number of components that have failed by time $t$. $N(t)$ is called a **renewal counting process**. In this section, we are interested in the behavior of $N(t)/t$ as $t$ becomes very large.

Let $X_j$ denote the lifetime of the $j$th component, then the time when the $n$th component fails is given by

$$S_n = X_1 + X_2 + \cdots + X_n, \tag{5.44}$$

where we assume that the $X_j$ are iid nonnegative random variables with $0 < E[X] = E[X_j] < \infty$. We say that $S_n$ is the time of the $n$th arrival or renewal, and we call the $X_j$'s the *interarrival or cycle times*. Figure 5.8 shows a realization of $N(t)$ and the associated sequence of interarrival times. The lines in the time axis indicate the arrival times. Note that $N(t)$ is a nondecreasing, integer-valued staircase function of time that increases without bound as $t$ approaches infinity.

Since the mean interarrival time is $E[X]$ seconds per event, we expect intuitively that $N(t)$ grows at a rate of $1/E[X]$ events per second. We will now use the strong law of large numbers to show this is the case. The average arrival rate in the first $t$ seconds is given by $N(t)/t$. We will show that with probability one, $N(t)/t \to 1/E[X]$ as $t \to \infty$.

Since $N(t)$ is the number of arrivals up to time $t$, then $S_{N(t)}$ is the time of the last arrival prior to time $t$, and $S_{N(t)+1}$ is the time of the first arrival after time $t$ (see Fig. 5.9). Therefore

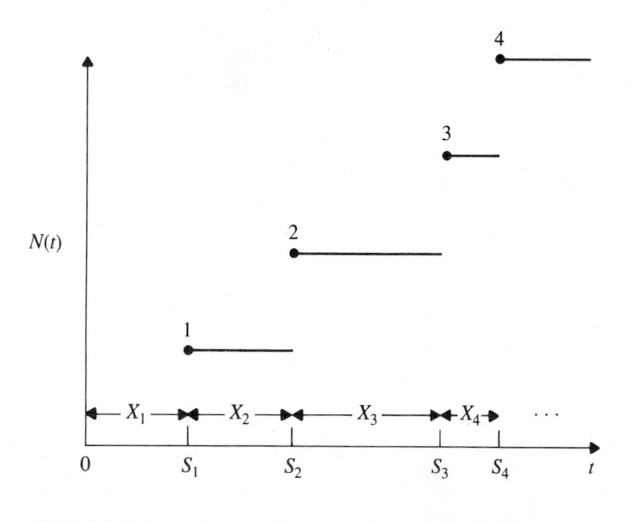


**FIGURE 5.8** A counting process and its interarrival times.

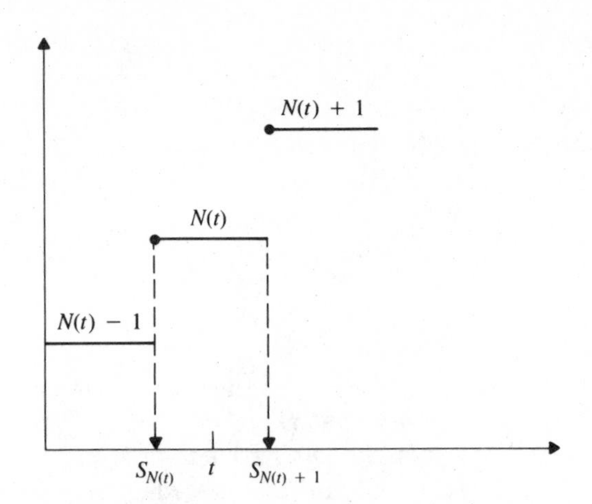


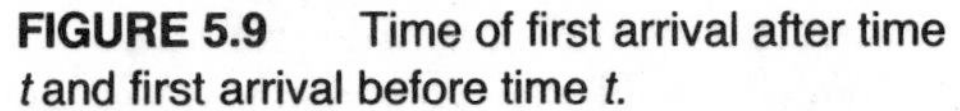

**FIGURE 5.9** Time of first arrival after time *t* and first arrival before time *t*.

$$S_{N(t)} \leq t < S_{N(t)+1}.$$

If we divide the above equation by $N(t)$, we obtain

$$\frac{S_{N(t)}}{N(t)} \leq \frac{t}{N(t)} < \frac{S_{N(t)+1}}{N(t)}. \tag{5.45}$$

The term on the left-hand side is the sample average interarrival time for the first $N(t)$ arrivals:

$$\frac{S_{N(t)}}{N(t)} = \frac{1}{N(t)} \sum_{j=1}^{N(t)} X_j.$$

As $t \to \infty$, $N(t)$ approaches infinity so the above sample average converges to $E[X]$, with probability one, by the strong law of large numbers. We now show that the term on the right-hand side also approaches $E[X]$:

$$\frac{S_{N(t)+1}}{N(t)} = \left(\frac{S_{N(t)+1}}{N(t)+1}\right)\left(\frac{N(t)+1}{N(t)}\right).$$

As $t \to \infty$, the first term on the right-hand side approaches $E[X]$ and the second term approaches 1 with probability one. Thus the lower and upper terms in Eq. (5.45) both approach $E[X]$ with probability one as $t$ approaches infinity. We have proved the following theorem:

**THEOREM 1: ARRIVAL RATE FOR iid INTERARRIVALS**

Let $N(t)$ be the counting process associated with the iid interarrival sequence $X_j$, with $0 < E[X_j] = E[X] < \infty$. Then with probability one,

$$\lim_{t\to\infty} \frac{N(t)}{t} \to \frac{1}{E[X]}. \tag{5.46}$$

### ■■ Example 5.18

Customers arrive at a service station with iid exponential interarrival times with mean $E[X_j] = 1/\alpha$. Find the long-term average arrival rate.

From Theorem 1, it immediately follows that with probability one,

$$\lim_{t\to\infty} \frac{N(t)}{t} = \frac{1}{\alpha^{-1}} = \alpha.$$

Thus $\alpha$ represents the long-term average arrival rate. ■■

---

### ■■ Example 5.19

Let $U_j$ be the "up" time during which a system is continuously functioning, and let $D_j$ be the "down" time required to repair the system when it breaks down. Find the long-term average rate at which repairs need to be done.

Define a repair cycle to consist of an "up" time followed by a "down" time, $X_j = U_j + D_j$, then the average cycle time is $E[U] + E[D]$. The number of repairs required by time $t$ is $N(t)$ and by Theorem 1, the rate at which repairs need to be done is

$$\lim_{t\to\infty} \frac{N(t)}{t} = \frac{1}{E[U] + E[D]}.$$

■■

---

## Long-Term Time Averages

Suppose that events occur at random with iid interevent times $X_j$, and that a cost $C_j$ is associated with each occurrence of an event. Let $C(t)$ be the cost incurred up to time $t$. We now determine the long term behavior of $C(t)/t$, that is, the long-term average rate at which costs are incurred.

We assume that the pairs $(X_j, C_j)$ form a sequence of iid random vectors, but that $X_j$ and $C_j$ need not be independent, that is, the cost associated with an event may depend on the associated interevent time. The total cost $C(t)$ incurred up to time $t$ is then the sum of costs associated with the $N(t)$ events that have occurred up to time $t$:

$$C(t) = \sum_{j=1}^{N(t)} C_j. \tag{5.47}$$

The time average of the cost up to time $t$ is $C(t)/t$, thus

$$\begin{aligned} \frac{C(t)}{t} &= \frac{1}{t}\sum_{j=1}^{N(t)} C_j \\ &= \frac{N(t)}{t}\left\{\frac{1}{N(t)}\sum_{j=1}^{N(t)} C_j\right\}. \end{aligned} \tag{5.48}$$

By Theorem 1, as $t \to \infty$, the first term on the right-hand side approaches

$1/E[X]$ with probability one. The expression inside the brackets is simply the sample mean of the first $N(t)$ costs. As $t \to \infty$, $N(t)$ approaches infinity so the second term approaches $E[C]$ with probability one, by the strong law of large numbers. Thus we have the following theorem:

**THEOREM 2**

Let $(X_j, C_j)$ be a sequence of iid interevent times and associated costs, with $0 < E[X_j] < \infty$ and $E[C_j] < \infty$, and let $C(t)$ be the cost incurred up to time $t$. Then, with probability one

$$\lim_{t\to\infty} \frac{C(t)}{t} = \frac{E[C]}{E[X]}. \tag{5.49}$$

The following series of examples demonstrate how Theorem 2 can be used to calculate long-term time averages.

### Example 5.20

Find the long-term proportion of time that the system is "up" in Example 5.19.

Let $I_U(t)$ be equal to one if the system is up at time $t$ and zero otherwise, then the long term proportion of time in which the system is up is

$$\lim_{t\to\infty} \frac{1}{t} \int_0^t I_U(t')\,dt',$$

where the integral is the total time the system is up in the time interval $[0, t]$.

Now define a cycle to consist of a system "up" time followed by a "down" time, then $X_j = U_j + D_j$, and $E[X] = E[U] + E[D]$. If we let the "cost" associated with each cycle be the "up" time $U_j$, then if $t$ is an instant when a cycle ends

$$\int_0^t I_U(t')\,dt' = \sum_{j=1}^{N(t)} U_j = C(t).$$

Thus $C(t)/t$ is the proportion of time that the system is "up" in the time interval $(0, t)$. By Theorem 2, the long-term proportion of time that the system is "up" is

$$\lim_{t\to\infty} \frac{C(t)}{t} = \frac{E[U]}{E[U] + E[D]}.$$

### Example 5.21

In the previous example, suppose that a cost $C_j$ is associated with each repair. Find the long-term average rate at which repair costs are incurred.

The mean interevent time is $E[U] + E[D]$, and the mean cost per repair is $E[C]$. Thus by Theorem 2, the long-term average repair cost rate is

$$\lim_{t\to\infty} \frac{C(t)}{t} = \frac{E[C]}{E[U] + E[D]}.$$

■■

## ■■ Example 5.22
### A Packet Voice Transmission System

A packet voice multiplexer can transmit up to $M$ packets every 10 msec period. Let $N$ be the number of packets input into the multiplexer every 10 msec. If $N \leq M$ the multiplexer transmits all $N$ packets, and if $N > M$ the multiplexer transmits $M$ packets and discards $(N - M)$ packets. Find the long-term proportion of packets discarded by the multiplexer.

Define a "cycle" by $X_j = N_j$, that is, the length of the "cycle" is equal to the number of packets produced in the $j$th interval. Define the cost in the $j$th cycle by $C_j = (N_j - M)^+ = \max(N_j - M, 0)$, that is, the number of packets that are discarded in the $j$th cycle. With these definitions, $t$ represents the first $t$ packets input into the multiplexer and $C(t)$ represents the number that had to be discarded. The long term proportion of packets discarded is then

$$\lim_{t\to\infty} \frac{C(t)}{t} = \frac{E[(N - M)^+]}{E[N]}$$

where

$$E[(N - M)^+] = \sum_{k=M}^{\infty} (k - M)p_k,$$

where $p_k$ is the pmf of $N$. This result was derived heuristically in Section 1.4, Eq. (1.10).

■■

## ■■ Example 5.23
### The Residual Lifetime

Let $X_1, X_2, \ldots$ be a sequence of interarrival times, and let the residual lifetime $r(t)$ be defined as the time from an arbitrary time instant $t$ until the next arrival as shown in Fig. 5.10. Find the long-term proportion of time that $r(t)$ exceeds $c$ seconds.

The amount of time that the residual lifetime exceeds $c$ in a cycle of length $X$ is $(X - c)^+$, that is, $X - c$ when the cycle is longer than $c$ seconds, and 0 when it is shorter than $c$ seconds. The long-term proportion of time that $r(t)$ exceeds $c$ seconds is obtained from Theorem 2 by defining

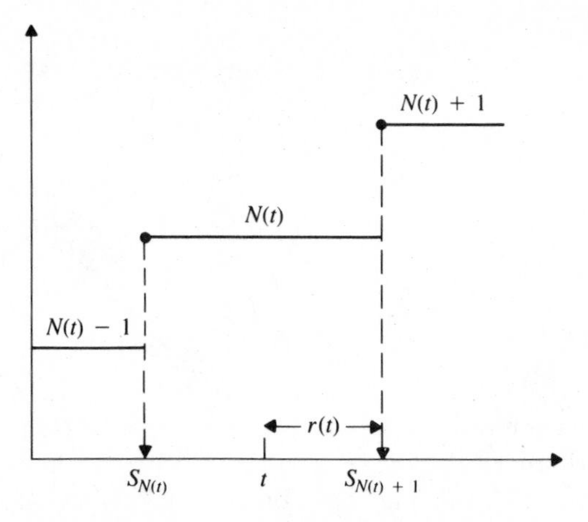

**FIGURE 5.10** Residual lifetime in a cycle.

the cost per cycle by $C_j = (X_j - c)^+$:

$$\begin{aligned}
\text{proportion of time } r(t) \text{ exceeds } c &= \frac{E[(X-c)^+]}{E[X]} \\
&= \frac{1}{E[X]}\int_0^\infty P[(X-c)^+ > x]\,dx \\
&= \frac{1}{E[X]}\int_0^\infty P[X > x + c]\,dx \\
&= \frac{1}{E[X]}\int_0^\infty \{1 - F_X(x+c)\}\,dx \\
&= \frac{1}{E[X]}\int_c^\infty \{1 - F_X(y)\}\,dy, \qquad (5.50)
\end{aligned}$$

where Eq. (3.59) was used for $E[(X-c)^+]$ in the second equality. This result is used extensively in reliability theory and in queueing theory. ■■

## *5.6

## A COMPUTER METHOD FOR EVALUATING THE DISTRIBUTION OF A RANDOM VARIABLE USING THE DISCRETE FOURIER TRANSFORM

In many situations we are forced to obtain the pmf or pdf of a random variable from its characteristic function using numerical methods because the inverse transform cannot be expressed in closed form. In the most common case, we are interested in finding the pmf/pdf corresponding to $\Phi_X(\omega)^n$, which corresponds to the characteristic function of the sum of $n$ iid random variables. In this section we introduce the discrete Fourier transform, which enables us to perform this numerical calculation in an efficient manner.

### Discrete Random Variables

First suppose that $X$ is an integer-valued random variable that takes on values in the set $\{0, 1, \ldots, N-1\}$. The characteristic function for this random variable is

$$\Phi_X(\omega) = \sum_{k=0}^{N-1} e^{j\omega k} p_k, \tag{5.51}$$

where $p_k = P[X = k]$ is the pmf. $\Phi_X(\omega)$ is a periodic function of $\omega$ with period $2\pi$ since $e^{(j(\omega+2\pi)k)} = e^{j\omega k}e^{jk2\pi} = e^{j\omega k}$.[3]

Consider the characteristic function at $N$ equally spaced values in the interval $[0, 2\pi)$:

$$c_m = \Phi_X\left(\frac{2\pi m}{N}\right) = \sum_{k=0}^{N-1} p_k e^{j2\pi km/N} \qquad m = 0, 1, \ldots, N-1. \tag{5.52}$$

Equation (5.52) defines the **discrete Fourier transform** (DFT) of the sequence $p_0, \ldots, p_{N-1}$. (The sign in the exponent in Eq. (5.52) is the opposite of that used in the usual definition of the DFT.) In general, the $c_m$'s are complex numbers. Note that if we extend the range of $m$ outside the range $\{0, N-1\}$ we obtain a periodic sequence consisting of a repetition of the basic sequence $c_0, \ldots, c_{N-1}$.

The sequence of $p_k$'s can be obtained from the sequence of $c_m$'s using the inverse DFT formula:

$$p_k = \frac{1}{N} \sum_{m=0}^{N-1} c_m e^{-j2\pi km/N} \qquad k = 0, 1, \ldots, N-1. \tag{5.53}$$

#### ■■ Example 5.24

A discrete random variable $X$ has pmf

$$p_0 = \frac{1}{2}, \qquad p_1 = \frac{3}{8}, \qquad \text{and} \qquad p_2 = \frac{1}{8}.$$

Find the characteristic function of $X$, the DFT for $N = 3$, and verify the inverse transform formula.

3. This follows from Euler's formula $e^{j\theta} = \cos\theta + \sin\theta$.

The characteristic function of $X$ is given by Eq. (5.51):

$$\Phi_X(\omega) = \frac{1}{2} + \frac{3}{8}e^{j\omega} + \frac{1}{8}e^{j2\omega}.$$

The DFT for $N = 3$ is given by the values of the characteristic function at $\omega = 2\pi m/3$, for $m = 0, 1, 2$:

$$c_0 = \Phi_X(0) = 1$$

$$c_1 = \Phi_X\left(\frac{2\pi}{3}\right) = \frac{1}{2} + \frac{3}{8}e^{j2\pi/3} + \frac{1}{8}e^{j4\pi/3}$$

$$= \frac{1}{2} + \frac{3}{8}(-.5 + j(.75)^{1/2}) + \frac{1}{8}(-.5 - j(.75)^{1/2})$$

$$= \frac{1}{4} + \frac{j(.75)^{1/2}}{4}$$

$$c_2 = \Phi_X\left(\frac{4\pi}{3}\right) = \frac{1}{2} + \frac{3}{8}e^{j4\pi/3} + \frac{1}{8}e^{j8\pi/3},$$

$$= \frac{1}{4} - \frac{j(.75)^{1/2}}{4}$$

where we have used Euler's formula to evaluate the complex exponentials.

We substitute the $c_j$'s into Eq. (5.53) to recover the pmf:

$$p_0 = \frac{1}{3}(c_0 + c_1 + c_2)$$

$$= \frac{1}{3}\left(1 + \frac{1}{4} + \frac{j(.75)^{1/2}}{4} + \frac{1}{4} - \frac{j(.75)^{1/2}}{4}\right)$$

$$= \frac{1}{2}$$

$$p_1 = \frac{1}{3}(c_0 + c_1 e^{-j2\pi/3} + c_2 e^{-j2\pi 2/3}) = \frac{3}{8}$$

$$p_2 = \frac{1}{3}(c_0 + c_1 e^{-j4\pi/3} + c_2 e^{-j4\pi 2/3}) = \frac{1}{8}.$$

■■

---

The range of the integer-valued random variable $X$ can be extended to the larger set $\{0, 1, \ldots, N-1, N, \ldots, L-1\}$ by defining a new pmf $p'_j$ given by

$$p'_j = \begin{cases} p_j & 0 \le j \le N-1 \\ 0 & N \le j \le L-1. \end{cases} \tag{5.54}$$

The characteristic function of the random variable, $\Phi_X(\omega)$ remains

ubstituting these values into Eq. (5.56) with $k = 1$ gives

$$P[Z = 1] = \frac{1}{3}\{d_0 + d_1 e^{-j2\pi/3} + d_2 e^{-j4\pi/3}\}$$
$$= \frac{1}{3}\left\{1 - \frac{1}{3}(e^{-j2\pi/3} + e^{-j4\pi/3})\right\}$$
$$= \frac{4}{9}.$$

We can verify this answer is correct by noting that

$$P[Z = 1] = P[\{X_1 = 0\} \cap (X_2 = 1\}] + P[\{X_1 = 1\} \cap \{X_2 = 0\}]$$
$$= \frac{1}{3}\frac{2}{3} + \frac{2}{3}\frac{1}{3} = \frac{4}{9}.$$

■■

In practice we are interested in using the DFT when the number of points in the pmf is large. An examination of Eq. (5.53) shows that the calculation of all $N$ points requires approximately $N^2$ multiplications of complex numbers. Thus if $N = 2^{10} = 1024$, approximately $10^6$ multiplica-

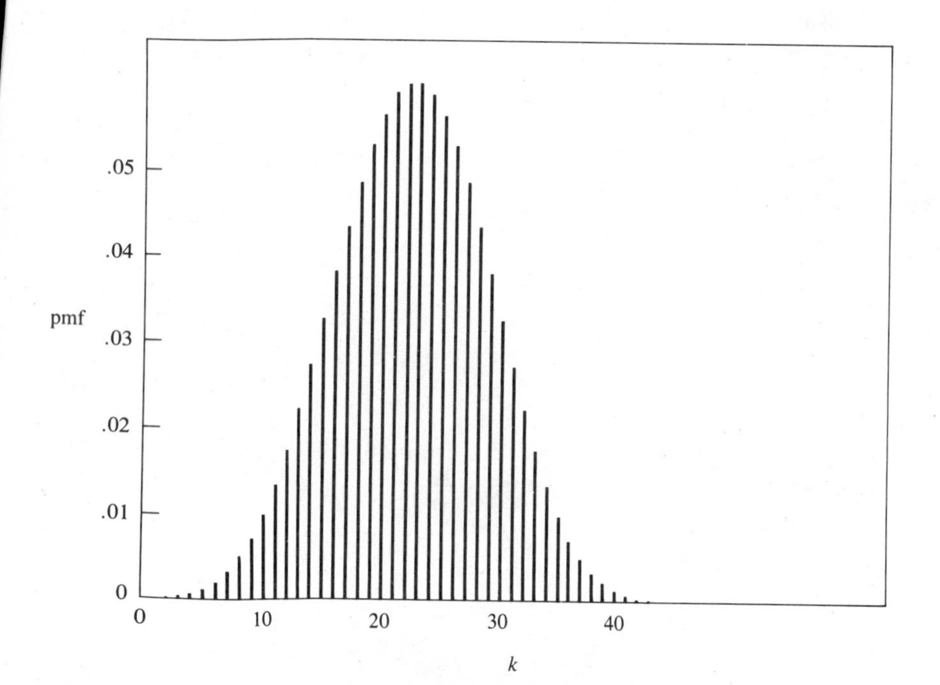


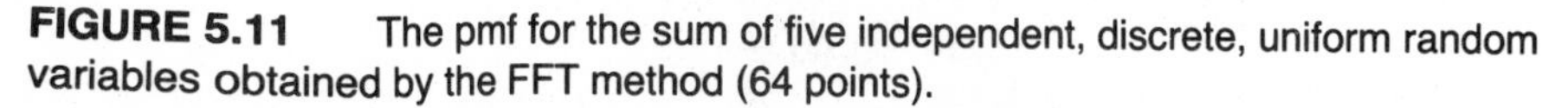

**FIGURE 5.11** The pmf for the sum of five independent, discrete, uniform random variables obtained by the FFT method (64 points).

unchanged, but the associated DFT now involves evaluati[illegible] different set of points:

$$c_m = \Phi_X\left(\frac{2\pi m}{L}\right) \qquad \text{for } m = 0, \ldots, L-1.$$

The inverse transform of the sequence in Eq. (5.55) then yield[illegible] Thus the pmf can be recovered using the DFT on $L \geq N$ sampl[illegible] as specified by Eq. (5.55). In essence, we have only padded t[illegible] $L - N$ zeros in Eq. (5.54).

The zero-padding method discussed above is required to ev[illegible] pmf of a sum of iid random variables. Suppose that

$$Z = X_1 + X_2 + \cdots + X_n,$$

where the $X_i$ are integer-valued iid random variables with char[illegible] function $\Phi_X(\omega)$. If the $X_i$ assume values from $\{0, 1, \ldots, N-1\}$[illegible] will assume values from $\{0, \ldots, n(N-1)\}$. The pmf of $Z$ is fou[illegible] the DFT evaluated at the $L = n(N-1) + 1$ points:

$$d_m = \Phi_Z\left(\frac{2\pi m}{L}\right) = \Phi_X\left(\frac{2\pi m}{L}\right)^n \qquad m = 0, \ldots, L-1,$$

since $\Phi_Z(\omega) = \Phi_X(\omega)^n$. Note that this requires evaluating the cha[illegible] istic function of $X$ at $L > N$ points. The pmf of $Z$ is then found from

$$P[Z = k] = \frac{1}{L}\sum_{m=0}^{L-1} d_m e^{-j2\pi km/L} \qquad k = 0, 1, \ldots, L-1.$$

### ■■ Example 5.25

Let $Z = X_1 + X_2$, where the $X_j$ are iid random variables with characteri[illegible] function:

$$\Phi_X(w) = \frac{1}{3} + \frac{2}{3}e^{j\omega}.$$

Find $P[Z = 1]$ using the DFT method.

$X$ assumes values from $\{0, 1\}$ and $Z$ from $\{0, 1, 2\}$, so $\Phi_Z(\omega) = \Phi_X(\omega)^{[illegible]}$ needs to be evaluated at three points:

$$d_m = \left\{\frac{1}{3} + \frac{2}{3}e^{j2\pi m/3}\right\}^2 \qquad m = 0, 1, 2.$$

These values are found to be

$$d_0 = 1, \qquad d_1 = -\frac{1}{3}, \qquad \text{and} \qquad d_2 = -\frac{1}{3}.$$

tions will be required. The popularity of the DFT method stems from the fact that algorithms, called **fast Fourier transform (FFT) algorithms,** have been developed that can carry out the above calculations in $N \log_2 N$ multiplications. For $N = 2^{10}$, $10^4$ multiplications will be required, a reduction by a factor of 100. Appendix 5.1 lists a FORTRAN program for one such algorithm.

### ■■ Example 5.26

Let $S_n = X_1 + \cdots + X_n$, where the $X_j$'s are iid random variables that are uniformly distributed in the set $\{0, 1, \ldots, 9\}$. The pmf for $S_n$, $n = 5$, was obtained using the FFT program in Appendix 5.1 and is shown in Fig. 5.11. ■■

---

So far, we have restricted $X$ to be an integer-valued random variable that takes on only a finite set of values $S_X = \{0, 1, \ldots, N-1\}$. We now consider the case where $S_X = \{0, 1, 2, \ldots\}$. Suppose that we know $\Phi_X(\omega)$, and that we obtain a pmf $p'_k$ from Eq. (5.53) using a finite set of sample points from $\Phi_X(\omega)$, $c_m = \Phi_X(2\pi m/N)$ for $m = 0, 1, \ldots, N-1$,

$$p'_k = \frac{1}{N} \sum_{m=0}^{N-1} c_m e^{-j2\pi km/N} \qquad k = 0, 1, \ldots, N-1. \tag{5.57}$$

To see what this calculation yields consider the points $c_m$ :

$$\begin{aligned}\Phi_X\left(\frac{2\pi m}{N}\right) &= \sum_{n=0}^{\infty} p_n e^{j2\pi mn/N} \\ &= (p_0 + p_N + \cdots)e^{j0} \\ &\quad + (p_1 + p_{N+1} + \cdots)e^{j2\pi m/N} \\ &\quad + \cdots \\ &\quad + (p_{N-1} + p_{2N-1} + \cdots)e^{j2\pi m(N-1)/N} \\ &= \sum_{k=0}^{N-1} p'_k e^{j2\pi km/N},\end{aligned} \tag{5.58}$$

where we have used the fact that $e^{j2\pi mn/N} = e^{j2\pi m(n+hN)/N}$, for $h$ an integer, to obtain the second equality and where for $k = 0, \ldots, N-1$,

$$p'_k = p_k + p_{N+k} + p_{2N+k} + \cdots. \tag{5.59}$$

Equation (5.57) states that the inverse transform of the points $c_m = \Phi_X(2\pi m/N)$ will yield $p'_0, \ldots, p'_{N-1}$, which are equal to the desired value $p_k$ plus the error

$$e_k = p_{N+k} + p_{2N+k} + \cdots.$$

Since the pmf must decay to zero as $k$ increases, the error term can be made small by making $N$ sufficiently large. The following example carries out an evaluation of the above error term in a case where the pmf is known. In practice, the pmf is not known so the appropriate value of $N$ is found by trial and error.

### Example 5.27

Suppose that $X$ is a geometric random variable. How large should $N$ be so that the percent error is 1%?

The error term for $p_k$ is given by

$$e_k = \sum_{h=1}^{\infty} p_{k+hN} = \sum_{h=1}^{\infty} (1-p)p^{k+hN} = (1-p)p^k \frac{p^N}{1-p^N}.$$

The percent error term for $p_k$ is

$$\frac{e_k}{p_k} = \frac{p^N}{1-p^N} = a \times 100\%.$$

By solving for $N$, we find that the error is less than $a = 0.01$ if

$$N > \frac{\log(a/1-a)}{\log p} \simeq \frac{-2.0}{\log_{10} p}.$$

Thus for example if $p = .1, .5, .9$, then the required $N$ is 2, 7, and 44, respectively. These numbers show how the required $N$ depends strongly on the rate of decay of the pmf.

### Continuous Random Variables

Let $X$ be a continuous random variable, and suppose that we are interested in finding the pdf of $X$ from $\Phi_X(\omega)$ using a numerical method. We can take the inverse Fourier transform formula and approximate it by a sum over intervals of width $\omega_0$:

$$f_X(x) = \frac{1}{2\pi}\int_{-\infty}^{\infty} \Phi_X(\omega)e^{-j\omega x}\,dw$$

$$\simeq \frac{1}{2\pi}\sum_{m=-M}^{M-1} \Phi_X(m\omega_0)e^{-jm\omega_0 x}\omega_0, \tag{5.60}$$

where the sum neglects the integral outside the range $[-M\omega_0, M\omega_0)$. The above sum takes on the form of a DFT if we consider the pdf in the range $[-2\pi/\omega_0, 2\pi/\omega_0)$ with $x = nd$, $d = 2\pi/N\omega_0$, and $N = 2M$:

$$f_X(nd) \simeq \frac{\omega_0}{2\pi}\sum_{m=-M}^{M-1} \Phi_X(m\omega_0)e^{-j2\pi nm/N} \qquad -M \le n \le M-1. \tag{5.61}$$

Equation (5.61) is a $2M$-point DFT of the sequence

$$c_m = \frac{\omega_0}{2\pi} \Phi_X(m\omega_0).$$

The FFT algorithm requires that $n$ range from 0 to $2M - 1$. Equation (5.61) can be cast into this form by recalling that the sequence $c_m$ is periodic with period $N$. An FFT algorithm will then calculate Eq. (5.61) if we input the sequence

$$c'_m = \begin{cases} c_m & 0 \leq m \leq M - 1 \\ c_{m-2M-1} & M < m \leq 2M - 1. \end{cases}$$

Three types of errors are introduced in approximating the pdf using Eq. (5.61). The first error involves approximating the integral by a sum. The second error results from neglecting the integral for frequencies outside the range $[-M\omega_0, M\omega_0)$. The third error results from neglecting the pdf outside the range $[-2\pi/\omega_0, 2\pi/\omega_0)$. The first and third errors are reduced by reducing $\omega_0$. The second error can be decreased by increasing $M$ while keeping $\omega_0$ fixed.

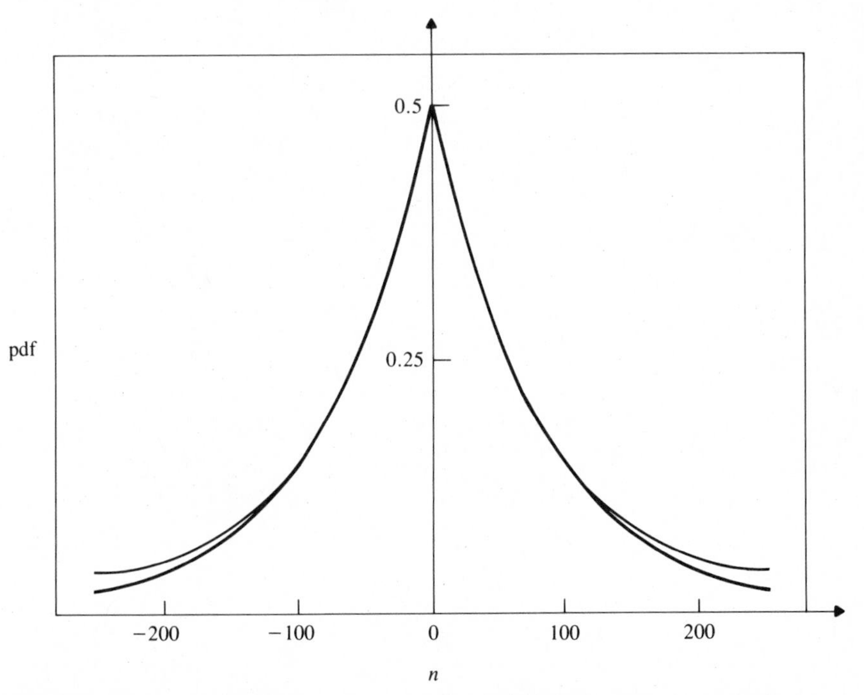


**FIGURE 5.12a** Comparison of exact pdf and pdf obtained by numerically inverting the characteristic function of a Laplacian random variable. Approximation using $\omega_0 = 1$ and $N = 512$.

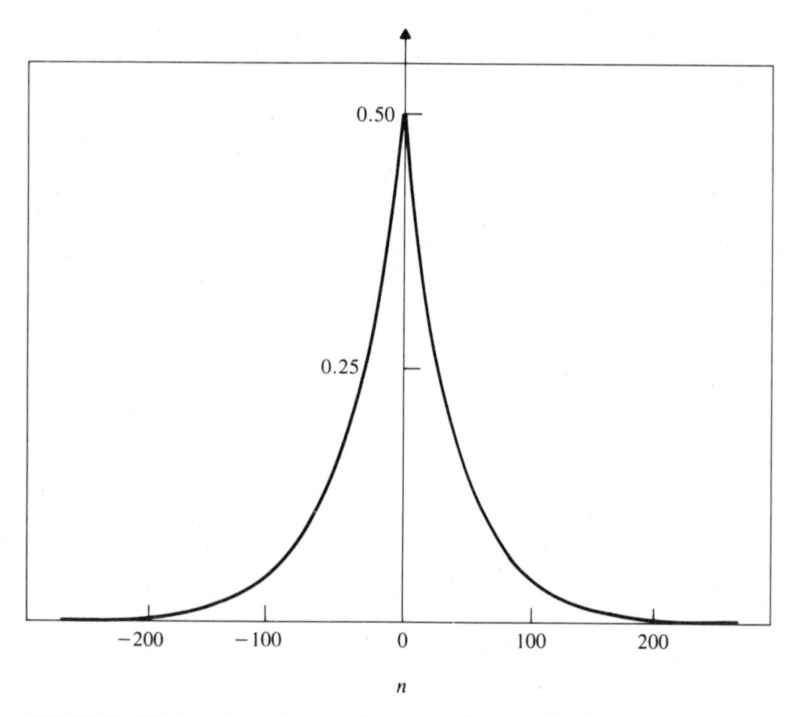


**FIGURE 5.12b** Comparison of exact pdf and pdf obtained by numerically inverting the characteristic function of a Laplacian random variable. Approximation using $\omega_0 = 1/2$ and $N = 512$.

## ■■ Example 5.28

The Laplacian random variable with parameter $\alpha = 1$ has characteristic function

$$\Phi_X(\omega) = \frac{1}{1 + \omega^2} \qquad -\infty < \omega < \infty.$$

Figures 5.12(a) and 5.12(b) compare the pdf with the approximation obtained using Eq. 5.61 with $N = 512$ points and two values of $\omega_0$. It can be seen that decreasing $\omega_0$ increases the accuracy of the approximation. ■■

---

## SUMMARY

- The expected value of a sum of random variables is always equal to the sum of the expected values of the random variables. In general, the variance of such a sum is not equal to the sum of the individual variances.

- The characteristic function of the sum of independent random variables is equal to the product of the characteristic functions of the individual random variables.
- The sample mean and the relative frequency estimators are used to estimate the expected value of random variables and the probabilities of events. The laws of large numbers state conditions under which these estimators approach the true values of the parameters they estimate as the number of samples become large.
- The central limit theorem states that the cdf of a sum of iid finite-mean, finite-variance random variables approaches that of a Gaussian random variable. This result allows us to approximate the pdf of sums of random variables by that of a Gaussian random variable.
- Confidence intervals are used to identify the region of values where a parameter of interest is likely to be, based on a series of observations. The confidence level indicates the probability with which the interval contains the parameter, and the width of the interval indicates the degree of accuracy with which the parameter can be pinpointed.
- A counting process counts the number of occurrences of an event in a certain time interval. When the time between occurrences of events are iid random variables, the strong law of large numbers enables us to obtain results concerning the rate at which events occur, and results concerning various long-term time averages.
- The discrete Fourier transform and the FFT algorithm allow us to compute numerically the pmf and pdf of random variables from their characteristic functions.

## CHECKLIST OF IMPORTANT TERMS

Arrival rate
Central limit theorem
Confidence interval
Confidence level
Counting process
Discrete Fourier transform
Fast Fourier transform
iid random variables
Method of batch means
Relative frequency
Sample mean
Sample variance
Strong law of large numbers
Student's $t$-distribution
Unbiased estimator
Weak law of large numbers

## ANNOTATED REFERENCES

See Chung [1, 220–233] for an insightful discussion of the laws of large numbers and the central limit theorem. Chapter 6 in Gnedenko [2] gives a detailed discussion of the laws of large numbers. Law and Kelton [3] give numerous examples on the computation of confidence intervals in practical situations. Chapter 7 in Ross [4] focusses on counting processes and their properties. Cadzow [5] gives a good introduction to the FFT algorithm.

1. K. L. Chung, *Elementary Probability Theory with Stochastic Processes,* Springer-Verlag, New York, 1975.
2. B. V. Gnedenko, *The Theory of Probability,* MIR Publishers, Moscow, 1976.
3. A. M. Law and W. D. Kelton, *Simulation Modeling and Analysis,* McGraw-Hill, New York, 1982.
4. S. M. Ross, *Introduction to Probability Models,* Academic Press, New York, 1985.
5. J. A. Cadzow, *Foundations of Digital Signal Processing and Data Analysis,* Macmillan, New York, 1987.
6. P. L. Meyer, *Introductory Probability and Statistical Applications,* Second Edition, Addison-Wesley Publishing Co., Reading, Mass., 1970.
7. J. W. Cooley, P. Lewis, and P. D. Welch, "The Fast Fourier Transform and Its Applications," *IEEE Transactions on Education,* vol. 12, pp. 27–34, March 1969.

## PROBLEMS

### Section 5.1
### Sums of Random Variables

1. Let $Z = X + Y + Z$, where $X$, $Y$, and $Z$ are zero-mean, unit-variance random variables with $\text{COV}(X, Y) = 1/4$ and $\text{COV}(Y, Z) = -1/4$ and $\text{COV}(X, Z) = 0$.
   a. Find the mean and variance of $Z$.
   b. Repeat Part a assuming $X$, $Y$, and $Z$ are uncorrelated random variables.
2. Let $X_1, \ldots, X_n$ be random variables with the same mean $\mu$ and with covariance function:

$$\text{COV}(X_i, X_j) = \begin{cases} \sigma^2 & \text{if } i = j \\ \rho\sigma^2 & \text{if } |i - j| = 1, \\ 0 & \text{otherwise,} \end{cases}$$

   where $|\rho| < 1$. Find the mean and variance of $S_n = X_1 + \cdots + X_n$.
3. Let $X_1, \ldots, X_n$ be random variables with the same mean $\mu$ and with

covariance function

$$\text{COV}(X_i, X_j) = \sigma^2 \rho^{-|i-j|},$$

where $|\rho| < 1$. Find the mean and variance of $S_n = X_1 + \cdots + X_n$.

4. Let $X$ and $Y$ be independent Cauchy random variables with parameters $\alpha$ and $\beta$ respectively. Let $Z = X + Y$.
   a. Find the characteristic function of $Z$.
   b. Find the pdf of $Z$ from the characteristic function found in Part a.
5. Let $S_k = X_1 + \cdots + X_k$, where the $X_i$'s are independent random variables, with $X_i$ a chi-square random variable with $n_i$ degrees of freedom. Show that $S_k$ is a chi-square random variable with $n = n_1 + \cdots + n_k$ degrees of freedom.
6. Let $S_n = X_1^2 + \cdots + X_n^2$ where the $X_i$'s are iid zero-mean, unit-variance Gaussian random variables.
   a. Show that $S_n$ is a chi-square random variable with $n$ degrees of freedom. *Hint*: See Example 3.26.
   b. Use the methods of Section 3.5 to find the pdf of
   $$T_n = \sqrt{X_1^2 + \cdots + X_n^2}.$$
   c. Show that $T_2$ is a Rayleigh random variable.
   d. Find the pdf for $T_3$. The random variable $T_3$ is used to model the speed of molecules in a gas. $T_3$ is said to have the *Maxwell distribution*.
7. Let $X$ and $Y$ be independent exponential random variables with parameters $\alpha$ and $\beta$ respectively. Let $Z = X + Y$.
   a. Find the characteristic function of $Z$.
   b. Find the pdf of $Z$ from the characteristic function found in Part a.
8. Let $Z = aX + bY$, where $X$ and $Y$ are independent random variables and $a$ and $b$ are arbitrary constants.
   a. Find the characteristic function of $Z$.
   b. Find the mean and variance of $Z$ by taking derivatives of the characteristic function found in Part a.
9. Let $M_n$ be the sample mean of $n$ iid random variables $X_j$. Find the characteristic function of $M_n$ in terms of the characteristic function of the $X_j$'s.
10. Let $S_k = X_1 + \cdots + X_k$, where the $X_i$'s are independent random variables, with $X_i$ a binomial random variable with parameters $n_i$ and $p$. Use the probability generating function to show that $S_n$ is a binomial random variable with parameters $n = n_1 + \cdots + n_k$ and $p$. Explain why this result is obvious.
11. Let $S_k = X_1 + \cdots + X_k$, where the $X_i$'s are independent random

variables, with $X_i$ a Poisson random variable with mean $\alpha_i$. Show that $S_k$ is a Poisson random variable with mean $\alpha = \alpha_1 + \cdots + \alpha_k$.

12. Let $X_1, X_2, \ldots$ be a sequence of independent integer-valued random variables, let $N$ be an integer-valued random variable independent of the $X_j$'s, and let

$$S = \sum_{k=1}^{N} X_k.$$

  a. Find the mean and variance of $S$.
  b. Show that

$$G_S(z) = E(z^S) = G_N(G_X(z)),$$

  where $G_X(z)$ is the generating function of each of the $X_k$'s.

13. Let the number of jobs arriving at a shop in a 1 hour period be a Poisson random variable with mean $L$. Each job requires $X_j$ seconds to complete, where the $X_j$'s are iid random variables that are equal to 3 minutes or 6 minutes with equal probability.
  a. Find the mean and variance of the total work $W$ (measured in minutes) arriving in a 1 hour period.
  b. Find the $G_W(z) = E[z^W]$.

14. Let the number of message transmissions by a computer in 1 hour be a binomial random variable with parameters $n$ and $p$. Suppose that the probability of a message transmission error is $\varepsilon$. Let $S$ be the number of transmission errors in a 1 hour period.
  a. Find the mean and variance of $S$.
  b. Find $G_S(z) = E[z^S]$.

### Section 5.2
### The Sample Mean and the Laws of Large Numbers

15. Suppose that the number of particle emissions by a radioactive mass in $t$ seconds is a Poisson random variable with mean $\lambda t$. Use the Chebyshev inequality to obtain a bound for the probability that $|N(t)/t - \lambda|$ exceeds $\varepsilon$.

16. Suppose that 10% of voters are in favor of certain legislation. A large number $n$ of voters are polled and a relative frequency estimate $f_A(n)$ for the above proportion is obtained. Use Eq. (5.20) to determine how many voters should be polled in order that the probability is at least .95 that $f_A(n)$ differs from 0.10 by less than 0.02.

17. A fair die is tossed 100 times. Use Eq. (5.20) to bound the probability that the total number of spots is between 300 and 400.

18. Let $X_i$ be a sequence of independent zero-mean, unit-variance

Gaussian random variables. Compare the bound given by Eq. (5.20) with the exact value obtained from the $Q$ function for $n = 10$ and $n = 100$.

19. Does the weak law of large numbers hold for the sample mean if the $X_j$'s have the covariance functions given in Problem 2?

20. Does the weak law of large numbers hold for the sample mean if the $X_j$'s have the covariance functions given in Problem 3?

21. (The **sample variance**) Let $X_1, \ldots, X_n$ be an iid sequence of random variables for which the mean and variance are unknown. The sample variance is defined as follows:

$$V_n^2 = \frac{1}{n-1} \sum_{j=1}^{n} (X_j - M_n)^2,$$

where $M_n$ is the sample mean.

a. Show that

$$\sum_{j=1}^{n} (X_j - \mu)^2 = \sum_{j=1}^{n} (X_j - M_n)^2 + n(M_n - \mu)^2.$$

b. Use the result in Part a to show that

$$E\left[k \sum_{j=1}^{n} (X_j - M_n)^2\right] = k(n-1)\sigma^2.$$

c. Use Part b to show that $E[V_n^2] = \sigma^2$. Thus $V_n^2$ is an unbiased estimator for the variance.

d. Find the expected value of the sample variance if $n - 1$ is replaced by $n$. Note that this is a biased estimator for the variance.

## Section 5.3
## The Central Limit Theorem

22. A fair coin is tossed 1000 times. Estimate the probability that the number of heads is between 400 and 600. Estimate the probability that the number is between 500 and 550.

23. Repeat Problem 16 using the central limit theorem.

24. A fair die is tossed 100 times. Use the central limit theorem to estimate the probability that the total number of spots is between 300 and 400. Compare the answer to the bound obtained in Problem 17.

25. The lifetime of a cheap lightbulb is an exponential random variable with mean 36 hours. Suppose that 16 lightbulbs are tested and their

lifetimes measured. Use the central limit theorem to estimate the probability that the sum of the lifetimes is less than 600 hours?

26. A student uses pens whose lifetime is an exponential random variable with mean 1 week. Use the central limit theorem to determine the minimum number of pens he should buy at the beginning of a 15 week semester, so that with probability .99 he does not run out of pens during the semester.

27. Let $S$ be the sum of 100 iid Poisson random variables with mean 0.2. Compare the exact value of $P[S = k]$ to an approximation given by the central limit theorem as in Eq. (5.30).

28. The number of messages arriving at a multiplexer is a Poisson random variable with mean 10 messages/second. Use the central limit theorem to estimate the probability that more than 650 messages arrive in one minute.

29. A binary transmission channel introduces bit errors with probability .15. Estimate the probability that there are 20 or fewer errors in 100 bit transmissions.

30. The sum of a list of 100 real numbers is to be computed. Suppose that numbers are rounded off to the nearest integer so that each number has an error that is uniformly distributed in the interval $(-0.5, 0.5)$. Use the central limit theorem to estimate the probability that the total error in the sum of the 100 numbers exceeds 6.

### Section *5.4
### Confidence Intervals

31. A voltage measurement consists of the sum of a constant unknown voltage and a Gaussian-distributed noise voltage of zero-mean and variance 10 microvolt$^2$. Thirty independent measurements are made and a sample mean of 100 microvolts is obtained. Find the corresponding 95% confidence interval.

32. Let $X_j$ be a Gaussian random variable with unknown mean $E[X] = \mu$ and variance 1.
    a. Find the width of the 95% confidence intervals for $\mu$ for $n = 4$, 16, 100.
    b. Repeat for 99% confidence intervals.

33. The lifetime of 225 lightbulbs is measured and the sample mean and sample variance are found to be 223 hours and 100 hours$^2$, respectively. Find a 95% confidence interval for the mean lifetime.

34. Let $X$ be a Gaussian random variable with unknown mean and

unknown variance. A set of 10 independent measurements of $X$ yields

$$\sum_{j=1}^{10} X_j = 350 \qquad \text{and} \qquad \sum_{j=1}^{10} X_j^2 = 12{,}645.$$

Find a 90% confidence interval for the mean of $X$.

35. Let $X$ be a Gaussian random variable with unknown mean and unknown variance. A set of 10 independent measurements of $X$ yields a sample mean of 57.3 and a sample variance of 23.2.
    a. Find the 90%, 95%, and 99% confidence intervals.
    b. Repeat Part a if a set of 20 measurements had yielded the above sample mean and sample variance.

36. A computer simulation program is used to produce 150 samples of a random variable. The samples are grouped into fifteen batches of ten samples each. The batch sample means are listed below:

| | | | | |
|---|---|---|---|---|
| 0.228 | −1.941 | 0.141 | 1.979 | −0.224 |
| 0.501 | −5.907 | −1.367 | −1.615 | −1.013 |
| −0.397 | −3.360 | −3.330 | −0.033 | −0.976 |

Find the 90% confidence interval for the sample mean.

37. A coin is flipped a total of 500 times, in 10 batches of 50 flips each. The number of heads in each of the batches is as follows:

    24, 27, 22, 24, 25, 24, 28, 26, 23, 26.

    Find the 95% confidence interval for the probability of heads $p$ using the method of batch means.

38. (Computer Exercise) In Section 5.4, the meaning of the confidence level $1 - \alpha$ is explained as follows: "If we were to compute confidence intervals a large number of times, we would find that approximately $(1 - \alpha) \times 100\%$ of the time, the computed intervals would contain the true value of the parameter." The following computer exercise is intended to check this statement.
    a. Assuming that the mean is unknown and that the variance is known, find the 90% confidence interval for the mean of a Gaussian random variable with $n = 10$.
    b. Write a computer program subroutine to generate zero-mean, unit-variance Gaussian random variables, and to compute the sample mean of 10 such random variables.
    c. Use the subroutine developed in Part b to obtain 500 sample means and associated confidence intervals. Find the proportion of confidence intervals that include the true mean (which by design is zero). Is this in agreement with the confidence level $1 - \alpha = .90$?

d. Repeat Part c using exponential random variables with mean one. Should the proportion of intervals including the true mean be given by $1 - \alpha$? Explain.

39. (Computer Exercise) Write a subroutine to generate random variables $X_j$ that are uniformly distributed in the interval $[-1, 1]$.

a. Suppose that 160 $X_j$'s are generated and that 90% confidence intervals for the sample mean are to be calculated. Find the confidence intervals for the mean using the following combinations:

4 batches of 40 samples each,
8 batches of 20 samples each,
16 batches of 10 samples each, and
32 batches of 5 samples each.

b. Write a computer program to perform the experiment in Part a 500 times. In each repetition of the experiment, compute the four confidence intervals defined in Part a. Calculate the proportion of time in which the above four confidence intervals include the true mean. Which of the above combinations of the batch size and number of batches are in better agreement with the results predicted by the confidence level? Explain why.

## Section *5.5
## Long-Term Arrival Rates and Associated Averages

40. The customer arrival times at a bus depot are iid exponential random variables with mean $T$. Suppose that buses leave as soon as $m$ seats are full. At what rate do buses leave the depot?

41. A faulty clock ticks forward every minute with probability $p$ and it does not tick forward with probability $1 - p$. What is the rate at which this clock moves forward?

42. a. Show that $\{N(t) \geq n\}$ and $\{S_n \leq t\}$ are equivalent events.
    b. Use Part a to find $P[N(t) \leq n]$ when the $X_j$ are iid exponential random variables with mean $1/\alpha$.

43. Explain why the following are not equivalent events:
    a. $\{N(t) \leq n\}$ and $\{S_n \geq t\}$.
    b. $\{N(t) > n\}$ and $\{S_n < t\}$.

44. A communication channel alternates between periods when it is errorfree and periods during which it introduces errors. Assuming that these periods are independent random variables of means $m_1$ and $m_2$, respectively, find the long-term proportion of time during which the channel is errorfree.

45. A worker works at a rate $r_1$ when the boss is around and at a rate $r_2$ when the boss is not present. Suppose that the sequence of durations of the time periods when the boss is present and absent are independent random variables with means $m_1$ and $m_2$, respectively. Find the long-term average rate at which the worker works.

46. A computer (repairman) continuously cycles through three tasks (machines). Suppose that each time the computer services task $i$, it spends time $X_i$ doing so.
    a. What is the long-term rate at which the computer cycles through the three tasks?
    b. What is the long-term proportion of time spent by the computer servicing task $i$?
    c. Repeat Parts a and b if a random time $W$ is required for the computer (repairman) to switch (walk) from one task (machine) to another.

47. Customers arrive at a phone booth and seize the phone for a random time $Y$, if the phone is free. If the phone is not free, the customers leave immediately. Suppose that the time between customer arrivals is an exponential random variable.
    a. Find the long-term rate at which customers seize the phone.
    b. Find the long-term proportion of customers that leave without seizing the phone.

48. The lifetime of a certain system component is an exponential random variable with mean $T$. Suppose that the component is replaced when it fails or when it reaches the age of $3T$ months.
    a. Find the long-term rate at which components are replaced.
    b. Find the long-term rate at which working components are replaced.

49. A data compression encoder segments a stream of information bits into patterns as shown below. Each pattern is then encoded into the codeword shown below.

| *pattern* | *codeword* | *probability* |
|---|---|---|
| 1 | 100 | .1 |
| 01 | 101 | .09 |
| 001 | 110 | .081 |
| 0001 | 111 | .0729 |
| 0000 | 0 | .6561 |

    a. If the information source produces a bit every millisecond, find the rate at which codewords are produced.
    b. Find the long-term ratio of encoded bits to information bits.

50. In Example 5.23 evaluate the proportion of time that the residual lifetime $r(t)$ exceeds $c$ seconds for the following cases:
    a. $X_j$ iid uniform random variables in the interval $[0, 2T]$.
    b. $X_j$ iid exponential random variables with mean $T$.
    c. $X_j$ iid Rayleigh random variables with mean $T$.
    d. Calculate and compare the mean residual time in each of the above three cases.

51. Let the age $a(t)$ of a cycle be defined as the time that has elapsed from the last arrival up to an arbitrary time instant $t$. Show that long-term proportion of time that $a(t)$ exceeds $c$ seconds is given by Eq. (5.50).

52. Suppose that the cost in each cycle grows at a rate proportional to the age $a(t)$ of the cycle, that is,

$$C_j = \int_0^{X_j} a(t')\,dt'.$$

    a. Show that $C_j = X_j^2/2$.
    b. Show that the long-term rate at which the cost grows is $E[X^2]/2E[X]$.
    c. Show that the result in Part b is also the long-term time average of $a(t)$, that is,

$$\lim_{t\to\infty} \frac{1}{t} \int_0^t a(t')\,dt' = \frac{E[X^2]}{2E[X]}.$$

    d. Explain why the average residual life is also given by the above expression.

53. Calculate the mean age and mean residual life in Problem 52 in the following cases:
    a. $X_j$ iid uniform random variables in the interval $[0, 2T]$,
    b. $X_j$ iid exponential random variables with mean $T$,
    c. $X_j$ iid Rayleigh random variables with mean $T$.

54. (The Regenerative Method) Suppose that a queueing system has the property that when a customer arrives and finds an empty system, the future behavior of the system is completely independent of the past. Define a cycle to consist as the time period between two consecutive customer arrivals to an empty system. Let $N_j$ be the number of customers served during the $j$th cycle and let $T_j$ be the total delay of all customers served during the $j$th cycle.
    a. Use Theorem 2 to show that the average customer delay is given by $E[T]/E[N]$, that is,

$$\lim_{n\to\infty} \frac{1}{n} \sum_{k=1}^{n} D_k = \frac{E[T]}{E[N]},$$

where $D_k$ is the delay of the $k$th customer.

b. How would you use this result to estimate the average delay in a computer simulation of queueing system?

### Section *5.6
### A Computer Method for Evaluating the Distribution of a Random Variable Using the Discrete Fourier Transform

55. Let the discrete random variable $X$ be uniformly distributed in the set $\{0, 1, 2\}$.
    a. Find the $N = 3$ DFT for $X$.
    b. Use the inverse DFT to recover $P[X = 1]$.
56. Let $S = X + Y$, where $X$ and $Y$ are iid random variables uniformly distributed in the set $\{0, 1, 2\}$.
    a. Find the $N = 5$ DFT for $S$.
    b. Use the inverse DFT to find $P[S = 2]$.
57. Let $X$ be a binomial random variable with parameter $n = 8$ and $p = 1/2$.
    a. Write an FFT program to obtain the pmf of $X$ from $\Phi_X(\omega)$.
    b. Use the FFT program to obtain the pmf of $Z = X + Y$ where $X$ and $Y$ are iid binomial random variables with $n = 8$ and $p = 1/2$.
58. Let $X_j$ be a discrete random variable that is uniformly distributed in the set $\{0, 1, \ldots, 9\}$. Use an FFT program to find the pmf of $S_n = X_1 + \cdots + X_n$ for $n = 5$ and $n = 10$. Plot your results and compare them to Figure 5.11
59. Let $X$ be the geometric random variable with parameter $p = 1/2$. Write an FFT program to evaluate Eq. (5.57) to compute $p_k'$ for $N = 8$ and $N = 16$. Compare the result to those given by Eq. (5.59).
60. Let $X$ be a Poisson random variable with mean $L = 5$.
    a. Use an FFT program to obtain the pmf from $\Phi_X(\omega)$. Find the value of $N$ for which the error in Eq. (5.57) is less than 1%.
    b. Let $S = X_1 + X_2 + \cdots + X_5$, where the $X_j$ are iid Poisson random variables with mean $L = 5$. Use an FFT program to compute the pmf of $S$ from $\Phi_X(\omega)$.
61. Let $X$ and $Y$ be independent binomial random variables with parameters $n_1 = 64$ and $p_1 = 1/2$ and $n_2 = 32$ and $p_2 = 1/4$, respectively. Use an FFT program to find the pmf of $S = X + Y$.
62. The probability generating function for the number $N$ of customers in a certain queueing system (the so-called $M/D/1$ system discussed in Chapter 9) is

$$G_N(z) = \frac{(1 - \rho)(1 - z)}{1 - ze^{\rho(1-z)}},$$

where $0 \le \rho \le 1$. Use an FFT program to obtain the pmf of $N$ for $\rho = 1/2$.

63. Use an FFT program to obtain approximately the pdf of a Laplacian random variable from its characteristic function. Use the same parameters as in Example 5.26 and compare your results to those shown in Figure 5.12.
64. Use an FFT program to obtain approximately the pdf of $Z = X + Y$, where $X$ and $Y$ are independent Laplacian random variables with parameters $\alpha = 1$ and $\alpha = 2$, respectively.
65. Use an FFT program to obtain approximately the pdf of a zero-mean, unit-variance Gaussian random variable from its characteristic function. Experiment with the values of $N$ and $\omega_0$ and compare the results given by the FFT with the exact values.
66. Figures 5.2 through 5.4 for the cdf of the sum of iid Bernoulli, uniform, and exponential random variables were obtained using an FFT program. Reproduce the results shown in these Figures.

## APPENDIX 5.1

### SUBROUTINE FFT(A, M, N)[4]

The subroutine argument $A$ is a complex vector of size $N = 2^M$. The DFT of the sequence $p_0, \ldots, p_{N-1}$ can be obtained by calling the subroutine with $A$ containing the entries $A_{k+1} = p_k$ for $k = 0, 1, \ldots, N-1$. The subroutine returns the DFT sequence in the vector $A$, with $A_{m+1} = C_m$ for $m = 0, 1, \ldots, N-1$.

The subroutine can also be used to compute the inverse DFT. The DFT is defined by

$$X_m = \sum_{k=0}^{N-1} x_k e^{j2\pi km/N},$$

and the inverse DFT is

$$x_k = \frac{1}{N} \sum_{m=0}^{N-1} X_m e^{-j2\pi km/N}.$$

If we take the complex conjugate of the above equation we obtain

$$x_k^* = \sum_{m=0}^{N-1} \left(\frac{X_m^*}{N}\right) e^{j2\pi km/N}.$$

4. After Cooley, Lewis, and Welch [7].

Thus

$$x_k = \left\{ \sum_{m=0}^{N-1} \left( \frac{X_m^*}{N} \right) e^{j2\pi km/N} \right\}^*.$$

Thus we can obtain $x_k$ by calling the subroutine with $A$ containing the sequence $X_m^*/N$, and then conjugating the sequence returned by the subroutine.

```
      SUBROUTINE FFT(A,M,N)
      COMPLEX A(N),U,W,T
      N = 2**M
      NV2 = N/2
      NM1 = N - 1
      J = 1
      DO 7 I = 1, NM1
         IF (I .GE. J) GO TO 5
         T = A(J)
         A(J) = A(I)
         A(I) = T
5        K = NV2
6        IF (K .GE. J) GO TO 7
         J = J - K
         K = K/2
         GO TO 6
7       J = J + K
        PI = 3.141592653589793
       DO 20 L = 1,M
         LE = 2**L
         LE1 = LE/2
         U = (1.0,0.)
         W = CMPLX(COS(PI/LE1), -SIN(PI/LE1))
         DO 20 J = 1,LE1
            DO 10 I = J,N,LE
            IP = I + LE1
            T = A(IP)*U
            A(IP) = A(I) - T
10          A(I) = A(I) + T
20     U = U*W
       RETURN
       END
```

CHAPTER 6

# Random Processes

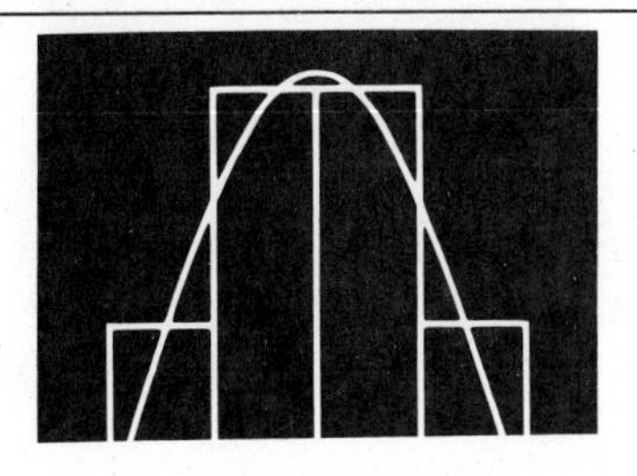

In certain random experiments, the outcome is a function of time or space. For example, in speech recognition systems, decisions are made on the basis of a voltage waveform corresponding to a speech utterance. In an image processing system, the intensity of the image varies over a rectangular region. In a queueing system, the number of customers in the system varies with time. In some situations, two or more functions of time may be of interest. For example, the temperature in a certain city and the demand placed on the local electric power utility vary together in time.

The random time functions in all of the above examples can be viewed as numerical quantities that evolve randomly in time or space. Thus what we really have is a family of random variables indexed by the time or space variable. In Section 6.1 we introduce the notion of a *random process* (or *stochastic process*), which is defined as an *indexed family of random variables.*

We are interested in specifying the joint behavior of the random variables within a family (i.e., the stochastic process at various time instants). In Section 6.2 we see that this is done by specifying joint distribution functions, as well as mean and covariance functions.

In Sections 6.3 and 6.4 we present examples of stochastic processes and show how models of complex processes can be developed from a few simple models. In Section 6.5 we introduce the class of stationary random processes that can be viewed as random processes in "steady state." In Section 6.6 we examine the properties of time averages of random processes and the problem of estimating the parameters of a random process.

## 6.1

## DEFINITION OF A RANDOM PROCESS

Consider a random experiment specified by the outcomes $\zeta$ from some sample space $S$, by the events defined on $S$, and by the probabilities on these events. Suppose that to every outcome $\zeta \in S$, we assign a function of time according to some rule:

$$X(t, \zeta) \qquad t \in I.$$

The graph of the function $X(t, \zeta)$ versus $t$, for $\zeta$ fixed, is called a **realization** or **sample path** of the random process. On the other hand, for each fixed $t_k$ from the index set $I$, $X(t_k, \zeta)$ is a random variable (see Fig. 6.1). Thus we have created an indexed family of random variables, $\{X(t, \zeta), t \in I\}$. This family is called a **random process.** We also refer to random processes as **stochastic processes.** We will usually suppress the $\zeta$ and use $X(t)$ to denote a random process.

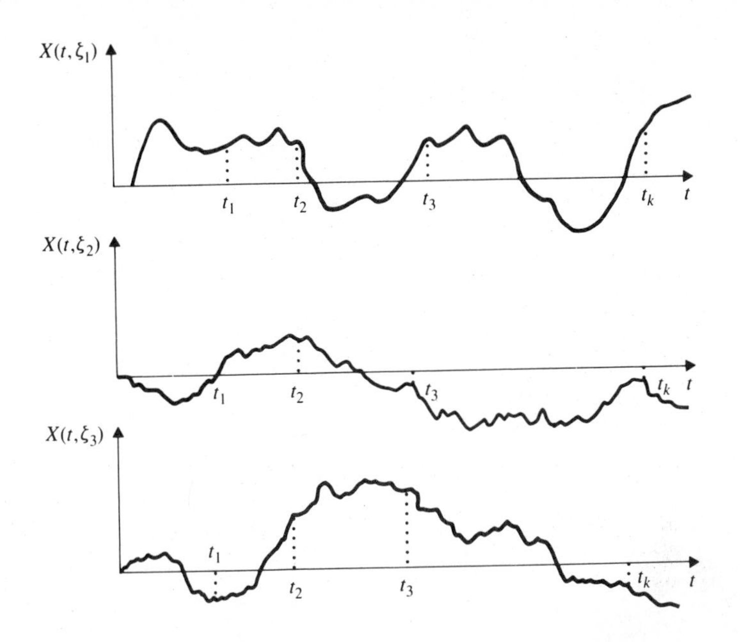


**FIGURE 6.1** Several realizations of a random process.

A stochastic process is said to be **discrete time** if the index set $I$ is a countable set (i.e., the set of integers or the set of nonnegative integers). When dealing with discrete-time processes, we will usually use $n$ to denote the time index and $X_n$ to denote the random process. A **continuous-time** stochastic process is one in which $I$ is continuous (i.e., the real line or the nonnegative real line).

The following example shows how we can imagine a stochastic process as resulting from nature selecting $\zeta$ at the beginning of time and gradually revealing it in time through $X(t, \zeta)$.

### ■■ Example 6.1

Let $\zeta$ be a number selected at random from the interval $S = [0, 1]$, and let $\cdot b_1 b_2 \cdots$ be the binary expansion of $\zeta$:

$$\zeta = \sum_{i=1}^{\infty} b_i 2^{-i} \qquad \text{where } b_i \in \{0, 1\}.$$

Define the discrete-time random process $X(n, \zeta)$ by

$$X(n, \zeta) = b_n \qquad n = 1, 2, \ldots.$$

The resulting process is sequence of binary numbers, with $X(n, \zeta)$ equal to the $n$th number in the binary expansion of $\zeta$. ■■

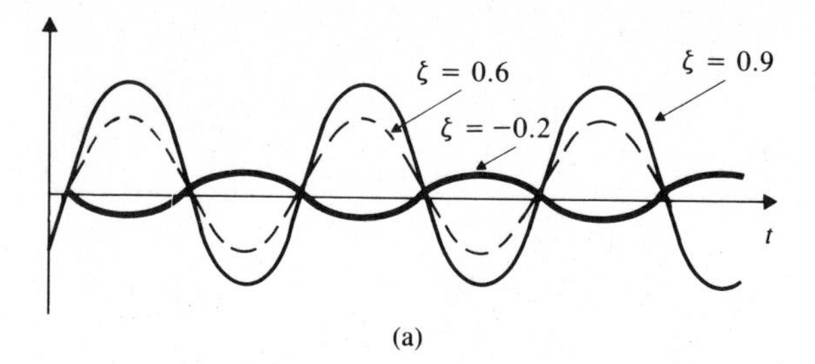


**FIGURE 6.2a** Sinusoid with random amplitude.

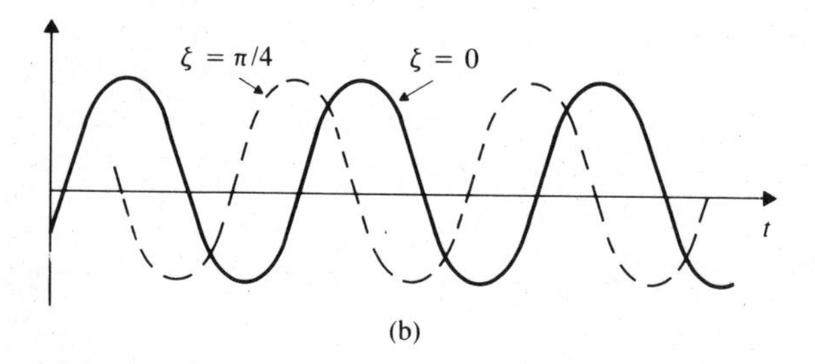


**FIGURE 6.2b** Sinusoid with random phase.

### Example 6.2

Let $\zeta$ be selected at random from the interval $[-1, 1]$. Define the continuous-time random process $X(t, \zeta)$ by

$$X(t, \zeta) = \zeta \cos(2\pi t) \qquad -\infty < t < \infty.$$

The realizations of this random process are sinusoids with amplitude $\zeta$ as shown in Fig. 6.2(a).

Let $\zeta$ be selected at random from the interval $(-\pi, \pi)$, and let $Y(t, \zeta) = \cos(2\pi t + \zeta)$. The realizations of $Y(t, \zeta)$ are time-shifted versions of $\cos 2\pi t$ as shown in Fig. 6.2(b).

---

The randomness in $\zeta$ induces randomness in the observed function $X(t, \zeta)$. In principle, one can deduce the probability of events involving a stochastic process at various instants of time from probabilities involving $\zeta$ by using the equivalent event method introduced in Chapter 3.

### Example 6.3

Find the following probabilities for the random process introduced in Example 6.1: $P[X(1, \zeta) = 0]$ and $P[X(1, \zeta) = 0 \text{ and } X(2, \zeta) = 1]$.

The probabilities are obtained by finding the equivalent events in terms of $\zeta$:

$$P[X(1, \zeta) = 0] = P\left[0 \le \zeta < \frac{1}{2}\right] = \frac{1}{2}.$$

$$P[X(1, \zeta) = 0 \text{ and } X(2, \zeta) = 1] = P\left[\frac{1}{4} \le \zeta < \frac{1}{2}\right] = \frac{1}{4}.$$

Clearly, the probability of any sequence of $k$ bits is $2^{-k}$. ■■

**■■ Example 6.4**

Find the pdf of $X_0 = X(t_0, \zeta)$ and $Y(t_0, \zeta)$ in Example 6.2.

If $t_0$ is such that $\cos(2\pi t_0) = 0$, then $X(t_0, \zeta) = 0$ for all $\zeta$ and the pdf of $X(t_0)$ is a delta function of unit weight at $x = 0$. Otherwise, $X(t_0, \zeta)$ is uniformly distributed in the interval $(-\cos 2\pi t_0, \cos 2\pi t_0)$ since $\zeta$ is uniformly distributed in $[-1, 1]$ (see Fig. 6.3a). Note that the pdf of $X(t_0, \zeta)$ depends on $t_0$.

The approach used in Example 3.28 can be used to show that $Y(t_0, \zeta)$ has an arcsine distribution:

$$f_Y(y) = \frac{1}{\pi\sqrt{1 - y^2}}, \qquad |y| < 1$$

(see Fig. 6.3b). Note that the pdf of $Y(t_0, \zeta)$ does not depend on $t_0$. ■■

In general, the sample paths of a stochastic process can be quite complicated and cannot be described by simple formulas. In addition, it is usually not possible to identify an underlying probability space for the family of observed functions of time. Thus the equivalent event approach for computing the probability of events involving $X(t, \zeta)$ in terms of the probabilities of events involving $\zeta$ does not prove useful in practice. In the

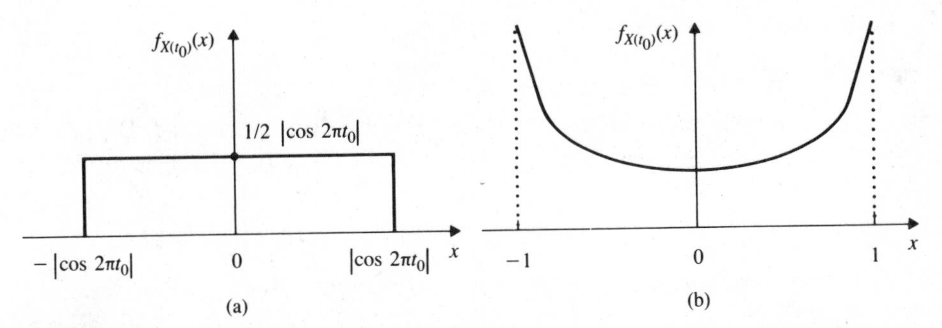


**FIGURE 6.3** (a) pdf of sinusoid with random amplitude, and (b) pdf of sinusoid with random phase.

next section we show an alternative method for specifying the probabilities of events involving a stochastic process.

## 6.2 SPECIFYING A RANDOM PROCESS

### Joint Distributions of Time Samples

Let $X_1, X_2, \ldots, X_k$ be the $k$ random variables obtained by sampling the random process $X(t, \zeta)$ at the times $t_1, t_2, \ldots, t_k$:

$$X_1 = X(t_1, \zeta), \quad X_2 = X(t_2, \zeta), \ldots, \quad X_k = X(t_k, \zeta),$$

as shown in Fig. 6.1. The joint behavior of the random process at these $k$ time instants is specified by the joint cumulative distribution of the *vector* random variable $(X_1, X_2, \ldots, X_k)$. The probabilities of any event involving the random process at all or some of these time instants can be computed from this cdf using the methods developed for vector random variables in Chapter 4. Thus, *a stochastic process is specified by the collection of kth order joint cumulative distribution functions*:

$$F_{X_1,\ldots,X_k}(x_1, x_2, \ldots, x_k) = P[X_1 \le x_1, X_2 \le x_2, \ldots, X_k \le x_k], \tag{6.1}$$

*for any k and any choice of sampling instants* $t_1, \ldots, t_k$. Note that the collection of cdf's must be consistent in the sense that lower order cdf's are obtained as marginals of higher order cdf's.

If the stochastic process is discrete-valued, then a collection of probability mass functions can be used to specify the stochastic process:

$$p_{X_1,\ldots,X_k}(x_1, x_2, \ldots, x_k) = P[X_1 = x_1, X_2 = x_2, \ldots, X_k = x_k] \tag{6.2}$$

If the stochastic process is continuous-valued, then a collection of probability density functions can be used instead:

$$f_{X_1,\ldots,X_k}(x_1, x_2, \ldots, x_k). \tag{6.3}$$

#### ■■ Example 6.5

Let $X_n$ be a sequence of independent, identically distributed Bernoulli random variables with $p = 1/2$. The joint pmf for any $k$ time samples is then

$$P[X_1 = x_1, X_2 = x_2, \ldots, X_k = x_k] = 2^{-k} \qquad x_i \in \{0, 1\} \text{ for all } i.$$

This binary random process is equivalent to the one discussed in Example 6.1.
■■

At first glance it does not appear that we have made much progress in specifying random processes because we are now confronted with the task of specifying a vast collection of joint cdf's! However, this approach works because most useful models of stochastic processes are obtained by elaborating on a few simple models, so the methods developed in the first five chapters of this book can be used to derive the required cdf's. We develop several important examples of this in Sections 6.3 and 6.4.

### The Mean, Autocorrelation, and Autocovariance Functions

The moments of time samples of a random process can be used to partially specify the random process because they summarize the information contained in the joint cdf's. The **mean $m_X(t)$** of a random process $X(t)$ is defined by

$$m_X(t) = E[X(t)] = \int_{-\infty}^{\infty} x f_{X(t)}(x)\, dx, \tag{6.4}$$

where $f_{X(t)}(x)$ is the pdf of $X(t)$. In general, $m_X(t)$ is a function of time. Trends in the behavior of $X(t)$ are reflected in the variation of $m_X(t)$ with time.

The **autocorrelation $R_X(t_1, t_2)$** of a random process $X(t)$ is defined as the joint moment of $X(t_1)$ and $X(t_2)$:

$$R_X(t_1, t_2) = E[X(t_1)X(t_2)] = \int_{-\infty}^{\infty}\int_{-\infty}^{\infty} xy f_{X(t_1),X(t_2)}(x, y)\, dx\, dy, \tag{6.5}$$

where $f_{X(t_1),X(t_2)}(x, y)$ is the second-order pdf of $X(t)$. In general, the autocorrelation is a function of $t_1$ and $t_2$.

The **autocovariance $C_X(t_1, t_2)$** of a random process $X(t)$ is defined as the covariance of $X(t_1)$ and $X(t_2)$:

$$C_X(t_1, t_2) = E[\{X(t_1) - m_X(t_1)\}\{X(t_2) - m_X(t_2)\}]. \tag{6.6}$$

From Eq. (4.70), the autocovariance can be expressed in terms of the autocorrelation and the means:

$$C_X(t_1, t_2) = R_X(t_1, t_2) - m_X(t_1)m_X(t_2). \tag{6.7}$$

Note that the **variance of $X(t)$** can be obtained from $C_X(t_1, t_2)$:

$$\text{VAR}[X(t)] = E[(X(t) - m_X(t))^2] = C_X(t, t). \tag{6.8}$$

The **correlation coefficient** of $X(t)$ is defined as the correlation coefficient of $X(t_1)$ and $X(t_2)$ (see Eq. (4.71):

$$\rho_X(t_1, t_2) = \frac{C_X(t_1, t_2)}{\sqrt{C_X(t_1, t_1)}\sqrt{C_X(t_2, t_2)}}. \tag{6.9}$$

Recall that the correlation coefficient is a measure of the extent to which a random variable can be predicted as a linear function of another. In

Chapter 7, we see that the autocovariance function and the autocorrelation function play a critical role in the design of linear methods for analyzing and processing random signals.

### ■■ Example 6.6
Sinusoid with Random Amplitude

Let $X(t) = A \cos 2\pi t$, where $A$ is some random variable (see Fig. 6.2a). The mean of $X(t)$ is found using Eq. (3.61)

$$m_X(t) = E[A \cos 2\pi t] = E[A] \cos 2\pi t.$$

Note that the mean varies with $t$. In particular, note that the process is always zero for values of $t$ where $\cos 2\pi t = 0$.

The autocorrelation is

$$\begin{aligned} R_X(t_1, t_2) &= E[A \cos 2\pi t_1 A \cos 2\pi t_2] \\ &= E[A^2] \cos 2\pi t_1 \cos 2\pi t_2, \end{aligned}$$

and the autocovariance is then

$$\begin{aligned} C_X(t_1, t_2) &= R_X(t_1, t_2) - m_X(t_1) m_X(t_2) \\ &= \{E[A^2] - E[A]^2\} \cos 2\pi t_1 \cos 2\pi t_2 \\ &= \text{VAR}[A] \cos 2\pi t_1 \cos 2\pi t_2. \end{aligned}$$

■■

### ■■ Example 6.7
Sinusoid with Random Phase

Let $X(t) = \cos(wt + \Theta)$ where $\Theta$ is uniformly distributed in the interval $(-\pi, \pi)$ (see Fig. 6.2b). The mean of $X(t)$ is found using Eq. (3.61)

$$m_X(t) = E[\cos(wt + \Theta)] = \frac{1}{2\pi} \int_{-\pi}^{\pi} \cos(wt + x)\, dx = 0.$$

The autocorrelation and autocovariance are then

$$\begin{aligned} C_X(t_1, t_2) = R_X(t_1, t_2) &= E[\cos(wt_1 + \Theta) \cos(wt_2 + \Theta)] \\ &= \frac{1}{2\pi} \int_{-\pi}^{\pi} \frac{1}{2} \{\cos(w(t_1 - t_2)) + \cos(w(t_1 + t_2) + 2x)\}\, dx \\ &= \frac{1}{2} \cos(w(t_1 - t_2)), \end{aligned}$$

where we used the identity $\cos(a)\cos(b) = \frac{1}{2}\cos(a + b) + \frac{1}{2}\cos(a - b)$. Note that $m_X(t)$ is a constant and that $C_X(t_1, t_2)$ depends only on $|t_1 - t_2|$.

■■

### Gaussian Random Processes

A random process $X(t)$ is a **Gaussian random process** if the samples $X_1 = X(t_1)$, $X_2 = X(t_2), \ldots, X_k = X(t_k)$ are jointly Gaussian random variables for all $k$, and all choices of $t_1, \ldots, t_k$. (Note that this definition applies for discrete-time and continuous-time processes.) Recall from Eq. (4.90) that the joint pdf of jointly Gaussian random variables is determined by the vector of means and by the covariance matrix:

$$f_{X_1,X_2,\ldots,X_k}(x_1, \ldots, x_k) = \frac{e^{-1/2(\boldsymbol{x}-\boldsymbol{m})^T K^{-1}(\boldsymbol{x}-\boldsymbol{m})}}{(2\pi)^{n/2}|K|^{1/2}}, \tag{6.10}$$

where

$$\mathbf{m} = \begin{bmatrix} m_X(t_1) \\ \vdots \\ m_X(t_k) \end{bmatrix} \qquad K = \begin{bmatrix} C_X(t_1, t_1) & C_X(t_1, t_2) & \ldots & C_X(t_1, t_k) \\ C_X(t_2, t_1) & C_X(t_2, t_2) & \ldots & C_X(t_2, t_k) \\ \vdots & \vdots & & \vdots \\ C_X(t_k, t_1) & \ldots & & C_X(t_k, t_k) \end{bmatrix}.$$

Gaussian random processes therefore have the special property that their joint pdf's are completely specified by the mean of the process $m_X(t)$ and by the covariance function $C_X(t_1, t_2)$.

#### ■■ Example 6.8
iid Gaussian Sequence

Let the discrete-time random process $X_n$ be a sequence of independent Gaussian random variables with mean $m$ and variance $\sigma^2$.

The covariance matrix for the times $t_1, \ldots, t_k$ is

$$\{C_X(t_i, t_j)\} = \{\sigma^2 \delta_{ij}\} = \sigma^2 I,$$

where $\delta_{ij} = 1$ when $i = j$ and 0 otherwise, and $I$ is the identity matrix. Thus the corresponding joint pdf is

$$\begin{aligned} f_{X_1,\ldots,X_k}(x_1, x_2, \ldots, x_k) &= \frac{1}{(2\pi\sigma^2)^{k/2}} \exp\left\{-\sum_{i=1}^{k} (x_i - m)^2/2\sigma^2\right\} \\ &= f_X(x_1)f_X(x_2)\ldots f_X(x_k). \end{aligned}$$

■■

### Multiple Random Processes

The joint behavior of two or more random processes is specified by the collection of joint distributions for all possible choices of time samples of the processes. For example, for a pair of random processes $X(t)$ and $Y(t)$ we must specify all possible joint density functions of $X(t_1), \ldots, X(t_k)$ and $Y(t'_1), \ldots, Y(t'_j)$ for all $k, j$, and all choices of $t_1, \ldots, t_k$ and $t'_1, \ldots, t'_j$.

The processes $X(t)$ and $Y(t)$ are said to be **independent** if the vector random variables $(X(t_1), \ldots, X(t_k))$ and $(Y(t'_1), \ldots, Y(t'_j))$ are independent for all $k, j$, and all choices of $t_1, \ldots, t_k$ and $t'_1, \ldots, t'_j$.

The **cross-correlation $R_{X,Y}(t_1, t_2)$** of $X(t)$ and $Y(t)$ is defined by

$$R_{X,Y}(t_1, t_2) = E[X(t_1)Y(t_2)]. \tag{6.12}$$

The processes $X(t)$ and $Y(t)$ are said to be **orthogonal** if

$$R_{X,Y}(t_1, t_2) = 0 \qquad \text{for all } t_1 \text{ and } t_2. \tag{6.13}$$

The **cross-covariance $C_{X,Y}(t_1, t_2)$** of $X(t)$ and $Y(t)$ is defined by

$$\begin{aligned} C_{X,Y}(t_1, t_2) &= E[\{X(t_1) - m_X(t_1)\}\{Y(t_2) - m_Y(t_2)\}] \\ &= R_{X,Y}(t_1, t_2) - m_X(t_1)m_Y(t_2). \end{aligned} \tag{6.14}$$

The processes $X(t)$ and $Y(t)$ are said to be **uncorrelated** if

$$C_{X,Y}(t_1, t_2) = 0 \qquad \text{for all } t_1 \text{ and } t_2. \tag{6.15}$$

## ■■ Example 6.9

Let $X(t) = \cos(wt + \Theta)$ and $Y(t) = \sin(wt + \Theta)$, where $\Theta$ is a random variable uniformly distributed in $[-\pi, \pi]$. Find the cross-covariance of $X(t)$ and $Y(t)$.

From Example 6.7 we know that $X(t)$ and $Y(t)$ are zero-mean. From Eq. (6.14), the cross-covariance is equal to the cross-correlation:

$$\begin{aligned} R_{X,Y}(t_1, t_2) &= E[\cos(wt_1 + \Theta)\sin(wt_2 + \Theta)] \\ &= E\left[-\frac{1}{2}\sin(w(t_1 - t_2)) + \frac{1}{2}\sin(w(t_1 + t_2) + 2\Theta)\right] \\ &= -\frac{1}{2}\sin(w(t_1 - t_2)), \end{aligned}$$

since $E[\sin(w(t_1 + t_2) + 2\Theta)] = 0$. ■■

## ■■ Example 6.10
Signal Plus Noise

Suppose we observe a process $Y(t)$, which consists of a desired signal $X(t)$ plus noise $N(t)$:

$$Y(t) = X(t) + N(t).$$

Find the cross-correlation between the observed signal and the desired signal assuming that $X(t)$ and $N(t)$ are independent random processes.

From Eq. (6.12), we have

$$\begin{aligned}R_{X,Y}(t_1, t_2) &= E[X(t_1)Y(t_2)]\\ &= E[X(t_1)\{X(t_2) + N(t_2)\}]\\ &= E[X(t_1)X(t_2)] + E[X(t_1)N(t_2)]\\ &= R_X(t_1, t_2) + E[X(t_1)]E[N(t_2)]\\ &= R_X(t_1, t_2) + m_X(t_1)m_N(t_2),\end{aligned}$$

where the third equality followed from the fact that $X(t)$ and $N(t)$ are independent. ■■

## 6.3

## EXAMPLES OF DISCRETE-TIME RANDOM PROCESSES

In this section we begin with the simplest class of random processes, independent identically distributed sequences, and gradually build up to more complex discrete-time random processes. In Section 6.4 we then use limiting arguments to develop models for continuous-time random processes. We derive the joint distributions and the moments of various random processes. The objective here is to introduce you to some important random processes and to show you how we develop complex models from simple models using the tools developed thus far.

### iid Random Processes

Let $X_n$ be a discrete-time random process consisting of a sequence of independent, identically distributed (iid) random variables with common cdf $F_X(x)$, mean $m$, and variance $\sigma^2$. The sequence $X_n$ is called the **iid random process.** The joint cdf for any time instants $n_1, \ldots, n_k$ is given by

$$\begin{aligned}F_{X_1,\ldots,X_k}(x_1, x_2, \ldots, x_k) &= P[X_1 \le x_1, X_2 \le x_2, \ldots, X_k \le x_k)\\ &= F_X(x_1)F_X(x_2)\ldots F_X(x_k),\end{aligned} \tag{6.16}$$

where for simplicity $X_k$ denotes $X_{n_k}$. Equation (6.16) implies that if $X_n$ is discrete-valued, the joint pmf factors into the product of individual pmf's, and if $X_n$ is continuous-valued, the joint pdf factors into the product of the individual pdf's.

The *mean of an iid process* is obtained from Eq. (6.4):

$$m_X(n) = E[X_n] = m \qquad \text{for all } n. \tag{6.17}$$

Thus, the mean is constant.

The autocovariance function is obtained from Eq. (6.6) as follows. If $n_1 \neq n_2$, then

$$C_X(n_1, n_2) = E[(X_{n_1} - m)(X_{n_2} - m)]$$
$$= E[(X_{n_1} - m)]E[(X_{n_2} - m)] = 0,$$

since $X_{n_1}$ and $X_{n_2}$ are independent random variables. If $n_1 = n_2 = n$, then

$$C_X(n_1, n_2) = E[(X_n - m)^2] = \sigma^2.$$

We can express the *autocovariance of the iid process* in compact form as follows:

$$C_X(n_1, n_2) = \sigma^2 \, \delta_{n_1,n_2}, \tag{6.18}$$

where $\delta_{n_1,n_2} = 1$ if $n_1 = n_2$ and 0 otherwise.

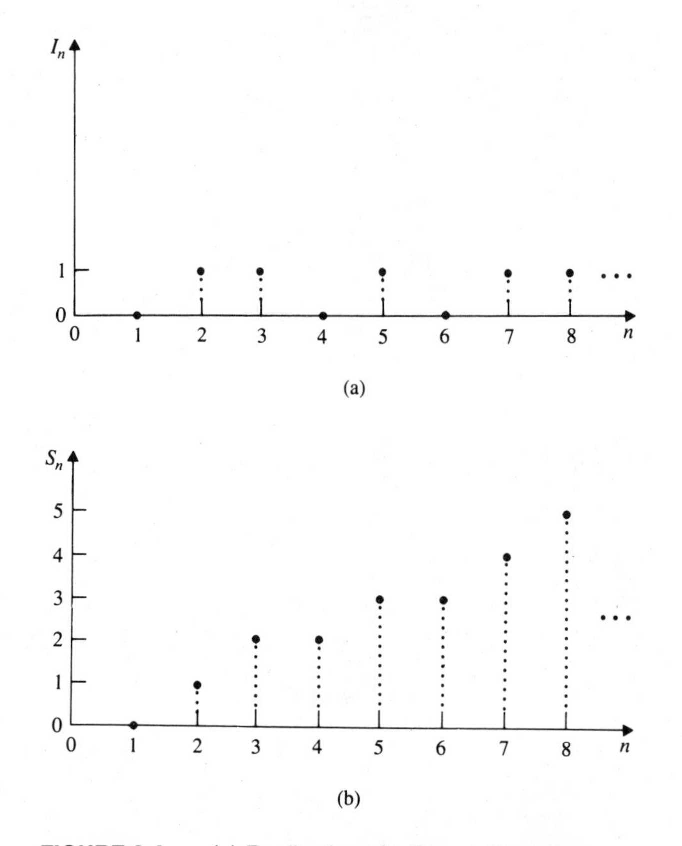


**FIGURE 6.4** (a) Realization of a Bernoulli process. $I_n = 1$ indicates that a light bulb fails and is replaced in day *n*. (b) Realization of a binomial process. $S_n$ denotes the number of light bulbs that have failed up to time *n*.

The autocorrelation function of the iid process is found from Eq. (6.7):

$$R_X(n_1, n_2) = C_X(n_1, n_2) + m^2. \tag{6.19}$$

### Example 6.11
Bernoulli Random Process

Let $I_n$ be a sequence of independent Bernoulli random variables. $I_n$ is then an iid random process taking on values from the set $\{0, 1\}$. A realization of such a process is shown in Fig. 6.4(a). For example, $I_n$ could be an indicator function for the event "a light bulb fails and is replaced on day $n$."

Since $I_n$ is a Bernoulli random variable, it has mean and variance

$$m_1(n) = p \qquad \text{VAR}[I_n] = p(1 - p).$$

The independence of the $I_n$'s makes probabilities easy to compute. For example, the probability that the first 4 bits in the sequence are 1001 is

$$\begin{aligned} &P[I_1 = 1, I_2 = 0, I_3 = 0, I_4 = 1] \\ &\quad = P[I_1 = 1]P[I_2 = 0]P[I_3 = 0]P[I_4 = 1] \\ &\quad = p^2(1 - p)^2. \end{aligned}$$

Similarly, the probability that the second bit is 0 and the seventh is 1 is

$$P[I_2 = 0, I_7 = 1] = P[I_2 = 0]P[I_7 = 1] = p(1 - p).$$

---

### Example 6.12

Let $D_n = 2I_n - 1$, where $I_n$ is the Bernoulli random process, then

$$D_n = \begin{cases} 1 & \text{if } I_n = 1 \\ -1 & \text{if } I_n = 0. \end{cases}$$

For example, $D_n$ might represent the change in position of a particle that moves along a straight line in jumps of $\pm 1$ every time unit. A realization of $D_n$ is shown in Fig. 6.5(a).

The mean of $D_n$ is

$$m_D(n) = E[D_n] = E[2I_n - 1] = 2E[I_n] - 1 = 2p - 1.$$

The variance of $D_n$ is found from Eqs. (3.69) and (3.70):

$$\text{VAR}[D_n] = \text{VAR}[2I_n - 1] = 2^2\,\text{VAR}[I_n] = 4p(1 - p).$$

The probabilities of events involving $D_n$ are computed as in Example 6.11.

---

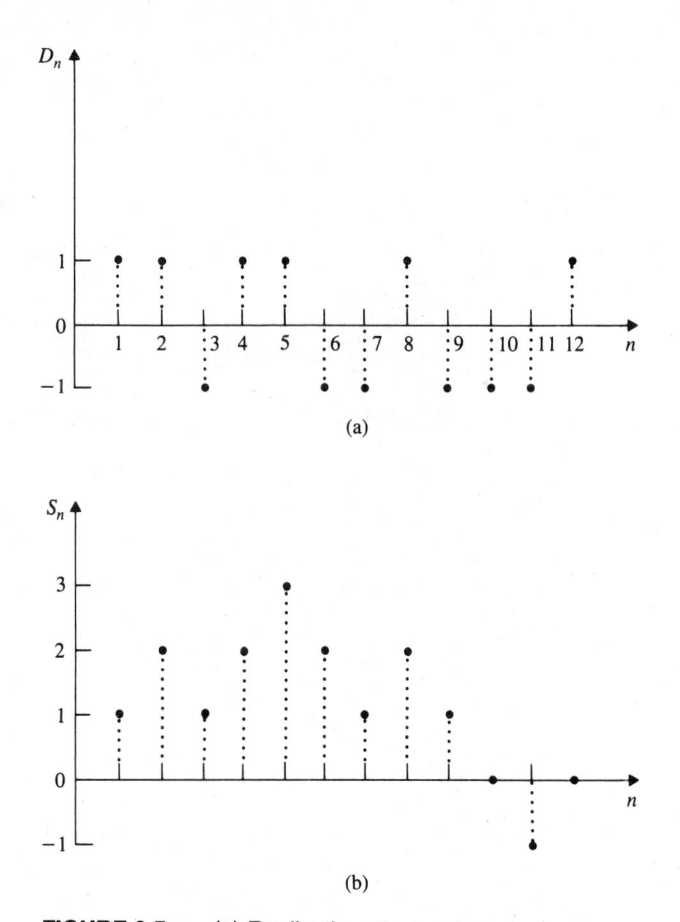


**FIGURE 6.5** (a) Realization of a random step process. $D_n = 1$ implies that the particle moves one step to the right at time $n$. $D_n = -1$ denotes a step to the left. (b) Realization of a random walk process. $S_n$ denotes the position of a particle at time $n$.

### Sum Processes: The Binomial Counting and Random Walk Processes

Many interesting random processes are obtained as the sum of a sequence of iid random variables, $X_1, X_2, \ldots$:

$$\begin{aligned} S_n &= X_1 + X_2 + \cdots + X_n \qquad n = 1, 2, \ldots \\ &= S_{n-1} + X_n. \end{aligned} \tag{6.20}$$

where $S_0 = 0$. We call $S_n$ the **sum process.** The pdf or pmf of $S_n$ is found using the convolution or characteristic-equation methods presented in Section 5.1. Note that $S_n$ depends on the "past," $S_1, \ldots, S_{n-1}$, only through

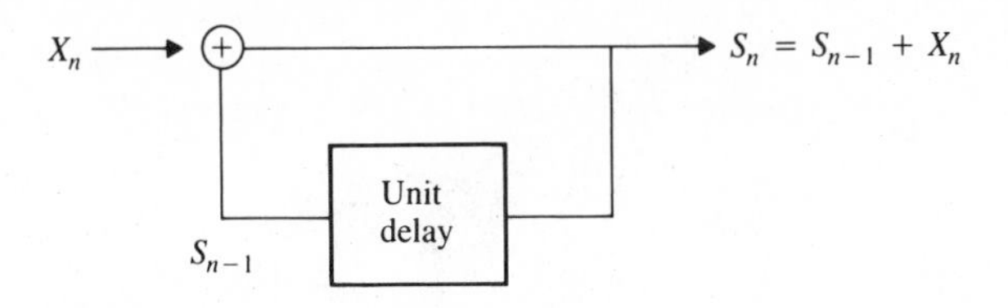


**FIGURE 6.6** The sum process $S_n = X_1 + \cdots + X_n$, $S_0 = 0$, can be generated in this way.

$S_{n-1}$, that is, $S_n$ is independent of the past when $S_{n-1}$ is known.[1] This can be seen clearly from Fig. 6.6, which shows a recursive procedure for computing $S_n$.

### ■■ Example 6.13
Binomial Counting Process

Let the $I_i$ be the sequence of independent Bernoulli random variables in Example 6.11, and let $S_n$ be the corresponding sum process. $S_n$ is then the **counting process** that gives the number of successes in the first $n$ Bernoulli trials. The sample function for $S_n$ corresponding to a particular sequence of $I_i$'s is shown in Fig. 6.4(b). If $I_n$ indicates that a light bulb fails and is replaced on day $n$, then $S_n$ denotes the number of light bulbs that have failed up to day $n$.

Since $S_n$ is the sum of $n$ independent Bernoulli random variables, $S_n$ is a binomial random variable with parameters $n$ and $p = P[I = 1]$:

$$P[S_n = j] = \binom{n}{j} p^j (1 - p)^{n-j} \qquad \text{for } 0 \leq j \leq n,$$

and zero otherwise. Thus $S_n$ has mean $np$ and variance $np(1 - p)$. Note that the mean of this process grows linearly with time. ■■

### ■■ Example 6.14
One-Dimensional Random Walk

Let $D_n$ be the iid process of $\pm 1$ random variables in Example 6.12, and let $S_n$ be the corresponding sum process. $S_n$ is then the position of the particle at time $n$. The random process $S_n$ is an example of a **one-dimensional random walk.** A sample function of $S_n$ is shown in Fig. 6.5(b).

The pmf of $S_n$ is found as follows. If there are $k$ $+1$s in the first $n$ trials, then there are $n - k$ $-1$s, and $S_n = k - (n - k) = 2k - n$. Conversely,

1. This is called the Markov property. Chapter 8 discusses in detail random processes with this property.

$S_n = j$ if the number of +1s is $k = (j + n)/2$. If $(j + n)/2$ is not an integer, then $S_n$ cannot equal $j$. Thus

$$P[S_n = 2k - n] = \binom{n}{k} p^k (1 - p)^{n-k} \qquad \text{for } k \in \{0, 1, \ldots, n\}.$$

■■

The sum process $S_n$ has **independent increments** in nonoverlapping time intervals. To see this consider two time intervals: $n_0 < n \le n_1$ and $n_2 < n \le n_3$, where $n_1 \le n_2$. The increments of $S_n$ in these disjoint time intervals are given by

$$\begin{aligned} S_{n_1} - S_{n_0} &= X_{n_0+1} + \cdots + X_{n_1} \\ S_{n_3} - S_{n_2} &= X_{n_2+1} + \cdots + X_{n_3}. \end{aligned} \tag{6.21}$$

The above increments do not have any of the $X_n$'s in common, so the independence of the $X_n$'s implies that the increments $(S_{n_1} - S_{n_0})$ and $(S_{n_3} - S_{n_2})$ are independent random variables.

For $n' > n$, the increment $S_{n'} - S_n$ is the sum of $n' - n$ iid random variables, so it has the same distribution as $S_{n'-n}$, the sum of the first $n' - n$ $X$'s, that is,

$$P[S_{n'} - S_n = y] = P[S_{n'-n} = y]. \tag{6.22}$$

Thus increments in intervals of the same length have the same distribution regardless of when the interval begins. For this reason, we say that $S_n$ has **stationary increments.**

## ■■ Example 6.15

The independent and stationary increments property is particularly easy to see for the binomial process since the increments in an interval is the number of successes in the corresponding trials. The independent increment property follows from the fact that the number of successes in disjoint time intervals are independent. The stationary increments property follows from the fact that the pmf for the increment in a time interval is the binomial pmf with the corresponding number of trials. ■■

The fact that the sum process $S_n$ has independent and stationary increments makes it easy to compute the joint pmf/pdf for any number of time instants. For simplicity, suppose that the $X_n$ are integer-valued, so $S_n$ is also integer-valued. We compute the joint pmf of $S_n$ at times $n_1$, $n_2$, and $n_3$:

$$\begin{aligned} &P[S_{n_1} = y_1, S_{n_2} = y_2, S_{n_3} = y_3] \\ &\qquad = P[S_{n_1} = y_1, S_{n_2} - S_{n_1} = y_2 - y_1, S_{n_3} - S_{n_2} = y_3 - y_2], \end{aligned} \tag{6.23}$$

since the process is equal to $y_1$, $y_2$, and $y_3$ at times $n_1$, $n_2$, and $n_3$, if and only if, it is equal to $y_1$ at time $n_1$, and the subsequent increments are $y_2 - y_1$, and $y_3 - y_2$. The independent increments property then implies that

$$P[S_{n_1} = y_1, S_{n_2} = y_2, S_{n_3} = y_3] = P[S_{n_1} = y_1]P[S_{n_2} - S_{n_1} = y_2 - y_1] \times P[S_{n_3} - S_{n_2} = y_3 - y_2]. \quad (6.24)$$

Finally, the stationary increments property implies that

$$P[S_{n_1} = y_1, S_{n_2} = y_2, S_{n_3} = y_3] = P[S_{n_1} = y_1]P[S_{n_2-n_1} = y_2 - y_1] P[S_{n_3-n_2} = y_3 - y_2].$$

Clearly, we can use this procedure to write the *joint pmf of* $S_n$ at any time instants $n_1 < n_2 < \cdots < n_k$ in terms of the pmf at the initial time instant and the pmf's of the subsequent increments:

$$P[S_{n_1} = y_1, S_{n_2} = y_2, \ldots, S_{n_k} = y_k] = P[S_{n_1} = y_1] \times P[S_{n_2-n_1} = y_2 - y_1] \cdots P[S_{n_k-n_{k-1}} = y_k - y_{k-1}] \quad (6.25)$$

If the $X_n$ are continuous-valued random variables, then it can be shown that the *joint density of* $S_n$ at times $n_1, n_2, \ldots, n_k$ is:

$$f_{S_{n_1}, S_{n_2}, \ldots, S_{n_k}}(y_1, y_2, \ldots, y_k) = f_{S_{n_1}}(y_1) f_{S_{n_2-n_1}}(y_2 - y_1) \cdots f_{S_{n_k-n_{k-1}}}(y_k - y_{k-1}). \quad (6.26)$$

## ■■ Example 6.16

Find the joint pmf for the binomial counting process at times $n_1$ and $n_2$.

Following the above approach we have

$$\begin{aligned} P[S_{n_1} = y_1, S_{n_2} = y_2] &= P[S_{n_1} = y_1]P[S_{n_2-n_1} = y_2 - y_1] \\ &= \binom{n_2 - n_1}{y_2 - y_1} p^{y_2-y_1}(1-p)^{n_2-n_1-y_2+y_1} \binom{n_1}{y_1} p^{y_1}(1-p)^{n_1-y_1} \\ &= \binom{n_2 - n_1}{y_2 - y_1}\binom{n_1}{y_1} p^{y_2}(1-p)^{n_2-y_2}. \end{aligned}$$

■■

## ■■ Example 6.17
### Sum of iid Gaussian Sequence

Let $X_n$ be a sequence of iid Gaussian random variables with zero mean and variance $\sigma^2$. Find the joint pdf of the corresponding sum process.

The sum process $S_n$ is also a Gaussian random process with mean zero and variance $n\sigma^2$ (see Example 5.3). The joint pdf of $S_n$ at times $n_1$ and $n_2$

is given by

$$f_{S_{n_1},S_{n_2}}(y_1, y_2) = f_{S_{n_2-n_1}}(y_2 - y_1)f_{S_{n_1}}(y_1)$$
$$= \frac{1}{\sqrt{2\pi(n_2 - n_1)\sigma^2}} e^{-(y_2-y_1)^2/[2(n_2-n_1)\sigma^2]} \frac{1}{\sqrt{2\pi n_1 \sigma^2}} e^{-y_1^2/2n_1\sigma^2}$$

■■

Since the sum process $S_n$ is the sum of $n$ iid random variables, it has *mean* and *variance*:

$$m_S(n) = E[S_n] = nE[X] = nm \tag{6.27}$$
$$\text{VAR}[S_n] = n\,\text{VAR}[X] = n\sigma^2. \tag{6.28}$$

The autocovariance of $S_n$ is

$$C_S(n, k) = E[(S_n - E[S_n])(S_k - E[S_k])]$$
$$= E[(S_n - nm)(S_k - km)]$$
$$= E\left[\left\{\sum_{i=1}^{n} (X_i - m)\right\}\left\{\sum_{j=1}^{k} (X_j - m)\right\}\right]$$
$$= \sum_{i=1}^{n}\sum_{j=1}^{k} E[(X_i - m)(X_j - m)].$$

The term inside the summation is the autocovariance $C_X(i,j)$ of the iid process $X_n$. Recall from Eq. (6.18) that $C_X(i,j)$ is equal to $\sigma^2$ when $i = j$ and zero otherwise. Thus, all but the $i = j$ terms in the double summation are zero, and the *autocovariance of the sum process* is

$$C_S(n, k) = \sum_{i=1}^{\min(n,k)} C_X(i, i) = \min(n, k)\sigma^2. \tag{6.29}$$

The property of independent increments allows us to compute the autocovariance in another way. Suppose $n \leq k$ so $n = \min(n, k)$, then

$$C_S(n, k) = E[(S_n - nm)(S_k - km)]$$
$$= E[(S_n - nm)\{(S_n - nm) + (S_k - km) - (S_n - nm)\}]$$
$$= E[(S_n - nm)^2] + E[(S_n - nm)(S_k - S_n - (k - n)m)].$$

Since $S_n$ and the increment $S_k - S_n$ are independent

$$C_S(n, k) = E[(S_n - nm)^2] + E[(S_n - nm)]E[(S_k - S_n - (k - n)m)]$$
$$= E[(S_n - nm)^2]$$
$$= \text{VAR}[S_n] = n\sigma^2,$$

since $E[S_n - nm] = 0$. Similarly if $k = \min(n, k)$, we would have obtained $k\sigma^2$. Thus we again obtain Eq. (6.29).

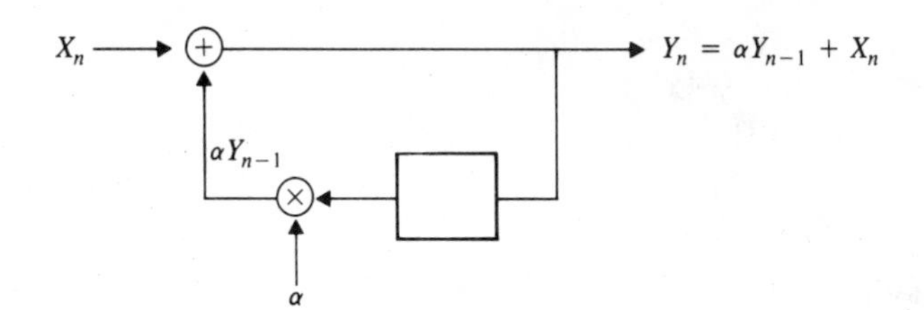


**FIGURE 6.7a** First-order autoregressive process.

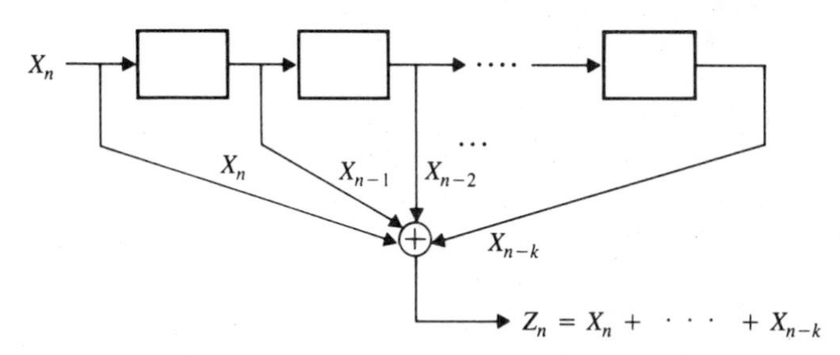


**FIGURE 6.7b** Moving average process.

### ■■ Example 6.18

Find the autocovariance of the one-dimensional random walk process discussed in Example 6.14.

From Example 6.12 and Eqs. (6.27) and (6.28), $S_n$ has mean $n(2p - 1)$ and variance $4np(1 - p)$. Thus its autocovariance is given by

$$C_s(n, k) = \min(n, k)4p(1 - p).$$ ■■

---

The sum process can be generalized in a number of ways. For example, the recursive structure in Fig. 6.6 can be modified as shown in Fig. 6.7(a). We then obtain first-order *autoregressive random processes,* which are of interest in time series analysis and in digital signal processing. If instead we use the structure shown in Fig. 6.7(b), we obtain *moving average processes*. We investigate these processes in Chapter 7.

## 6.4 EXAMPLES OF CONTINUOUS-TIME RANDOM PROCESSES

In this section we show how continuous-time random processes can be obtained as the limit of discrete-time processes. We introduce the Poisson process and the Brownian motion random process, which are two important examples of this limiting procedure. We also introduce several random processes that are derived from the Poisson process.

### Poisson Process

Consider a situation in which events occur at random instants of time at an average rate of $\lambda$ events per second. For example, an event could represent the arrival of a customer to a service station or the breakdown of a component in some system. Let $N(t)$ be the number of event occurrences in the time interval $[0, t]$. $N(t)$ is then a nondecreasing, integer-valued, continuous-time random process as shown in Fig. 6.8.

Suppose that the interval $[0, t]$ is divided into $n$ subintervals of very short duration $\delta = t/n$. Assume that the following two conditions hold:

1. The probability of more than one event occurrence in a subinterval is negligible compared to the probability of observing one or zero events.
2. Whether or not an event occurs in a subinterval is independent of the outcomes in other subintervals.

The first assumption implies that the outcome in each subinterval can be viewed as a Bernoulli trial. The second assumption implies that these Bernoulli trials are independent. Thus the two assumptions together imply that the counting process $N(t)$ can be approximated by the binomial counting process that counts the number of successes in the $n$ Bernoulli trials.

If the probability of an event occurrence in each subinterval is $p$, then the expected number of event occurrences in the interval $[0, t]$ is $np$. Since events occur at a rate of $\lambda$ events per second, the average number of events

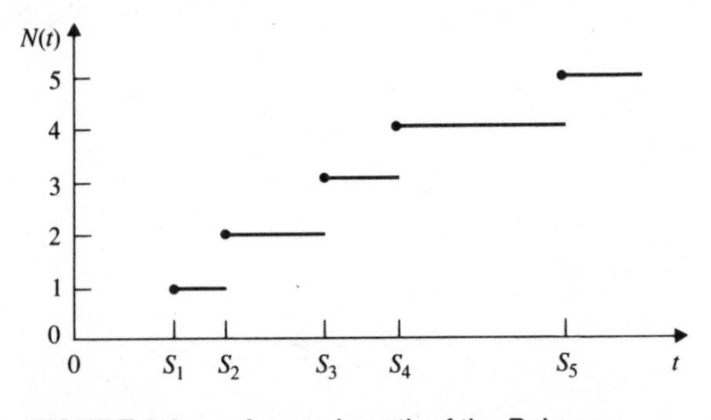


**FIGURE 6.8** A sample path of the Poisson counting process. The event occurrence times are denoted by $S_1, S_2, \ldots$. The $j$th interevent time is denoted by $X_j = S_j - S_{j-1}$.

in the interval $[0, t]$ is also $\lambda t$. Thus we must have that

$$\lambda t = np.$$

If we now let $n \to \infty$ (i.e., $\delta \to 0$) and $p \to 0$ while $np = \lambda t$ remains fixed, then from Eq. (3.31) the binomial distribution approaches a Poisson distribution with parameter $\lambda t$. We therefore conclude that *the number of event occurrences $N(t)$ in the interval $[0, t]$ has a Poisson distribution with mean $\lambda t$*:

$$P[N(t) = k] = \frac{(\lambda t)^k}{k!} e^{-\lambda t} \qquad \text{for } k = 0, 1, \ldots. \tag{6.30}$$

for this reason $N(t)$ is called the **Poisson process.**

The process $N(t)$ inherits the property of independent and stationary increments from the underlying binomial process. Thus *the distribution for the number of event occurrences in any interval of length t is given by Eq. (6.30).* The independent and stationary increments property allows us to write the joint pmf for $N(t)$ at any number of points. For example, for $t_1 < t_2$

$$\begin{aligned} P[N(t_1) = i, N(t_2) = j] &= P[N(t_1) = i]P[N(t_2) - N(t_1) = j - i] \\ &= P[N(t_1) = i]P[N(t_2 - t_1) = j - i] \\ &= \frac{(\lambda t_1)^i e^{-\lambda t_1}}{i!} \frac{(\lambda(t_2 - t_1))^{j-i} e^{-\lambda(t_2 - t_1)}}{(j - i)!}. \end{aligned} \tag{6.31}$$

## ■■ Example 6.19

Inquiries arrive at a recorded message device according to a Poisson process of rate 15 inquiries per minute. Find the probability that in a 1 minute period, 3 inquiries arrive during the first 10 seconds and 2 inquiries arrive during the last 15 seconds.

The arrival rate in seconds is $\lambda = 15/60 = 1/4$ inquiries per second. Writing time in seconds, the probability of interest is

$$P[N(10) = 3 \text{ and } N(60) - N(45) = 2].$$

By applying first the independent increments property, and then the stationary increments property, we obtain

$$\begin{aligned} &P[N(10) = 3 \text{ and } N(60) - N(45) = 2] \\ &\quad = P[N(10) = 3]P[N(60) - N(45) = 2] \\ &\quad = P[N(10) = 3]P[N(60 - 45) = 2] \\ &\quad = \frac{(10/4)^3 e^{-10/4}}{3!} \frac{(15/4)^2 e^{-15/4}}{2!}. \end{aligned}$$

■■

Consider the time $T$ between event occurrences in a Poisson process. Again suppose that the time interval $[0, t]$ is divided into $n$ subintervals of length $\delta = t/n$. The probability that the interevent time $T$ exceeds $t$ seconds is equivalent to no event occurring in $t$ seconds (or in $n$ Bernoulli trials):

$$\begin{aligned} P[T > t] &= P[\text{no events in } t \text{ seconds}] \\ &= (1 - p)^n \\ &= \left(1 - \frac{\lambda t}{n}\right)^n \\ &\to e^{-\lambda t} \qquad \text{as } n \to \infty. \end{aligned} \tag{6.32}$$

Equation (6.32) implies that $T$ is an exponential random variable with parameter $\lambda$. Since the time between event occurrences in the underlying binomial process are independent geometric random variables, it follows that the sequence of interevent times in a Poisson process are independent random variables. We therefore conclude that the *interevent times in a Poisson process form an iid sequence of exponential random variables with mean* $1/\lambda$.

Another quantity of interest is the time $S_n$ at which the $n$th event occurs in a Poisson process. Let $T_j$ denote the iid exponential interarrival times, then

$$S_n = T_1 + T_2 + \cdots + T_n.$$

In Example 5.5, we saw that the sum of $n$ iid exponential random variables has an Erlang distribution. Thus the pdf of $S_n$ is

$$f_{S_n}(y) = \frac{(\lambda y)^{n-1}}{(n-1)!} \lambda e^{-\lambda y} \qquad y \geq 0. \tag{6.33}$$

## ■■ Example 6.20

Find the mean and variance of the time until the arrival of the tenth inquiry in Example 6.19.

The arrival rate is $\lambda = 1/4$ inquiries per second, so the interarrivals times are exponential random variables with parameter $\lambda$. From Table 3.2, the mean and variance of an interarrival time are then $1/\lambda$ and $1/\lambda^2$, respectively. The time of the tenth arrival is the sum of ten such iid random variables, thus

$$E[S_{10}] = 10E[T] = \frac{10}{\lambda} = 40 \text{ seconds}$$

$$\text{VAR}[S_{10}] = 10\,\text{VAR}[T] = \frac{10}{\lambda^2} = 160 \text{ seconds}^2.$$

■■

In applications where the Poisson process models customer interarrival times, it is customary to say that arrivals occur "at random." We now explain what is meant by this statement. Suppose that we are *given* that only one arrival occurred in an interval $[0, t]$, and let $X$ be the arrival time of the single customer. For $0 < x < t$, let $N(x)$ be the number of events up to time $x$, and let $N(t) - N(x)$ be the increment in the interval $(x, t]$, then

$$\begin{aligned}
P[X \le x] &= P[N(x) = 1 \mid N(t) = 1] \\
&= \frac{P[N(x) = 1 \text{ and } N(t) = 1]}{P[N(t) = 1]} \\
&= \frac{P[N(x) = 1 \text{ and } N(t) - N(x) = 0]}{P[N(t) = 1]} \\
&= \frac{P[N(x) = 1]P[N(t) - N(x) = 0]}{P[N(t) = 1]} \\
&= \frac{\lambda x e^{-\lambda x} e^{-\lambda(t-x)}}{\lambda t e^{-\lambda t}} \\
&= \frac{x}{t}.
\end{aligned} \tag{6.34}$$

Equation (6.32) implies that given that one arrival has occurred in the interval $[0, t]$, then the customer arrival time is uniformly distributed in the interval $[0, t]$. It is in this sense that customer arrivals times occur "at random." It can be shown that *if the number of arrivals in the interval $[0, t]$ is k, then the individual arrival times are distributed independently and uniformly in the interval.*

### ■■ Example 6.21

Suppose that two customers arrive at a shop during a 2 minute period. Find the probability that both customers arrived during the first minute.

The arrival times of the customers are independent and uniformly distributed in the two minute interval. Each customer arrives during the first minute with probability 1/2. Thus the probability that both arrive during the first minute is $(1/2)^2 = 1/4$. This answer can be verified by showing that $P[N(1) = 2 \mid N(2) = 2] = 1/4$. ■■

## Random Telegraph Signal and Other Processes Derived from the Poisson Process

Many processes are derived from the Poisson process. In this section, we present two examples of such random processes.

## Example 6.22
Random Telegraph Signal

Consider a random process $X(t)$ that assumes the values $\pm 1$. Suppose that $X(0) = \pm 1$ with probability 1/2, and suppose that $X(t)$ then changes polarity with each occurrence of an event in a Poisson process of rate $\alpha$. Figure 6.9 shows a sample function of $X(t)$.

The pmf of $X(t)$ is given by

$$P[X(t) = \pm 1] = P[X(t) = \pm 1 \mid X(0) = 1]P[X(0) = 1] + P[X(t) = \pm 1 \mid X(0) = -1]P[X(0) = -1]. \quad (6.35)$$

The conditional pmf's are found by noting that $X(t)$ will have the same polarity as $X(0)$ only when an even number of events occur in the interval $(0, t]$. Thus

$$\begin{aligned} P[X(t) = \pm 1 \mid X(0) = \pm 1] &= P[N(t) = \text{even integer}] \\ &= \sum_{j=0}^{\infty} \frac{(\alpha t)^{2j}}{(2j)!} e^{-\alpha t} \\ &= e^{-\alpha t} \frac{1}{2} \{e^{\alpha t} + e^{-\alpha t}\} \\ &= \frac{1}{2}\{1 + e^{-2\alpha t}\}. \end{aligned} \quad (6.36)$$

$X(t)$ and $X(0)$ will differ in sign if the number of events in $t$ is odd:

$$\begin{aligned} P[X(t) = \pm 1 \mid X(0) = \mp 1] &= \sum_{j=0}^{\infty} \frac{(\alpha t)^{2j+1}}{(2j+1)!} e^{-\alpha t} \\ &= e^{-\alpha t} \frac{1}{2} \{e^{\alpha t} - e^{-\alpha t}\} \\ &= \frac{1}{2}\{1 - e^{-2\alpha t}\}. \end{aligned} \quad (6.37)$$

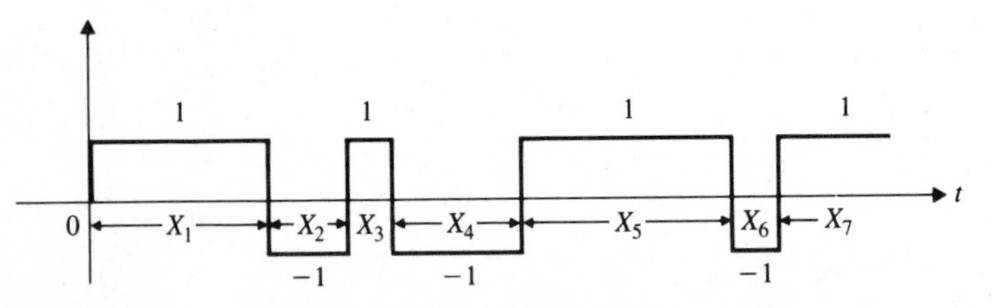


**FIGURE 6.9** Sample path of a random telegraph signal. The times between transitions $X_j$ are iid exponential random variables.

We obtain the pmf for $X(t)$ by substituting into Eq. (6.35):

$$P[X(t) = 1] = \frac{1}{2}\frac{1}{2}\{1 + e^{-2\alpha t}\} + \frac{1}{2}\frac{1}{2}\{1 - e^{-2\alpha t}\} = \frac{1}{2}$$

$$P[X(t) = -1] = 1 - P[X(t) = 1] = \frac{1}{2}. \tag{6.38}$$

Thus the random telegraph signal is equally likely to be $\pm 1$ at any time $t > 0$.

The mean and variance of $X(t)$ are

$$m_X(t) = 1P[X(t) = 1] + (-1)P[X(t) = -1] = 0$$
$$\text{VAR}[X(t)] = E[X(t)^2] = (1)^2P[X(t) = 1] + (-1)^2P[X(t) = -1] = 1.$$

The autocovariance of $X(t)$ is found as follows:

$$\begin{aligned} C_X(t_1, t_2) &= E[X(t_1)X(t_2)] \\ &= 1P[X(t_1) = X(t_2)] + (-1)P[X(t_1) \neq X(t_2)] \\ &= \frac{1}{2}\{1 + e^{-2\alpha|t_2-t_1|}\} - \frac{1}{2}\{1 - e^{-2\alpha|t_2-t_1|}\} \\ &= e^{-2\alpha|t_2-t_1|}. \end{aligned} \tag{6.39}$$

Thus time samples of $X(t)$ become less and less correlated as the time between them increases. ■■

---

The Poisson process and the random telegraph processes are examples of the continuous-time Markov chain processes that are discussed in Chapter 8.

### ■■ Example 6.23
Filtered Poisson Impulse Train

The Poisson process is zero at $t = 0$ and increases by one unit at the random arrival times $S_j$, $j = 1, 2, \ldots$. Thus the Poisson process can be expressed as the sum of randomly shifted step functions:

$$N(t) = \sum_{i=1}^{\infty} u(t - S_i) \qquad N(0) = 0,$$

where the $S_i$ are the arrival times.

Since the integral of a delta function $\delta(t - S)$ is a step function $u(t - S)$, we can view $N(t)$ as the result of integrating a train of delta functions that occur at times $S_n$ as shown in Fig. 6.10(a):

$$Z(t) = \sum_{i=1}^{\infty} \delta(t - S_i).$$

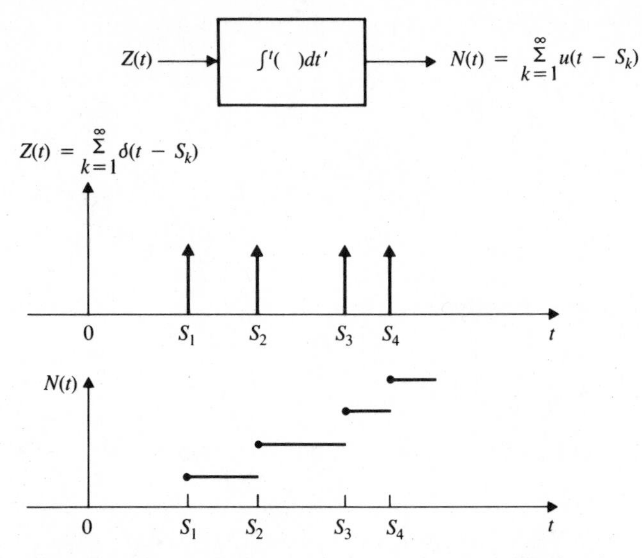


**FIGURE 6.10a** Poisson process as integral of train of delta functions.

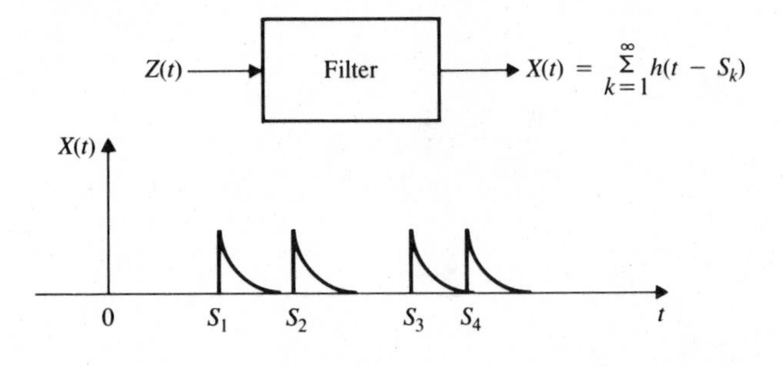


**FIGURE 6.10b** Filtered train of delta functions.

We can obtain other continuous-time processes by replacing the step function by another function $h(t)$[2] as shown in Fig. 6.10(b):

$$X(t) = \sum_{i=1}^{\infty} h(t - S_i). \tag{6.40}$$

For example, $h(t)$ could represent the current pulse that results when a photoelectron hits a detector. $X(t)$ is then the total current flowing at time $t$. $X(t)$ is called a **shot noise** process. ■■

2. This is equivalent to passing $Z(t)$ through a linear system whose response to a delta function is $h(t)$.

The following example shows how the properties of the Poisson process can be used to evaluate averages involving the filtered process.

### ■■ Example 6.24

Find the expected value of the shot noise process $X(t)$.

We condition on $N(t)$, the number of impulses that have occurred up to time $t$:

$$E[X(t)] = E[E[X(t) \mid N(t)]].$$

Suppose $N(t) = k$, then

$$\begin{aligned} E[X(t) \mid N(t) = k] &= E\left[\sum_{j=1}^{k} h(t - S_j)\right] \\ &= \sum_{j=1}^{k} E[h(t - S_j)]. \end{aligned}$$

Since the arrival times, $S_1, \ldots, S_k$, when the impulses occurred are independent, uniformly distributed in the interval $[0, t]$,

$$E[h(t - S_j)] = \int_0^t h(t - s) \frac{ds}{t} = \frac{1}{t} \int_0^t h(u)\, du.$$

Thus

$$E[X(t) \mid N(t) = k] = \frac{k}{t} \int_0^t h(u)\, du,$$

and

$$E[X(t) \mid N(t)] = \frac{N(t)}{t} \int_0^t h(u)\, du.$$

Finally, we obtain

$$\begin{aligned} E[X(t)] &= E[E[X(t) \mid N(t)]] \\ &= \frac{E[N(t)]}{t} \int_0^t h(u)\, du \\ &= \lambda \int_0^t h(u)\, du, \end{aligned} \tag{6.41}$$

where we used the fact that $E[N(t)] = \lambda t$. Note that $E[X(t)]$ approaches a constant value as $t$ becomes large if the above integral is finite. ■■

### Brownian Motion Process

Suppose that the symmetric random walk process (i.e., $p = 1/2$) of Example 6.14 takes steps of magnitude $h$ every $\delta$ seconds. We obtain a continuous-time process by letting $X_\delta(t)$ be the accumulated sum up to time $t$. At time $t$, the process will have taken $n = [t/\delta]$ jumps, so it is equal to

$$X_\delta(t) = h(D_1 + D_2 + \cdots + D_{[t/\delta]}) = hS_n. \tag{6.42}$$

The mean and variance of $X_\delta(t)$ are

$$E[X_\delta(t)] = hE[S_n] = 0$$
$$\text{VAR}[X_\delta(t)] = h^2 n \,\text{VAR}[D_n] = h^2 n,$$

where we used the fact that $\text{VAR}[D_n] = 4p(1-p) = 1$ since $p = 1/2$.

Suppose that we simultaneously shrink the size of the jumps and the time between jumps. In particular let $\delta \to 0$ and $h \to 0$ with $h = \sqrt{\alpha\delta}$, and let $X(t)$ be the resulting process.

$X(t)$ then has *mean* and *variance* given by

$$E[X(t)] = 0 \tag{6.43}$$
$$\text{VAR}[X(t)] = (\sqrt{\alpha\delta})^2(t/\delta) = \alpha t. \tag{6.44}$$

Thus we obtain a continuous-time process $X(t)$ that begins at the origin, has zero mean for all time, but has a variance that increases linearly with time. Figure 6.11 shows a sample function of the process. $X(t)$ is called the **Brownian motion** random process because it is used to model the motion of particles suspended in a fluid that move under the rapid and random impact of neighboring particles.

As $\delta \to 0$, Eq. (6.42) implies that $X(t)$ approaches the sum of an infinite number of random variables since $n = [t/\delta] \to \infty$. By the central limit theorem the *pdf of* $X(t)$ therefore approaches that of a Gaussian random variable with mean zero and variance $\alpha t$:

$$f_{X(t)}(x) = \frac{1}{\sqrt{2\pi\alpha t}} e^{-x^2/2\alpha t}. \tag{6.45}$$

$X(t)$ inherits the property of independent and stationary increments

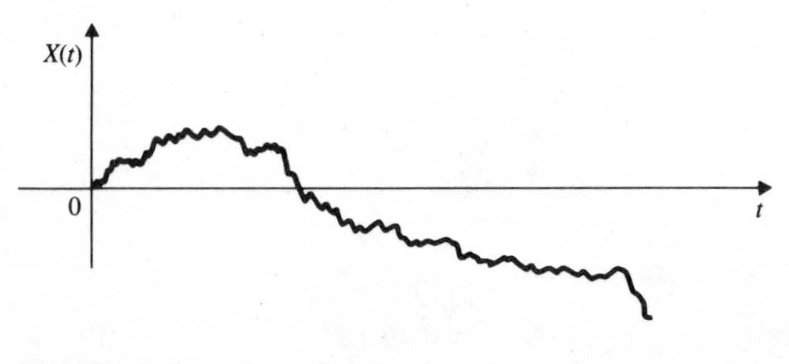

**FIGURE 6.11** Sample function of Brownian motion process.

from the random walk process from which it is derived. As a result, the *joint pdf of* $X(t)$ at several times $t_1, t_2, \ldots, t_k$ can be obtained by using Eq. (6.26):

$$f_{X(t_1),\ldots,X(t_k)}(x_1, \ldots, x_k) = f_{X(t_1)}(x_1) f_{X(t_2-t_1)}(x_2 - x_1) \cdots f_{X(t_k-t_{k-1})}(x_k - x_{k-1})$$

$$= \frac{\exp\left\{-\frac{1}{2}\left[\frac{x_1^2}{\alpha t_1} + \frac{(x_2 - x_1)^2}{\alpha(t_2 - t_1)} + \cdots + \frac{(x_k - x_{k-1})^2}{\alpha(t_k - t_{k-1})}\right]\right\}}{\sqrt{(2\pi\alpha)^k t_1(t_2 - t_1)\cdots(t_k - t_{k-1})}} \quad (6.46)$$

The independent increments property and the same sequence of steps that led to Eq. (6.29) can be used to show that the *autocovariance of* $X(t)$ is given by

$$C_X(t_1, t_2) = \alpha \min(t_1, t_2). \quad (6.47)$$

### ■■ Example 6.25

It can be shown that Eq. (6.46) implies that Brownian motion process is a Gaussian random process. For example, consider the $k = 2$ case and the times $t_1 < t_2$. The mean vector is zero since $m_X(t) = 0$. The covariance matrix and its inverse are

$$K = \begin{vmatrix} \alpha t_1 & \alpha t_1 \\ \alpha t_1 & \alpha t_2 \end{vmatrix} \qquad K^{-1} = \frac{1}{\alpha t_1(t_2 - t_1)} \begin{vmatrix} t_2 & -t_1 \\ -t_1 & t_1 \end{vmatrix}.$$

It is then easy to verify that

$$-\frac{1}{2}\mathbf{x}^T K^{-1}\mathbf{x} = -\frac{1}{2}\left\{\frac{x_1^2}{\alpha t_1} + \frac{(x_2 - x_1)^2}{\alpha(t_2 - t_1)}\right\}$$

and that

$$|K| = \alpha^2 t_1(t_2 - t_1).$$

These expressions are in agreement with Eq. (6.10) for $k = 2$. ■■

## 6.5 STATIONARY RANDOM PROCESSES

Many random processes have the property that the nature of the randomness in the process does not change with time. An observation of the process in the time interval $(t_0, t_1)$ exhibits the same type of random behavior as an observation in some other time interval $(t_0 + \tau, t_1 + \tau)$. This leads us to postulate that the probabilities of samples of the process do not depend on the instant when we begin taking observations, that is,

probabilities involving samples taken at times $t_1, \ldots, t_k$ will not differ from those taken at $t_1 + \tau, \ldots, t_k + \tau$.

If we are dealing with random processes that began at $t = -\infty$, then the above condition can be stated precisely as follows. *A discrete-time or continuous-time random process $X(t)$ is* **stationary** *if the joint distribution of any set of samples does not depend on the placement of the time origin.* This means that the joint cdf of $X(t_1), X(t_2), \ldots, X(t_k)$ is the same as that of $X(t_1 + \tau), X(t_2 + \tau), \ldots, X(t_k + \tau)$:

$$F_{X(t_1),\ldots,X(t_k)}(x_1, \ldots, x_k) = F_{X(t_1+\tau),\ldots,X(t_k+\tau)}(x_1, \ldots, x_k), \tag{6.48}$$

for all time shifts $\tau$, all $k$, and all choices of sample times $t_1, \ldots, t_k$. If a process begins at some definite time (i.e., $n = 0$ or $t = 0$), then we say it is stationary if its joint distributions do not change under time shifts to the right.

Two processes $X(t)$ and $Y(t)$ are said to be **jointly stationary** if the joint cdf's of $X(t_1), \ldots, X(t_k)$ and $Y(t'_1), \ldots, Y(t'_j)$ do not depend on the placement of the time origin for all $k$ and $j$ and all choices of sampling times $t_1, \ldots, t_k$ and $t'_1, \ldots, t'_j$.

The *first order cdf of a stationary random process must be independent of time* since by Eq. (6.48)

$$F_{X(t)}(x) = F_{X(t+\tau)}(x) = F_X(x) \qquad \text{all } t, \tau. \tag{6.49}$$

This implies that the mean and variance of $X(t)$ are constant and independent of time:

$$m_X(t) = E[X(t)] = m \qquad \text{for all } t \tag{6.50}$$

$$\text{VAR}[X(t)] = E[(X(t) - m)^2] = \sigma^2 \qquad \text{for all } t. \tag{6.51}$$

The *second order cdf of a stationary random process can depend only on the time difference between the samples* and not on the particular time of the samples since by Eq. (6.48)

$$F_{X(t_1),X(t_2)}(x_1, x_2) = F_{X(0),X(t_2-t_1)}(x_1, x_2) \qquad \text{for all } t_1, t_2. \tag{6.52}$$

This implies that the autocorrelation and the autocovariance of $X(t)$ can depend only on $t_2 - t_1$:

$$R_X(t_1, t_2) = R_X(t_2 - t_1) \qquad \text{for all } t_1, t_2 \tag{6.53}$$

$$C_X(t_1, t_2) = C_X(t_2 - t_1) \qquad \text{for all } t_1, t_2. \tag{6.54}$$

### ■■ Example 6.26
iid Random Process

Show that the iid random process is stationary.

The joint cdf for the samples at any $k$ time instants, $t_1, \ldots, t_k$ is

$$F_{X(t_1),\ldots,X(t_k)}(x_1, x_2, \ldots, x_k) = F_X(x_1)F_X(x_2) \cdots F_X(x_k)$$

for all $k$, $t_1, \ldots, t_k$. Thus Eq. (6.48) is satisfied, and so the iid random process is stationary. ■■

### ■■ Example 6.27

Is the sum process a discrete-time stationary process?

The sum process is defined by $S_n = X_1 + X_2 + \cdots + X_n$, where the $X_i$ are an iid sequence. The process has mean and variance

$$m_S(n) = nm \qquad \text{VAR}[S_n] = n\sigma^2,$$

where $m$ and $\sigma^2$ are the mean and variance of the $X_n$. It can be seen that the mean and variance are not constant but grow linearly with the time index $n$. Therefore the sum process cannot be a stationary process. ■■

### ■■ Example 6.28

Random Telegraph Signal

Show that the random telegraph signal discussed in Example 6.22 is a stationary random process when $P[X(0) = \pm 1] = 1/2$. Show that $X(t)$ settles into stationary behavior as $t \to \infty$ even if $P[X(0) = \pm 1] \neq 1/2$.

We need to show that the following two joint pmf's are equal:

$$P[X(t_1) = a_1, \ldots, X(t_k) = a_k] \\ = P[X(t_1 + \tau) = a_1, \ldots, X(t_k + \tau) = a_k],$$

for any $k$, any $t_1 < \cdots < t_k$, and any $a_j = \pm 1$. The independent increments property of the Poisson process implies that

$$P[X(t_1) = a_1, \ldots, X(t_k) = a_k] = P[X(t_1) = a_1] \\ \times P[X(t_2) = a_2 \,|\, X(t_1) = a_1] \cdots P[X(t_k) = a_k \,|\, X(t_{k-1}) = a_{k-1}],$$

since the values of the random telegraph at the times $t_1, \ldots, t_k$ is determined by the number of occurrences of events of the Poisson process in the time intervals $(t_j, t_{j+1})$. Similarly,

$$P[X(t_1 + \tau) = a_1, \ldots, X(t_k + \tau) = a_k] \\ = P[X(t_1 + \tau) = a_1]P[X(t_2 + \tau) = a_2 \,|\, X(t_1 + \tau) = a_1] \cdots \\ \times P[X(t_k + \tau) = a_k \,|\, X(t_{k-1} + \tau) = a_{k-1}].$$

The corresponding transition probabilities in the previous two equations are equal since:

$$P[X(t_{j+1}) = a_{j+1} \,|\, X(t_j) = a_j] = \begin{cases} \frac{1}{2}\{1 + e^{-2\alpha(t_{j+1} - t_j)}\} & \text{if } a_j = a_{j+1} \\ \frac{1}{2}\{1 - e^{-2\alpha(t_{j+1} - t_j)}\} & \text{if } a_j \neq a_{j+1} \end{cases} \\ = P[X(t_{j+1} + \tau) = a_{j+1} \,|\, X(t_j + \tau) = a_j].$$

Thus the two joint probabilities differ only in the first term, namely, $P[X(t_1) = a_1]$ and $P[X(t_1 + \tau) = a_1]$.

From Example 6.22 we know that if $P[X(0) = \pm 1] = 1/2$ then $P[X(t) = \pm 1] = 1/2$, for all $t$. Thus $P[X(t_1) = a_1] = 1/2$, $P[X(t_1 + \tau) = a_1] = 1/2$, and

$$P[X(t_1) = a_1, \ldots, X(t_k) = a_k] = P[X(t_1 + \tau) = a_1, \ldots, X(t_k + \tau) = a_k].$$

Thus we conclude that the process is stationary when $P[X(0) = \pm 1] = 1/2$.

If $P[X(0) = \pm 1] \neq 1/2$, then the two joint pmf's are not equal because $P[X(t_1) = a_1] \neq P[X(t_1 + \tau) = a_1]$. Let's see what happens if we know that the process started at a specific value, say $X(0) = 1$, that is, $P[X(0) = 1] = 1$. The pmf for $X(t)$ is obtained from Eqs. (6.35) through (6.37):

$$P[X(t) = a] = P[X(t) = a \mid X(0) = 1]1$$
$$= \begin{cases} \dfrac{1}{2}\{1 + e^{-2\alpha t}\} & \text{if } a = 1 \\ \dfrac{1}{2}\{1 - e^{-2\alpha t}\} & \text{if } a = -1. \end{cases}$$

For very small $t$, the probability that $X(t) = 1$ is close to 1; but as $t$ increases, the probability that $X(t) = 1$ becomes 1/2. Therefore as $t_1$ becomes large, $P[X(t_1) = a_1] \to 1/2$ and $P[X(t_1 + \tau) = a_1] = 1/2$ and the two joint pmf's become equal. In other words, the process "forgets" the initial condition and it settles down into "steady state," that is, stationary behavior. ■■

### Wide-Sense Stationary Random Processes

In many situations we cannot determine whether a random process is stationary, but we can determine whether the mean is a constant:

$$m_X(t) = m \qquad \text{for all } t, \tag{6.55}$$

and whether the autocovariance (or equivalently the autocorrelation) is a function of $t_1 - t_2$ only:

$$C_X(t_1, t_2) = C_X(t_1 - t_2) \qquad \text{for all } t_1, t_2. \tag{6.56}$$

*A discrete-time or continuous-time random process $X(t)$ is* **wide-sense stationary** (WSS) *if it satisfies Eqs. (6.55) and (6.56).* Similarly, we say that the processes $X(t)$ and $Y(t)$ are **jointly wide-sense stationary** if they are both wide-sense stationary and if their cross-covariance depends only on $t_1 - t_2$. When $X(t)$ is wide-sense stationary, we write

$$C_X(t_1, t_2) = C_X(\tau) \qquad \text{and} \qquad R_X(t_1, t_2) = R_X(\tau),$$

where $\tau = t_1 - t_2$.

*All stationary random processes are wide-sense stationary* since they satisfy Eqs. (6.55) and (6.56). The following example shows that some wide-sense stationary processes are not stationary.

### Example 6.29

Let $X_n$ consist of two interleaved sequences of independent random variables. For $n$ even, $X_n$ assumes the values $\pm 1$ with probability 1/2; for $n$ odd, $X_n$ assumes the values 1/3 and $-3$ with probabilities 9/10 and 1/10, respectively. $X_n$ is not stationary since its pmf varies with $n$. It is easy to show that $X_n$ has mean

$$m_X(n) = 0 \qquad \text{for all } n$$

and covariance function

$$C_X(i, j) = \begin{cases} E[X_i]E[X_j] = 0 & \text{for } i \neq j \\ E[X_i^2] = 1 & \text{for } i = j. \end{cases}$$

$X_n$ is therefore wide-sense stationary.

We will see in Chapter 7 that the autocorrelation function of wide-sense stationary processes plays a crucial role in the design of linear signal processing algorithms. We now develop several results that enable us to deduce properties of a WSS process from properties of its autocorrelation function.

First, the autocorrelation function at $\tau = 0$ gives the **average power** (second moment) of the process:

$$R_X(0) = E[X(t)^2] \qquad \text{for all } t. \tag{6.57}$$

Second, the autocorrelation function is an even function of $\tau$ since

$$R_X(\tau) = E[X(t + \tau)X(t)] = E[X(t)X(t + \tau)] = R_X(-\tau). \tag{6.58}$$

Third, the autocorrelation function is a measure of the rate of change of a random process in the following sense. Consider the change in the process from time $t$ to $t + \tau$:

$$\begin{aligned} P[|X(t + \tau) - X(t)| > \varepsilon] &= P[(X(t + \tau) - X(t))^2 > \varepsilon^2] \\ &\leq \frac{E[(X(t + \tau) - X(t))^2]}{\varepsilon^2} \\ &= \frac{2\{R_X(0) - R_X(\tau)\}}{\varepsilon^2}, \end{aligned} \tag{6.59}$$

where we used the Markov inequality, Eq. (3.72) to obtain the upper bound. Equation (6.59) states that if $R_X(0) - R_X(\tau)$ is small, that is, $R_X(\tau)$ drops off slowly, then the probability of a large change in $X(t)$ in $\tau$ seconds is small.

Fourth, the autocorrelation function is maximum at $\tau = 0$. We use the following inequality:[3]

$$E[XY]^2 \leq E[X^2]E[Y^2], \tag{6.60}$$

for any two random variables $X$ and $Y$. If we apply this equation to $X(t + \tau)$ and $X(t)$, we obtain

$$R_X(\tau)^2 = E[X(t + \tau)X(t)]^2 \leq E[X^2(t + \tau)]E[X^2(t)] = R_X(0)^2.$$

Thus

$$|R_X(\tau)| \leq R_X(0). \tag{6.61}$$

Fifth, if $R_X(0) = R_X(d)$, then $R_X(\tau)$ is periodic with period $d$ and $X(t)$ is mean-square periodic, that is, $E[(X(t + d) - X(t))^2] = 0$. If we apply Eq. (6.60) to $X(t + \tau + d) - X(t + \tau)$ and $X(t)$, we obtain

$$E[(X(t + \tau + d) - X(t + \tau))X(t)]^2$$
$$\leq E[(X(t + \tau + d) - X(t + \tau))^2]E[X^2(t)],$$

which implies that

$$\{R_X(\tau + d) - R_X(\tau)\}^2 \leq 2\{R_X(0) - R_X(d)\}R_X(0).$$

Thus $R_X(d) = R_X(0)$ implies that the right-hand side of the equation is zero, and thus that $R_X(\tau + d) = R_X(\tau)$ for all $\tau$. Repeated applications of this result imply that $R_X(\tau)$ is periodic with period $d$. The fact that $X(t)$ is mean-square periodic follows from

$$E[(X(t + d) - X(t))^2] = 2\{R_X(0) - R_X(d)\} = 0.$$

Sixth, let $X(t) = m + N(t)$, where $N(t)$ is a zero-mean process for which $R_N(\tau) \to 0$ as $\tau \to \infty$, then

$$R_X(\tau) = E[(m + N(t))^2] = m^2 + 2mE[N(t)] + R_N(\tau)$$
$$= m^2 + R_N(\tau) \to m^2 \qquad \text{as } \tau \to \infty.$$

In other words, $R_X(\tau)$ approaches the square of the mean of $X(t)$ as $\tau \to \infty$.

In summary, the autocorrelation function can have three types of components: (1) a component that approaches zero as $\tau \to \infty$; (2) a periodic component; and (3) a component due to a nonzero mean.

---

3. The proof of the bound is essentially the same as that used to show that $|\rho| \leq 1$ in Section 4.7.

## ■■ Example 6.30

Figure 6.12 shows several typical autocorrelation functions. Figure 6.12(a) shows the autocorrelation function for the random telegraph signal $X(t)$ (see Eq. 6.39):

$$R_X(\tau) = e^{-2\alpha|\tau|} \qquad \text{for all } \tau.$$

$X(t)$ is zero-mean and $R_X(\tau) \to 0$ as $|\tau| \to \infty$.

Figure 6.12(b) shows the autocorrelation function for a sinusoid $Y(t)$ with amplitude $a$ and random phase (see Example 6.7):

$$R_Y(\tau) = \frac{a^2}{2}\cos(2\pi f_0 \tau) \qquad \text{for all } \tau.$$

$Y(t)$ is zero-mean and $R_Y(\tau)$ is periodic with period $1/f_0$.

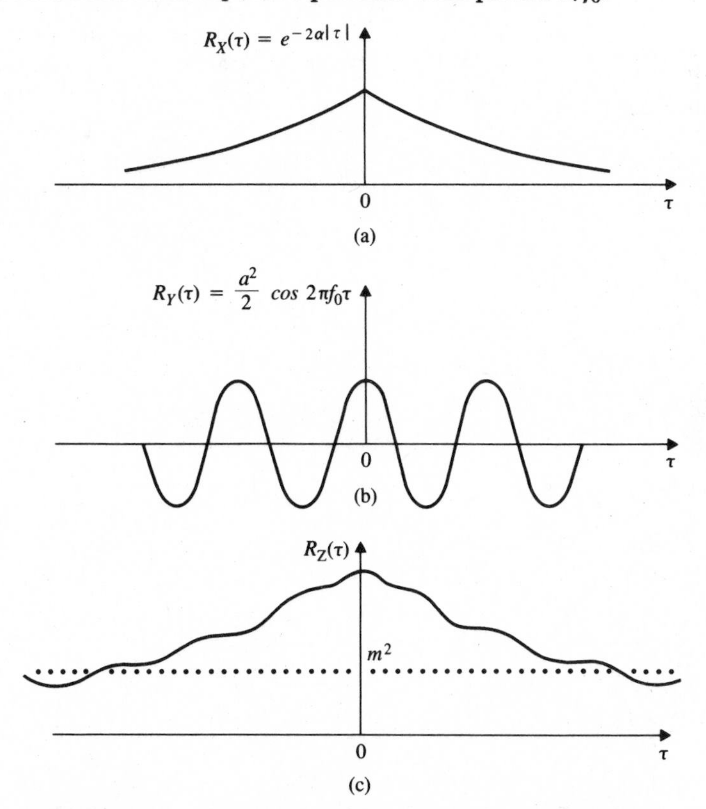


**FIGURE 6.12** (a) Autocorrelation function of a random telegraph signal. (b) Autocorrelation function of a sinusoid with random phase. (c) Autocorrelation function of a random process that has nonzero mean, a periodic component, and a "random" component.

Figure 6.12(c) shows the autocorrelation function for the process $Z(t) = X(t) + Y(t) + m$, where $X(t)$ is the random telegraph process, $Y(t)$ is a sinusoid with random phase, and $m$ is a constant. If we assume that $X(t)$ and $Y(t)$ are independent processes, then

$$\begin{aligned} R_Z(\tau) &= E[\{X(t+\tau) + Y(t+\tau) + m\}\{X(t) + Y(t) + m\}] \\ &= R_X(\tau) + R_Y(\tau) + m^2. \end{aligned}$$

■■

### Wide-Sense Stationary Gaussian Random Processes

*If a Gaussian random process is wide-sense stationary, then it is also stationary.* Recall from Section 6.2, Eq. (6.10), that the joint pdf of a Gaussian random process is completely determined by the mean $m_X(t)$ and the autocovariance $C_X(t_1, t_2)$. If $X(t)$ is wide-sense stationary, then its mean is a constant $m$ and its autocovariance depends only on the difference of the sampling times, $t_i - t_j$. It then follows that the joint pdf of $X(t)$ depends only on this set of differences, and hence it is invariant with respect to time shifts. Thus the process is also stationary.

The above result makes WSS Gaussian random processes particularly easy to work with since all the information required to specify the joint pdf is contained in $m$ and $C_X(\tau)$.

#### ■■ Example 6.31
A Gaussian Moving Average Process

Let $X_n$ be an iid sequence of Gaussian random variables with zero mean and variance $\sigma^2$, and let $Y_n$ be the average of two consecutive values of $X_n$:

$$Y_n = \frac{X_n + X_{n-1}}{2}.$$

The mean of $Y_n$ is zero since $E[X_i] = 0$ for all $i$. The covariance is

$$\begin{aligned} C_Y(i,j) = E[Y_i Y_j] &= \frac{1}{4} E[(X_i + X_{i-1})(X_j + X_{j-1})] \\ &= \frac{1}{4}\{E[X_i X_j] + E[X_i X_{j-1}] + E[X_{i-1} X_j] + E[X_{i-1} X_{j-1}]\} \\ &= \begin{cases} \frac{1}{2}\sigma^2 & \text{if } i = j \\ \frac{1}{4}\sigma^2 & \text{if } |i - j| = 1 \\ 0 & \text{otherwise.} \end{cases} \end{aligned}$$

We see that $Y_n$ has a constant mean and a covariance function that depends only on $|i - j|$, thus $Y_n$ is a wide-sense stationary process. $Y_n$ is a Gaussian random variable since it is defined by a linear function of Gaussian random variables (see Section 4.8). Thus the joint pdf of $Y_n$ is given by Eq. (6.10) with zero mean vector and with entries of the covariance matrix specified by $C(i, j)$ above.

## 6.6

## TIME AVERAGES OF RANDOM PROCESSES AND ERGODIC THEOREMS

At some point, the parameters of a random process must be obtained through measurement. The results from Chapter 5 suggest that we repeat the random experiment that gives rise to the random process a large number of times and take the arithmetic average of the quantities of interest. For example, to estimate the mean $m_X(t)$ of a random process $X(t, s)$, we repeat the random experiment and take the following average:

$$\hat{m}_X(t) = \frac{1}{N} \sum_{i=1}^{N} X(t, s_i), \tag{6.62}$$

where $N$ is the number of repetitions of the experiment, and $X(t, s_i)$ is the realization observed in the $i$th repetition.

In some situations, we are interested in estimating the mean or autocorrelation functions from the **time average** of a single realization, that is,

$$\langle X(t) \rangle_T = \frac{1}{2T} \int_{-T}^{T} X(t, s)\, dt. \tag{6.63}$$

An **ergodic theorem** states conditions under which a time average converges as the observation interval becomes large. In this section, we are interested in ergodic theorems that state when time averages converge to the ensemble average (expected value).

The strong law of large numbers, presented in Chapter 5, is one of the most important ergodic theorems. It states that if $X_n$ is an iid discrete-time random process with finite mean $E[X_n] = m$, then the time average of the samples converges to the ensemble average with probability one:

$$P\left[\lim_{n \to \infty} \frac{1}{n} \sum_{i=1}^{n} X_i = m\right] = 1. \tag{6.64}$$

This result allows us to estimate $m$ by taking the time average of a single realization of the process. We are interested in obtaining results of this type for a larger class of random processes, that is, for non-iid, discrete-time random processes, and for continuous-time random processes.

The following example shows that in general time averages do not converge to ensemble averages.

### ■■ Example 6.32

Let $X(t) = A$ for all $t$, where $A$ is a zero-mean, unit-variance random variable. Find the limiting value of the time average.

The mean of the process is $m_X(t) = E[X(t)] = E[A] = 0$. However, Eq. (6.63) gives

$$\langle X(t) \rangle_T = \frac{1}{2T} \int_{-T}^{T} A\, dt = A.$$

Thus the time-average mean does not always converge to $m_X(t) = 0$. ■■

Consider the estimate given by Eq. (6.63) for $E[X(t)] = m_X(t)$. The estimate yields a single number, so obviously it only makes sense to consider processes for which $m_X(t) = m$, a constant. We now develop an ergodic theorem for the time average of wide-sense stationary processes.

Let $X(t)$ be a WSS process. The expected value of $\langle X(t) \rangle_T$ is

$$E[\langle X(t) \rangle_T] = E\left[\frac{1}{2T} \int_{-T}^{T} X(t)\, dt\right] = \frac{1}{2T} \int_{-T}^{T} E[X(t)]\, dt = m. \tag{6.65}$$

The expected value and the integral can be interchanged since the integral is the limit of a sum, and the expected value of a sum is the sum of the expected values. Equation (6.65) states that $\langle X(t) \rangle_T$ is an unbiased estimator for $m$.

Consider the variance of $\langle X(t) \rangle_T$:

$$\begin{aligned}
\text{VAR}[\langle X(t) \rangle_T] &= E[(\langle X(t) \rangle_T - m)^2] \\
&= E\left[\left\{\frac{1}{2T} \int_{-T}^{T} (X(t) - m)\, dt\right\}\left\{\frac{1}{2T} \int_{-T}^{T} (X(t') - m)\, dt'\right\}\right] \\
&= \frac{1}{4T^2} \int_{-T}^{T} \int_{-T}^{T} E[(X(t) - m)(X(t') - m)]\, dt\, dt' \\
&= \frac{1}{4T^2} \int_{-T}^{T} \int_{-T}^{T} C_X(t, t')\, dt\, dt'.
\end{aligned} \tag{6.66}$$

Since the process $X(t)$ is WSS, Eq. (6.66) becomes

$$\text{VAR}[\langle X(t) \rangle_T] = \frac{1}{4T^2} \int_{-T}^{T} \int_{-T}^{T} C_X(t - t')\, dt\, dt' \tag{6.67}$$

Figure 6.13 shows the region of integration for this integral. The integrand

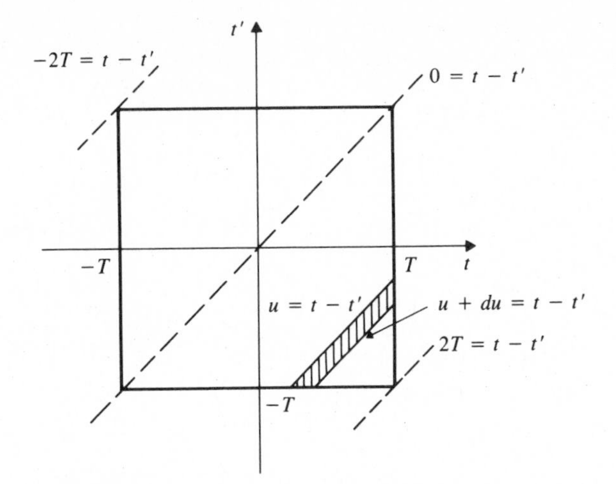


**FIGURE 6.13** Region of integration for integral in Eq. (6.66).

is constant along the line $u = t - t'$ for $-2T < u < 2T$, so we can evaluate the integral as the sums of infinitesimal strips as shown in the figure. It can be shown that each strip has area $(2T - |u|)\,du$, so the contribution of each strip to the integral is $(2T - |u|)C_X(u)\,du$. Thus

$$\begin{aligned} \text{VAR}[\langle X(t)\rangle_T] &= \frac{1}{4T^2}\int_{-2T}^{2T}(2T - |u|)C_X(u)\,du \\ &= \frac{1}{2T}\int_{-2T}^{2T}\left(1 - \frac{|u|}{2T}\right)C_X(u)\,du. \end{aligned} \tag{6.68}$$

Therefore, $\langle X(t)\rangle_T$ will approach $m$ in the mean-square sense, that is, $E[(\langle X(t)\rangle_T - m)^2] \to 0$, if the expression in Eq. (6.68) approaches zero with increasing $T$. We have just proved the following ergodic theorem.

**THEOREM**

Let $X(t)$ be a WSS process with $m_X(t) = m$, then

$$\lim_{T\to\infty}\langle X(t)\rangle_T = m$$

in the mean-square sense, if and only if

$$\lim_{T\to\infty}\frac{1}{2T}\int_{-2T}^{2T}\left(1 - \frac{|u|}{2T}\right)C_X(u)\,du = 0.$$

In keeping with engineering usage, we say that a WSS process is **mean-ergodic** if it satisfies the conditions of the above theorem.

The above theorem can be used to obtain ergodic theorems for the time average of other quantities. For example, if we replace $X(t)$ by $Y(t + \tau)Y(t)$ in Eq. (6.63), we obtain a time-average estimate for the autocorrelation function of the process $Y(t)$:

$$\langle Y(t+\tau)Y(t)\rangle_T = \frac{1}{2T}\int_{-T}^{T} Y(t+\tau)Y(t)\,dt. \tag{6.69}$$

It is easily shown that $E[\langle Y(t+\tau)Y(t)\rangle_T] = R_Y(\tau)$ if $Y(t)$ is WSS. The above ergodic theorem then implies that the time average autocorrelation converges to $R_Y(\tau)$ in the mean square sense if the term in Eq. (6.68) with $X(t)$ replaced by $Y(t)Y(t+\tau)$ converges to zero.

### ■■ Example 6.33

Is the random telegraph process mean-ergodic?

The covariance function for the random telegraph process is $C_X(\tau) = e^{-2\alpha|\tau|}$, so the variance of $\langle X(t)\rangle_T$ is

$$\begin{aligned} \text{VAR}[\langle X(t)\rangle_T] &= \frac{2}{2T}\int_0^{2T}\left(1 - \frac{u}{2T}\right)e^{-2\alpha u}\,du \\ &< \frac{1}{T}\int_0^{2T} e^{-2\alpha u}\,du = \frac{1 - e^{-4\alpha T}}{2\alpha T} \end{aligned}$$

The bound approaches zero as $T \to \infty$, so $\text{VAR}[\langle X(t)\rangle_T] \to 0$. Therefore the process is mean-ergodic. ■■

---

If the random process under consideration is discrete-time, then the time-average estimate for the mean and the autocorrelation functions of $X_n$ are given by

$$\langle X_n\rangle_T = \frac{1}{2T+1}\sum_{n=-T}^{T} X_n \tag{6.70}$$

$$\langle X_{n+k}X_n\rangle_T = \frac{1}{2T+1}\sum_{n=-T}^{T} X_{n+k}X_n. \tag{6.71}$$

If $X_n$ is a WSS random process, then $E[\langle X_n\rangle_T] = m$, and so $\langle X_n\rangle_T$ is an unbiased estimate for $m$. It is easy to show that the variance of $\langle X_n\rangle_T$ is

$$\text{VAR}[\langle X_n\rangle_T] = \frac{1}{2T+1}\sum_{k=-2T}^{2T}\left(1 - \frac{|k|}{2T+1}\right)C_X(k). \tag{6.72}$$

Therefore, $\langle X_n\rangle_T$ approaches $m$ in the mean-square sense and is mean-ergodic if the expression in Eq. (6.72) approaches zero with increasing $T$.

## SUMMARY

- A random process or stochastic process is an indexed family of random variables that is specified by the set of joint distributions of any number and choice of random variables in the family. The mean, autocovariance, and autocorrelation functions summarize some of the information contained in the joint distributions of pairs of time samples.

- The mean and covariance functions completely specify all joint distributions of a Gaussian random process.

- The sum process of an iid sequence has the property of stationary and independent increments, which facilitates the evaluation of the joint pdf/pmf of the process at any set of time instants. The binomial and random processes are sum processes. The Poisson and Brownian motion processes are obtained as limiting forms of these sum processes.

- The Poisson process has independent, stationary increments that are Poisson distributed. The interarrival times in a Poisson process are iid exponential random variables.

- A random process is stationary if its joint distributions are independent of the choice of time origin. If a random process is stationary, then $m_X(t)$ is constant, and $R_X(t_1, t_2)$ depends only on $t_1 - t_2$.

- A random process is wide-sense stationary (WSS) if its mean is constant and if its autocorrelation and autocovariance depends only on $t_1 - t_2$. A WSS process need not be stationary.

## CHECKLIST OF IMPORTANT TERMS

Autocorrelation function
Autocovariance function
Bernoulli random process
Binomial counting process
Brownian motion process
Cross-correlation function
Gaussian random process
iid random process
Independent increments
Independent random processes
Mean-ergodic random process
Mean function
Orthogonal random processes
Poisson process
Random process
Random telegraph signal
Random walk process
Realization or sample path
Stationary increments
Stationary random process
Stochastic process
Sum random process
Uncorrelated random processes
WSS random process

## ANNOTATED REFERENCES

References [1] through [4] can be consulted for further reading on random processes. Reference [5] provides many examples of autocorrelation functions in the context of electrical engineering.

1. A. Papoulis, *Probability, Random Variables, and Stochastic Processes,* McGraw-Hill, 1965.
2. W. B. Davenport, *Probability and Random Processes: An Introduction for Applied Scientists and Engineers,* McGraw-Hill, 1970.
3. H. Stark and J. W. Woods, *Probability, Random Processes, and Estimation Theory for Engineers,* Prentice-Hall, 1986.
4. R. M. Gray and L. D. Davisson, *Random Processes, A Mathematical Approach for Engineers,* Prentice-Hall, 1986.
5. G. R. Cooper and C. D. MacGillem, *Probabilistic Methods of Signal and System Analysis,* Holt, Reinhart & Winston, 1986.

## PROBLEMS

### Sections 6.1 & 6.2
### Definition and Specification of a Stochastic Process

1. In Example 6.1, find the joint pmf for $X_1$ and $X_2$. Why are $X_1$ and $X_2$ independent?
2. A discrete-time random process $X_n$ is defined as follows. A fair coin is tossed. If the outcome is heads, $X_n = 1$ for all $n$; if the outcome is tails, $X_n = -1$ for all $n$.
   a. Sketch some sample paths of the process.
   b. Find the pmf for $X_n$.
   c. Find the joint pmf for $X_n$ and $X_{n+k}$.
   d. Find the mean and autocovariance functions of $X_n$.
3. A discrete-time random process $X_n$ is defined as follows. A fair coin is tossed. If the outcome is heads, $X_n = (-1)^n$ for all $n$; if the outcome is tails, $X_n = (-1)^{n+1}$ for all $n$.
   a. Sketch some sample paths of the process.
   b. Find the pmf for $X_n$.
   c. Find the joint pmf for $X_n$ and $X_{n+k}$.
   d. Find the mean and autocovariance functions of $X_n$.
4. A discrete-time random process is defined by $X_n = s^n$, for $n \geq 0$, where $s$ is selected at random from the interval $(0, 1)$.
   a. Sketch some sample paths of the process.

b. Find the cdf of $X_n$.
c. Find the joint cdf for $X_n$ and $X_{n+1}$.
d. Find the mean and autocovariance functions of $X_n$.

5. Let $g(t)$ be the rectangular pulse shown in Fig. P6.1.

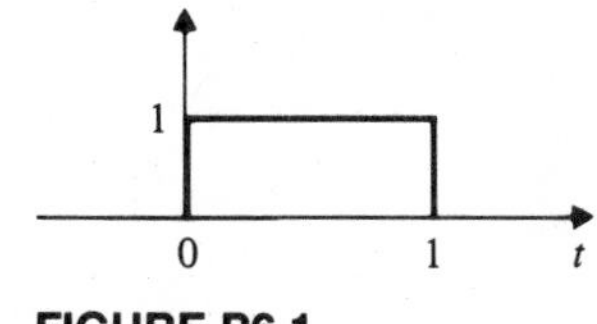

FIGURE P6.1

The random process $X(t)$ is defined as

$$X(t) = Ag(t),$$

where $A$ assumes the values $\pm 1$ with equal probability.

a. Find the pmf of $X(t)$.
b. Find $m_X(t)$.
c. Find the joint pmf of $X(t)$ and $X(t + d)$.
d. Find $C_X(t, t + d)$, $d > 0$.

6. A random process is defined by

$$Y(t) = g(t - T),$$

where $g(t)$ is the rectangular pulse of Problem 5, and $T$ is a uniformly-distributed random variable in the interval $(0, 1)$.

a. Find the pmf of $Y(t)$.
b. Find $m_X(t)$.

7. A random process is defined by

$$X(t) = g(t - T),$$

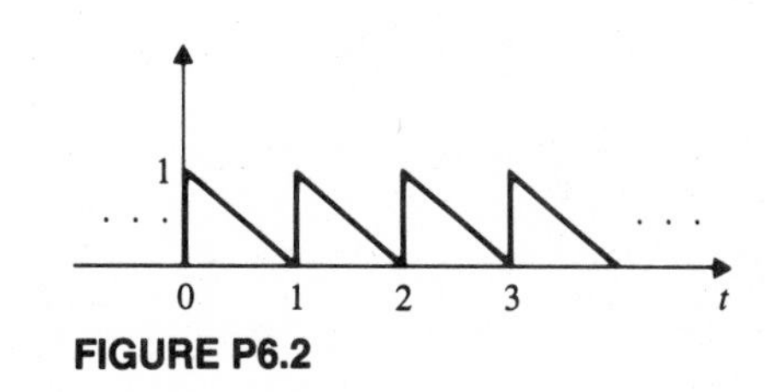


FIGURE P6.2

where $T$ is a uniform random variable in the interval $(0, 1)$ and $g(t)$ is the periodic triangular waveform shown in Fig. P6.2.

a. Find the cdf of $X(t)$ for $0 < t < 1$.
b. Find $m_X(t)$.

8. Let $Y(t) = g(t - T)$ as in Problem 6, but let $T$ be an exponentially distributed random variable with parameter $\alpha$.
   a. Find the pmf of $Y(t)$.
   b. Find the joint pmf of $Y(t)$ and $Y(t + d)$. Consider two cases: $d > 2$, and $0 < d < 2$.
   c. Find $m_Y(t)$ and $C_Y(t, t + d)$ for $d > 2$ and $0 < d < 2$.
9. Let $Z(t) = At + B$, where $A$ and $B$ are independent random variables.
   a. Find the pdf of $Z(t)$.
   b. Find $m_Z(t)$.
10. The random process $H(t)$ is defined as the "hard-limited" version of $X(t)$:
$$H(t) = \begin{cases} +1 & \text{if } X(t) \geq 0 \\ -1 & \text{if } X(t) < 0. \end{cases}$$
   a. Find the pdf, mean, and autocovariance of $H(t)$ if $X(t)$ is the sinusoid with a random amplitude presented in Example 6.2.
   b. Find the pdf, mean, and autocovariance of $H(t)$ if $X(t)$ is the sinusoid with random phase presented in Example 6.7.
   c. Find a general expression for the mean of $H(t)$ in terms of the cdf of $X(t)$.
   d. Find an expression for the cross-correlation between $H(t)$ and $X(t)$.
11. Find an expression for $E[|X_{t_2} - X_{t_1}|^2]$ in terms of autocorrelation function.
12. Are uncorrelated random processes orthogonal? Are orthogonal random processes uncorrelated? Explain.
13. Let $X(t)$ and $Y(t)$ be jointly Gaussian random processes. Explain the relation between the conditions of independence, uncorrelatedness, and orthogonality of $X(t)$ and $Y(t)$.
14. Let $X(t)$ be a zero-mean Gaussian random process with autocovariance function given by
$$C_X(t_1, t_2) = \sigma^2 e^{-|t_1 - t_2|}.$$
Find the joint pdf of $X(t)$ and $X(t + s)$.
15. The random process $Z(t)$ is defined by
$$Z(t) = Xt + Y,$$
where $X$ and $Y$ are a pair of random variables with means $m_X$, $m_Y$, variances, $\sigma_X^2$, $\sigma_Y^2$ and correlation coefficient $\rho_{X,Y}$.
   a. Find the mean and autocovariance of $Z(t)$.
   b. Find the pdf of $Z(t)$ if $X$ and $Y$ are jointly Gaussian random variables.

16. Let $X(t) = A\cos wt + B\sin wt$, where $A$ and $B$ are iid Gaussian random variables with zero mean and variance $\sigma^2$.
    a. Find the mean and autocovariance of $X(t)$.
    b. Find the joint pdf of $X(t)$ and $X(t+s)$.
17. Let $Y(t) = X(t) - aX(t+d)$, where $X(t)$ is a Gaussian random process.
    a. Find the mean and autocovariance of $Y(t)$.
    b. Find the pdf of $Y(t)$.
    c. Find the joint pdf of $Y(t)$ and $Y(t+s)$.
18. Let $X(t)$ and $Y(t)$ be independent Gaussian random processes with zero means and the same covariance function $C(t_1, t_2)$. Define the following "amplitude-modulated signal" by

$$Z(t) = X(t)\cos wt + Y(t)\sin wt.$$

    a. Find the mean and autocovariance of $Z(t)$.
    b. Find the pdf of $Z(t)$.
19. Let $X(t)$ be a zero-mean Gaussian random process with autocovariance function given by $C_X(t_1, t_2)$. If $X(t)$ is the input to a "square law detector," then the output is

$$Y(t) = X(t)^2.$$

Find the mean and autocovariance of the output $Y(t)$.

## Section 6.3
## Examples of Discrete-Time Random Processes

20. Let $Y_n$ be the process that results when individual 1s in a Bernoulli process are erased with probability $\alpha$. Find the pmf of $S'_n$, the counting process for $Y_n$. Does $Y_n$ have independent and stationary increments?
21. Let $S_n$ denote a binomial counting process.
    a. Show that $P[S_n = j, S_{n'} = i] \neq P[S_n = j]P[S_{n'} = i]$.
    b. Find $P[S_{n_2} = j \mid S_{n_1} = i]$, where $n_2 > n_1$.
    c. Show that $P[S_{n_2} = j \mid S_{n_1} = i,\ S_{n_0} = k] = P[S_{n_2} = j \mid S_{n_1} = i]$, where $n_2 > n_1 > n_0$.
22. Find $P[S_n = 0]$ for the random walk process.
23. Consider the following *moving average* processes:

$$Y_n = \frac{1}{2}(X_n + X_{n-1}) \qquad X_0 = 0$$

$$Z_n = \frac{2}{3}X_n + \frac{1}{3}X_{n-1} \qquad X_0 = 0.$$

a. Flip a coin 10 times to obtain a realization of a Bernoulli random process $X_n$. Find the resulting realizations of $Y_n$ and $Z_n$.
b. Repeat Part a with $X_n$ given by the random step process introduced in Example 6.12.
c. Find the mean, variance, and covariance of $Y_n$ and $Z_n$ if $X_n$ is a Bernoulli random process. Are the sample means of $Y_n$ and $Z_n$ in Part a close to their respective means?
d. Repeat Part c if $X_n$ is the random step process.

24. Consider the following *autoregressive processes*:

$$W_n = 2W_{n-1} + X_n \qquad W_0 = 0$$

$$Z_n = \frac{1}{2}Z_{n-1} + X_n \qquad Z_o = 0.$$

a. Flip a coin 10 times to obtain a realization of the Bernoulli process. Find the resulting realizations of $W_n$ and $Z_n$. What trends do the processes exhibit? Is the sample mean meaningful for either of these processes?
b. Express $W_n$ and $Z_n$ in terms of $X_n, X_{n-1}, \ldots, X_1$ and then find $E[W_n]$ and $E[Z_n]$. Do these results agree with the trends observed in Part a?
c. Do $W_n$ or $Z_n$ have independent increments? stationary increments?

25. Let $M_n$ be the discrete time process defined as the sequence of sample means of an iid sequence:

$$M_n = \frac{X_1 + X_2 + \cdots + X_n}{n}.$$

a. Find the mean, variance, and covariance of $M_n$.
b. Does $M_n$ have independent increments? stationary increments?

26. Find the pdf of the processes defined in Problem 23 if the $X_n$ are an iid sequence of zero-mean, unit-variance Gaussian random variables.

27. Let $X_n$ consist of an iid sequence of Cauchy random variables.

a. Find the pdf of the sum process $S_n$. *Hint*: Use the characteristic function method.
b. Find the joint pdf of $S_n$ and $S_{n+k}$.

28. Let $X_n$ consist of an iid sequence of Poisson random variables with mean $\alpha$.

a. Find the pmf of the sum process $S_n$.
b. Find the joint pmf of $S_n$ and $S_{n+k}$.

29. Let $X_n$ be an iid sequence of zero-mean, unit-variance Gaussian random variables.

a. Find the pdf of $M_n$ defined in Problem 25.
b. Find the joint pdf of $M_n$ and $M_{n+k}$. *Hint*: Use the independent increments property of $S_n$.

30. Suppose that an experiment has three possible outcomes, say 0, 1, and 2, and suppose that these occur with probabilities $p_0$, $p_1$, and $p_2$. Consider a sequence of independent repetitions of the experiment, and let $X_j(n)$ be the indicator function for outcome $j$. The vector

$$\mathbf{X}(n) = (X_0(n), X_1(n), X_2(n))$$

then constitutes a vector-valued Bernoulli random process. Consider the counting process for $\mathbf{X}(n)$:

$$\mathbf{S}(n) = \mathbf{X}(n) + \mathbf{X}(n-1) + \cdots + \mathbf{X}(1) \qquad \mathbf{S}(0) = 0.$$

a. Show that $\mathbf{S}(n)$ has a multinomial distribution.
b. Show that $\mathbf{S}(n)$ has independent increments, then find the joint pmf of $\mathbf{S}(n)$ and $\mathbf{S}(n+k)$.
c. Show that components $S_j(n)$ of the vector process are binomial counting processes.

## Section 6.4
## Examples of Continuous-Time Random Processes

31. Suppose that a secretary receives calls that arrive according to a Poisson process with rate 10 calls per hour. What is the probability that no calls go unanswered if the secretary is away from the office for the first and last 15 minutes of an hour?

32. Customers arrive at a soft drink dispensing machine according to a Poisson process with rate $\lambda$. Suppose that each time a customer deposits money, the machine dispenses a soft drink with probability $p$. Find the pmf for the number of soft drinks dispensed in time $t$. (Assume the machine holds an infinite number of soft drinks.)

33. Noise impulses occur on a telephone line according to a Poisson process of rate $\lambda$.
    a. Find the probability that no impulses occur during the transmission of a message that is $t$ seconds long.
    b. Suppose that the message is encoded so that the errors caused by a single impulse can be corrected. What is the probability that a $t$-second message is either error-free or correctable?

34. Messages arrive at a computer from two telephone lines according to independent Poisson processes of rates $\lambda_1$ and $\lambda_2$ respectively.
    a. Find the probability that a message arrives first on line 1.
    b. Find the pdf for the time until a message arrives on either line.

c. Find the pmf for $N(t)$, the total number of messages that arrive in an interval of length $t$.

d. Generalize the result of Part c for the "merging" of $k$ independent Poisson processes of rates $\lambda_1, \ldots, \lambda_k$, respectively:

$$N(t) = N_1(t) + \cdots + N_k(t).$$

35. Find $P[N(t-d) = j \mid N(t) = k]$ with $d > 0$ where $N(t)$ is a Poisson process with rate $\lambda$.

36. Suppose that the time required to service a customer in a queueing system is an exponential random variable with parameter $\beta$. If customers arrive at the system according to a Poisson process with parameter $\lambda$, find the pmf for the number of customers that arrive during one customer's service time. *Hint*: Condition on the service time.

37. Is the difference of two independent Poisson random processes also a Poisson process?

38. Let $N(t)$ be a Poisson random process with parameter $\lambda$. Suppose that each time an event occurs, a coin is flipped and the outcome (heads or tails) is recorded. Let $N_1(t)$ and $N_2(t)$ denote the number of heads and tails recorded up to time $t$, respectively. Assume that $p$ is the probability of heads.

a. Find $P[N_1(t) = j, N_2(t) = k \mid N(t) = k + j]$.

b. Use Part a to show that $N_1(t)$ and $N_2(t)$ are independent Poisson random processes of rates $p\lambda$ and $(1-p)\lambda$, respectively:

$$P[N_1(t) = j, N_2(t) = k] = \frac{(p\lambda)^j}{j!}e^{-p\lambda}\frac{((1-p)\lambda)^k}{k!}e^{-(1-p)\lambda}.$$

39. Customers arrive at a soft drink dispensing machine according to a Poisson process with rate $\lambda$. Let $N(t)$ be the number of customer arrivals up to time $t$. Suppose that each time a customer deposits money, the machine dispenses a random number of soft drinks. In particular assume that this random number is a Poisson random variable with parameter 1. (Assume the machine holds an infinite number of soft drinks.) Let $X(t)$ be the number of drinks dispensed up to time $t$.

a. Find $P[X(t) = j \mid N(t) = n]$.

b. Find $P[X(t) = j]$.

40. Let $X(t)$ denote the random telegraph signal, and $Y(t)$ be a process derived from $X(t)$ as follows: Each time $X(t)$ changes polarity, $Y(t)$ changes polarity with probability $p$.

a. Find the $P[Y(t) = \pm 1]$.

b. Find the autocovariance function of $Y(t)$. Compare it to that of $X(t)$.

41. Let $Y(t)$ be the random signal obtained by switching between the values 0 and 1 according to the events in a Poisson process of rate $\lambda$. Compare the pmf and autocovariance of $Y(t)$ with that of the random telegraph signal.

42. Let $Z(t)$ be the random signal obtained by switching between the values 0 and 1 according to the events in a counting process $N(t)$. Let

$$P[N(t) = k] = \frac{1}{1 + \lambda t}\left(\frac{\lambda t}{1 + \lambda t}\right)^k \qquad k = 0, 1, 2, \ldots.$$

   a. Find the pmf of $Z(t)$.
   b. Find $m_Z(t)$.

43. In the filtered Poisson process (Eq. 6.41), let $h(t)$ be a pulse of unit amplitude as shown in Fig. P6.3.
   a. Show that $X(t)$ is then the increment in the Poisson process in the interval $(t - T, t]$.
   b. Find the mean and autocorrelation function of $X(t)$.

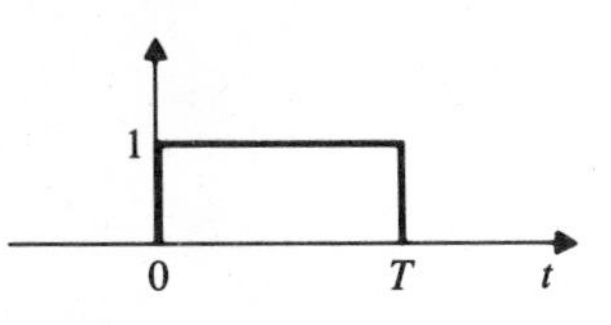


**FIGURE P6.3**

44. a. Find the second moment and variance of the shot noise process discussed in Example 6.24.
   b. Find the mean and variance of the shot noise process if $h(t) = e^{-\beta t}$ for $t \geq 0$.

45. Messages arrive at a message center according to a Poisson process of rate $\lambda$. Every hour the messages that have arrived during the previous hour are forwarded to their destination. Find the mean of the total time waited by all the messages that arrive during the hour. *Hint*: Condition on the number of arrivals.

46. Let $Y(t) = X(t) + \mu t$, where $X(t)$ is the Brownian motion process.
   a. Find the pdf of $Y(t)$.
   b. Find the joint pdf of $Y(t)$ and $Y(t + s)$.

47. Let $Y(t) = X^2(t)$, where $X(t)$ is the Brownian motion process.
   a. Find the pdf of $Y(t)$.
   b. Find the conditional pdf of $Y(t_2)$ given $Y(t_1)$.

48. Let $Z(t) = X(t) - aX(t - s)$, where $X(t)$ is the Brownian motion process.

a. Find the pdf of $Z(t)$.
b. Find $m_Z(t)$.

49. Let $Z(t) = X(t) + X(t - s)$, where $X(t)$ is the Brownian motion process.
    a. Find the pdf of $Z(t)$.
    b. Find $m_Z(t)$.

## Section 6.5
## Stationary Random Processes

50. Is the random amplitude sinusoid in Example 6.2 a stationary random process? Is it wide-sense stationary?
51. Is the random phase sinusoid in Example 6.7 a stationary random process? Is it wide-sense stationary?
52. Let $X(t) = g(t - T)$, where $g(t)$ is the periodic waveform introduced in Problem 7, and $T$ is a uniformly distributed random variable in the interval $(0, 1)$. Is $X(t)$ a stationary random process? Is $X(t)$ wide-sense stationary?
53. Let $X(t)$ be defined by

$$X(t) = A \cos wt + B \sin wt,$$

where $A$ and $B$ are independent, zero-mean random variables.
    a. Show that $X(t)$ is wide-sense stationary.
    b. Show that $X(t)$ is not stationary. *Hint*: Consider $E[X^3(t)]$.

54. Consider the following moving average process:

$$Y_n = \frac{1}{2}(X_n + X_{n-1}) \qquad X_0 = 0.$$

    a. Is $Y_n$ a stationary random process if $X_n$ is an iid process?
    b. Is $Y_n$ a stationary random process if $X_n$ is a stationary process?

55. Let $Z_n$ be the autoregressive random process discussed in Problem 24:

$$Z_n = \frac{1}{2}Z_{n-1} + X_n \qquad Z_0 = 0,$$

where $X_n$ is a zero-mean iid process.
    a. Find the autocovariance of $Z_n$ and determine whether $Z_n$ is wide-sense stationary.
    b. Does $Z_n$ eventually settle down into stationary behavior?
    c. Find the pdf of $Z_n$ if $X_n$ is an iid sequence of zero-mean, unit-variance Gaussian random variables. What is the pdf of $Z_n$ as $n \to \infty$?

56. Let $Y(t) = X(t) - aX(t + s)$, where $X(t)$ is a wide-sense stationary random process.
    a. Determine whether $Y(t)$ is also a wide-sense stationary random process.
    b. Find the pdf of $Y(t)$ if $X(t)$ is also a Gaussian random process.
57. Let $X(t)$ and $Y(t)$ be independent, wide-sense stationary random processes with zero means and the same covariance function $C_X(\tau)$. Let $Z(t)$ be defined by

$$Z(t) = aX(t) + bY(t).$$

    a. Determine whether $Z(t)$ is also wide-sense stationary.
    b. Determine the pdf of $Z(t)$ if $X(t)$ and $Y(t)$ are also Gaussian random processes.
58. Let $X(t)$ and $Y(t)$ be independent, wide-sense stationary random processes with zero means and the same covariance function $C_X(\tau)$. Let $Z(t)$ be given by

$$Z(t) = X(t)\cos wt + Y(t)\sin wt.$$

    a. Determine whether $Z(t)$ is a wide-sense stationary random process.
    b. Find the pdf of $Z(t)$ if $X(t)$ and $Y(t)$ are also Gaussian random processes.
59. Let $X(t)$ be a zero-mean wide-sense stationary random process with autocovariance function given by $C_X(\tau)$. The output of a "square law detector" is

$$Y(t) = X(t)^2.$$

    a. Determine whether $Y(t)$ is also a wide-sense stationary random process.
    b. Find the pdf of $Y(t)$ if $X(t)$ is also a Gaussian random process.
60. A WSS process $X(t)$ has mean 1 and autocorrelation function given in Fig. P6.4.
    a. Find the mean component of $R_X(\tau)$.
    b. Find the periodic component of $R_X(\tau)$.
    c. Find the remaining component of $R_X(\tau)$.

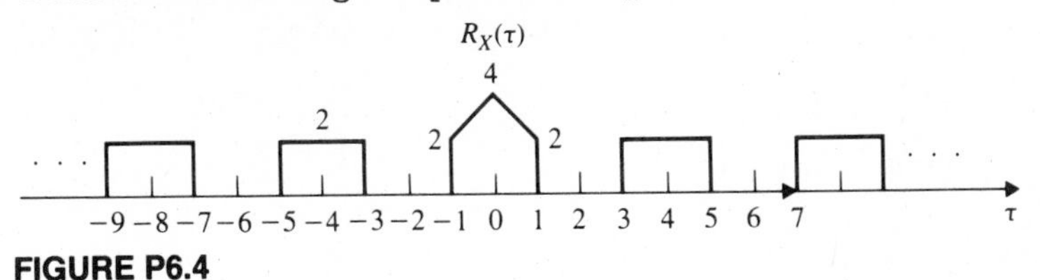


**FIGURE P6.4**

## Section 6.6
## Time Averages of Random Processes and Ergodic Theorems

61. Find the variance of the time average given in Example 6.32.
62. A WSS random process $X(t)$ has

$$R_X(\tau) = \begin{cases} 0 & |\tau| > 1 \\ A(1 - |\tau|) & |\tau| \leq 1. \end{cases}$$

Is $X(t)$ mean-ergodic?

63. Let $X(t) = A\cos(2\pi ft)$, where $A$ is a random variable with mean $m$ and variance $\sigma^2$.
    a. Evaluate $\langle X(t)\rangle_T$, find its limit as $T \to \infty$, and compare to $m_X(t)$.
    b. Evaluate $\langle X(t + \tau)X(t)\rangle$, find its limit as $T \to \infty$, and compare to $R_X(t + \tau, t)$.
64. Repeat the previous problem with $X(t) = A\cos(2\pi ft + \Theta)$, where $A$ is as before, $\Theta$ is a random variable uniformly distributed in $(0, 2\pi)$, and $A$ and $\Theta$ are independent random variables.
65. Find an exact expression for $\text{VAR}[\langle X(t)\rangle_T]$ in Example 6.33.
66. The WSS random process $X_n$ has mean $m$ and autocovariance $C_X(k) = (1/2)^{|k|}$. Is $X_n$ mean-ergodic?
67. Show that a WSS random process is mean-ergodic if

$$\int_{-\infty}^{\infty} |C(u)|\, du < \infty.$$

68. Let $\langle X^2(t)\rangle_T$ denote a time-average estimate for the mean power of a WSS random process. Under what conditions is this time average a valid estimate for $E[X^2(t)]$?
69. Under what conditions is the time average $\langle X(t + \tau)X(t)\rangle_T$ a valid estimate for the autocorrelation $R_X(\tau)$ of a WSS random process $X(t)$?
70. Let $Y(t)$ be the indicator function for the event $\{a < X(t) \leq b\}$, that is,

$$Y(t) = \begin{cases} 1 & \text{if } X(t) \in (a, b] \\ 0 & \text{otherwise} \end{cases}$$

    a. Show that $\langle Y(t)\rangle_T$ is the proportion of time in the time interval $(-T, T)$ that $X(t) \in (a, b]$.
    b. Find $E[\langle Y(t)\rangle_T]$.
    c. Under what conditions does $\langle Y(t)\rangle_T \to P[a < X(t) \leq b]$.
    d. How can $\langle Y(t)\rangle_T$ be used to estimate $P[X(t) \leq x]$?
71. Repeat problem 70 for the time average of the discrete-time $Y_n$, which is defined as the indicator for the event $\{a < X_n \leq b\}$.

72. Define $Z_n = u(a - X_n)$, where $u(x)$ is the unit step function, that is, $Z_n = 1$ if and only if $X_n \leq a$.
   a. Show that the time average $\langle Z_n \rangle_N$ is the proportion of $Z_n$'s that are less than $a$ in the first $N$ samples.
   b. Show that if the process is ergodic (in some sense), then this time average is equal to $F_X(a) = P[X \leq a]$.

CHAPTER 7

# Analysis and Processing of Random Signals

In this chapter we introduce methods for analyzing and processing random signals. In Section 7.1 we introduce the notion of power spectral density, which allows us to view random processes in the frequency domain. Sections 7.2 and 7.3 discuss two important methods for processing random signals: filtering and modulation. Section 7.4 considers the design of optimum linear systems, which can be used for prediction of the future of a random process or for recovering signals corrupted by noise. Finally Section 7.5 discusses the problem of estimating the power spectral density from observations of the random process.

## 7.1 POWER SPECTRAL DENSITY

The Fourier series and the Fourier transform allow us to view nonrandom time functions as the weighted sum or integral of sinusoidal functions. A time function that varies slowly has the weighting concentrated at the low frequency sinusoidal components. A time function that varies rapidly has the weighting concentrated at higher frequency components. Thus the rate at which a nonrandom time function varies is related to the weighting function of the Fourier series or transform. This weighting function is called the "spectrum" of the time function.

The notion of a time function as being composed of sinusoidal components is also very useful for random processes. However, since a sample function of a random process can be viewed as being selected from an ensemble of allowable time functions, the weighting function or "spectrum" for a random process must refer in some way to the average rate of change of the ensemble of allowable time functions. Equation (6.59) shows that, for wide-sense stationary processes, the autocorrelation function $R_X(\tau)$ is an appropriate measure for the average rate of change of a random process. Indeed if a random process changes slowly with time, then it remains correlated with itself for a long period of time, and $R_X(\tau)$ decreases slowly as a function of $\tau$. On the other hand, a rapidly varying random process, quickly becomes uncorrelated with itself, and $R_X(\tau)$ decreases rapidly with $\tau$.

We now present the **Einstein-Wiener-Khinchin theorem,** which states that the power spectral density of a wide-sense stationary random process is given by the Fourier transform of the autocorrelation function.[1]

1. This result is usually called the Wiener-Khinchin theorem after Norbert Wiener and A. Ya. Khinchin who proved the result in the early 1930s. Recently it has been discovered that this result was stated by Albert Einstein in a 1914 paper (see Reference 10).

### Continuous-Time Random Processes

Let $X(t)$ be a continuous-time WSS random process with mean $m_X$ and autocorrelation function $R_X(\tau)$. The **power spectral density of $X(t)$** is given by the Fourier transform of the autocorrelation function:

$$\begin{aligned} S_X(f) &= \mathcal{F}\{R_X(\tau)\} \\ &= \int_{-\infty}^{\infty} R_X(\tau)e^{-j2\pi f\tau}\,d\tau. \end{aligned} \tag{7.1}$$

A table of Fourier transforms and its properties is given in the Appendix 2.

For real-valued random processes, the autocorrelation function is an even function of $\tau$:

$$R_X(\tau) = R_X(-\tau). \tag{7.2}$$

Substitution into Eq. (7.1) implies that

$$\begin{aligned} S_X(f) &= \int_{-\infty}^{\infty} R_X(\tau)\{\cos 2\pi f\tau - j \sin 2\pi f\tau\}\,d\tau \\ &= \int_{-\infty}^{\infty} R_X(\tau) \cos 2\pi f\tau\,d\tau, \end{aligned} \tag{7.3}$$

since the integral of the product of an even function ($R_X(\tau)$) and an odd function ($\sin 2\pi f\tau$) is zero. Equation (7.3) implies that $S_X(f)$ *is real-valued and an even function of* $f$. Later we show that $S_X(f)$ *is non-negative*:

$$S_X(f) \geq 0 \qquad \text{for all } f. \tag{7.4}$$

The autocorrelation function can be recovered from the power spectral density by applying the inverse Fourier transform formula to Eq. (7.1):

$$\begin{aligned} R_X(\tau) &= \mathcal{F}^{-1}\{S_X(f)\} \\ &= \int_{-\infty}^{\infty} S_X(f)e^{j2\pi f\tau}\,df. \end{aligned} \tag{7.5}$$

In electrical engineering it is customary to refer to the second moment of $X(t)$ as the **average power of $X(t)$**.[2] Equation (7.5) together with Eq. (6.57) give

$$E[X^2(t)] = R_X(0) = \int_{-\infty}^{\infty} S_X(f)\,df. \tag{7.6}$$

Equation (7.6) states that the average power of $X(t)$ is obtained by integrating $S_X(f)$ over all frequencies. This is consistent with the fact that $S_X(f)$ is the "density of power" of $X(t)$ at the frequency $f$.

2. If $X(t)$ is a voltage or current developed across a 1-ohm resistor, then $X^2(t)$ is the instantaneous power absorbed by the resistor.

Since the autocorrelation and autocovariance functions are related by $R_X(\tau) = C_X(\tau) + m_X^2$, the power spectral density is also given by

$$\begin{aligned} S_X(f) &= \mathcal{F}\{C_X(\tau) + m_X^2\} \\ &= \mathcal{F}\{C_X(\tau)\} + m_X^2\,\delta(f), \end{aligned} \tag{7.7}$$

where we have used the fact that the Fourier transform of a constant is a delta function. We say the $m_X$ is the "*dc*" component of $X(t)$.

The notion of power spectral density can be generalized to two jointly wide-sense stationary processes. The **cross-power spectral density** $\boldsymbol{S_{X,Y}(f)}$ is defined by

$$S_{X,Y}(f) = \mathcal{F}\{R_{X,Y}(\tau)\}, \tag{7.8}$$

where $R_{X,Y}(\tau)$ is the cross-correlation between $X(t)$ and $Y(t)$:

$$R_{X,Y}(\tau) = E[X(t+\tau)Y(t)]. \tag{7.9}$$

In general, $S_{X,Y}(f)$ is a complex function of $f$ even if $X(t)$ and $Y(t)$ are both real-valued.

### ■■ Example 7.1
Random Telegraph Signal

Find the power spectral density of the random telegraph signal.

In Example 6.22, the autocorrelation function of the random telegraph process was found to be

$$R_X(\tau) = e^{-2\alpha|\tau|}$$

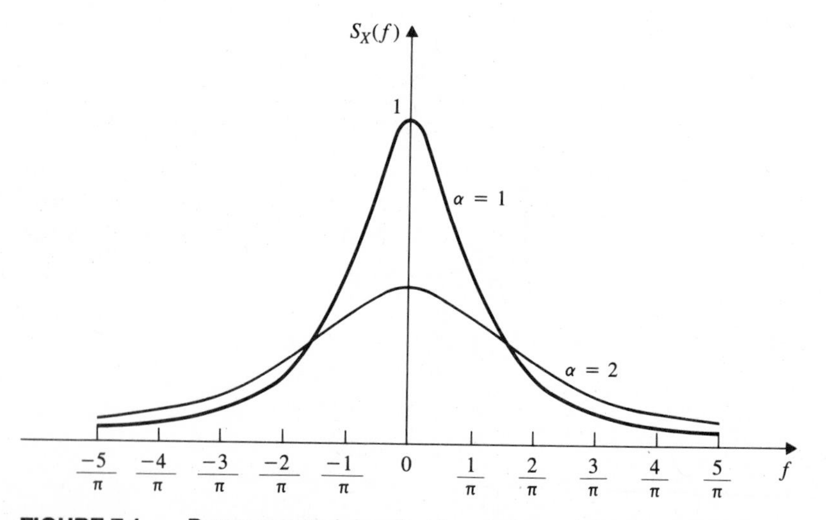


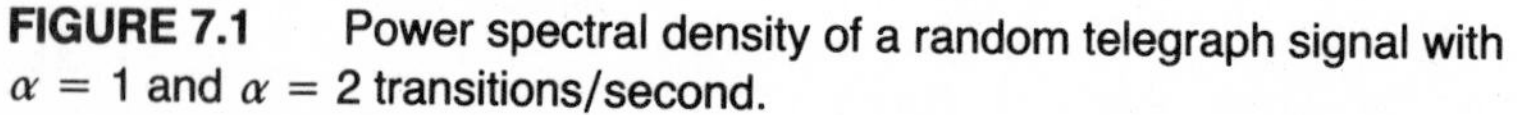

**FIGURE 7.1** Power spectral density of a random telegraph signal with $\alpha = 1$ and $\alpha = 2$ transitions/second.

where $\alpha$ is the average transition rate of the signal. Therefore, the power spectral density of the process is:

$$S_X(f) = \int_{-\infty}^{0} e^{2\alpha\tau} e^{-j2\pi f\tau}\, d\tau + \int_{0}^{\infty} e^{-2\alpha\tau} e^{-j2\pi f\tau}\, d\tau$$

$$= \frac{1}{2\alpha - j2\pi f} + \frac{1}{2\alpha + j2\pi f}$$

$$= \frac{4\alpha}{4\alpha^2 + 4\pi^2 f^2}. \tag{7.10}$$

Figure 7.1 shows the power spectral density for $\alpha = 1$ and $\alpha = 2$ transitions per second. The process changes two times more quickly when $\alpha = 2$; it can be seen from the figure that the power spectral density for $\alpha = 2$ has greater high-frequency content. ■■

---

### ■■ Example 7.2
Sinusoid with Random Phase

Let $X(t) = a \cos(2\pi f_0 t + \Theta)$ where $\Theta$ is uniformly distributed in the interval $(0, 2\pi)$. Find $S_X(f)$.

From Example 6.7, the autocorrelation for $X(t)$ is

$$R_X(\tau) = \frac{a^2}{2} \cos 2\pi f_0 \tau.$$

Thus, the power spectral density is

$$S_X(f) = \frac{a^2}{2} \mathscr{F}\{\cos 2\pi f_0 \tau\}$$

$$= \frac{a^2}{4} \delta(f - f_0) + \frac{a^2}{4} \delta(f + f_0). \tag{7.11}$$

where we have used the table of Fourier transforms in Appendix B. The signal has average power $R_X(0) = a^2/2$. All of this power is concentrated at the frequencies $\pm f_0$, so the power density at these frequencies is infinite. ■■

---

### ■■ Example 7.3
White Noise

The power spectral density of a WSS white noise process whose frequency components are limited to the range $-W \le f \le W$ is shown in Fig. 7.2(a). The process is said to be "white" in analogy to white light, which contains all frequencies in equal amounts. The average power in this process is

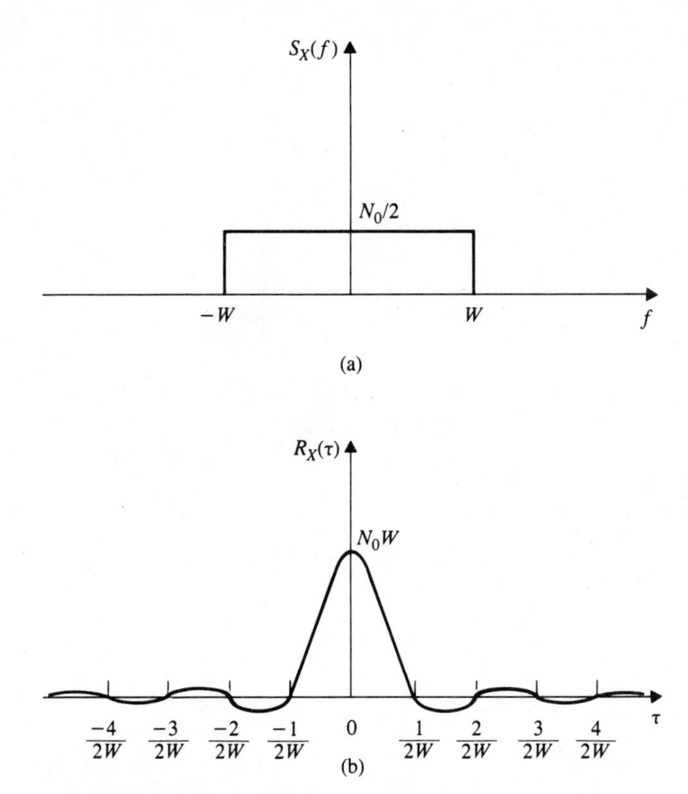


**FIGURE 7.2** Bandlimited white noise: (a) power spectral density, (b) autocorrelation function.

obtained from Eq. (7.6):

$$E[X^2(t)] = \int_{-W}^{W} \frac{N_0}{2} df = N_0 W. \tag{7.12}$$

The autocorrelation for this process is obtained from Eq. (7.5):

$$\begin{aligned} R_X(\tau) &= \frac{1}{2} N_0 \int_{-W}^{W} e^{-j2\pi f \tau} df \\ &= \frac{1}{2} N_0 \frac{e^{-j2\pi W\tau} - e^{j2\pi W\tau}}{-j2\pi\tau} \\ &= \frac{N_0 \sin(2\pi W\tau)}{2\pi\tau}. \end{aligned} \tag{7.13}$$

$R_X(\tau)$ is shown in Fig. 7.2(b). Note that $X(t)$ and $X(t + \tau)$ are uncorrelated at $\tau = \pm k/2W$, $k = 1, 2, \ldots$.

The term white noise usually refers to a random process $W(t)$ whose power spectral density is $N_0/2$ for *all* frequencies:

$$S_W(f) = \frac{N_0}{2} \qquad \text{for all } f. \tag{7.14}$$

Equation (7.12) with $W = \infty$ shows that such a process must have infinite average power. By taking the limit $W \to \infty$ in Eq. (7.13), we find that the autocorrelation of such a process approaches

$$R_W(\tau) = \frac{N_0}{2}\delta(\tau). \tag{7.15}$$

■■

---

### ■■ Example 7.4
Sum of Two Processes

Find the power spectral density of $Z(t) = X(t) + Y(t)$, where $X(t)$ and $Y(t)$ are jointly WSS processes.

The autocorrelation of $Z(t)$ is

$$\begin{aligned} R_Z(t) &= E[Z(t+\tau)Z(t)] = E[(X(t+\tau) + Y(t+\tau))(X(t) + Y(t))] \\ &= R_X(\tau) + R_{YX}(\tau) + R_{XY}(\tau) + R_Y(\tau). \end{aligned}$$

The power spectral density is then

$$\begin{aligned} S_Z(f) &= \mathcal{F}\{R_X(\tau) + R_{YX}(\tau) + R_{XY}(\tau) + R_Y(\tau)\} \\ &= S_X(f) + S_{YX}(f) + S_{XY}(f) + S_Y(f). \end{aligned} \tag{7.16}$$

■■

---

### ■■ Example 7.5

Let $Y(t) = X(t - d)$, where $d$ is a constant delay and where $X(t)$ is WSS. Find $R_{YX}(\tau)$, $S_{YX}(f)$, $R_Y(\tau)$ and $S_Y(f)$.

The definitions of $R_{YX}(\tau)$, $S_{YX}(f)$, and $R_Y(\tau)$ give

$$R_{YX}(\tau) = E[Y(t+\tau)X(t)] = E[X(t+\tau-d)X(t)] = R_X(\tau - d) \tag{7.17}$$

The time-shifting property of the Fourier transform gives:

$$\begin{aligned} S_{YX}(f) &= \mathcal{F}\{R_X(\tau - d)\} = S_X(f)e^{-j2\pi fd} \\ &= S_X(f)\cos(2\pi fd) - jS_X(f)\sin(2\pi fd). \end{aligned} \tag{7.18}$$

Finally:

$$R_Y(\tau) = E[Y(t+\tau)Y(t)] = E[X(t+\tau-d)X(t-d)] = R_X(\tau). \tag{7.19}$$

Equation (7.19) implies that

$$S_Y(f) = \mathscr{F}\{R_Y(\tau)\} = \mathscr{F}\{R_X(\tau)\} = S_X(f). \tag{7.20}$$

Note from Eq. (7.18) that the cross-power spectral density is complex. Note from Eq. (7.20) that $S_X(f) = S_Y(f)$ despite the fact that $X(t) \neq Y(t)$. Thus, $S_X(f) = S_Y(f)$ *does not imply that* $X(t) = Y(t)$. ■■

### Discrete-Time Random Processes

Let $X_n$ be a discrete-time WSS random process with mean $m_X$ and autocorrelation function $R_X(k)$. The **power spectral density of $X_n$** is defined as the Fourier transform of the autocorrelation sequence

$$\begin{aligned} S_X(f) &= \mathscr{F}\{R_X(k)\} \\ &= \sum_{k=-\infty}^{\infty} R_X(k) e^{-j2\pi fk}. \end{aligned} \tag{7.21}$$

Note that we need only consider frequencies in the range $-1/2 < f \leq 1/2$, since $S_X(f)$ is periodic in $f$ with period 1. As in the case of continuous random processes, $S_X(f)$ can be shown to be a real-valued, nonnegative, even function of $f$.

The inverse Fourier transform formula applied to Eq. (7.21) implies that[3]

$$R_X(k) = \int_{-1/2}^{1/2} S_X(f) e^{j2\pi fk}\, df. \tag{7.22}$$

The **cross-power spectral density $S_{X,Y}(f)$** of two jointly WSS discrete-time processes $X_n$ and $Y_n$ is defined by

$$S_{X,Y}(f) = \mathscr{F}\{R_{X,Y}(k)\}, \tag{7.23}$$

where $R_{X,Y}(k)$ is the cross-correlation between $X_n$ and $Y_n$:

$$R_{X,Y}(k) = E[X_{n+k}Y_n]. \tag{7.24}$$

#### ■■ Example 7.6
White Noise

Let the process $X_n$ be a sequence of uncorrelated random variables with zero mean and variance $\sigma_X^2$. Find $S_X(f)$.

The autocorrelation of this process is

$$R_X(k) = \begin{cases} \sigma_X^2 & k = 0 \\ 0 & k \neq 0. \end{cases}$$

3. You can view $R_X(k)$ as the coefficients of the Fourier series of the periodic function $S_X(f)$.

The power spectral density of the process is found by substituting $R_X(k)$ into Eq. (7.21):

$$S_X(f) = \sigma_X^2 \qquad -\frac{1}{2} < f < \frac{1}{2}. \tag{7.25}$$

Thus the process $X_n$ contains all possible frequencies in equal measure. ■■

---

## ■■ Example 7.7
Moving Average Process

Let the process $Y_n$ be defined by

$$Y_n = X_n + \alpha X_{n-1}, \tag{7.26}$$

where $X_n$ is the white noise process of Example 7.6. Find $S_Y(f)$.

It is easily shown that the mean and autocorrelation of $Y_n$ are given by

$$E[Y_n] = 0,$$

and

$$E[Y_n Y_{n+k}] = \begin{cases} (1+\alpha^2)\sigma_X^2 & k = 0 \\ \alpha\sigma_X^2 & k = \pm 1 \\ 0 & \text{otherwise.} \end{cases} \tag{7.27}$$

The power spectral density is then

$$\begin{aligned} S_Y(f) &= (1+\alpha^2)\sigma_X^2 + \alpha\sigma_X^2\{e^{j2\pi f} + e^{-j2\pi f}\} \\ &= \sigma_X^2\{(1+\alpha^2) + 2\alpha \cos 2\pi f\}. \end{aligned} \tag{7.28}$$

$S_Y(f)$ is shown in Fig. 7.3 for $\alpha = -1$. ■■

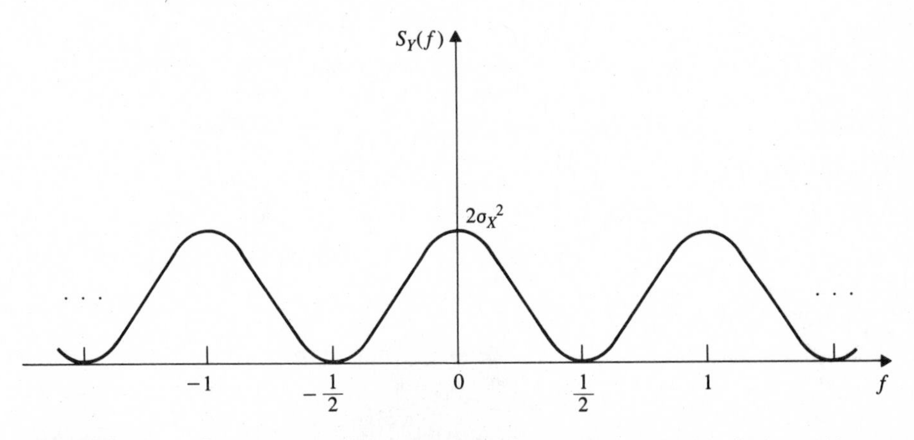


**FIGURE 7.3** Power spectral density of moving-average process discussed in Example 7.7.

**■■ Example 7.8**
Signal Plus Noise

Let the observation $Z_n$ be given by

$$Z_n = X_n + Y_n,$$

where $X_n$ is the signal we wish to observe, $Y_n$ is a white noise process with power $\sigma_Y^2$, and $X_n$ and $Y_n$ are independent random processes. Suppose further that $X_n = A$ for all $n$, where $A$ is a random variable with zero mean and variance $\sigma_A^2$. Thus $Z_n$ represents a sequence of noisy measurements of the random variable $A$. Find the power spectral density of $Z_n$.

The mean and autocorrelation of $Z_n$ are

$$E[Z_n] = E[A] + E[Y_n] = 0$$

and

$$\begin{aligned} E[Z_n Z_{n+k}] &= E[(X_n + Y_n)(X_{n+k} + Y_{n+k})] \\ &= E[X_n X_{n+k}] + E[X_n]E[Y_{n+k}] + E[X_{n+k}]E[Y_n] + E[Y_n Y_{n+k}] \\ &= E[A^2] + R_Y(k). \end{aligned}$$

Thus $Z_n$ is also a WSS process.

The power spectral density of $Z_n$ is then

$$S_Z(f) = E[A^2]\delta(f) + S_Y(f),$$

where we have used the fact that the Fourier transform of a constant is a delta function. ■■

## Power Spectral Density as a Time Average

In the above discussion, we simply stated that the power spectral density is given as the Fourier transform of the autocorrelation without supplying a proof. We now show how the power spectral density arises naturally when we take Fourier transforms of realizations of random processes.

Let $X_0, \ldots, X_{k-1}$ be $k$ observations from the discrete-time, WSS process $X_n$. Let $\tilde{x}_k(f)$ denote the discrete Fourier transform of this sequence

$$\tilde{x}_k(f) = \sum_{m=0}^{k-1} X_m e^{-j2\pi f m}. \tag{7.29}$$

Note that $\tilde{x}_k(f)$ is a complex-valued random variable. The magnitude squared of $\tilde{x}_k(f)$ is a measure of the "energy" at the frequency $f$. If we divide this energy by the total "time" $k$, we obtain an estimate for the "power" at the frequency $f$:

$$\tilde{p}_k(f) = \frac{1}{k}|\tilde{x}_k(f)|^2. \tag{7.30}$$

$\tilde{p}_k(f)$ is called the **periodogram estimate** for the power spectral density.

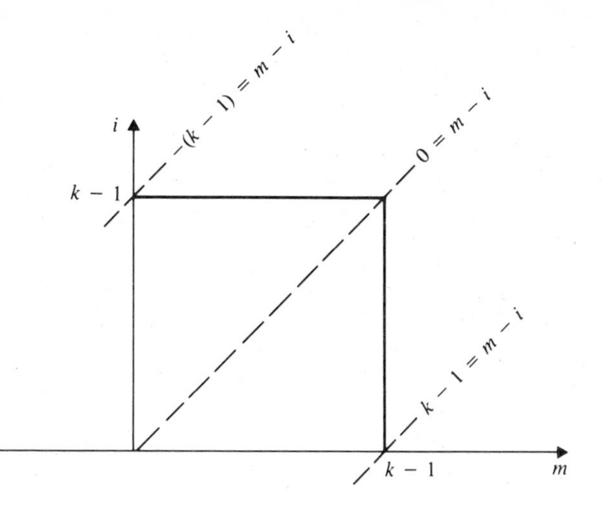


**FIGURE 7.4** Range of summation in Eq. (7.31).

Consider the expected value of the periodogram estimate:

$$\begin{aligned}
E[\tilde{p}_k(f)] &= \frac{1}{k} E[\tilde{x}_k(f)\tilde{x}_k^*(f)] \\
&= \frac{1}{k} E\left[\sum_{m=0}^{k-1} X_m e^{-j2\pi fm} \sum_{i=0}^{k-1} X_i e^{j2\pi fi}\right] \\
&= \frac{1}{k} \sum_{m=0}^{k-1} \sum_{i=0}^{k-1} E[X_m X_i] e^{-j2\pi f(m-i)} \\
&= \frac{1}{k} \sum_{m=0}^{k-1} \sum_{i=0}^{k-1} R_X(m - i) e^{-j2\pi f(m-i)}.
\end{aligned} \tag{7.31}$$

Figure 7.4 shows the range of the double summation in Eq. (7.31). Note that all the terms along the diagonal $m' = m - i$ are equal, that $m'$ ranges from $-(k - 1)$ to $k - 1$, and that there are $k - |m'|$ terms along the diagonal $m' = m - i$. Thus Eq. (7.31) becomes

$$\begin{aligned}
E[\tilde{p}_k(f)] &= \frac{1}{k} \sum_{m'=-(k-1)}^{k-1} \{k - |m'|\} R_X(m') e^{-j2\pi fm'} \\
&= \sum_{m'=-(k-1)}^{k-1} \left\{1 - \frac{|m'|}{k}\right\} R_X(m') e^{-j2\pi fm'}.
\end{aligned} \tag{7.32}$$

Comparison of Eq. (7.32) to Eq. (7.21) shows that the mean of the periodogram estimate is not equal to $S_X(f)$ for two reasons. First, Eq. (7.21) does not have the term in brackets in Eq. (7.32). Second, the limits of the summation in Eq. (7.32) are not $\pm\infty$. We say that $\tilde{p}_k(f)$ is a "biased" estimator for $S_X(f)$. However as $k \to \infty$, we see that the term in brackets approaches one, and that the limits of the summation approach $\pm\infty$. Thus

$$E[\tilde{p}_k(f)] \to S_X(f) \qquad \text{as } k \to \infty, \tag{7.33}$$

that is, the mean of the periodogram estimate does indeed approach $S_X(f)$. Note that Eq. (7.33) shows that $S_X(f)$ is nonnegative for all $f$, since $\tilde{p}_k(f)$ is nonnegative for all $f$.

In order to be useful, the variance of the periodogram estimate should also approach zero. The answer to this question involves looking more closely at the problem of power spectral density estimation. We defer this topic to Section 7.5.

All of the above results hold for a continuous-time WSS random process $X(t)$ after appropriate changes are made from summations to integrals. The **periodogram estimate for $S_X(f)$,** for an observation in the interval $0 < t < T$, is defined by

$$\tilde{p}_T(f) = \frac{1}{T}|\tilde{x}_T(f)|^2, \tag{7.34}$$

where

$$\tilde{x}_T(f) = \int_0^T X(t')e^{-j2\pi f t'}\,dt'. \tag{7.35}$$

The same derivation that led to Eq. (7.32) can be used to show that the mean of the periodogram estimate is given by

$$E[\tilde{p}_T(f)] = \int_{-T}^{T}\left\{1 - \frac{|\tau|}{T}\right\}R_X(\tau)e^{-j2\pi f\tau}\,d\tau. \tag{7.36}$$

It then follows that

$$E[\tilde{p}_T(f)] \to S_X(f) \qquad \text{as } T \to \infty. \tag{7.37}$$

## 7.2

## RESPONSE OF LINEAR SYSTEMS TO RANDOM SIGNALS

Many applications involve the processing of random signals (i.e., random processes) in order to achieve certain ends. For example, in prediction, we are interested in predicting future values of a signal in terms of past values. In filtering and smoothing, we are interested in recovering signals that have been corrupted by noise. In modulation, we are interested in converting low-frequency information signals into high-frequency transmission signals that propagate more readily through various transmission media.

Signal processing involves converting a signal from one form into another. Thus a signal processing method is simply a transformation or mapping from one time function into another function. If the input to the transformation is a random process, then the output will also be a random process. In the next two sections, we are interested in determining the statistical properties of the output process when the input is a wide-sense stationary random process.

### Continuous-Time Systems

Consider a **system** in which an input signal $x(t)$ is mapped into the output signal $y(t)$ by the transformation

$$y(t) = T[x(t)].$$

The system is **linear** if superposition holds, that is,

$$T[\alpha x_1(t) + \beta x_2(t)] = \alpha T[x_1(t)] + \beta T[x_2(t)],$$

where $x_1(t)$ and $x_2(t)$ are arbitrary input signals, and $\alpha$ and $\beta$ are arbitrary constants. Let $y(t)$ be the response to input $x(t)$, then the system is said to be **time-invariant** if the response to $x(t - \tau)$ is $y(t - \tau)$. The **impulse response** $h(t)$ of a linear, time-invariant system is defined by

$$h(t) = T[\delta(t)].$$

The response of the system to an arbitrary input $x(t)$ is then

$$y(t) = h(t) * x(t) = \int_{-\infty}^{\infty} h(s)x(t - s)\,ds = \int_{-\infty}^{\infty} h(t - s)x(s)\,ds. \quad (7.38)$$

Therefore a linear, time-invariant system is completely specified by its impulse response. The impulse response $h(t)$ can also be specified by giving its Fourier transform, the **transfer function** of the system:

$$H(f) = \mathcal{F}\{h(t)\} = \int_{-\infty}^{\infty} h(t)e^{-j2\pi ft}\,dt. \quad (7.39)$$

A system is said to be **causal** if the response at time $t$ depends only on past values of the input, that is, if $h(t) = 0$ for $t < 0$.

If the input to a linear, time-invariant system is a random process $X(t)$ as shown in Fig. 7.5, then the output of the system is the random process

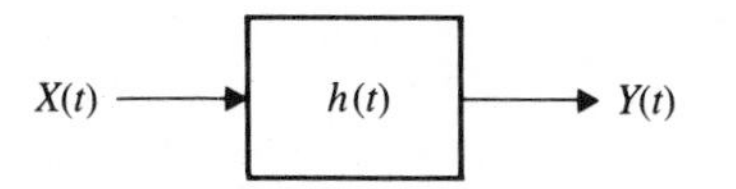


**FIGURE 7.5** A linear system with a random input signal.

given by

$$Y(t) = \int_{-\infty}^{\infty} h(s)X(t - s)\,ds = \int_{-\infty}^{\infty} h(t - s)X(s)\,ds. \tag{7.40}$$

We now show that if $X(t)$ is a wide-sense stationary process, then $Y(t)$ is also wide-sense stationary.[4]

The mean of $Y(t)$ is given by

$$E[Y(t)] = E\left[\int_{-\infty}^{\infty} h(s)X(t - s)\,ds\right] = \int_{-\infty}^{\infty} h(s)E[X(t - s)]\,ds.$$

The expected value and the integral can be interchanged since the integral is the limit of a sum, and the expected value of a sum is equal to the sum of expected values. Now $m_X = E[X(t - \tau)]$ since $X(t)$ is wide-sense stationary, so

$$E[Y(t)] = m_X \int_{-\infty}^{\infty} h(\tau)\,d\tau = m_X H(0), \tag{7.41}$$

where $H(f)$ is the transfer function of the system. Thus the mean of the output $Y(t)$ is the constant $m_Y = H(0)m_X$.

The autocorrelation of $Y(t)$ is given by

$$\begin{aligned} E[Y(t)Y(t + \tau)] &= E\left[\int_{-\infty}^{\infty} h(s)X(t - s)\,ds \int_{-\infty}^{\infty} h(r)X(t + \tau - r)\,dr\right] \\ &= \int_{-\infty}^{\infty}\int_{-\infty}^{\infty} h(s)h(r)E[X(t - s)X(t + \tau - r)]\,ds\,dr \\ &= \int_{-\infty}^{\infty}\int_{-\infty}^{\infty} h(s)h(r)R_X(\tau + s - r)\,ds\,dr, \end{aligned} \tag{7.42}$$

where we have used the fact that $X(t)$ is wide-sense stationary. The expression on the right-hand side of Eq. (7.42) depends only on $\tau$. Thus the autocorrelation of $Y(t)$ depends only on $\tau$, and since the $E[Y(t)]$ is a constant, we conclude that $Y(t)$ is a wide-sense stationary process.

We are now ready to compute the power spectral density of the output of a linear, time-invariant system. Taking the transform of $R_Y(\tau)$ as given in Eq. (7.42), we obtain

$$\begin{aligned} S_Y(f) &= \int_{-\infty}^{\infty} R_Y(\tau)e^{-j2\pi f\tau}\,d\tau \\ &= \int_{-\infty}^{\infty}\int_{-\infty}^{\infty}\int_{-\infty}^{\infty} h(s)h(r)R_X(\tau + s - r)e^{-j2\pi f\tau}\,ds\,dr\,d\tau. \end{aligned}$$

4. Equation (7.40) supposes that the input was applied at an infinite time in the past. If the input is applied at $t = 0$ then $Y(t)$ is not wide-sense stationary. However it becomes wide-sense stationary as the response reaches "steady state" (see Problem 22).

Change variables, letting $u = \tau + s - r$:

$$
\begin{aligned}
S_Y(f) &= \int_{-\infty}^{\infty}\int_{-\infty}^{\infty}\int_{-\infty}^{\infty} h(s)h(r)R_X(u)e^{-j2\pi f(u-s+r)}\,ds\,dr\,du \\
&= \int_{-\infty}^{\infty} h(s)e^{j2\pi fs}\,ds \int_{-\infty}^{\infty} h(r)e^{-j2\pi fr}\,dr \int_{-\infty}^{\infty} R_X(u)e^{-j2\pi fu}\,du \\
&= H^*(f)H(f)S_X(f) \\
&= |H(f)|^2\,S_X(f),
\end{aligned}
\tag{7.43}
$$

where we have used the definition of the transfer function. Equation (7.43) relates the input and output power spectral densities to the system transfer function. Note that $R_Y(\tau)$ can also be found by computing Eq. (7.43) and then taking the inverse Fourier transform.

Equations (7.41) through (7.43) only enable us to determine the mean and autocorrelation function of the output process $Y(t)$. In general this is not enough to determine probabilities of events involving $Y(t)$. However, if the input process is a Gaussian WSS random process, then it can be shown that the output process will also be a Gaussian WSS random process. (This should not be surprising since the convolution integral in Eq. 7.40 is the limit of a sum, and since we have shown in Section 4.8 that a linear transformation of jointly Gaussian random vectors is also Gaussian.) Thus the mean and autocorrelation function provided by Eqs. (7.41) through (7.43) are enough to determine all joint pdf's involving the Gaussian random process $Y(t)$.

The cross-correlation between the input and output processes is also of interest:

$$
\begin{aligned}
R_{Y,X}(\tau) &= E[Y(t+\tau)X(t)] \\
&= E\left[X(t)\int_{-\infty}^{\infty} X(t+\tau-r)h(r)\,dr\right] \\
&= \int_{-\infty}^{\infty} E[X(t)X(t+\tau-r)]h(r)\,dr \\
&= \int_{-\infty}^{\infty} R_X(\tau-r)h(r)\,dr \\
&= R_X(\tau) * h(\tau).
\end{aligned}
\tag{7.44}
$$

By taking the Fourier transform, we obtain the cross-power spectral density:

$$
S_{Y,X}(f) = H(f)S_X(f). \tag{7.45a}
$$

Since $R_{X,Y}(\tau) = R_{Y,X}(-\tau)$, we have that

$$S_{X,Y}(f) = S^*_{Y,X}(f) = H^*(f)S_X(f), \tag{7.45b}$$

where * denotes the complex conjugate.

### ■■ Example 7.9
Filtered White Noise

Find the power spectral density of the output of a linear, time-invariant system whose input is a white noise process.

Let $X(t)$ be the input process with power spectral density

$$S_X(f) = \frac{N_0}{2} \qquad \text{for all } f.$$

The power spectral density of the output $Y(t)$ is then

$$S_Y(f) = |H(f)|^2 \frac{N_0}{2}. \tag{7.46}$$

Thus the transfer function completely determines the shape of the power spectral density of the output process. ■■

---

Example 7.9 provides us with a method for generating WSS processes with arbitrary power spectral density $S_Y(f)$. We simply need to filter white noise through a filter with transfer function $H(f) = \sqrt{S_Y(f)}$. In general this filter will be noncausal. We can usually, but not always, obtain a *causal* filter with transfer function $H(f)$ such that $S_Y(f) = H(f)H^*(f)$. For example, if $S_Y(f)$ is a rational function, that is, if it consists of the ratio of two polynomials, then it is easy to factor $S_X(f)$ into the above form as shown in the next example. Furthermore any power spectral density can be approximated by a rational function. Thus filtered white noise can be used to synthesize WSS random processes with arbitrary power spectral densities, and hence arbitrary autocorrelation functions.

### ■■ Example 7.10

Find the impulse response of a causal filter that can be used to generate a random process with output power spectral density:

$$S_Y(f) = \frac{A}{\alpha^2 + 4\pi^2 f^2}.$$

This power spectral density factors as follows:

$$S_Y(f) = \frac{\sqrt{A}}{(\alpha - j2\pi f)} \frac{\sqrt{A}}{(\alpha + j2\pi f)}.$$

If we let the filter transfer function be $H(f) = \sqrt{A}/(\alpha + j2\pi f)$, then the impulse response is

$$h(t) = \sqrt{A}\, e^{-\alpha t} \qquad \text{for } t \geq 0,$$

which is the response of a causal system. Thus if we filter white noise with power spectral density $N_0/2 = 1$ using the above filter, we obtain a process with the desired power spectral density. ■■

### ■■ Example 7.11
Ideal Filters

Let $Z(t) = X(t) + Y(t)$, where $X(t)$ and $Y(t)$ are independent random processes with power spectral densities shown in Fig. 7.6(a). Find the output if $Z(t)$ is input into an ideal lowpass filter with transfer function shown in Fig. 7.6(b). Find the output if $Z(t)$ is input into an ideal bandpass filter with transfer function shown in Fig. 7.6(c).

The power spectral density of the output $W(t)$ of the lowpass filter is

$$S_W(f) = |H_{LP}(f)|^2 S_X(f) + |H_{LP}(f)|^2 S_Y(f) = S_X(f),$$

since $H_{LP}(f) = 1$ for the frequencies where $S_X(f)$ is nonzero, and $H_{LP}(f) = 0$ where $S_Y(f)$ is nonzero. Thus $W(t)$ has the same power spectral density as $X(t)$. As indicated in Example 7.5, this does not imply that $W(t) = X(t)$.

To show that $W(t) = X(t)$, in the mean square sense, consider $D(t) = W(t) - X(t)$. It is easily shown that

$$R_D(\tau) = R_W(\tau) - R_{WX}(\tau) - R_{XW}(\tau) + R_X(\tau).$$

The corresponding power spectral density is

$$\begin{aligned} S_D(f) &= S_W(f) - S_{WX}(f) - S_{XW}(f) + S_X(f) \\ &= |H_{LP}(f)|^2 S_X(f) - H_{LP}(f) S_X(f) - H_{LP}^*(f) S_X(f) + S_X(f) \\ &= 0. \end{aligned}$$

Therefore $R_D(\tau) = 0$ for all $\tau$, and $W(t) = X(t)$ in the mean square sense since:

$$E[(W(t) - X(t))^2] = E[D^2(t)] = R_D(0) = 0.$$

Thus we have shown that the lowpass filter removes $Y(t)$ and passes $X(t)$. Similarly, the bandpass filter removes $X(t)$ and passes $Y(t)$. ■■

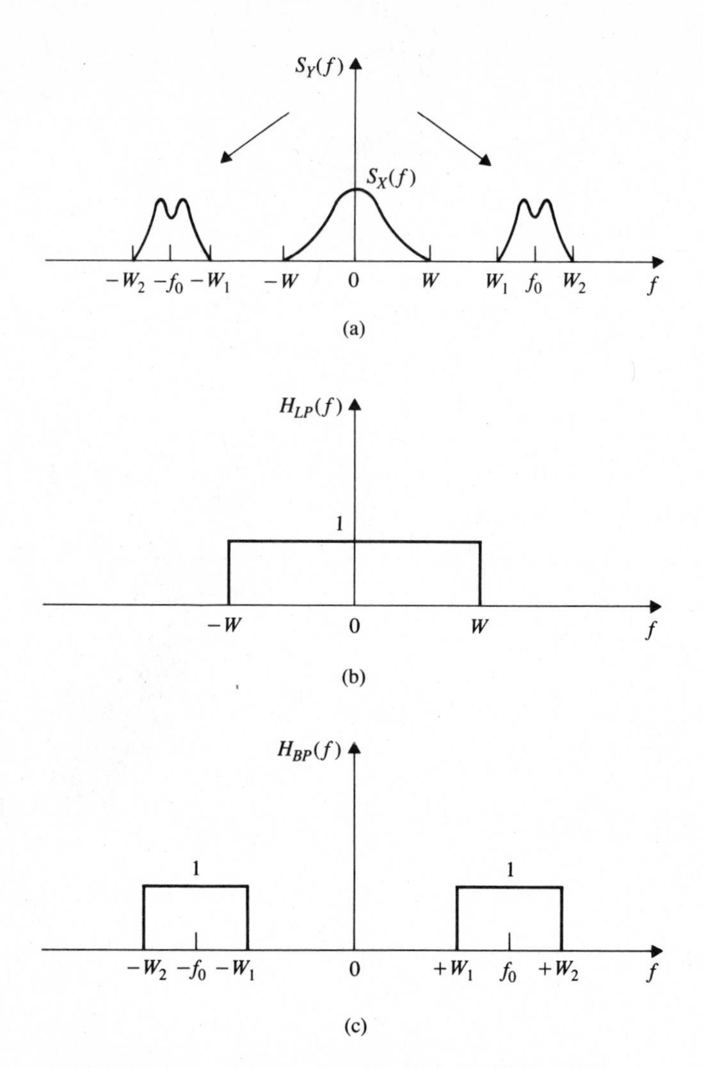


**FIGURE 7.6** (a) Input signal to filters is $X(t) + Y(t)$, (b) lowpass filter, (c) bandpass filter.

## Example 7.12

A random telegraph signal is passed through an RC lowpass filter which has transfer function

$$H(f) = \frac{\beta}{\beta + j2\pi f},$$

where $\beta = 1/RC$ is the time-constant of the filter. Find the power spectral density and autocorrelation of the output.

In Example 7.1, the power spectral density of the random telegraph signal with transition rate $\alpha$ was found to be

$$S_X(f) = \frac{4\alpha}{4\alpha^2 + 4\pi^2 f^2}.$$

From Eq. (7.43) we have

$$S_Y(f) = \left(\frac{\beta^2}{\beta^2 + 4\pi^2 f^2}\right)\left(\frac{4\alpha}{4\alpha^2 + 4\pi^2 f^2}\right)$$
$$= \frac{4\alpha\beta^2}{\beta^2 - 4\alpha^2}\left\{\frac{1}{4\alpha^2 + 4\pi^2 f^2} - \frac{1}{\beta^2 + 4\pi^2 f^2}\right\}.$$

$R_Y(\tau)$ is found by inverting the above expression:

$$R_Y(\tau) = \frac{1}{\beta^2 - 4\alpha^2}\{\beta^2 e^{-2\alpha|\tau|} - 2\alpha\beta e^{-\beta|\tau|}\}.$$

■■

---

### Discrete-Time Systems

The results obtained above for continuous-time signals also hold for discrete-time signals after appropriate changes are made from integrals to summations.

Let the **unit-sample response** $h_n$ be the response of a discrete-time, linear, time-invariant system to a unit-sample input $\delta_n$:

$$\delta_n = \begin{cases} 1 & n = 0 \\ 0 & n \neq 0. \end{cases} \tag{7.47}$$

The response of the system to an arbitrary input random process $X_n$ is then given by

$$Y_n = h_n * X_n = \sum_{j=-\infty}^{\infty} h_j X_{n-j} = \sum_{j=-\infty}^{\infty} h_{n-j} X_j. \tag{7.48}$$

Thus discrete-time, linear, time-invariant systems are determined by the unit-sample response $h_n$. The **transfer function** of such a system is defined by

$$H(f) = \sum_{i=-\infty}^{\infty} h_i e^{-j2\pi f i}. \tag{7.49}$$

The derivation from the previous section can be used to show that if $X_n$ is a wide-sense stationary process, then $Y_n$ is also wide-sense stationary. The mean of $Y_n$ is given by

$$m_Y = m_X \sum_{j=-\infty}^{\infty} h_j = m_X H(0). \tag{7.50}$$

The autocorrelation of $Y_n$ is given by

$$R_Y(k) = \sum_{j=-\infty}^{\infty} \sum_{i=-\infty}^{\infty} h_j h_i R_X(k + j - i). \tag{7.51}$$

By taking the Fourier transform of $R_Y(k)$ it is readily shown that the power spectral density of $Y_n$ is

$$S_Y(f) = |H(f)|^2 S_X(f). \tag{7.52}$$

This is the same equation that was found for continuous-time systems.

Finally, we note that if the input process $X_n$ is a Gaussian WSS random process, then the output process $Y_n$ is also a Gaussian WSS random whose statistics are completely determined by the mean and autocorrelation function provided by Eqs. (7.50) through (7.52).

**■■ Example 7.13**
Filtered White Noise

Let $X_n$ be a white noise sequence with zero mean and average power $\sigma_X^2$. If $X_n$ is the input to a linear, time-invariant system with transfer function $H(f)$, then the output process $Y_n$ has power spectral density:

$$S_Y(f) = |H(f)|^2 \sigma_X^2 \tag{7.53}$$

■■

---

Equation (7.53) provides us with a method for generating discrete-time random processes with arbitrary power spectral densities or autocorrelation functions. If the power spectral density can be written as a rational function of $z = e^{j2\pi f}$ in Eq. (7.21), then a causal filter can be found to generate a process with the power spectral density. Note that this is a generalization of the methods presented in Section 4.9 for generating vector random variables with arbitrary covariance matrix.

**■■ Example 7.14**
First-order Autoregressive Process

A first-order autoregressive process $Y_n$ with zero mean is defined by

$$Y_n = \alpha Y_{n-1} + X_n, \tag{7.54}$$

where $X_n$ is a zero-mean white noise input random process with average power $\sigma_x^2$. Find the power spectral density of $Y_n$.

The unit-sample response can be determined from Eq. (7.54):

$$h_n = \begin{cases} 0 & n < 0 \\ 1 & n = 0 \\ \alpha^n & n > 0. \end{cases}$$

Note we require $|\alpha| < 1$ for the system to be stable.[5] Therefore the transfer function is

$$H(f) = \sum_{n=0}^{\infty} \alpha^n e^{-j2\pi fn} = \frac{1}{1 - \alpha e^{-j2\pi f}}.$$

Equation (7.52) then gives

$$\begin{aligned} S_Y(f) &= \frac{\sigma_X^2}{(1 - \alpha e^{-j2\pi f})(1 - \alpha e^{j2\pi f})} \\ &= \frac{\sigma_X^2}{1 + \alpha^2 - (\alpha e^{-j2\pi f} + \alpha e^{j2\pi f})} \\ &= \frac{\sigma_X^2}{1 + \alpha^2 - 2\alpha \cos 2\pi f}. \end{aligned}$$

■■

---

### ■■ Example 7.15
ARMA Random Process

An **autoregressive moving average (ARMA) process** is defined by

$$Y_n = \sum_{i=1}^{q} \alpha_i Y_{n-i} + \sum_{i'=0}^{p} \beta_{i'} W_{n-i'}, \tag{7.55}$$

where $W_n$ is a WSS, white noise input process. Moving average processes and autoregressive processes are obtained as special cases of ARMA processes if the appropriate coefficients are set to zero. It can be shown that the transfer function of the linear system defined by the above equation is

$$H(f) = \frac{\sum_{i'=0}^{p} \beta_{i'} e^{-j2\pi f}}{\sum_{i=0}^{q} \alpha_i e^{-j2\pi f}}.$$

The power spectral density of the ARMA process is

$$S_Y(f) = |H(f)|^2 \sigma_W^2.$$

ARMA models are used extensively in random time series analysis and in signal processing. ■■

---

5. A system is said to be **stable** if $\sum_n |h_n| < \infty$. The response of a stable system to any bounded input is also bounded.

## 7.3

## AMPLITUDE MODULATION BY RANDOM SIGNALS

Many of the transmission media used in communication systems can be modeled as linear systems and their behavior can be specified by a transfer function $H(f)$, which passes certain frequencies and rejects others. Quite often the information signal $A(t)$ (i.e., a speech or music signal) is not at the frequencies that propagate well. The purpose of a **modulator** is to map the information signal $A(t)$ into a transmission signal $X(t)$ that is in a frequency range that propagates well over the desired medium. At the receiver, we need to perform an inverse mapping to recover $A(t)$ from $X(t)$. In this section, we discuss two of the amplitude modulation methods.

Let $A(t)$ be a WSS random process that represents an information signal. In general $A(t)$ will be "lowpass" in character, that is, its power spectral density will be concentrated at low frequencies as shown in Fig. 7.7(a). An **amplitude modulation** (AM) system produces a transmission

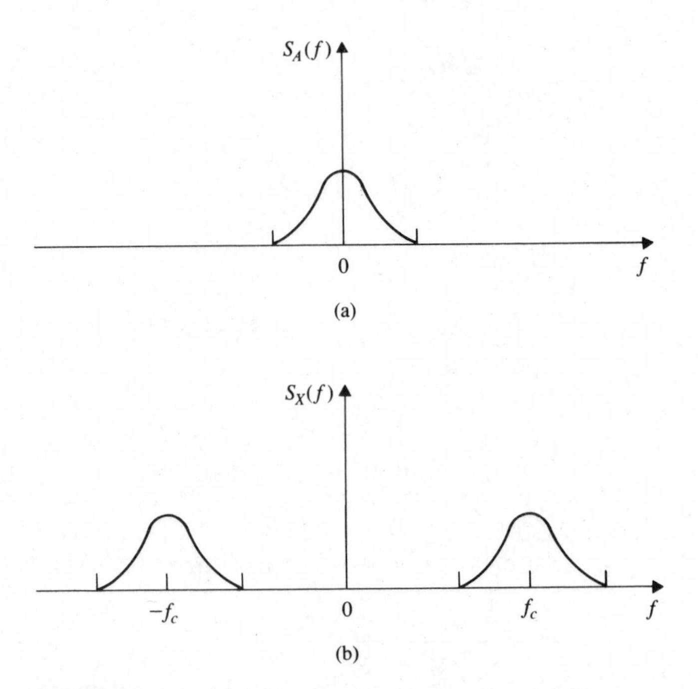


**FIGURE 7.7** (a) A lowpass information signal; (b) an amplitude modulated signal.

signal by multiplying $A(t)$ by a "carrier" signal $\cos(2\pi f_c t + \Theta)$:

$$X(t) = A(t)\cos(2\pi f_c t + \Theta), \tag{7.56}$$

where we assume $\Theta$ is a random variable that is uniformly distributed in the interval $(0, 2\pi)$, and $\Theta$ and $A(t)$ are independent.

The autocorrelation of $X(t)$ is

$$\begin{aligned} E[X(t+\tau)X(t)] &= E[A(t+\tau)\cos(2\pi f_c(t+\tau)+\Theta)A(t)\cos(2\pi f_c t+\Theta)] \\ &= E[A(t+\tau)A(t)]E[\cos(2\pi f_c(t+\tau)+\Theta)\cos(2\pi f_c t+\Theta)] \\ &= R_A(\tau)E\left[\frac{1}{2}\cos(2\pi f_c\tau) + \frac{1}{2}\cos(2\pi f_c(2t+\tau)+2\Theta)\right] \\ &= \frac{1}{2}R_A(\tau)\cos(2\pi f_c\tau), \end{aligned} \tag{7.57}$$

where we used the fact that $E[\cos(2\pi f_c(2t+\tau)+2\Theta)] = 0$ (see Example 6.7). Thus $X(t)$ is also a wide-sense stationary random process.

The power spectral density of $X(t)$ is

$$\begin{aligned} S_X(f) &= \mathcal{F}\left\{\frac{1}{2}R_A(\tau)\cos(2\pi f_c\tau)\right\} \\ &= \frac{1}{4}S_A(f+f_c) + \frac{1}{4}S_A(f-f_c), \end{aligned} \tag{7.58}$$

where we used the table of Fourier transforms in the Appendix B. Figure 7.7(b) shows $S_X(f)$. It can be seen that the power spectral density of the information signal has been shifted to the regions around $\pm f_c$. $X(t)$ is an example of a **bandpass signal.** Bandpass signals are characterized as having their power spectral density concentrated about some frequency much greater than zero.

The transmission signal is demodulated by multiplying it by the carrier signal and lowpass filtering as shown in Fig. 7.8. Let

$$Y(t) = X(t)\,2\cos(2\pi f_c t + \Theta). \tag{7.59}$$

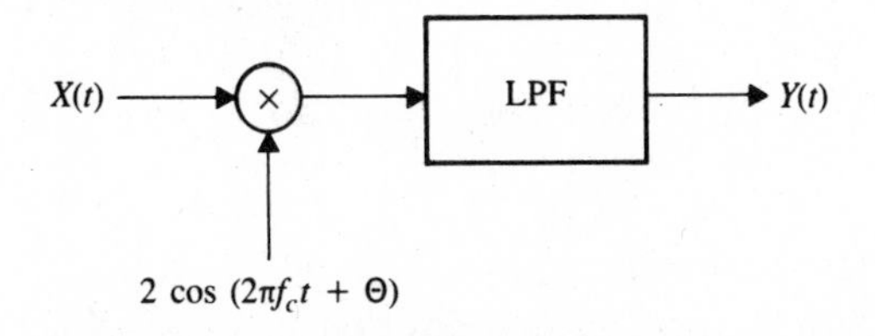


FIGURE 7.8 AM demodulator.

Proceeding as above, we find that

$$S_Y(f) = \frac{1}{2} S_X(f + f_c) + \frac{1}{2} S_X(f - f_c)$$
$$= \frac{1}{2}\{S_A(f + 2f_c) + S_A(f)\} + \frac{1}{2}\{S_A(f) + S_A(f - 2f_c)\}.$$

The ideal lowpass filter passes $S_A(f)$ and blocks $S_A(f \pm 2f_c)$, thus the output of the lowpass filter has power spectral density

$$S_Y(f) = S_A(f).$$

In fact, from Example 7.11 we know the output is the original information signal, $A(t)$.

The modulation method in Eq. (7.56) can only produce bandpass signals for which $S_X(f)$ is locally symmetric about $f_c$, $S_X(f_c + \delta f) = S_X(f_c - \delta f)$ for $|\delta f| < W$, as in Fig. 7.7(b). The method cannot yield real-valued transmission signals whose power spectral density lack this symmetry, such as shown in Fig. 7.9(a). The following **quadrature amplitude modulation** (QAM) method can be used to produce such signals:

$$X(t) = A(t)\cos(2\pi f_c t + \Theta) + B(t)\sin(2\pi f_c t + \Theta), \tag{7.60}$$

where $A(t)$ and $B(t)$ are real-valued, jointly wide-sense stationary random processes, and we require that

$$R_A(\tau) = R_B(\tau) \tag{7.61a}$$

$$R_{B,A}(\tau) = -R_{A,B}(\tau) \tag{7.61b}$$

Note that Eq. (7.61a) implies that $S_A(f) = S_B(f)$, a real-valued, even function of $f$ as shown in Fig. 7.9(b). Note also that Eq. (7.61b) implies that $S_{B,A}(f)$ is a purely imaginary, odd function of $f$ as also shown in Fig. 7.9(c) (see Problem 43).

Proceeding as before, we can show that $X(t)$ is a wide-sense stationary random process with autocorrelation function

$$R_X(\tau) = R_A(\tau)\cos(2\pi f_c \tau) + R_{B,A}(\tau)\sin(2\pi f_c \tau) \tag{7.62}$$

and power spectral density

$$S_X(f) = \frac{1}{2}\{S_A(f - f_c) + S_A(f + f_c)\} + \frac{1}{2j}\{S_{BA}(f - f_c) - S_{BA}(f + f_c)\}. \tag{7.63}$$

The resulting power spectral density is as shown in Fig. 7.9(a). Thus QAM can be used to generate real-valued bandpass signals with arbitrary power spectral density.

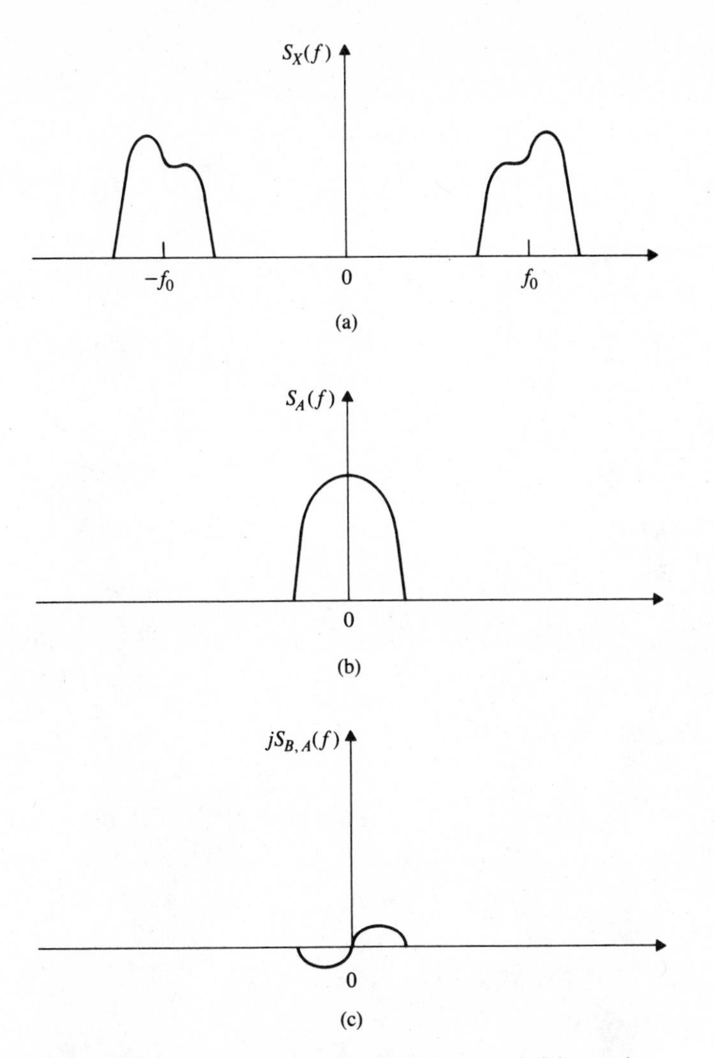


**FIGURE 7.9** (a) A general bandpass signal, (b) a real-valued even function of *f*, (c) an imaginary odd function of *f*.

Bandpass random signals, such as those in Fig. 7.9(a), arise in communication systems when wide-sense stationary white noise is filtered by bandpass filters. Let $N(t)$ be such a process with power spectral density $S_N(f)$. It can be shown that $N(t)$ can be represented by

$$N(t) = N_c(t)\cos(2\pi f_c t + \Theta) - N_s(t)\sin(2\pi f_c t + \Theta), \tag{7.64}$$

where $N_c(t)$ and $N_s(t)$ are jointly wide sense stationary processes with

$$S_{N_c}(f) = S_{N_s}(f) = \{S_N(f - f_c) + S_N(f + f_c)\}_L \tag{7.65}$$

and

$$S_{N_c,N_s}(f) = j\{S_N(f - f_c) - S_N(f + f_c)\}_L, \tag{7.66}$$

where the subscript $L$ denotes the lowpass portion of the expression in brackets. In words, every real-valued bandpass process can be treated as if it had been generated by a QAM modulator.

### ■■ Example 7.16
Demodulation of Noisy Signal

The received signal in an AM system is

$$Y(t) = A(t)\cos(2\pi f_c t + \Theta) + N(t)$$

where $N(t)$ is a bandlimited white noise process with spectral density

$$S_N(f) = \begin{cases} \dfrac{N_0}{2} & |f \pm f_c| < W \\ 0 & \text{elsewhere.} \end{cases}$$

Find the signal to noise ratio of the recovered signal.

Equation (7.64) allows us to represent the received signal by

$$Y(t) = \{A(t) + N_c(t)\}\cos(2\pi f_c t + \Theta) - N_s(t)\sin(2\pi f_c t + \Theta).$$

The demodulator in Fig. 7.8 is used to recover $A(t)$. After multiplication by $2\cos(2\pi f_c t + \Theta)$, we have

$$\begin{aligned} 2Y(t)\cos(2\pi f_c t + \Theta) &= \{A(t) + N_c(t)\}2\cos^2(2\pi f_c t + \Theta) \\ &\quad - N_s(t)2\cos(2\pi f_c t + \Theta)\sin(2\pi f_c t + \Theta) \\ &= \{A(t) + N_c(t)\}(1 + \cos(4\pi f_c t + 2\Theta)) \\ &\quad - N_s(t)\sin(4\pi f_c t + 2\Theta). \end{aligned}$$

After lowpass filtering, the recovered signal is:

$$A(t) + N_c(t).$$

The power in the signal and noise components, respectively, are

$$\sigma_A^2 = \int_{-W}^{W} S_A(f)\,df$$

$$\sigma_{N_c}^2 = \int_{-W}^{W} S_{N_c}(f)\,df = \int_{-W}^{W} \left(\frac{N_0}{2} + \frac{N_0}{2}\right) df = 2WN_o.$$

The output signal to noise ratio is then

$$\text{SNR} = \frac{\sigma_A^2}{2WN_o}.$$

■■

# 7.4

## OPTIMUM LINEAR SYSTEMS

Many problems can be posed in the following way. We observe a discrete-time, zero-mean process $X_\alpha$ over a certain time interval $I = \{t - a, \ldots, t + b\}$, and we are required to use the $a + b + 1$ resulting observations $\{X_{t-a}, \ldots, X_t, \ldots, X_{t+b}\}$ to obtain an estimate $Y_t$ for some other (presumably related) zero-mean process $Z_t$. The estimate $Y_t$ is required to be linear as shown in Fig. 7.10:

$$Y_t = \sum_{\beta=t-a}^{t+b} h_{t-\beta} X_\beta = \sum_{\beta=-b}^{a} h_\beta X_{t-\beta}. \tag{7.67}$$

The figure of merit for the estimator is the mean square error

$$E[e_t^2] = E[(Z_t - Y_t)^2] \tag{7.68}$$

and we seek to find the **optimum filter,** which is characterized by the impulse response $h_\beta$ that minimizes the mean square error.

Examples 7.17 and 7.18 show that different choices of $Z_t$ and $X_\alpha$ and of observation interval correspond to different estimation problems.

### ■■ Example 7.17
Filtering and Smoothing Problems

Let the observations be the sum of a "desired signal" $Z_\alpha$ plus unwanted "noise" $N_\alpha$:

$$X_\alpha = Z_\alpha + N_\alpha \qquad \alpha \in I.$$

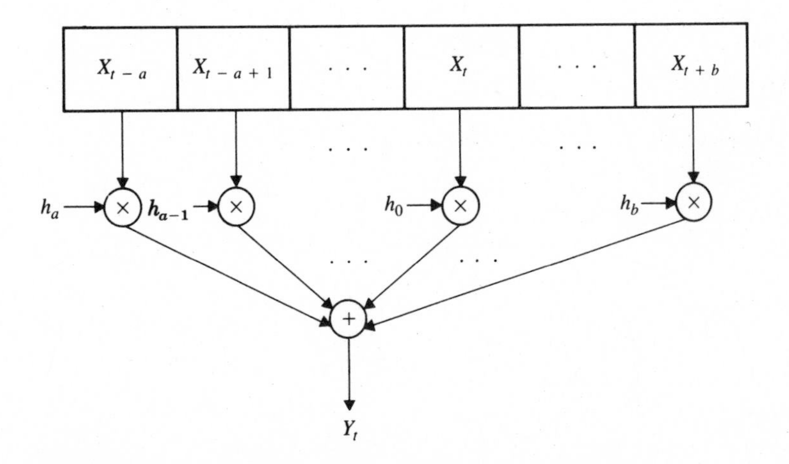

**FIGURE 7.10** A linear system for producing an estimate $Y_t$.

We are interested in estimating the desired signal at time $t$. The relation between $t$ and the observation interval $I$ gives rise to a variety of estimation problems.

If $I = (-\infty, t)$, that is, $a = \infty$ and $b = 0$, then we have a *filtering* problem where we estimate $Z_t$ in terms of noisy observations of the past and present. If $I = (t - a, t)$, then we have a filtering problem in which we estimate $Z_t$ in terms of the $a + 1$ most recent noisy observations.

If $I = (-\infty, \infty)$, that is $a = b = \infty$, then we have a *smoothing* problem where we are attempting to recover the signal from its entire noisy version. There are applications where this makes sense, for example, if the entire realization $X_\alpha$ has been recorded and the estimate $Z_t$ is obtained by "playing back" $X_\alpha$. ■■

---

### ■■ Example 7.18
Prediction

Suppose we want to predict $Z_t$ in terms of its recent past: $\{Z_{t-a}, \ldots, Z_{t-1}\}$. The general estimation problem becomes this *prediction* problem if we let the observation $X_\alpha$ be the past $a$ values of the signal $Z_\alpha$, that is,

$$X_\alpha = Z_\alpha \qquad t - a \leq \alpha \leq t - 1.$$

The estimate $Y_t$ is then a linear prediction of $Z_t$ in terms of its most recent values. ■■

---

### The Orthogonality Condition

It can be shown that the optimum filter must satisfy the **orthogonality condition** (see Eqs. 4.82a and 4.82b), which states that the error $e_t$ must be orthogonal to all the observations $X_\alpha$, that is,

$$\begin{aligned} 0 &= E[e_t X_\alpha] \qquad \text{for all } \alpha \in I \\ &= E[(Z_t - Y_t)X_\alpha] = 0, \end{aligned} \tag{7.69}$$

or equivalently

$$E[Z_t X_\alpha] = E[Y_t X_\alpha] \qquad \text{for all } \alpha \in I. \tag{7.70}$$

If we substitute Eq. (7.67) into Eq. (7.70) we find

$$\begin{aligned} E[Z_t X_\alpha] &= E\left[\sum_{\beta=-b}^{a} h_\beta X_{t-\beta} X_\alpha\right] \qquad \text{for all } \alpha \in I \\ &= \sum_{\beta=-b}^{a} h_\beta E[X_{t-\beta} X_\alpha] \\ &= \sum_{\beta=-b}^{a} h_\beta R_X(t - \alpha - \beta) \qquad \text{for all } \alpha \in I. \end{aligned} \tag{7.71}$$

Equation (7.71) shows that $E[Z_t X_\alpha]$ depends only on $t - \alpha$, and thus $X_\alpha$ and $Z_t$ are jointly wide-sense stationary processes. Therefore, we can rewrite Eq. (7.71) as follows:

$$R_{Z,X}(t - \alpha) = \sum_{\beta=-b}^{a} h_\beta R_X(t - \beta - \alpha) \qquad t - a \leq \alpha \leq t + b.$$

Finally, letting $m = t - \alpha$, we obtain the following key equation:

$$R_{Z,X}(m) = \sum_{\beta=-b}^{a} h_\beta R_X(m - \beta) \qquad -b \leq m \leq a. \tag{7.72}$$

*The optimum linear filter must satisfy the set of $a + b + 1$ linear equations given by Eq.* (7.72).

In the above derivation we deliberately used the notation $Z_t$ instead of $Z_n$ to suggest that the same development holds for the continuous-time version of the problem. In particular, if we seek to estimate $Z(t)$ from a set of observations $X(\alpha)$, $t - a \leq \alpha \leq t + b$, then the filter $h(\beta)$ that minimizes the mean square error is specified by

$$R_{Z,X}(\tau) = \int_{-b}^{a} h(\beta) R_X(\tau - \beta)\, d\beta \qquad -b \leq \tau \leq a. \tag{7.73}$$

Thus in the time-continuous case we obtain an integral equation instead of a set of linear equations. The analytic solution of this integral equation can be quite difficult, but the equation can be solved numerically by approximating the integral by a summation.[6]

We now determine the mean square error of the optimum filter. First we note that for the optimum filter, the error $e_t$ and the estimate $Y_t$ are orthogonal since

$$E[e_t Y_t] = E\left[e_t \sum h_{t-\beta} X_\beta\right] = \sum h_{t-\beta} E[e_t X_\beta] = 0,$$

where the terms inside the last summation are 0 because of Eq. (7.69). Since $e_t = Z_t - Y_t$, the mean square error is then

$$\begin{aligned} E[e_t^2] &= E[e_t(Z_t - Y_t)] \\ &= E[e_t Z_t], \end{aligned}$$

6. Equation (7.73) can be solved in a very elegant manner by using the Karhumen-Loeve expansion. For example, see Chapter 10 in Stark and Woods (1986).

since $e_t$ and $Y_t$ are orthogonal. Substituting for $e_t$, yields

$$\begin{aligned}E[e_t^2] &= E[(Z_t - Y_t)Z_t] = E[Z_tZ_t] - E[Y_tZ_t] \\ &= R_Z(0) - E[Z_tY_t] \\ &= R_Z(0) - E\left[Z_t \sum_{\beta=-b}^{a} h_\beta X_{t-\beta}\right] \\ &= R_Z(0) - \sum_{\beta=-b}^{a} h_\beta R_{Z,X}(\beta).\end{aligned} \tag{7.74}$$

Similarly, it can be shown that the mean square error of the optimum filter in the continuous-time case is

$$E[e^2(t)] = R_Z(0) - \int_{-b}^{a} h(\beta) R_{Z,X}(\beta)\, d\beta. \tag{7.75}$$

The following theorem summarizes the above results. The theorem is stated for the discrete-time case, but the results hold for the continuous-time case after appropriate changes are made from summations to integrals. (See Eqs. 7.73 and 7.75.)

**THEOREM**

Let $X_t$ and $Z_t$ be zero-mean, jointly wide-sense stationary processes, and let $Y_t$ be an estimate for $Z_t$ of the form

$$Y_t = \sum_{\beta=t-a}^{t+b} h_{t-\beta} X_\beta = \sum_{\beta=-b}^{a} h_\beta X_{t-\beta}$$

The filter that minimizes $E[(Z_t - Y_t)^2]$ satisfies the equation

$$R_{Z,X}(m) = \sum_{\beta=-b}^{a} h_\beta R_X(m - \beta) \qquad -b \le m \le a$$

and has mean square error given by

$$E[(Z_t - Y_t)^2] = R_Z(0) - \sum_{\beta=-b}^{a} h_\beta R_{Z,X}(\beta).$$

### ■■ Example 7.19
Filtering of Signal Plus Noise

Suppose we are interested in estimating the signal $Z_n$ from the $p + 1$ most recent noisy observations:

$$X_\alpha = Z_\alpha + N_\alpha \qquad \alpha \in I = \{n - p, \ldots, n - 1, n\}.$$

Find the set of linear equations for the optimum filter if $Z_\alpha$ and $N_\alpha$ are independent random processes.

For this choice of observation interval, Eq. (7.72) becomes

$$R_{Z,X}(m) = \sum_{\beta=0}^{p} h_\beta R_X(m - \beta) \qquad m \in \{0, 1, \ldots, p\}. \tag{7.76}$$

The cross-correlation terms in Eq. (7.76) are given by

$$R_{Z,X}(m) = E[Z_n X_{n-m}] = E[Z_n(Z_{n-m} + N_{n-m})] = R_Z(m).$$

The autocorrelation terms are given by

$$\begin{aligned} R_X(m-\beta) &= E[X_{n-\beta}X_{n-m}] = E[(Z_{n-\beta} + N_{n-\beta})(Z_{n-m} + N_{n-m})] \\ &= R_Z(m-\beta) + R_{Z,N}(m-\beta) + R_{N,Z}(m-\beta) + R_N(m-\beta) \\ &= R_Z(m-\beta) + R_N(m-\beta), \end{aligned}$$

since $Z_\alpha$ and $N_\alpha$ are independent random processes. Thus Eq. (7.76) for the optimum filter becomes

$$R_Z(m) = \sum_{\beta=0}^{p} h_\beta\{R_Z(m-\beta) + R_N(m-\beta)\} \quad m \in \{0, 1, \ldots, p\}. \tag{7.77}$$

This set of $p+1$ linear equations in $p+1$ unknowns $h_\beta$ is solved by matrix inversion. ■■

### ■■ Example 7.20

Find the set of equations for the optimum filter in Example 7.19 if $Z_\alpha$ is a first-order autoregressive process with average power $\sigma_Z^2$ and parameter $r$, $|r| < 1$, and $N_\alpha$ is a white noise process with average power $\sigma_N^2$.

The autocorrelation for a first-order autoregressive process is given by

$$R_Z(m) = \sigma_Z^2 r^{|m|} \qquad m = 0, \pm 1, \pm 2, \ldots.$$

(See Problem 36.) The autocorrelation for the white noise process is

$$R_N(m) = \sigma_N^2 \delta(m).$$

Substituting $R_Z(m)$ and $R_N(m)$ into Eq. (7.77) yields the following set of linear equations:

$$\sigma_Z^2 r^{|m|} = \sum_{\beta=0}^{p} h_\beta(\sigma_Z^2 r^{|m-\beta|} + \sigma_N^2 \delta(m-\beta)) \qquad m \in \{0, \ldots, p\}. \tag{7.78}$$

If we divide both sides of Eq. (7.78) by $\sigma_Z^2$ and let $\Gamma = \sigma_N^2/\sigma_Z^2$, we obtain the following matrix equation:

$$\begin{bmatrix} 1+\Gamma & r & r^2 & \cdots & r^p \\ r & 1+\Gamma & r & \cdots & r^{p-1} \\ r^2 & r & 1+\Gamma & \cdots & r^{p-2} \\ \cdot & \cdot & \cdot & \cdots & \cdot \\ r^p & r^{p-1} & r^{p-2} & \cdots & 1+\Gamma \end{bmatrix} \begin{bmatrix} h_0 \\ h_1 \\ \cdot \\ \cdot \\ h_p \end{bmatrix} = \begin{bmatrix} 1 \\ r \\ \cdot \\ \cdot \\ r^p \end{bmatrix} \tag{7.79}$$

Note that when the noise power is zero, i.e., $\Gamma = 0$, then the solution is

$h_0 = 1$, $h_j = 0$ $j = 1, \ldots, p$, that is, no filtering is required to obtain $Z_n$. See Problem 44 for an example involving the numerical solution of Eq. (7.79). ■■

---

**■■ Example 7.21**
Prediction

Consider the prediction problem in which we wish to predict $Z_n$ in terms of $\{Z_{n-p}, \ldots, Z_{n-1}\}$:

$$Y_n = \sum_{\beta=1}^{p} h_\beta Z_{n-\beta}. \tag{7.80}$$

Find the set of equations that specify the optimum filter, and find an expression for the resulting mean square error.

For this problem $X_\alpha = Z_\alpha$, so Eq. (7.72) becomes

$$R_Z(m) = \sum_{\beta=1}^{p} h_\beta R_Z(m - \beta) \qquad m \in \{1, \ldots, p\}. \tag{7.81}$$

Equation (7.74) for the mean-squared error becomes

$$E[e_n^2] = R_Z(0) - \sum_{\beta=1}^{p} h_\beta R_Z(\beta). \tag{7.82}$$

In Example 4.43 in Chapter 4, we solved a $p = 2$ example of this prediction problem. ■■

---

### Estimation Using the Entire Realization of the Observed Process

Suppose that $Z_t$ is to be estimated using the entire realization of $X_t$, that is, $I = (-\infty, \infty)$. Equations (7.72) and (7.73) become

$$R_{Z,X}(m) = \sum_{\beta=-\infty}^{\infty} h_\beta R_X(m - \beta) \qquad \text{for all } m \tag{7.83a}$$

$$R_{Z,X}(\tau) = \int_{-\infty}^{\infty} h(\beta) R_X(\tau - \beta)\, d\beta \qquad \text{for all } \tau. \tag{7.83b}$$

The Fourier transform of the first equation, and the Fourier transform of the second equation both yield the same expression:

$$S_{Z,X}(f) = H(f) S_X(f),$$

which is readily solved for the transfer function of the optimum filter:

$$H(f) = \frac{S_{Z,X}(f)}{S_X(f)}. \tag{7.84}$$

The impulse response of the optimum filter is then obtained by taking the appropriate inverse transform. In general the filter obtained from Eq. (7.84) will be noncausal, that is, its impulse response is nonzero for $t < 0$. We already indicated that there are applications where this makes sense, namely, in situations where the entire realization $X_\alpha$ is recorded and the estimate $Z_t$ is obtained in "nonreal time" by "playing back" $X_\alpha$.

### ■■ Example 7.22
Infinite Smoothing

Find the transfer function for the optimum filter for estimating $Z(t)$ from $X(\alpha) = Z(\alpha) + N(\alpha)$, $\alpha \in (-\infty, \infty)$, where $Z(\alpha)$ and $N(\alpha)$ are independent, zero-mean random processes.

The cross-correlation between the observation and the desired signal is

$$\begin{aligned} R_{Z,X}(\tau) &= E[Z(t+\tau)X(t)] = E[Z(t+\tau)(Z(t)+N(t))] \\ &= E[Z(t+\tau)Z(t)] + E[Z(t+\tau)N(t)] \\ &= R_Z(\tau), \end{aligned}$$

since $Z(t)$ and $N(t)$ are zero-mean, independent random processes. The cross-power spectral density is then

$$S_{Z,X}(f) = S_Z(f). \tag{7.85}$$

The autocorrelation of the observation process is

$$\begin{aligned} R_X(\tau) &= E[(Z(t+\tau) + N(t+\tau))(Z(t)+N(t))] \\ &= R_Z(\tau) + R_N(\tau). \end{aligned}$$

The corresponding power spectral density is

$$S_X(f) = S_Z(f) + S_N(f). \tag{7.86}$$

Substituting Eqs. (7.85) and (7.86) into Eq. (7.84) gives

$$H(f) = \frac{S_Z(f)}{S_Z(f) + S_N(f)}. \tag{7.87}$$

Note that the optimum filter $H(f)$ is nonzero only at the frequencies where $S_Z(f)$ is nonzero, that is, where the signal has power content. By dividing the numerator and denominator of Eq. (7.87) by $S_Z(f)$, we see that $H(f)$ emphasizes the frequencies where the ratio of signal to noise power density is large. ■■

### *Estimation Using Causal Filters

Now, suppose that $Z_t$ is to be estimated using only the past and present of $X_\alpha$, that is, $I = (-\infty, t)$. Equations (7.72) and (7.73) become

$$R_{Z,X}(m) = \sum_{\beta=0}^{\infty} h_\beta R_X(m - \beta) \qquad \text{for all } m \tag{7.88a}$$

$$R_{Z,X}(\tau) = \int_0^\infty h(\beta) R_X(\tau - \beta)\, d\beta \qquad \text{for all } \tau. \tag{7.88b}$$

Equations (7.88a) and (7.88b) are called the **Wiener-Hopf equations,** and though similar in appearance to Eqs. (7.83a) and (7.83b) are considerably more difficult to solve.

First, let us consider the special case where the observation process is white, that is, for the discrete-time case $R_X(m) = \delta_m$. Equation (7.88a) is then

$$R_{Z,X}(m) = \sum_{\beta=0}^{\infty} h_\beta\, \delta_{m-\beta} = h_m. \qquad m \geq 0 \tag{7.89}$$

Thus in this special case, the optimum causal filter has coefficients given by

$$h_m = \begin{cases} 0 & m < 0 \\ R_{Z,X}(m) & m \geq 0. \end{cases}$$

The corresponding transfer function is

$$H(f) = \sum_{m=0}^{\infty} R_{Z,X}(m) e^{-j2\pi fm}. \tag{7.90}$$

Note Eq. (7.90) is *not* $S_{Z,X}(f)$ since the limits of the Fourier transform in Eq. 7.90 do not extend from $-\infty$ to $+\infty$. However, $H(f)$ can be obtained from $S_{Z,X}(f)$ by finding $h_m = \mathscr{F}^{-1}[S_{Z,X}(f)]$, keeping the causal part (i.e., $h_m$ for $m \geq 0$) and setting the noncausal part to 0.

We now show how the solution of the above special case can be used to solve the general case. It can be shown that under very general conditions, the power spectral density of a random process can be factored into the form

$$S_X(f) = |G(f)|^2 = G(f)G^*(f), \tag{7.91}$$

where $G(f)$ and $1/G(f)$ are *causal* filters.[7] This suggests that we can find the optimum filter in two steps as shown in Fig. 7.11. First, we pass the observation process through a "whitening" filter with transfer function

7. The method for factoring $S_X(f)$ as specified by Eq. (7.91) is called **spectral factorization.** See Example 7.10 and the references at the end of the chapter.

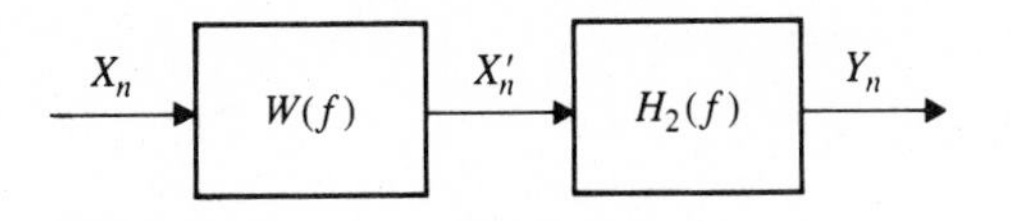


**FIGURE 7.11** Whitening filter approach for solving Wiener-Hopf equations.

$W(f) = 1/G(f)$ to produce a white noise process $X'_n$ since

$$S_{X'}(f) = |W(f)|^2 S_X(f) = \frac{|G(f)|^2}{|G(f)|^2} = 1 \qquad \text{for all } f.$$

Second, we find the best estimator for $Z_n$ using the whitened observation process $X'_n$ as given by Eq. 7.90. The filter that results from the tandem combination of the whitening filter and the estimation filter is the solution to the Wiener-Hopf equations.

The transfer function of the second filter in Fig. 7.11 is

$$H_2(f) = \sum_{m=0}^{\infty} R_{Z,X'}(m) e^{-j2\pi fm} \tag{7.92}$$

by Eq. (7.90). To evaluate Eq. (7.92) we need to find

$$\begin{aligned} R_{Z,X'}(k) &= E[Z_{n+k} X'_n] \\ &= \sum_{i=0}^{\infty} w_i E[Z_{n+k} X_{n-i}] \\ &= \sum_{i=0}^{\infty} w_i R_{Z,X}(k+i), \end{aligned} \tag{7.93}$$

where $w_i$ is the impulse response of the whitening filter. The Fourier transform of Eq. (7.93) gives an expression that is easier to work with:

$$S_{Z,X'}(f) = W^*(f) S_{Z,X}(f) = \frac{S_{Z,X}(f)}{G^*(f)}. \tag{7.94}$$

The inverse Fourier transform of Eq. (7.94) yields the desired $R_{Z,X'}(k)$, which can then be substituted into Eq. (7.92) to obtain $H_2(f)$.

In summary, the optimum filter is found using the following procedure:

1. Factor $S_X(f)$ as in Eq. (7.91) and obtain a causal whitening filter $W(f) = 1/G(f)$.
2. Find $R_{Z,X'}(k)$ from Eq. (7.93) or from Eq. (7.94).
3. $H_2(f)$ is then given by Eq. (7.92).

4. The optimum filter is then

$$H(f) = W(f)H_2(f). \tag{7.95}$$

This procedure is valid for the continuous-time version of the optimum causal filter problem, after appropriate changes are made from summations to integrals. The following example considers a continuous-time problem.

### Example 7.23
Wiener Filter

Find the optimum causal filter for estimating a signal $Z(t)$ from the observation $X(t) = Z(t) + N(t)$, where $Z(t)$ and $N(t)$ are independent random processes, $N(t)$ is zero-mean white noise with noise density 1, and $Z(t)$ has power spectral density

$$S_Z(f) = \frac{2}{1 + 4\pi^2 f^2}.$$

The optimum filter in this problem is called the **Wiener filter.**

The cross-power spectral density between $Z(t)$ and $X(t)$ is

$$S_{Z,X}(f) = S_Z(f),$$

since the signal and noise are independent random processes. The power spectral density for the observation process is

$$\begin{aligned} S_X(f) &= S_Z(f) + S_N(f) \\ &= \frac{3 + 4\pi^2 f^2}{1 + 4\pi^2 f^2} \\ &= \left(\frac{j2\pi f + \sqrt{3}}{j2\pi f + 1}\right)\left(\frac{-j2\pi f + \sqrt{3}}{-j2\pi f + 1}\right). \end{aligned}$$

If we let

$$G(f) = \frac{j2\pi f + \sqrt{3}}{j2\pi f + 1},$$

then it is easy to verify that $W(f) = 1/G(f)$ is the whitening causal filter.

Next we evaluate Eq. (7.94):

$$\begin{aligned} S_{Z,X'}(f) &= \frac{S_{Z,X}(f)}{G^*(f)} = \frac{2}{1 + 4\pi^2 f^2} \frac{1 - j2\pi f}{\sqrt{3} - j2\pi f} \\ &= \frac{2}{(1 + j2\pi f)(\sqrt{3} - j2\pi f)} \\ &= \frac{c}{1 + j2\pi f} + \frac{c}{\sqrt{3} - j2\pi f}, \end{aligned} \tag{7.96}$$

where $c = 2/(1 + \sqrt{3})$. If we take the inverse Fourier transform of $S_{Z,X'}(f)$, we obtain

$$R_{Z,X'}(\tau) = \begin{cases} ce^{-\tau} & \tau > 0 \\ ce^{\sqrt{3}\tau} & \tau < 0. \end{cases}$$

Equation (7.92) states that $H_2(f)$ is given by the Fourier transform of the $\tau > 0$ portion of $R_{Z,X'}(\tau)$:

$$H_2(f) = \mathcal{F}\{ce^{-\tau}u(\tau)\} = \frac{c}{1 + j2\pi f}$$

Note that we could have gotten this result directly from Eq. (7.96) by noting that only the first term gives rise to the positive-time (i.e., causal) component.

The optimum filter is then

$$H(f) = \frac{1}{G(f)} H_2(f) = \frac{c}{\sqrt{3} + j2\pi f}.$$

The impulse response of this filter is

$$h(t) = ce^{-\sqrt{3}t} \qquad t > 0.$$

■■

## *7.5 A COMPUTER METHOD FOR ESTIMATING THE POWER SPECTRAL DENSITY

Let $X_0, \ldots, X_{k-1}$ be $k$ observations of the discrete-time, zero-mean, wide-sense stationary process $X_n$. The periodogram estimate for $S_X(f)$ is defined as

$$\tilde{p}_k(f) = \frac{1}{k} |\tilde{x}_k(f)|^2, \tag{7.97}$$

where $\tilde{x}_k(f)$ is obtained as a Fourier transform of the observation sequence:

$$\tilde{x}_k(f) = \sum_{m=0}^{k-1} X_m e^{-j2\pi fm}. \tag{7.98}$$

In Section 7.1 we showed that the expected value of the periodogram estimate is

$$E[\tilde{p}_k(f)] = \sum_{m'=-(k-1)}^{k-1} \left\{1 - \frac{|m'|}{k}\right\} R_X(m')e^{-j2\pi fm'} \tag{7.99}$$

so $\tilde{p}_k(f)$ is a biased estimator for $S_X(f)$. However as $k \to \infty$,

$$E[\tilde{p}_k(f)] \to S_X(f) \tag{7.100}$$

so the mean of the periodogram estimate approaches $S_X(f)$.

Before proceeding to find the variance of the periodogram estimate, we note that the periodogram estimate is equivalent to taking the Fourier transform of an estimate for the autocorrelation sequence; that is,

$$\tilde{p}_k(f) = \sum_{m=-(k-1)}^{k-1} \hat{r}_k(m)e^{-j2\pi fm}, \tag{7.101}$$

where the estimate for the autocorrelation is

$$\hat{r}_k(m) = \frac{1}{k} \sum_{n=0}^{k-|m|-1} X_n X_{n+m}. \tag{7.102}$$

See Problem 61.

We might expect that as we increase the number of samples $k$, the periodogram estimate converges to $S_X(f)$. This does not happen. Instead, we find that $\tilde{p}_k(f)$ fluctuates wildly about the true spectral density, and

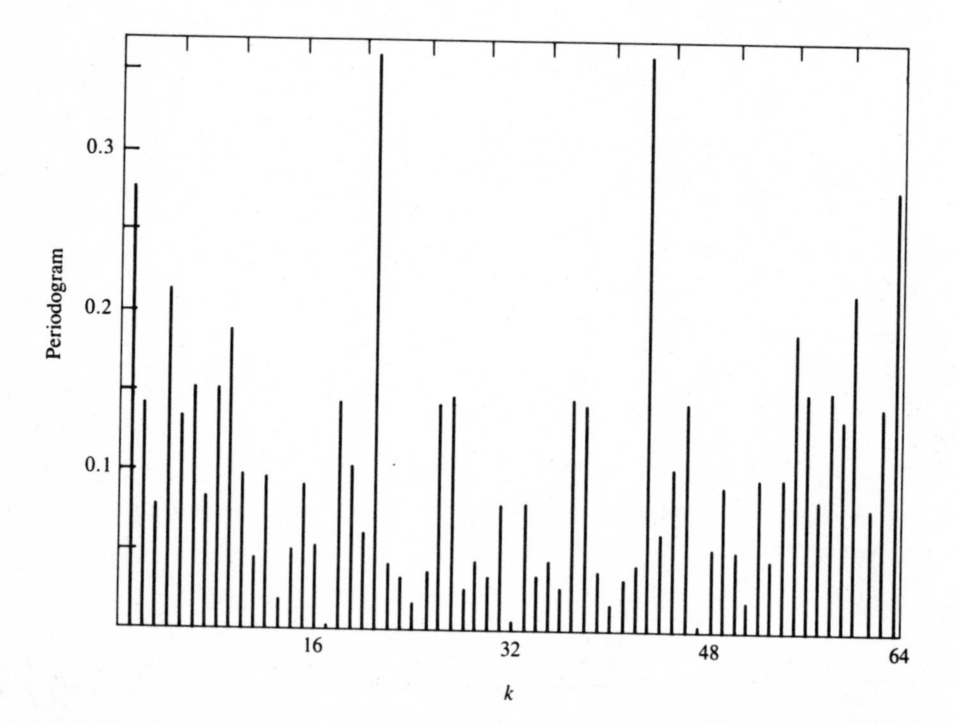


**FIGURE 7.12** Periodogram for 64 samples of white noise sequence $X_n$ iid uniform in (0, 1), $S_X(f) = \sigma_X^2 = 1/12 = 0.083$.

that this random variation does not decrease with increased $k$ (see Fig. 7.12). To see why this happens, in the next section we compute the statistics of the periodogram estimate for a white noise Gaussian random process. We find that the estimates given by the periodogram have a variance that does *not* approach zero as the number of samples is increased. This explains the lack of improvement in the estimate as $k$ is increased. Furthermore, we show that the periodogram estimates are uncorrelated at uniformly spaced frequencies in the interval $-1/2 \leq f < 1/2$. This explains the erratic appearance of the periodogram estimate as a function of $f$. In the final section, we obtain another estimate for $S_X(f)$ whose variance does approach zero as $k$ increases.

### Variance of Periodogram Estimate

Following the approach of Jenkins and Watts (1968, pp. 230–233), we consider the periodogram of samples of a white noise process with $S_X(f) = \sigma_X^2$ at the frequencies $f = n/k$, $-k/2 \leq n < k/2$, which will cover the frequency range $-1/2 \leq f < 1/2$. (In practice these are the frequencies we would evaluate if we were using the FFT algorithm to compute $\tilde{x}_k(f)$.) First we rewrite Eq. (7.98) at $f = n/k$ as follows:

$$\begin{aligned}\tilde{x}_k\left(\frac{n}{k}\right) &= \sum_{m=0}^{k-1} X_m\left(\cos\left(\frac{2\pi mn}{k}\right) - j\sin\left(\frac{2\pi mn}{k}\right)\right) \\ &= A_k(n) - jB_k(n) \qquad -k/2 \leq n < k/2,\end{aligned} \tag{7.103}$$

where

$$A_k(n) = \sum_{m=0}^{k-1} X_m \cos\left(\frac{2\pi mn}{k}\right) \tag{7.104}$$

and

$$B_k(n) = \sum_{m=0}^{k-1} X_m \sin\left(\frac{2\pi mn}{k}\right). \tag{7.105}$$

Then it follows that the periodogram estimate is

$$\tilde{p}_k\left(\frac{n}{k}\right) = \frac{1}{k}\left|\tilde{x}_k\left(\frac{n}{k}\right)\right|^2 = \frac{1}{k}\{A_k^2(n) + B_k^2(n)\} \tag{7.106}$$

We find the variance of $\tilde{p}_k(n/k)$ from the statistics of $A_k(n)$ and $B_k(n)$.

The random variables $A_k(n)$ and $B_k(n)$ are defined as linear functions of the jointly Gaussian random variables $X_0, \ldots, X_{k-1}$. Therefore $A_k(n)$ and $B_k(n)$ are also jointly Gaussian random variables. If we take the expected value of Eqs. (7.104) and (7.105) we find

$$E[A_k(n)] = 0 = E[B_k(n)] \qquad \text{for all } n. \tag{7.107}$$

Note also that the $n = -k/2$ and $n = 0$ terms are different in that

$$B_k(-k/2) = 0 = B_k(0) \tag{7.108a}$$

$$A_k(-k/2) = \sum_{i=0}^{k-1} (-1)^i X_i \qquad A_k(0) = \sum_{i=0}^{k-1} X_i. \tag{7.108b}$$

The correlation between $A_k(n)$ and $A_k(m)$ (for $n$, $m$ not equal to $-k/2$ or 0) is

$$\begin{aligned} E[A_k(n)A_k(m)] &= \sum_{i=0}^{k-1}\sum_{l=0}^{k-1} E[X_iX_l]\cos\left(\frac{2\pi ni}{k}\right)\cos\left(\frac{2\pi ml}{k}\right) \\ &= \sigma_X^2 \sum_{i=0}^{k-1} \cos\left(\frac{2\pi ni}{k}\right)\cos\left(\frac{2\pi mi}{k}\right) \\ &= \sigma_X^2 \sum_{i=0}^{k-1} \frac{1}{2}\cos\left(\frac{2\pi(n-m)i}{k}\right) + \sigma_X^2 \sum_{i=0}^{k-1} \frac{1}{2}\cos\left(\frac{2\pi(n+m)i}{k}\right), \end{aligned}$$

where we used the fact that $E[X_iX_1] = \sigma_X^2\delta_{il}$ since the noise is white. The second summation is equal to zero, and the first summation is zero except when $n = m$. Thus

$$E[A_k(n)A_k(m)] = \frac{1}{2}k\sigma_X^2\,\delta_{nm} \qquad \text{for all } n, m \neq -k/2, 0. \tag{7.109a}$$

It can similarly be shown that

$$E[B_k(n)B_k(m)] = \frac{1}{2}k\sigma_X^2\,\delta_{nm} \qquad n, m \neq 0 - k/2, 0 \tag{7.109b}$$

$$E[A_k(n)B_k(m)] = 0 \qquad \text{for all } n, m. \tag{7.109c}$$

When $n = -k/2$ or 0, we have

$$E[A_k(n)A_k(m)] = k\sigma_X^2\,\delta_{nm} \qquad \text{for all } m. \tag{7.109d}$$

Equations (7.109a) through (7.109d) imply that $A_k(n)$ and $B_k(m)$ are uncorrelated random variables. Since $A_k(n)$ and $B_k(n)$ are jointly Gaussian random variables, this implies that they are zero mean, *independent* Gaussian random variables.

We are now ready to find the statistics of the periodogram estimates at the frequencies $f = n/k$. Equation (7.106) gives

$$\begin{aligned} \tilde{p}_k\left(\frac{n}{k}\right) &= \frac{1}{k}\{A_k^2(n) + B_k^2(n)\} \qquad n \neq -k/2, 0 \\ &= \frac{1}{2}\sigma_X^2\left\{\frac{A_k^2(n)}{(1/2)k\sigma_X^2} + \frac{B_k^2(n)}{(1/2)k\sigma_X^2}\right\}. \end{aligned} \tag{7.110}$$

The quantity in brackets is the sum of the squares of two zero-mean, unit variance independent Gaussian random variables. This is a chi-square random variable with two degrees of freedom (see Problem 6 in Chapter 5). From Table 3.2, we see that a chi-square random variable with $v$ degrees of freedom has variance $2v$. Thus the expression in the brackets has variance 4, and the periodogram estimate $\hat{p}_k(n/k)$ has variance

$$\text{VAR}\left[\tilde{p}_k\left(\frac{n}{k}\right)\right] = \left(\frac{1}{2}\sigma_X^2\right)^2 4 = \sigma_X^4 = S_X(f)^2. \tag{7.111a}$$

For $n = -k/2$ and $n = 0$,

$$\tilde{p}_k\left(\frac{n}{k}\right) = \sigma_X^2\left\{\frac{A_k^2(n)}{k\sigma_X^2}\right\}.$$

The quantity in brackets is a chi-square random variable with one degree of freedom and variance 2, so the variance of the periodogram estimate is

$$\text{VAR}\left[\tilde{p}_k\left(\frac{n}{k}\right)\right] = 2\sigma_X^4 \qquad n = -k/2, 0. \tag{7.111b}$$

Thus we conclude from Eq. (7.111a) and (7.111b) that *the variance of the periodogram estimate is proportional to the square of the power spectral density and does not approach zero as $k$ increases*. In addition, Eqs. (7.109a) through (7.109d) imply that *the periodogram estimates at the frequencies $f = -n/k$ are uncorrelated random variables*. A more detailed analysis (Jenkins and Watts, 1968, p. 238) shows that for arbitrary $f$

$$\text{VAR}[\tilde{p}_k(f)] = S_X(f)^2\left\{1 + \left(\frac{\sin(2\pi fk)}{k\sin(2\pi f)}\right)^2\right\}. \tag{7.112}$$

Thus variance of the periodogram estimate does not approach zero as the number of samples is increased.

The above discussion has only considered the spectrum estimation for a white noise, Gaussian random process, but the general conclusions are also valid for nonwhite, non-Gaussian processes. If the $X_i$ are not Gaussian, we note from Eqs. (7.104) and (7.105) that $A_k$ and $B_k$ are approximately Gaussian by the central limit theorem if $k$ is large. Thus the periodogram estimate is then approximately a chi-square random variable.

If the process $X_i$ is not white, then it can be viewed as filtered white noise:

$$X_n = h_n * W_n,$$

where $S_W(f) = \sigma_W^2$ and $|H(f)|^2 S_W(f) = S_X(f)$. The periodograms of $X_n$ and $W_n$ are related by

$$\frac{1}{k}\left|\tilde{x}_k\left(\frac{n}{k}\right)\right|^2 = \frac{1}{k}\left|H\left(\frac{n}{k}\right)\right|^2\left|\tilde{w}_k\left(\frac{n}{k}\right)\right|^2. \tag{7.113}$$

Thus

$$\left|\tilde{w}_k\left(\frac{n}{k}\right)\right|^2 = \frac{|\tilde{x}_k(n/k)|^2}{|H(n/k)|^2}. \tag{7.114}$$

From our previous results, we know that $|\tilde{w}_k(n/k)|^2/k$ is a chi-square random variable with variance $\sigma_W^4$. This implies that

$$\text{VAR}\left[\frac{|\tilde{x}_k(n/k)|^2}{k}\right] = \left|H\left(\frac{n}{k}\right)\right|^4 \sigma_W^4 = S_X(f)^2. \tag{7.115}$$

Thus we conclude that the variance of the periodogram estimate for nonwhite noise is also proportional to $S_X(f)^2$.

### Smoothing of Periodogram Estimate

A fundamental result in probability theory is that the sample mean of a sequence of *independent* realizations of a random variable approaches the true mean with probability one. We obtain an estimate for $S_X(f)$ that goes to zero with the number of observations $k$ by taking the average of $N$ *independent* periodograms on samples of size $k$:

$$\langle \tilde{p}_k(f) \rangle_N = \frac{1}{N} \sum_{i=1}^{N} \tilde{p}_{k,i}(f), \tag{7.116}$$

where $\{\tilde{p}_{k,i}(f)\}$ are $N$ independent periodograms computed using separate sets of $k$ samples each. Figures 7.13 and 7.14 show the $N = 10$ and $N = 50$ smoothed periodograms corresponding to the unsmoothed periodogram of Fig. 7.12. It is evident that the variance of the power spectrum estimates is decreasing with $N$.

The mean of the smoothed estimator is

$$\begin{aligned} E\langle \tilde{p}_k(f) \rangle_N &= \frac{1}{N} \sum_{i=1}^{N} E[\tilde{p}_{k,i}(f)] = E[\tilde{p}_k(f)] \\ &= \sum_{m'=-(k-1)}^{k-1} \left\{1 - \frac{|m'|}{k}\right\} R_X(m') e^{-j2\pi f m'}, \end{aligned} \tag{7.117}$$

where we have used Eq. (7.32). Thus the smoothed estimator has the same mean as the periodogram estimate on a sample of size $k$.

The variance of the smoothed estimator is

$$\begin{aligned} \text{VAR}[\langle \tilde{p}_k(f) \rangle_N] &= \frac{1}{N^2} \sum_{i=1}^{N} \text{VAR}[\tilde{p}_{k,i}(f)] \\ &= \frac{1}{N} \text{VAR}[\tilde{p}_k(f)] \\ &= \frac{1}{N} S_X(f)^2. \end{aligned}$$

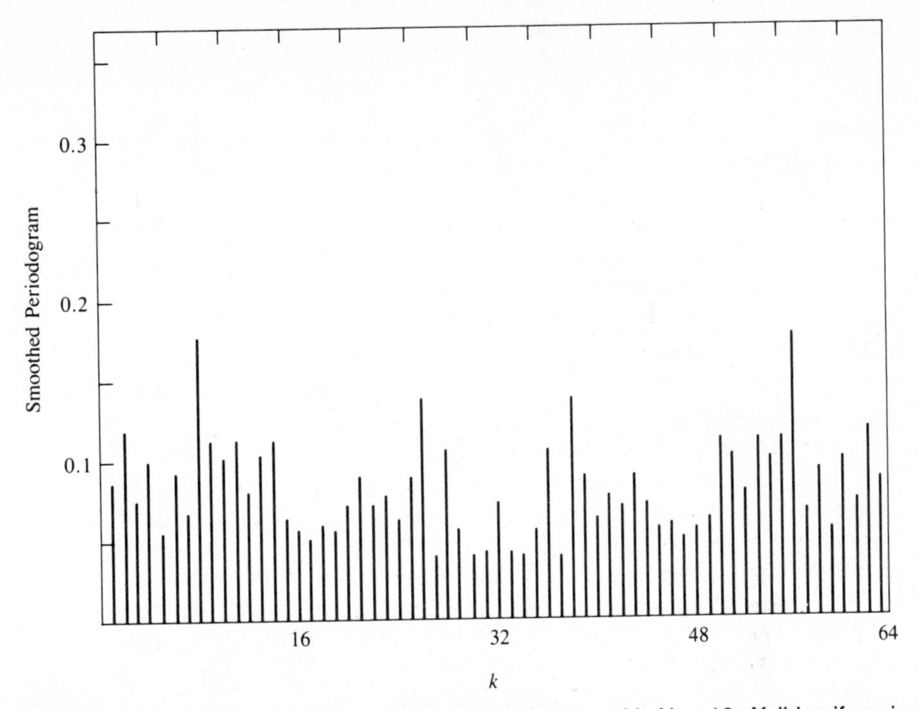


**FIGURE 7.13** Sixty-four point smoothed periodogram with $N = 10$, $X_n$ iid uniform in $(0, 1)$, $S_X(f) = 1/12 = 0.083$.

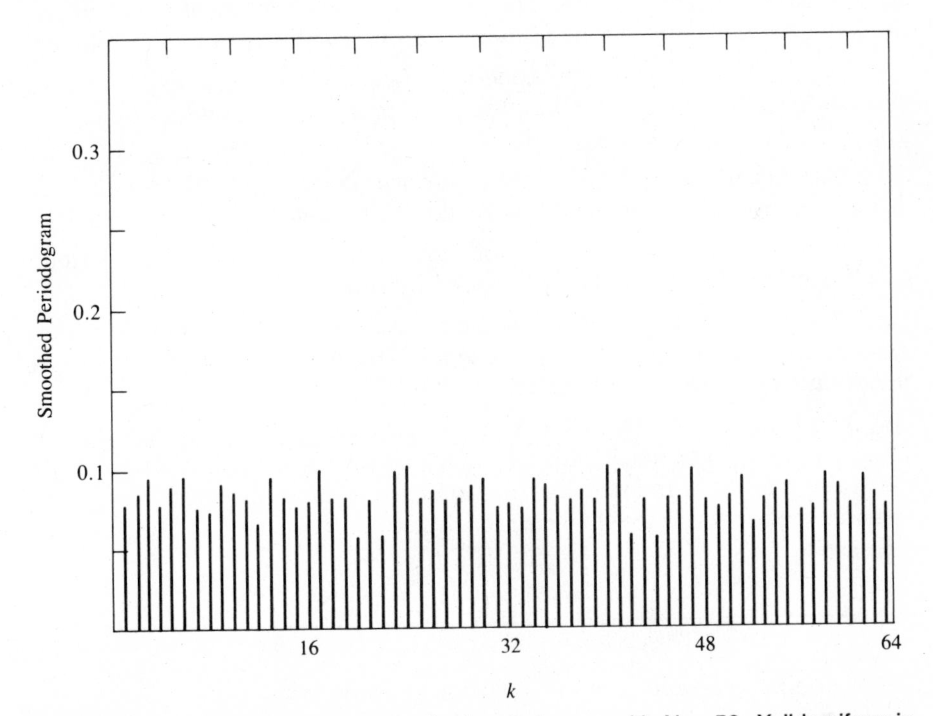


**FIGURE 7.14** Sixty-four point smoothed periodogram with $N = 50$, $X_n$ iid uniform in $(0, 1)$, $S_X(f) = 1/12 = 0.083$.

Thus the variance of the smoothed estimator can be reduced by increasing $N$, the number of periodograms used in Eq. (7.116).

In practice, a sample set of size $Nk$, $X_0, \ldots, X_{N_{k-1}}$, is divided into $N$ blocks and a separate periodogram is computed for each block. The smoothed estimate is then the average over the $N$ periodograms. This method is called **Bartlett's smoothing procedure.** Note that, in general, the resulting periodograms are not independent because the underlying blocks are not independent. Thus this smoothing procedure must be viewed as an approximation to the computation and averaging of independent periodograms.

The choice of $k$ and $N$ are determined by the desired frequency resolution and variance of the estimate. The blocksize $k$ determines the number of frequencies for which the spectral density is computed (i.e., the frequency resolution). The variance of the estimate is controlled by the number of periodograms $N$. The actual choice of $k$ and $N$ depend on the nature of the signal being investigated.

## SUMMARY

- The power spectral density of a WSS process is the Fourier transform of its autocorrelation function. The power spectral density of a real-valued random process is a real-valued, nonnegative, even function of frequency.
- The output of a linear, time-invariant system is a WSS random process if its input is a WSS random process that is applied an infinite time in the past.
- The output of a linear, time-invariant system is a Gaussian WSS random process if its input is a Gaussian WSS random process.
- Wide-sense stationary random processes with arbitrary rational power spectral density can be generated by filtering white noise.
- The orthogonality condition can be used to obtain equations for linear systems that minimize mean square error. These systems arise in filtering, smoothing, and prediction problems.
- The variance of the periodogram estimate for the power spectral density does not approach zero as the number of samples is increased. An average of several independent periodograms is required to obtain an estimate whose variance does approach zero as the number of samples is increased.

## CHECKLIST OF IMPORTANT TERMS

Amplitude modulation
ARMA process
Autoregressive process
Causal system

Cross-power spectral density
Filtering
Impulse response
Linear system
Moving average process
Periodogram
Power spectral density
Prediction
Smoothed periodogram
Smoothing
Transfer function
White noise

## ANNOTATED REFERENCES

References [1] through [5] contain good discussions of the notion of power spectral density and of the response of linear systems to random inputs. References [5] and [6] give accessible introductions to the spectral factorization problem. References [6] through [8] discuss linear filtering and power spectrum estimation in the context of digital signal processing. Reference [9] discusses the basic theory underlying power spectrum estimation.

1. A. Papoulis, *Probability, Random Variables, and Stochastic Processes,* McGraw-Hill, New York, 1965.
2. W. B. Davenport, *Probability and Random Processes: An Introduction for Applied Scientists and Engineers,* McGraw-Hill, New York, 1970.
3. H. Stark and J. W. Woods, *Probability, Random Processes, and Estimation Theory for Engineers,* Prentice-Hall, Englewood Cliffs, N.J., 1986.
4. R. M. Gray and L. D. Davisson, *Random Processes, A Mathematical Approach for Engineers,* Prentice-Hall, Englewood Cliffs, N.J., 1986.
5. G. R. Cooper and C. D. MacGillem, *Probabilistic Methods of Signal and System Analysis,* Holt, Reinhart & Winston, New York, 1986.
6. J. A. Cadzow, *Foundations of Digital Signal Processing and Data Analysis,* Macmillan, New York, 1987.
7. N. S. Jayant and P. Noll, *Digital Coding of Waveforms,* Prentice-Hall, Englewood Cliffs, N.J., 1984.
8. M. Kunt, *Digital Signal Processing,* Artech House, Dedham, Mass., 1986.
9. G. M. Jenkins and D. G. Watts, *Spectral Analysis and Its Applications,* Holden-Day, San Francisco, 1968.
10. A. Einstein, "Method for the Determination of the Statistical Values of Observations Concerning Quantities Subject to Irregular Observations," reprinted in *IEEE ASSP Magazine,* October 1987, p. 6.

## PROBLEMS

### Section 7.1
### Power Spectral Density

1. Let $g(x)$ denote the triangular function shown in Fig. P7.1.
   a. Find the power spectral density corresponding to $R_X(\tau) = g(\tau/T)$.
   b. Find the autocorrelation corresponding to the power spectral density $S_X(f) = g(f/W)$.

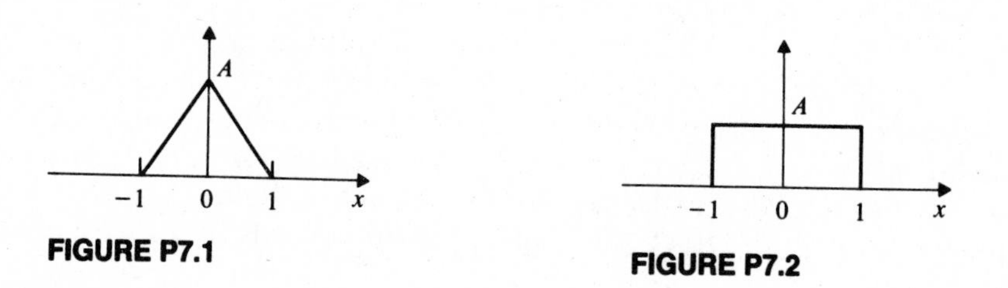


FIGURE P7.1

FIGURE P7.2

2. Let $p(x)$ be the rectangular function shown in Fig. P7.2. Is $R_X(\tau) = p(\tau/T)$ a valid autocorrelation function?
3. Find the power spectral density $S_Y(f)$ of a random process with autocorrelation function $R_X(\tau)\cos(2\pi f_0 \tau)$, where $R_X(\tau)$ is itself an autocorrelation function.
4. Find the autocorrelation function corresponding to the power spectral density shown in Fig. P7.3.

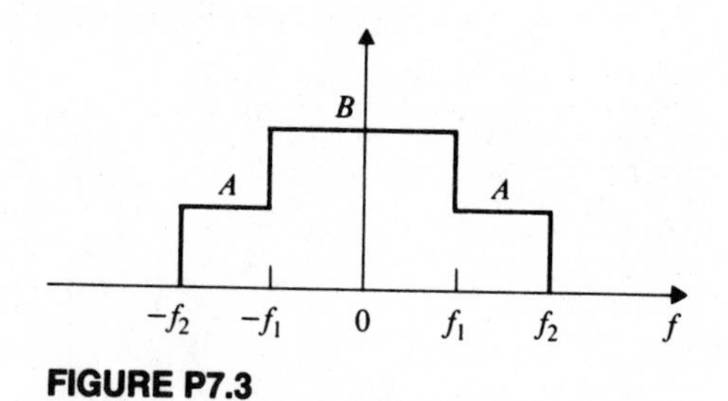


FIGURE P7.3

5. Let $Z(t) = X(t) + Y(t)$. Under what conditions does $S_Z(f) = S_X(f) + S_Y(f)$?
6. Show that
   a. $R_{X,Y}(\tau) = R_{Y,X}(-\tau)$.
   b. $S_{X,Y}(f) = S^*_{Y,X}(f)$.

7. Let $Y(t) = X(t) - X(t - d)$.
   a. Find $R_{X,Y}(\tau)$ and $S_{X,Y}(f)$.
   b. Find $R_Y(\tau)$ and $S_Y(f)$.
8. Let $X(t)$ and $Y(t)$ be independent wide-sense stationary random processes, and define

   $$Z(t) = X(t)Y(t).$$

   a. Show that $Z(t)$ is wide-sense stationary.
   b. Find $R_Z(\tau)$ and $S_Z(f)$.
9. Let $R_X(k) = 4(1/2)^{|k|} + 16(1/4)^{|k|}$. Find $S_X(f)$.
10. Let $R_X(k) = (1 - |k|/N)$, for $|k| < N$ and 0 elsewhere. Find $S_X(f)$.
11. Let $X_n = \cos(2\pi f_0 n + \Theta)$, where $\Theta$ is a uniformly distributed random variable in the interval $(0, 2\pi)$. Find $S_X(f)$.
12. Let $D_n = X_n - X_{n-d}$, where $d$ is an integer constant, and $X_n$ is a zero-mean, WSS random process.
    a. Find $R_D(k)$ and $S_D(f)$ in terms of $R_X(k)$ and $S_X(f)$.
    b. Find $E[D_n^2]$.
13. Find $R_D(k)$ and $S_D(f)$ in Problem 12 if $X_n$ is the moving average process of Example 7.7 with $\alpha = 1$.
14. Let $X_n$ be a zero-mean, bandlimited white noise random process with $S_X(f) = 1$ for $|f| < f_c$ and 0 elsewhere, where $f_c < 1/2$.
    a. Find $R_X(k)$.
    b. Find $R_X(k)$ when $f_c = 1/4$.
15. Let $W_n$ be a zero-mean white noise sequence, and let $X_n$ be a random process that is independent of $W_n$.
    a. Show that $Y_n = W_n X_n$ is a white sequence, and find $\sigma_Y^2$.
    b. Suppose $X_n$ is a Gaussian random process with autocorrelation $R_X(k) = (1/2)^{|k|}$. Specify the joint pmf's for $Y_n$.
16. Evaluate the periodogram estimate for the random process $X(t) = a\cos(2\pi f_0 t + \Theta)$, where $\Theta$ is a uniformly distributed random variable in the interval $(0, 2\pi)$. What happens as $T \to \infty$?

### Section 7.2
### Response of Linear Systems to Random Signals

17. Let $X(t)$ be a differentiable WSS random process, and define

    $$Y(t) = \frac{d}{dt}X(t).$$

    Find an expression for $S_Y(f)$ and $R_Y(\tau)$. *Hint*: For this system, $H(f) = j2\pi f$.

18. Let $Y(t)$ be the derivative of $X(t)$, a bandlimited white noise process as in Example 7.3.
    a. Find $S_Y(f)$ and $R_Y(\tau)$.
    b. What is the average power of the output?
19. Let $Y(t)$ be a short-term integration of $X(t)$:

$$Y(t) = \frac{1}{T}\int_{t-T}^{t} X(t')\,dt'.$$

Find $S_Y(f)$ in terms of $S_X(f)$.

20. In Problem 19, let $R_X(\tau) = (1 - |\tau|/T)$ for $|\tau| < T$ and zero elsewhere.
    a. Find $S_Y(f)$.
    b. Find $E[Y^2(t)]$.
21. The input into a filter is zero-mean white noise with noise power density $N_0/2$. The filter has transfer function:

$$H(f) = \frac{1}{1 + j2\pi f}.$$

    a. Find $S_{Y,X}(f)$ and $R_{Y,X}(\tau)$.
    b. Find $S_Y(f)$ and $R_Y(\tau)$.
    c. What is the average power of the output?
22. A WSS process $X(t)$ is applied to a linear system at $t = 0$. Find the mean and autocorrelation function of the output process. Show that the output process becomes WSS as $t \to \infty$.
23. Let $Y(t)$ be the output of a linear system with impulse response $h(t)$ and input $X(t)$. Find $R_{Y,X}(\tau)$ when the input is white noise. Explain how this result can be used to estimate the impulse response of a linear system.
24. A WSS Gaussian random process $X(t)$ is applied to two linear systems as shown in Fig. P7.4. Find the joint pdf of $Y(t_1)$ and $W(t_2)$.

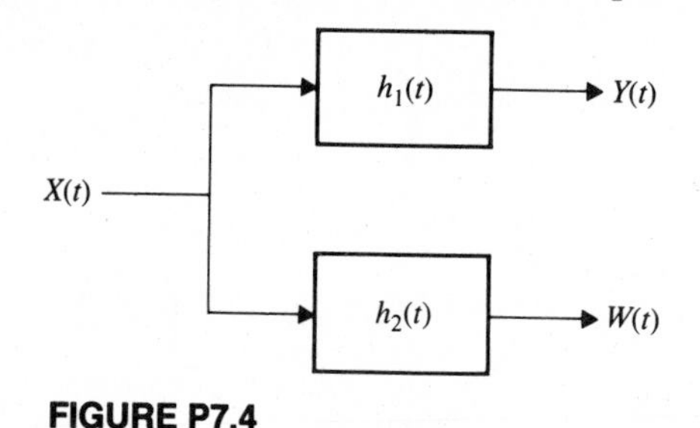


**FIGURE P7.4**

25. Let $Y(t) = h(t) * X(t)$ and $Z(t) = X(t) - Y(t)$ as shown in Fig. P7.5.

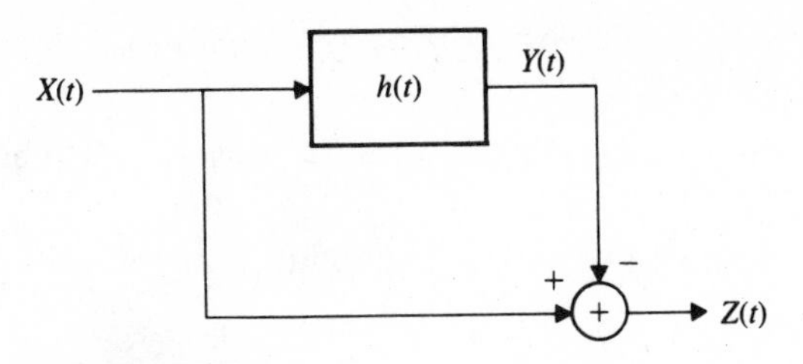


FIGURE P7.5

a. Find $S_Z(f)$ in terms of $S_X(f)$.
b. Find $E[Z^2(t)]$.

26. Let $Y(t)$ be the output of a linear system with impulse response $h(t)$ and input $X(t) + N(t)$. Let $Z(t) = X(t) - Y(t)$.
 a. Find $S_Z(f)$.
 b. Find $S_Z(f)$ if we are given that $X(t)$ and $N(t)$ are independent random processes.
27. A random telegraph signal is passed through an ideal lowpass filter with cutoff frequency $W$. Find the power spectral density of the difference between the input and output of the filter. Find the average power of the difference signal.
28. Let $Y(t) = a\cos(2\pi f_c t + \Theta) + N(t)$ be passed through an ideal bandpass filter that passes the frequencies $|f - f_c| < W/2$. Assume that $\Theta$ is uniformly distributed in $(0, 2\pi)$. Find the ratio of signal power to noise power at the output of the filter.
29. A zero-mean white noise sequence is input into a cascade of two systems (see Fig. P7.6). System 1 has impulse response $h_n = (1/2)^n u(n)$ and System 2 has impulse response $g_n = (1/4)^n u(n)$ where $u(n) = 1$ for $n \geq 0$ and 0 elsewhere.
 a. Find $S_Y(f)$ and $S_Z(f)$.
 b. Find $R_{W,Y}(k)$ and $R_{W,Z}(k)$ and $S_{W,Y}(f)$ and $S_{W,Z}(k)$.
 c. Find $E[Z_n^2]$.

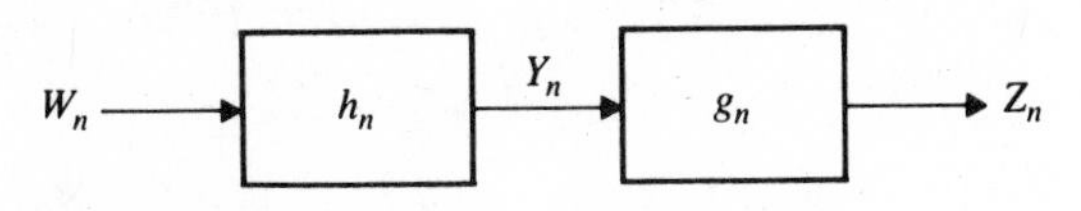


FIGURE P7.6

30. Let $Y_n = X_n + \beta X_{n-1}$ where $X_n$ is a zero-mean, first-order autoregressive process with autocorrelation $R_X(k) = \sigma^2\alpha^k$, $|\alpha| < 1$.
 a. Find $R_{Y,X}(k)$ and $S_{Y,X}(f)$.
 b. Find $S_Y(f)$, $R_Y(k)$ and $E[Y_n^2]$.
 c. For what value of $\beta$ is $Y_n$ a white noise process.

31. Let $Y_n = (X_{n+1} + X_n + X_{n-1})/3$ be a "smoothed" version of $X_n$. Find $R_Y(k)$, $S_Y(f)$, and $E[Y_n^2]$.
32. Suppose $X_n$ is a Gaussian white noise process in Problem 31. Find the joint pmf for $(Y_n, Y_{n+1}, Y_{n+2})$.
33. A moving average process $X_n$ is produced as follows:

$$X_n = W_n + \alpha_1 W_{n-1} + \cdots + \alpha_p W_{n-p},$$

where $W_n$ is a zero-mean, white noise process.

a. Show that $R_X(k) = 0$ for $|k| > p$.
b. Find $R_X(k)$ by computing $E[X_{n+k}X_n]$, then find $S_X(f) = \mathcal{F}\{R_X(k)\}$.
c. Find the impulse response $h_n$ of the linear system that defines the moving average process. Find the corresponding transfer function $H(f)$, and then $S_X(f)$. Compare your answer to Part b.

34. Consider the second-order autoregressive process defined by

$$Y_n = \frac{3}{4}Y_{n-1} - \frac{1}{8}Y_{n-2} + W_n,$$

where $W_n$ is a zero-mean, white noise process.

a. Verify that the unit-sample response is $h_n = 2(1/2)^n - (1/4)^n$ for $n \geq 0$, and 0 otherwise.
b. Find the transfer function.
c. Find $S_Y(f)$ and $R_Y(k) = \mathcal{F}^{-1}\{S_Y(f)\}$.

35. Suppose the autoregressive process defined in Problem 34 is the input to the following moving average system:

$$Z_n = Y_n - \frac{1}{4}Y_{n-1}.$$

a. Find $S_Z(f)$ and $R_Z(k)$.
b. Explain why $Z_n$ is a first-order autoregressive process.
c. Find a moving average system that will produce a white noise sequence when $Z_n$ is the input.

36. An autoregressive process $Y_n$ is produced as follows:

$$Y_n = \alpha_1 Y_{n-1} + \cdots + \alpha_q Y_{n-q} + W_n,$$

where $W_n$ is a zero-mean, white noise process.

a. Show that the autocorrelation of $Y_n$ satisfies the following set of equations:

$$R_Y(0) = \sum_{i=1}^{q} \alpha_i R_Y(i) + R_W(0)$$

$$R_Y(k) = \sum_{i=1}^{q} \alpha_i R_Y(k - i).$$

b. Use these recursive equations to compute the autocorrelation of the process in Example 7.20.

**Section 7.3**
**Amplitude Modulation by Random Signals**

37. Plot the power spectral density of the amplitude modulated signal $Y(t)$ in Example 7.16 assuming $f_c > W$ and $f_c < W$.

38. Suppose that a random telegraph signal with transition rate $\alpha$ is the input signal in an amplitude modulation system. Plot the power spectral density of the modulated signal assuming $f_c = \alpha/\pi$ and $f_c = 10\alpha/\pi$.

39. Let the input to an amplitude modulation system be $2\cos(2\pi f_1 t + \Phi)$, where $\Phi$ is uniformly distributed in $(-\pi, \pi)$. Find the power spectral density of the modulated signal assuming $f_c > f_1$.

40. Find the signal-to-noise ratio in the recovered signal in Example 7.16 if $S_N(f) = \alpha f^2$ for $|f + f_c| < W$ and zero elsewhere.

41. The input signals to a QAM system are independent random processes with power spectral densities shown in Fig. P7.7.
Sketch the power spectral density of the QAM signal.

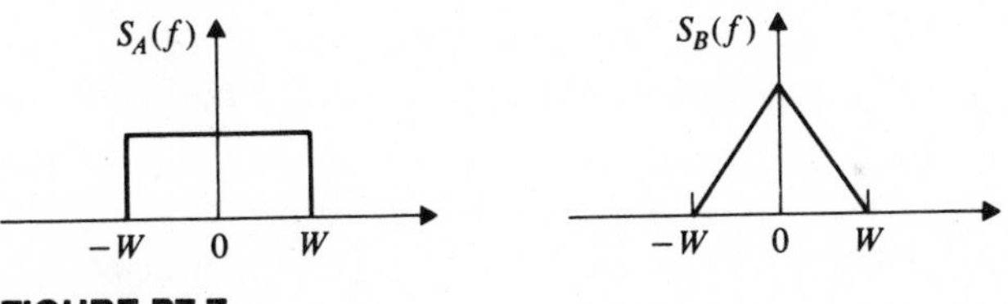

**FIGURE P7.7**

42. Under what conditions does the receiver shown in Fig. P7.8 recover the input signals to a QAM signal?

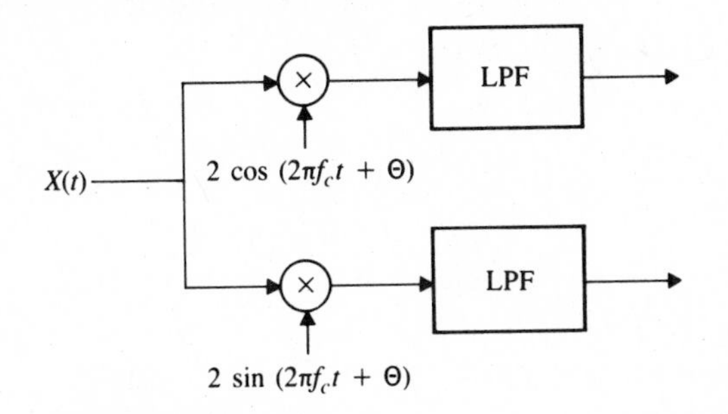


**FIGURE P7.8**

43. Show that Equation (7.61b) implies that $S_{BA}(f)$ is a purely imaginary, odd function of $f$.

## Section 7.4
## Optimum Linear Filters

44. Let $X_\alpha = Z_\alpha + N_\alpha$ as in Example 7.20, where $Z_\alpha$ is a first-order autoregressive process with $R_Z(k) = 4(1/2)^{|k|}$ and $N_\alpha$ is white noise with $\sigma_N^2 = 1$.
    a. Find the optimum $p = 1$ filter for estimating $Z_\alpha$.
    b. Find the mean square error of the resulting filter.
45. Let $X_\alpha = Z_\alpha + N_\alpha$ as in Example 7.19, where $Z_\alpha$ has $R_Z(k) = \sigma_Z^2(r_1)^{|k|}$ and $N_\alpha$ has $R_N(k) = \sigma_N^2 r_2^{|k|}$ where $r_1$ and $r_2$ are less than one in magnitude.
    a. Find the equation for the optimum filter for estimating $Z_\alpha$.
    b. Write the matrix equation for the filter coefficients.
    c. Solve the $p = 2$ case, if $\sigma_Z^2 = 4$, $r_1 = 1/2$, $\sigma_N^2 = 1$, $r_2 = 1/10$.
    d. Find the mean square error for the optimum filter in Part c.
46. Let $X_\alpha = Z_\alpha + N_\alpha$ as in Example 7.19, where $Z_\alpha$ is the first-order moving average process of Example 7.7, and $N_\alpha$ is white noise.
    a. Find the equation for the optimum filter for estimating $Z_\alpha$.
    b. For the $p = 2$ case, write and solve the matrix equation for the filter coefficients.
    c. Find the mean square error for the optimum filter in Part c.
47. Let $X_\alpha = Z_\alpha + N_\alpha$ as in Example 7.17, and suppose that an estimator for $Z_\alpha$ uses observations from the following times instants $I = \{n - p, \ldots, n, \ldots, n + p\}$.
    a. Find the equation for the optimum filter.
    b. Write the matrix equation for the $2p + 1$ filter coefficients.
    c. Solve the $p = 1$ case in Part *b* if $Z_\alpha$ and $N_\alpha$ are as in Problem 44.
    d. Find the mean square error in Part c.
48. Consider the predictor introduced in Example 7.21.
    a. Find the optimum predictor coefficients in the $a = 2$ case when $R_Z(k) = 4(1/2)^{|k|}$.
    b. Find the mean square error in Part b.
49. Let $X(t)$ be a WSS, continuous-time process.
    a. Use the orthogonality principle to find the best estimator for $X(t)$ of the form

    $$\hat{X}(t) = aX(t_1) + bX(t_2),$$

    where $t_1$ and $t_2$ are two given time instants.
    b. Find the mean square error of the optimum estimator.
50. Find the optimum filter and its mean square error in Problem 49 if $t_1 = t - d$ and $t_2 = t + d$.

51. Find the optimum filter and its mean square error in Problem 49 if $t_1 = t - d$ and $t_2 = t - 2d$, and $R_X(\tau) = e^{-\alpha|\tau|}$. Compare the performance of this filter to the performance of the optimum filter of the form $\hat{X}(t) = aX(t - d)$.

52. Show that the mean square error for the optimum infinite smoothing filter in Example 7.22 is

$$E[e^2(t)] = \int_{-\infty}^{\infty} \frac{S_Z(f)S_N(f)}{S_Z(f) + S_N(f)} df$$

where $e(t) = Z(t) - Y(t)$. *Hint*: $E[e^2(t)] = R_e(0)$.

53. Solve the infinite-smoothing problem in Example 7.22 if $Z(t)$ is the random telegraph signal and $N(t)$ is white noise. What is the resulting mean square error?

54. Solve the infinite-smoothing problem in Example 7.22 if $Z(t)$ is bandlimited white noise of density $N_1/2$ and $N(t)$ is (infinite-bandwidth) white noise of noise density $N_0/2$. What is the resulting mean square error?

55. Solve the infinite-smoothing problem in Example 7.22 if $Z(t)$ and $N(t)$ are as given in Example 7.23. Find the resulting mean square error.

56. Let $X_n = Z_n + N_n$, where $Z_n$ and $N_n$ are independent, zero-mean random processes.
    a. Find the smoothing filter given by Eq. (7.84) when $Z_n$ is a first-order autoregressive process and $N_n$ is white noise.
    b. Find an expression for the resulting mean square error.

57. Find the Wiener filter in Example 7.23 if $N(t)$ is white noise of noise density $N_0/2$ and $Z(t)$ has power spectral density

$$S_Z(f) = \frac{4\alpha}{4\alpha^2 + 4\pi^2 f^2}.$$

58. Find the mean square error for the Wiener filter found in Example 7.23. Compare this to the mean square error of the infinite-smoothing filter found in Problem 55.

59. Suppose we wish to estimate (predict) $X(t + d)$ by

$$\hat{X}(t + d) = \int_0^{\infty} h(\tau)X(t - \tau)\, d\tau.$$

    a. Show that the optimum filter must satisfy

$$R_X(\tau + d) = \int_0^{\infty} h(x)R_X(\tau - x)\, dx \qquad \tau \geq 0.$$

b. Use the Wiener-Hopf method to find the optimum filter when $R_X(\tau) = e^{-\alpha|\tau|}$.

*60. Let $X_n = Z_n + N_n$, where $Z_n$ and $N_n$ are independent random processes, $N_n$ is a white noise process with $\sigma_N^2 = 1$, and $Z_n$ is a first-order autoregressive process with $R_Z(k) = 4(1/2)^{|k|}$. Find the optimum filter for estimating $Z_n$ from $X_n, X_{n-1}, \ldots$.

## Section 7.5
## A Computer Method for Estimating the Power Spectral Density

61. Verify Eqs. (7.101) and (7.102).

62. Write a computer program that generates a sequence $X_n$ of iid random variables that are uniformly distributed in $(0, 1)$.
    a. Compute several 64-point periodograms and verify the random behavior of the periodogram as a function of $f$. Does the periodogram vary about the true power spectral density?
    b. Compute the smoothed periodogram based on 10, 20, and 50 independent periodograms. Compare the smoothed periodograms to the true power spectral density.

63. Repeat Problem 62 with $X_n$ a first-order autoregressive process with autocorrelation:
    a. $R_X(k) = (.9)^{|k|}$.
    b. $R_X(k) = (1/2)^{|k|}$.
    c. $R_X(k) = (.1)^{|k|}$.

64. Consider the following estimator for the autocorrelation function

$$\hat{r}'_k(m) = \frac{1}{k - |m|} \sum_{n=0}^{k-|m|-1} X_n X_{n+m}.$$

Show that if we estimate the power spectrum of $X_n$ by the Fourier transform of $\hat{r}'_k(m)$, the resulting estimator has mean

$$E[\tilde{p}_k(f)] = \sum_{m'=-(k-1)}^{k-1} R_X(m') e^{-j2\pi f m'}.$$

Why is the estimator biased?

CHAPTER 8

# Markov Chains

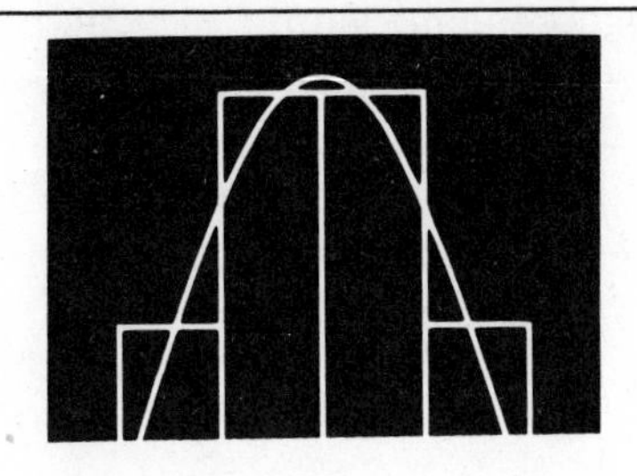

In general, the random variables within the family defining a stochastic process are not independent, and in fact can be statistically dependent in very complex ways. In this chapter we introduce the class of Markov random processes that have a simple form of dependence and that are quite useful in modeling problems found in practice. We concentrate on integer-valued Markov processes, which are called Markov chains. We develop methods for finding the transient and the long-term behavior of these random processes. In Chapter 9, we see that Markov chains form the basis for many queueing system models.

## 8.1

## MARKOV PROCESSES

A random process $X(t)$ is a **Markov process** if the future of the process given the present is independent of the past, that is, if for arbitrary times $t_1 < t_2 < \cdots < t_k < t_{k+1}$;

$$\begin{aligned} P[X(t_{k+1}) &= x_{k+1} \mid X(t_k) = x_k, \ldots, X(t_1) = x_1] \\ &= P[X(t_{k+1}) = x_{k+1} \mid X(t_k) = x_k] \end{aligned} \tag{8.1}$$

if $X(t)$ is discrete-valued, and

$$\begin{aligned} P[a < X(t_{k+1}) \le b \mid X(t_k) &= x_k, \ldots, X(t_1) = x_1] \\ &= P[a < X(t_{k+1}) \le b \mid X(t_k) = x_k] \end{aligned} \tag{8.2.a}$$

if $X(t)$ is continuous-valued. If the samples of $X(t)$ are jointly continuous, then Eq. (8.2a) is equivalent to

$$f_{X(t_{k+1})}(x_{k+1} \mid X(t_k) = x_k, \ldots, X(t_1) = x_1) = f_{X(t_{k+1})}(x_{k+1} \mid X(t_k) = x_k). \tag{8.2b}$$

We refer to Eqs. (8.1) and (8.2) as the **Markov property.** In the above expression $t_k$ is the "present," $t_{k+1}$ is the "future," and $t_1, \ldots, t_{k-1}$ is the "past." Thus in Markov processes, pmf's and pdf's that are conditioned on several time instants always reduce to a pmf/pdf that is conditioned only on the most recent time instant. For this reason we refer to the value of $X(t)$ at time $t$ as the **state** of the process at time $t$.

### Example 8.1

Consider the sum process discussed in Section 6.3:

$$S_n = X_1 + X_2 + \cdots + X_n = S_{n-1} + X_n,$$

where the $X_i$'s are an iid sequence of random variables and where $S_0 = 0$.

$S_n$ is Markov process since

$$\begin{aligned} P[S_{n+1} = s_{n+1} \mid S_n = s_n, \ldots, S_1 = s_1] &= P[X_{n+1} = s_{n+1} - s_n] \\ &= P[S_{n+1} = s_{n+1} \mid S_n = s_n]. \end{aligned}$$

■■

### ■■ Example 8.2

Consider the moving average of a Bernoulli sequence:

$$Y_n = \frac{1}{2}(X_n + X_{n-1}),$$

where the $X_i$ are an independent Bernoulli sequence with $p = 1/2$. We now show that $Y_n$ is not a Markov process.

The pmf of $Y_n$ is

$$P[Y_n = 0] = P[X_n = 0, X_{n-1} = 0] = \frac{1}{4},$$

$$P\left[Y_n = \frac{1}{2}\right] = P[X_n = 0, X_{n-1} = 1] + P[X_n = 1, X_{n-1} = 0] = \frac{1}{2},$$

and

$$P[Y_n = 1] = P[X_n = 1, X_{n-1} = 1] = \frac{1}{4}.$$

Now consider the following conditional probability for two consecutive values of $Y_n$:

$$\begin{aligned} P\left[Y_n = 1 \mid Y_{n-1} = \frac{1}{2}\right] &= \frac{P[Y_n = 1, Y_{n-1} = 1/2]}{P[Y_{n-1} = 1/2]} \\ &= \frac{P[X_n = 1, X_{n-1} = 1, X_{n-2} = 0]}{1/2} = \frac{(1/2)^3}{1/2} = \frac{1}{4}. \end{aligned}$$

Now suppose we have additional knowledge about the past:

$$P\left[Y_n = 1 \mid Y_{n-1} = \frac{1}{2}, Y_{n-2} = 1\right] = \frac{P[Y_n = 1, Y_{n-1} = 1/2, Y_{n-2} = 1]}{P[Y_{n-1} = 1/2, Y_{n-2} = 1]} = 0,$$

since no sequence of $X_n$'s leads to the sequence 1, 1/2, 1. Thus

$$P\left[Y_n = 1 \mid Y_{n-1} = \frac{1}{2}, Y_{n-2} = 1\right] \neq P\left[Y_n = 1 \mid Y_{n-1} = \frac{1}{2}\right],$$

and the process is not Markov. ■■

### Example 8.3

The Poisson process is a continuous-time Markov process since

$$\begin{aligned}P[N(t_{k+1}) = j \mid N(t_k) = i, N(t_{k-1}) = x_{k-1}, \ldots, N(t_1) = x_1] \\ = P[j - i \text{ events in } t_{k+1} - t_k \text{ seconds}] \\ = P[N(t_{k+1}) = j \mid N(t_k) = i].\end{aligned}$$

### Example 8.4

The random telegraph signal of Example 6.22 is a continuous-time Markov process since

$$\begin{aligned}P[X(t_{k+1}) = a \mid X(t_k) = b, \ldots, X(t_1) = x_1] \\ = P[\text{even (odd) number of jumps in } t_{k+1} \\ - t_k \text{ seconds if } a = b (a \neq b)] \\ = P[X(t_{k+1}) = a \mid X(t_k) = b]\end{aligned}$$

An integer-valued Markov random process is called a **Markov chain.** In the remainder of this chapter we concentrate on Markov chains.

If $X(t)$ is a Markov chain, then the joint pmf for three arbitrary time instants is

$$\begin{aligned}P[X(t_3) = x_3, X(t_2) = x_2, X(t_1) = x_1] \\ = P[X(t_3) = x_3 \mid X(t_2) = x_2, X(t_1) = x_1]P[X(t_2) = x_2, X(t_1) = x_1] \\ = P[X(t_3) = x_3 \mid X(t_2) = x_2]P[X(t_2) = x_2, X(t_1) = x_1] \\ = P[X(t_3) = x_3 \mid X(t_2) = x_2]P[X(t_2) = x_2 \mid X(t_1) = x_1]P[X(t_1) = x_1],\end{aligned}$$

where we have used the definition of conditional probability and the Markov property. In general, the joint pmf for $k + 1$ arbitrary time instants is

$$\begin{aligned}P[X(t_{k+1}) = x_{k+1}, X(t_k) = x_k, \ldots, X(t_1) = x_1] \\ = P[X(t_{k+1}) = x_{k+1} \mid X(t_k) = x_k]P[X(t_k) = x_k \mid X(t_{k-1}) = x_{k-1}] \cdots \\ \times P[X(t_2) = x_2 \mid X(t_1) = x_1]P[X(t_1) = x_1]. \quad (8.3)\end{aligned}$$

Thus the *joint pmf of $X(t)$ at arbitrary time instants is given by the product of the pmf of the initial time instant and the probabilities for the subsequent state transitions*. Clearly, the state transition probabilities determine the statistical behavior of a Markov chain.

## 8.2

## DISCRETE-TIME MARKOV CHAINS

Let $X_n$ be a discrete-time integer-valued Markov chain that starts at $n = 0$ with pmf

$$p_j(0) \triangleq P[X_0 = j] \qquad j = 0, 1, 2, \ldots. \tag{8.4}$$

From Eq. (8.3), the joint pmf for the first $n + 1$ values of the process is

$$P[X_n = i_n, \ldots, X_0 = i_0]$$
$$= P[X_n = i_n \mid X_{n-1} = i_{n-1}] \cdots P[X_1 = i_1 \mid X_0 = i_0] P[X_0 = i_0]. \tag{8.5}$$

Thus the joint pmf for a particular sequence is simply the product of the probability for the initial state and the probabilities for the subsequent one-step state transitions.

We will assume that the one-step state transition probabilities are fixed and do not change with time, that is,

$$P[X_{n+1} = j \mid X_n = i] = p_{ij} \qquad \text{for all } n. \tag{8.6}$$

$X_n$ is said to be **homogeneous** in time. The joint pmf for $X_n, \ldots, X_0$ is then given by

$$P[X_n = i_n, \ldots, X_0 = i_0] = p_{i_{n-1}, i_n} \cdots p_{i_0, i_1} p_{i_0}(0) \tag{8.7}$$

Thus $X_n$ is completely specified by the *initial pmf* $p_i(0)$ and the *matrix of one-step transition probabilities* $P$:

$$P = \begin{bmatrix} p_{00} & p_{01} & p_{02} & \cdots \\ p_{10} & p_{11} & p_{12} & \cdots \\ \cdot & \cdot & \cdot & \\ p_{i0} & p_{i1} & \cdots & \\ \cdot & \cdot & \cdots & \end{bmatrix} \tag{8.8}$$

Note that each row of $P$ must add to one since

$$1 = \sum_j P[X_{n+1} = j \mid X_n = i] = \sum_j p_{ij}. \tag{8.9}$$

### ■■ Example 8.5
Two-State Markov Chain

A Markov model for packet speech assumes that if the $n$th packet contains silence, then the probability of silence in the next packet is $1 - \alpha$ and the probability of speech activity is $\alpha$. Similarly, if the $n$th packet contains

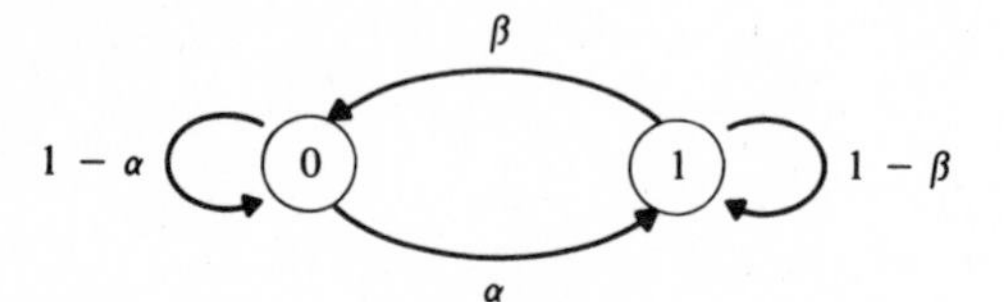

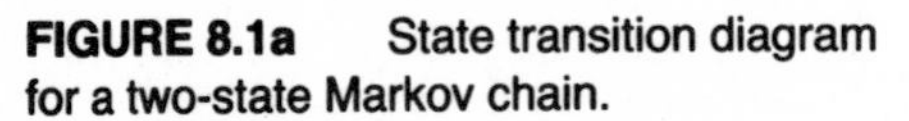

FIGURE 8.1a State transition diagram for a two-state Markov chain.

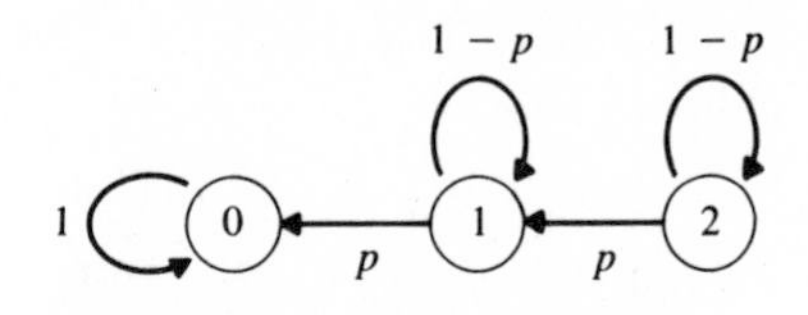


FIGURE 8.1b State transition diagram for Markov chain for light bulb inventory.

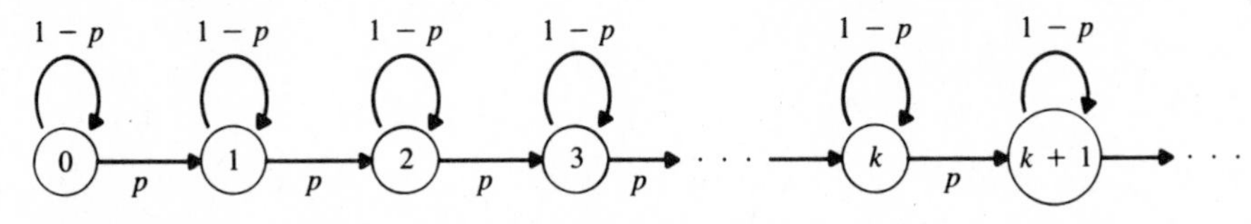


FIGURE 8.1c State transition diagram for binomial counting process.

speech activity, then the probability of speech activity in the next packet is $1 - \beta$ and the probability of silence is $\beta$.

Let $X_n$ be the indicator function for speech activity in a packet at time $n$, then $X_n$ is a two-state Markov chain with the state transition diagram shown in Fig. 8.1(a), and transition probability matrix

$$P = \begin{bmatrix} 1-\alpha & \alpha \\ \beta & 1-\beta \end{bmatrix}. \tag{8.10}$$

■■

## ■■ Example 8.6

On day 0 a house has two new light bulbs in reserve. The probability that the house will need a single new light bulb during day $n$ is $p$, and the probability that it will not need any is $q = 1 - p$. Let $Y_n$ be the number of new light bulbs left in the house at the end of day $n$. $Y_n$ is a Markov chain with state transition diagram shown in Fig. 8.1(b), and transition probability matrix

$$P = \begin{bmatrix} 1 & 0 & 0 \\ p & q & 0 \\ 0 & p & q \end{bmatrix}.$$

■■

## ■■ Example 8.7

Let $S_n$ be the binomial counting process introduced in Example 6.13. In one step, $S_n$ can either stay the same or increase by one. The state transition diagram is shown in Fig. 8.1(c), and the transition probability matrix is given by

$$P = \begin{bmatrix} 1-p & p & 0 & 0 & \cdots \\ 0 & 1-p & p & 0 & \cdots \\ 0 & 0 & 1-p & p & \cdots \\ \cdot & \cdot & \cdots & & \cdots \end{bmatrix}.$$

■■

### The *n*-step Transition Probabilities

To evaluate the joint pmf for arbitrary time instants [see Eq. (8.3)], we need to know the transition probabilities for an arbitrary number of steps. Let $P(n) = \{p_{ij}(n)\}$ *be the matrix of n-step transition probabilities,* where

$$p_{ij}(n) = P[X_{n+k} = j \mid X_k = i] \qquad n \geq 0,\ i,j \geq 0. \tag{8.11}$$

Note that $P[X_{n+k} = j \mid X_k = i] = P[X_n = j \mid X_0 = i]$ for all $n \geq 0$ and $k \geq 0$, since the transition probabilities do not depend on time.

First, consider the two-step transition probabilities. The probability of going from state $i$ at $t = 0$, passing through state $k$ at $t = 1$, and ending at state $j$ at $t = 2$ is

$$\begin{aligned} P[X_2 = j, X_1 = k \mid X_0 = i] &= \frac{P[X_2 = j, X_1 = k, X_0 = i]}{P[X_0 = i]} \\ &= \frac{P[X_2 = j \mid X_1 = k]P[X_1 = k \mid X_0 = i]P[X_0 = i]}{P[X_0 = i]} \\ &= P[X_2 = j \mid X_1 = k]P[X_1 = k \mid X_0 = i] \\ &= p_{ik}(1)p_{kj}(1). \end{aligned}$$

Note that $p_{ik}(1)$ and $p_{kj}(1)$ are components of $P$, the one-step transition probability matrix. We obtain $p_{ij}(2)$, the probability of going from $i$ at $t = 0$ to $j$ at $t = 2$, by summing over all possible intermediate states $k$:

$$p_{ij}(2) = \sum_k p_{ik}(1)p_{kj}(1) \qquad \text{for all } i, j. \tag{8.12}$$

The set of equations given by Eq. (8.12) state that the matrix $P(2)$ is obtained by multiplying the one-step transition probability matrices

$$P(2) = P(1)P(1) = P^2. \tag{8.13a}$$

Using the same arguments as above, we find that $P(n)$ is found by multiplying $P(n-1)$ by $P$:

$$P(n) = P(n-1)P. \tag{8.13b}$$

Equations (8.13a) and (8.13b) together imply that

$$P(n) = P^n, \tag{8.14}$$

that is, the *n-step transition probability matrix is the nth power of the one-step transition probability matrix.*

### The State Probabilities

Now consider the state probabilities at time $n$. Let $\mathbf{p}(n) = \{p_j(n)\}$ denote the row vector of state probabilities at time $n$. The probability $p_j(n)$ is related to $\mathbf{p}(n-1)$ by

$$\begin{aligned} p_j(n) &= \sum_i P[X_n = j \mid X_{n-1} = i] P[X_{n-1} = i] \\ &= \sum_i p_{ij} p_i(n-1). \end{aligned} \tag{8.15a}$$

Equation (8.15a) states that $\mathbf{p}(n)$ is obtained by multiplying the row vector $\mathbf{p}(n-1)$ by the matrix $P$:

$$\mathbf{p}(n) = \mathbf{p}(n-1)P. \tag{8.15b}$$

Similarly, $p_j(n)$ is related to $\mathbf{p}(0)$ by

$$\begin{aligned} p_j(n) &= \sum_i P[X_n = j \mid X_0 = i] P[X_0 = i] \\ &= \sum_i p_{ij}(n) p_i(0), \end{aligned} \tag{8.16a}$$

and in matrix notation

$$\mathbf{p}(n) = \mathbf{p}(0)P(n) = \mathbf{p}(0)P^n \qquad n = 1, 2, \ldots. \tag{8.16b}$$

Thus the *state pmf at time n is obtained by multiplying the initial state pmf by $P^n$.*

### ■■ Example 8.8

To find the $n$-step transition probability in Example 8.6, note that

$$p_{22}(n) = P[\text{no new light bulbs needed in } n \text{ days}] = q^n$$
$$p_{21}(n) = P[\text{1 light bulb needed in } n \text{ days}] = npq^{n-1}$$
$$p_{20}(n) = 1 - p_{22}(n) - p_{21}(n).$$

The other terms in $P(n)$ are found in similar fashion, thus

$$P(n) = \begin{bmatrix} 1 & 0 & 0 \\ 1 - q^n & q^n & 0 \\ 1 - q^n - npq^{n-1} & npq^{n-1} & q^n \end{bmatrix}.$$

Note that if $q < 1$ then, as $n \to \infty$,

$$P(n) \to \begin{bmatrix} 1 & 0 & 0 \\ 1 & 0 & 0 \\ 1 & 0 & 0 \end{bmatrix}.$$

As a result, the state pmf $\mathbf{p}(n) = (p_0(n), p_1(n), p_2(n))$ approaches

$$\begin{aligned} \mathbf{p}(n) &= (p_0(0), p_1(0), p_2(0))P(n) \\ &= (0, 0, 1)P(n) \\ &\to (0, 0, 1)\begin{bmatrix} 1 & 0 & 0 \\ 1 & 0 & 0 \\ 1 & 0 & 0 \end{bmatrix} = (1, 0, 0), \end{aligned}$$

where $(p_0(0), p_1(0), p_2(0))$ is the row vector of initial state probabilities and $(p_0(0), p_1(0), p_2(0)) = (0, 0, 1)$ since we start with two light bulbs. In words, the above equation states that we eventually run out of light bulbs!

■■

---

## ■■ Example 8.9

Let $\alpha = 1/10$ and $\beta = 1/5$ in Example 8.5. Find $P(n)$ for $n = 2$, 4, 8, and 16.

$$P^2 = \begin{bmatrix} .9 & .1 \\ .2 & .8 \end{bmatrix}^2 = \begin{bmatrix} .83 & .17 \\ .34 & .66 \end{bmatrix}$$

$$P^4 = \begin{bmatrix} .83 & .17 \\ .34 & .66 \end{bmatrix}^2 = \begin{bmatrix} .7467 & .2533 \\ .5066 & .4934 \end{bmatrix}$$

and similarly

$$P^8 = \begin{bmatrix} .6859 & .3141 \\ .6282 & .3718 \end{bmatrix} \qquad P^{16} = \begin{bmatrix} .6678 & .3322 \\ .6644 & .3356 \end{bmatrix}.$$

There is a clear trend here: It appears that as $n \to \infty$,

$$P^n \to \begin{bmatrix} 2/3 & 1/3 \\ 2/3 & 1/3 \end{bmatrix}.$$

Indeed it can be shown with a little linear algebra (Anton, 1981, p. 276) that

$$P^n = \frac{1}{\alpha + \beta}\begin{bmatrix} \beta & \alpha \\ \beta & \alpha \end{bmatrix} + \frac{(1 - \alpha - \beta)^n}{\alpha + \beta}\begin{bmatrix} \alpha & -\alpha \\ -\beta & \beta \end{bmatrix},$$

which clearly approaches

$$\frac{1}{\alpha + \beta}\begin{bmatrix} \beta & \alpha \\ \beta & \alpha \end{bmatrix} = \begin{bmatrix} 2/3 & 1/3 \\ 2/3 & 1/3 \end{bmatrix}.$$

■■

### ■■ Example 8.10

Let the initial state probabilities in the Example 8.9 be

$$P[X_0 = 0] = p_0(0) \quad \text{and} \quad P[X_0 = 1] = 1 - p_0(0).$$

Find the state probabilities as $n \to \infty$.

The state probability vector at time $n$ is

$$\mathbf{p}(n) = (p_0(0), 1 - p_0(0))P^n.$$

As $n \to \infty$, we have that

$$\mathbf{p}(n) \to (p_0(0), 1 - p_0(0))\begin{bmatrix} 2/3 & 1/3 \\ 2/3 & 1/3 \end{bmatrix} = \left[\frac{2}{3}, \frac{1}{3}\right].$$

We see that the state probabilities do not depend on the initial state probabilities as $n \to \infty$. ■■

---

### Steady State Probabilities

Example 8.10 is typical of Markov chains that settle into stationary behavior after the process has been running for a long time. As $n \to \infty$, the $n$-step transition probability matrix approaches a matrix in which all the rows are equal to the same pmf, that is,

$$p_{ij}(n) \to \pi_j \qquad \text{for all } i. \tag{8.17}$$

As $n \to \infty$, Eq. (8.16a) becomes

$$p_j(n) \to \sum_i \pi_j p_i(0) = \pi_j.$$

Thus as $n \to \infty$, the probability of state $j$ approaches a constant independent of time and of the initial state probabilities

$$p_j(n) \to \pi_j \qquad \text{for all } j. \tag{8.18}$$

We say that the system reaches "equilibrium" or "steady state."

We can find the pmf $\boldsymbol{\pi} \triangleq \{\pi_j\}$ in Eq. (8.18) (when it exists) by noting that as $n \to \infty$, $p_j(n) \to \pi_j$ and $p_i(n-1) \to \pi_i$ so Eq. (8.15a) approaches

$$\pi_j = \sum_i p_{ij}\pi_i, \tag{8.19a}$$

which in matrix notation is

$$\boldsymbol{\pi} = \boldsymbol{\pi}P. \tag{8.19b}$$

In general, Eq. (8.19b) has $n - 1$ linearly independent equations. The additional equation needed is provided by

$$\sum_i \pi_i = 1. \tag{8.19c}$$

We refer to $\boldsymbol{\pi}$ as the **stationary state pmf** of the Markov chain. If we start the Markov chain with initial state pmf $\mathbf{p}(0) = \boldsymbol{\pi}$, then by Eqs. (8.16b) and (8.19b) we have that the state probability vector

$$\mathbf{p}(n) = \boldsymbol{\pi}P^n = \boldsymbol{\pi} \qquad \text{for all } n.$$

The resulting process is a stationary random process as defined in Section 6.5, since the probability of the sequence of states $i_0, i_1, \ldots, i_n$ starting at time $k$ is, by Eq. (8.5),

$$\begin{aligned} P[X_{n+k} &= i_n, \ldots, X_k = i_0] \\ &= P[X_{n+k} = i_n \mid X_{n+k-1} = i_{n-1}] \cdots P[X_{1+k} = i_1 \mid X_k = i_0]P[X_k = i_0] \\ &= P[X_{n+k} = i_n \mid X_{n+k-1} = i_{n-1}] \cdots P[X_{1+k} = i_1 \mid X_k = i_0]\pi_{i_0} \\ &= p_{i_{n-1},i_n} \cdots p_{i_0,i_1}\pi_{i_0}, \end{aligned}$$

which is independent of the initial time $k$. Thus the probabilities are independent of the choice of time origin, and the process is stationary.

### ■■ Example 8.11

Find the stationary state pmf in Example 8.5.

Equation (8.19a) gives

$$\pi_0 = (1 - \alpha)\pi_0 + \beta\pi_1$$
$$\pi_1 = \alpha\pi_0 + (1 - \beta)\pi_1,$$

which imply that $\alpha\pi_0 = \beta\pi_1 = \beta(1 - \pi_0)$ since $\pi_0 + \pi_1 = 1$. Thus

$$\pi_0 = \frac{\beta}{\alpha + \beta} = \frac{2}{3} \qquad \pi_1 = \frac{\alpha}{\alpha + \beta} = \frac{1}{3}.$$

■■

---

It is worth noting at this point that not all Markov chains settle into stationary behavior. For example, the binomial counting process (Example 6.13) with $p > 0$ grows steadily so that for any fixed $j$, $p_j(n) \to 0$ as $n \to \infty$. In Section 8.4 we take a closer look at the long-term behavior of Markov chains and determine conditions under which Markov chains have a steady state.

## 8.3 CONTINUOUS-TIME MARKOV CHAINS

In Section 8.2 we saw that the transition probability matrix determines the behavior of a discrete-time Markov chain. In this section we see that the same is true for continuous-time Markov chains.

The joint pmf for $k + 1$ arbitrary time instants of a Markov chain is given by Eq. (8.3):

$$P[X(t_{k+1}) = x_{k+1}, X(t_k) = x_k, \ldots, X(t_1) = x_1]$$
$$= P[X(t_{k+1}) = x_{k+1} \mid X(t_k) = x_k] \cdots \times P[X(t_2) = x_2 \mid X(t_1) = x_1] P[X(t_1) = x_1]. \quad (8.20)$$

This result holds regardless of whether the process is discrete-time or continuous-time. In the continuous-time case, Eq. (8.20) requires that we know the transition probabilities from an arbitrary time $s$ to an arbitrary time $s + t$:

$$P[X(s + t) = j \mid X(s) = i] \qquad t \geq 0.$$

We assume here that the transition probabilities depend only on the difference between the two times:

$$P[X(s + t) = j \mid X(s) = i] = P[X(t) = j \mid X(0) = i] = p_{ij}(t) \qquad t \geq 0, \text{ all } s. \quad (8.21)$$

We say that $X(t)$ is **homogeneous** in time.

Let $P(t) = \{p_{ij}(t)\}$ *denote the matrix of transition probabilities in an interval of length* $t$. Since $p_{ii}(0) = 1$ and $p_{ij}(0) = 0$ for $i \neq j$, we have

$$P(0) = I, \quad (8.22)$$

where $I$ is the identity matrix.

### ■■ Example 8.12
Poisson Process

For the Poisson process, the transition probabilities satisfy

$$p_{ij}(t) = P[j - i \text{ events in } t \text{ seconds}]$$
$$= p_{0,j-i}(t)$$
$$= \frac{(\alpha t)^{j-i}}{(j - i)!} e^{-\alpha t} \qquad j \geq i.$$

Therefore

$$P(t) = \begin{bmatrix} e^{-\alpha t} & \alpha t e^{-\alpha t} & (\alpha t)^2 e^{-\alpha t}/2! & \cdot & \cdots \\ 0 & e^{-\alpha t} & \alpha t e^{-\alpha t} & (\alpha t)^2 e^{-\alpha t}/2! & \cdots \\ 0 & 0 & e^{-\alpha t} & \alpha t e^{-\alpha t} & \cdots \\ \cdot & \cdot & \cdot & \cdot & \cdots \end{bmatrix}.$$

As $t$ approaches zero, $e^{-\alpha t} \approx 1 - \alpha t$. Thus for a small time interval $\delta$,

$$P(\delta) \approx \begin{bmatrix} 1 - \alpha\delta & \alpha\delta & 0 & \cdots \\ 0 & 1 - \alpha\delta & \alpha\delta & \cdots \\ 0 & 0 & 1 - \alpha\delta & \cdots \\ \cdot & \cdot & \cdot & \cdots \end{bmatrix}$$

where all terms of order $\delta^2$ or higher have been neglected. Thus the probability of more than one transition in a very short time interval is negligible. ■■

---

**■■ Example 8.13**
Random Telegraph

In the random telegraph example, the process $X(t)$ changes with each occurrence of an event in a Poisson process. The transition probabilities are as follows:

$$P[X(t) = a \mid X(0) = a] = \frac{1}{2}\{1 + e^{-2\alpha t}\}$$

$$P[X(t) = a \mid X(0) = b] = \frac{1}{2}\{1 - e^{-2\alpha t}\} \qquad \text{if } a \neq b.$$

Thus the transition probability matrix is

$$P(t) = \begin{bmatrix} 1/2\{1 + e^{-2\alpha t}\} & 1/2\{1 - e^{-2\alpha t}\} \\ 1/2\{1 - e^{-2\alpha t}\} & 1/2\{1 + e^{-2\alpha t}\} \end{bmatrix}.$$

■■

---

### State Occupancy Times

Since the random telegraph signal changes polarity with each occurrence of an event in a Poisson process, it follows that the time spent in each state is an exponential random variable. It turns out that this is a property of all continuous-time Markov chains, that is: *$X(t)$ remains at a given value (state) for an exponentially-distributed random time.* To see why, let $T_i$ be the time spent in a state $i$. The probability of spending more than $t$ seconds in this state is then

$$P[T_i > t].$$

Now suppose that the process has already been in state $i$ for $s$ seconds, the probability of spending $t$ more seconds in this state is

$$P[T_i > t + s \mid T_i > s] = P[T_i > t + s \mid X(s') = i, 0 \leq s' \leq s],$$

since the $\{T_i > s\}$ implies that the system has been in state $i$ during the time interval $(0, s)$. The Markov property implies that if $X(s) = i$, then the past is irrelevant and we can view the system as being restarted in state $i$ at time $s$:

$$P[T_i > t + s \mid T_i > s] = P[T_i > t]. \tag{8.23}$$

Only the exponential random variable satisfies this property (see Section 3.4). Thus the time spent in state $i$ is an exponential random variable with some mean $1/v_i$:

$$P[T_i > t] = e^{-v_i t}. \tag{8.24}$$

The **mean state occupancy time** $1/v_i$ will usually be different for each state.

The above result provides us with another way of looking at continuous-time Markov chains. Each time a state, say $i$, is entered, an exponentially distributed state occupancy time $T_i$ is selected. When the time is up, the next state $j$ is selected according to a *discrete-time* Markov chain, with transition probabilities $\tilde{q}_{ij}$. Then the new state occupancy time is selected according to $T_j$, and so on.[1] We call $\tilde{q}_{ij}$ an **embedded Markov chain.**

### ■■ Example 8.14

The random telegraph signal in Example 8.13 spends an exponentially distributed time with mean $1/\alpha$ in each state. When a transition occurs, the transition is always from the present state to the only other state, thus the embedded Markov chain is

$$\tilde{q}_{00} = 0 \qquad \tilde{q}_{01} = 1$$
$$\tilde{q}_{10} = 1 \qquad \tilde{q}_{11} = 0.$$

■■

## Transition Rates and Time-Dependent State Probabilities

Consider the transition probabilities in a very short time interval of duration $\delta$ seconds. The probability that the process remains in state $i$ during the interval is:

$$\begin{aligned} P[T_i > \delta] &= e^{-v_i \delta} \\ &= 1 - \frac{v_i \delta}{1!} + \frac{v_i^2 \delta^2}{2!} - \cdots \\ &= 1 - v_i \delta + o(\delta) \end{aligned}$$

where $o(\delta)$ denotes terms that become negligible relative to $\delta$ as $\delta$

1. This view of Markov chains is useful in setting up computer simulation models of Markov chain processes.

approaches zero.[2] The exponential distributions of the state occupancy times imply that the probability of two or more transitions in an interval of duration $\delta$ is $o(\delta)$. Thus for small $\delta$, $p_{ii}(\delta)$ is approximately equal to the probability that the process remains in state $i$ for $\delta$ seconds:

$$\begin{aligned} p_{ii}(\delta) &= P[T_i > \delta] + o(\delta) \\ &= 1 - v_i\delta + o(\delta) \end{aligned}$$

or equivalently,

$$1 - p_{ii}(\delta) = v_i\delta + o(\delta). \tag{8.25}$$

We call $v_i$ the *rate at which the process $X(t)$ leaves state $i$.*

Once the process leaves state $i$, it will enter state $j$ with probability $\tilde{q}_{ij}$. Thus

$$\begin{aligned} p_{ij}(\delta) &= (1 - p_{ii}(\delta))\tilde{q}_{ij} \\ &= v_i\tilde{q}_{ij}\,\delta + o(\delta) \\ &= \gamma_{ij}\,\delta + o(\delta) \end{aligned} \tag{8.26a}$$

We call $\gamma_{ij} = v_i\,\tilde{q}_{ij}$ the *rate at which the process $X(t)$ enters state $j$ from state $i$.* For completeness, we define $\gamma_{ii} = -v_i$, so that by Eq. (8.25)

$$p_{ii}(\delta) - 1 = \gamma_{ii}\,\delta + o(\delta). \tag{8.26b}$$

If we divide both sides of Eqs. (8.26a) and (8.26b) by $\delta$ and take the limit $\delta \to 0$, we obtain

$$\lim_{\delta \to 0} \frac{p_{ij}(\delta)}{\delta} = \gamma_{ij} \qquad i \neq j \tag{8.27a}$$

and

$$\lim_{\delta \to 0} \frac{p_{ii}(\delta) - 1}{\delta} = \gamma_{ii}, \tag{8.27b}$$

since

$$\lim_{\delta \to 0} \frac{o(\delta)}{\delta} = 0,$$

because $o(\delta)$ is of order higher than $\delta$.

We are now ready to develop a set of equations for finding the state probabilities at time $t$, which will be denoted by

$$p_j(t) \triangleq P[X(t) = j].$$

---

2. A function $g(h)$ is said to be $o(h)$ if $\lim_{h\to 0} g(h)/h = 0$.

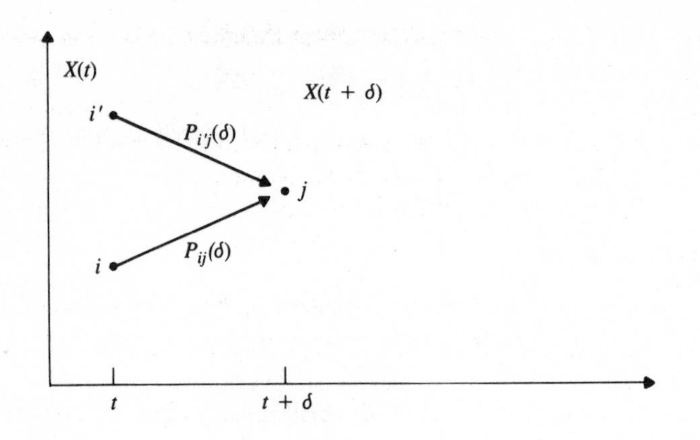


FIGURE 8.2 Transitions into state $j$.

For $\delta > 0$, we have (see Fig. 8.2)

$$\begin{aligned} p_j(t+\delta) &= P[X(t+\delta) = j] \\ &= \sum_i P[X(t+\delta) = j \mid X(t) = i] P[X(t) = i] \\ &= \sum_i p_{ij}(\delta) p_i(t). \end{aligned} \tag{8.28}$$

If we subtract $p_j(t)$ from both sides, we obtain

$$p_j(t+\delta) - p_j(t) = \sum_{i \neq j} p_{ij}(\delta) p_i(t) + (p_{jj}(\delta) - 1) p_j(t). \tag{8.29}$$

If we divide by $\delta$, apply Eqs. (8.27a) and (8.27b) and let $\delta \to 0$, we obtain

$$p_j'(t) = \sum_i \gamma_{ij} p_i(t) \tag{8.30}$$

Equation (8.30) is a form of the **Chapman-Kolmogorov equations** for continuous-time Markov chains. To find $p_j(t)$ we need to solve this system of differential equations with initial conditions specified by the initial state pmf $\{p_j(0), j = 0, 1, \ldots\}$.

Note that if we solve Eq. (8.30) under the assumption that the state at time zero was $i$, that is, with initial condition $p_i(0) = 1$ and $p_j(0) = 0$ for all $j \neq i$, then the solution is actually $p_{ij}(t)$ the $ij$ component of $P(t)$. Thus Eq. (8.30) can also be used to find the transition probability matrix.

### ■■ Example 8.15
A Simple Queueing System

A queueing system alternates between two states. In state 0, the system is idle and waiting for a customer to arrive. This idle time is an exponential random variable with mean $1/\alpha$. In state 1, the system is busy servicing a customer. The time in the busy state is an exponential random variable with mean $1/\beta$. Find the state probabilities $p_0(t)$ and $p_1(t)$ in terms of the initial state probabilities $p_0(0)$ and $p_1(0)$.

The system moves from state 0 to state 1 at a rate $\alpha$, and from state 1 to state 0 at a rate $\beta$:

$$\gamma_{00} = -\alpha \qquad \gamma_{01} = \alpha$$
$$\gamma_{10} = \beta \qquad \gamma_{11} = -\beta.$$

Equation (8.30) then gives

$$p_0'(t) = -\alpha p_0(t) + \beta p_1(t)$$
$$p_1'(t) = \alpha p_0(t) - \beta p_1(t).$$

Since $p_0(t) + p_1(t) = 1$, the first equation becomes

$$p_0'(t) = -\alpha p_0(t) + \beta(1 - p_0(t)),$$

which is a first order differential equation:

$$p_0'(t) + (\alpha + \beta)p_0(t) = \beta \qquad p_0(0) = p_0.$$

The general solution of this equation is

$$p_0(t) = \frac{\beta}{\alpha + \beta} + Ce^{-(\alpha+\beta)t}.$$

We obtain $C$ by setting $t = 0$ and solving in terms of $p_0(0)$; then we find

$$p_0(t) = \frac{\beta}{\alpha + \beta} + \left(p_0(0) - \frac{\beta}{\alpha + \beta}\right)e^{-(\alpha+\beta)t}$$

and

$$p_1(t) = \frac{\alpha}{\alpha + \beta} + \left(p_1(0) - \frac{\alpha}{\alpha + \beta}\right)e^{-(\alpha+\beta)t}.$$

Note that as $t \to \infty$

$$p_0(t) \to \frac{\beta}{\alpha + \beta} \qquad \text{and} \qquad p_1(t) \to \frac{\alpha}{\alpha + \beta}.$$

Thus as $t \to \infty$, the state probabilities approach constant values that are independent of the initial state probabilities. ■■

## ■■ Example 8.16
### The Poisson Process

Find the state probabilities for the Poisson process.

The Poisson process moves only from state $i$ to state $i + 1$ at a rate $\alpha$.

Thus

$$\gamma_{ii} = -\alpha \qquad \text{and} \qquad \gamma_{i,i+1} = \alpha.$$

Equation (8.30) then gives

$$p_o'(t) = -\alpha p_o(t) \qquad \text{for } j = 0$$
$$p_j'(t) = -\alpha p_j(t) + \alpha p_{j-1}(t) \qquad \text{for } j \geq 1.$$

The initial condition for the Poisson process is $p_0(0) = 1$, so the solution for the $j = 0$ equation is

$$p_0(t) = e^{-\alpha t}.$$

The equation for $j = 1$ is

$$p_1'(t) = -\alpha p_1(t) + \alpha e^{-\alpha t} \qquad p_1(0) = 0,$$

which is also a first order differential equation for which the solution is

$$p_1(t) = \frac{\alpha t}{1!} e^{-\alpha t}.$$

It can be shown by an induction argument that the solution of the state $j$ equation is

$$p_j(t) = \frac{(\alpha t)^j}{j!} e^{-\alpha t}.$$

Note that for any $j$, $p_j(t) \to 0$ as $t \to \infty$. Thus for the Poisson process, the probability of any finite state approaches zero as $t \to \infty$. This is consistent with the fact that the process grows steadily with time. ■■

### Steady State Probabilities and Global Balance Equations

As $t \to \infty$, the state probabilities in the two-state queueing system in Example 8.15 converge to a pmf that does not depend on the initial conditions. This is typical of systems that reach "equilibrium" or "steady state." For such system $p_j(t) \to p_j$ and $p_j'(t) \to 0$, so Eq. (8.30) becomes

$$0 = \sum_i \gamma_{ij} p_i \qquad \text{for all } j, \tag{8.31a}$$

or equivalently recalling that $\gamma_{jj} = -\nu_j$

$$\nu_j p_j = \sum_{i \neq j} \gamma_{ij} p_i \qquad \text{for all } j. \tag{8.31b}$$

Equation (8.31b) can be rewritten as follows:

$$p_j \left( \sum_{i \neq j} \gamma_{ji} \right) = \sum_{i \neq j} \gamma_{ij} p_i \tag{8.31c}$$

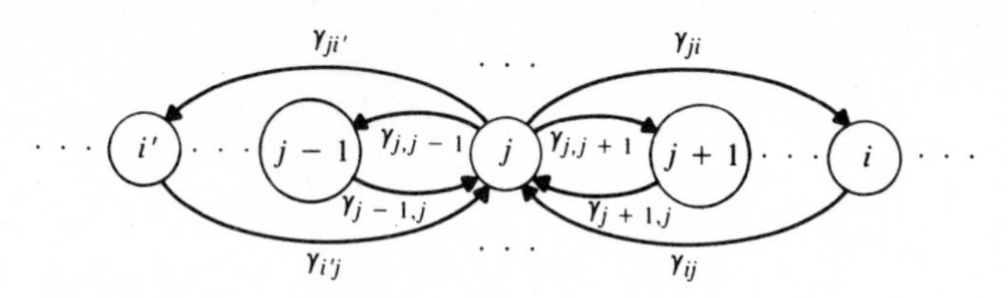


**FIGURE 8.3** Global balance of probability flows.

since

$$v_j = \sum_{i \neq j} \gamma_{ji}.$$

The system of linear equations given by Eqs. (8.31b) or (8.31c) are called the **global balance equations.** These equations state that at equilibrium, the rate of probability flow out of state $j$, namely $v_j p_j$, is equal to the rate of flow into state $j$ as shown in Fig. 8.3. By solving this set of linear equations we can obtain the stationary state pmf of the system (when it exists).

We refer to $\mathbf{p} = \{p_i\}$ as the **stationary state pmf** of the Markov chain. Since $\mathbf{p}$ satisfies Eq. (8.30), if we start the Markov chain with initial state pmf given by $\mathbf{p}$, then the state probabilities will be

$$p_i(t) = p_i \qquad \text{for all } t.$$

The resulting process is a stationary random process as defined in Section 6.5, since the probability of the sequence of states $i_0, i_1, \ldots, i_n$ at times $t < t_1 + t < \cdots < t_n + t$ is, by Eq. (8.20):

$$\begin{aligned} P[X(t) &= i_0, X(t_1 + t) = i_1, \ldots, X(t_n + t) = i_n] \\ &= P[X(t_n + t) = i_n \mid X(t_{n-1} + t) = i_{n-1}] \cdots \\ &\quad \times P[X(t_1 + t) = i_1 \mid X(t) = i_0] P[X(t) = i_0]. \end{aligned}$$

The transition probabilities depend only on the difference between the associated times. Thus the above joint probability depends on the choice of origin only through $P[X(t) = i_0]$. But $P[X(t) = i_0] = p_{i_0}$ for all $t$. Thus we conclude that the above joint probability is independent of the choice of time origin and thus that the process is stationary.

## ■■ Example 8.17

Find the stationary state pmf for the two-state queueing system discussed in Example 8.15.

Equation (8.31b) for this system gives

$$\alpha p_0 = \beta p_1 \qquad \text{and} \qquad \beta p_1 = \alpha p_0.$$

Noting that $p_0 + p_1 = 1$, we obtain

$$p_0 = \frac{\beta}{\alpha + \beta} \qquad \text{and} \qquad p_1 = \frac{\alpha}{\alpha + \beta}.$$

■■

---

### ■■ Example 8.18
The M/M/1 Single-Server Queueing System

Consider a queueing system in which customers are served one at a time in order of arrival. The time between customer arrivals is exponentially distributed with rate $\lambda$, and the time required to service a customer is exponentially distributed with rate $\mu$. Find the steady state pmf for the number of customers in the system.

The state transition rates are as follows. Customers arrive at a rate $\lambda$, so

$$\gamma_{i,i+1} = \lambda \qquad i = 0, 1, 2, \ldots.$$

When the system is nonempty, customers depart at the rate $\mu$. Thus

$$\gamma_{i,i-1} = \mu \qquad i = 1, 2, 3, \ldots.$$

The transition rate diagram is shown in Fig. 8.4. The global balance equations are

$$\lambda p_0 = \mu p_1 \qquad \text{for } j = 0 \tag{8.32a}$$

$$(\lambda + \mu)p_j = \lambda p_{j-1} + \mu p_{j+1} \qquad \text{for } j = 1, 2, \ldots. \tag{8.32b}$$

We can rewrite Eq. (8.32b) as follows:

$$\lambda p_j - \mu p_{j+1} = \lambda p_{j-1} - \mu p_j \qquad \text{for } j = 1, 2, \ldots,$$

which implies that

$$\lambda p_{j-1} - \mu p_j = \text{constant} \qquad \text{for } j = 1, 2, \ldots. \tag{8.33}$$

Equation (8.33) with $j = 1$ and Eq. (8.32a) together imply that

$$\text{constant} = \lambda p_0 - \mu p_1 = 0.$$

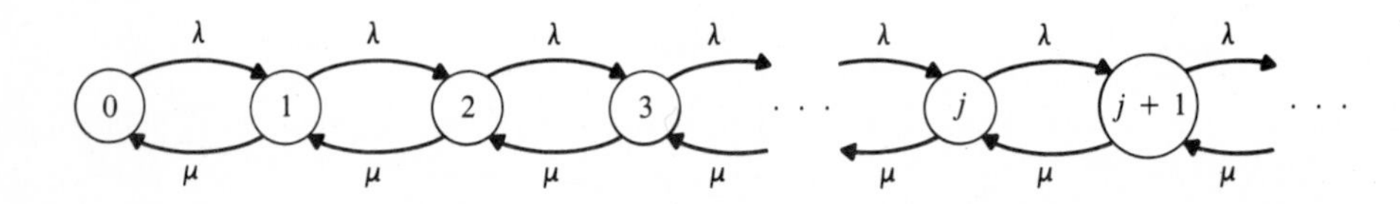


**FIGURE 8.4** Transition rate diagram for M/M/1 queueing system.

Thus Eq. (8.33) becomes

$$\lambda p_{j-1} = \mu p_j,$$

or equivalently,

$$p_j = \rho p_{j-1} \qquad j = 1, 2, \ldots$$

and by a simple induction argument

$$p_j = \rho^j p_0,$$

where $\rho = \lambda/\mu$. We obtain $p_0$ by noting that the sum of the probabilities must be one:

$$1 = \sum_{j=0}^{\infty} p_j = (1 + \rho + \rho^2 + \cdots)p_0 = \frac{1}{1-\rho} p_0,$$

where the series converges if and only if $\rho < 1$.

Thus

$$p_j = (1 - \rho)\rho^j \qquad j = 0, 1, 2, \ldots. \tag{8.34}$$

This queueing system is discussed in detail in Section 9.3.

The condition for the existence of a steady state solution has a simple explanation. The condition $\rho < 1$ is equivalent to

$$\lambda < \mu,$$

that is, the rate at which customers arrive must be less than the rate at which the system can process them. Otherwise the queue builds up without limit as time progresses. ■■

---

## ■■ Example 8.19
### A Birth-and-Death Process

A birth-and-death process is a Markov chain in which only transitions between adjacent states occur as shown in Fig. 8.5. The single server queueing system discussed in Example 8.18 is an example of a birth-and-death process.

The global balance equations for a general birth-and-death process are

$$\lambda_0 p_0 = \mu_1 p_1 \qquad j = 0 \tag{8.35a}$$

$$\lambda_j p_j - \mu_{j+1} p_{j+1} = \lambda_{j-1} p_{j-1} - \mu_j p_j \qquad j = 1, 2, \ldots. \tag{8.35b}$$

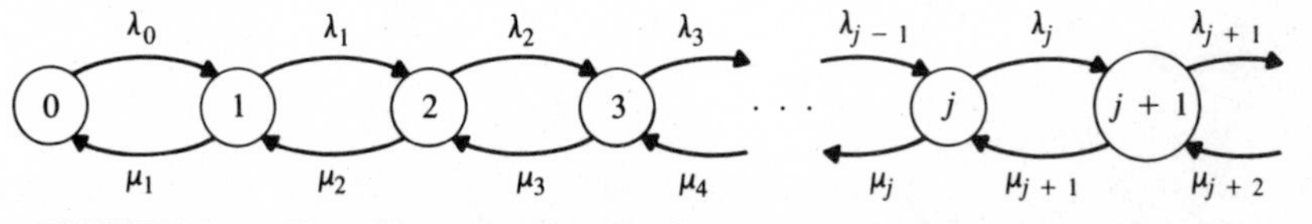


**FIGURE 8.5** Transition rate diagram for general birth-and-death process.

As in the previous example, it then follows that

$$p_j = r_j p_{j-1} \qquad j = 1, 2, \ldots$$

and

$$p_j = r_j r_{j-1} \cdots r_1 p_0 \qquad j = 1, 2, \ldots, \tag{8.36}$$

where $r_j = (\lambda_{j-1})/\mu_j$. If we define

$$R_j = r_j r_{j-1} \cdots r_1 \qquad \text{and} \qquad R_0 = 1,$$

then $p_0$ is found from

$$1 = \left(\sum_{j=0}^{\infty} R_j\right) p_0.$$

If the series in the above equation converges, then the stationary pmf is given by

$$p_j = \frac{R_j}{\sum_{i=0}^{\infty} R_i} \tag{8.37}$$

If the series does not converge, then a stationary pmf does not exist, and $p_j = 0$ for all $j$. In Chapter 9, we see that many useful queueing systems can be modeled by birth-and-death processes. ■■

## *8.4

## CLASSES OF STATES, RECURRENCE PROPERTIES, AND LIMITING PROBABILITIES

In this section we take a closer look at the relation between the behavior of a Markov chain and its transition probability matrix. First we see that the states of a discrete-time Markov chain can be divided into one or more separate classes and that these classes can be of several types. We then show that the long-term behavior of a Markov chain is related to the types of its state classes. Finally, we use these results to relate the long-term behavior of continuous-time Markov chains to that of its embedded Markov chain. Figure 8.6 summarizes the types of classes to which a state can belong and identifies the associated long-term behavior.

### Classes of States

We say that **state $j$ is accessible from state $i$** if for some $n \geq 0$, $p_{ij}(n) > 0$, that is, if there is a sequence of transitions from $i$ to $j$ that has nonzero probability. We say that **states $i$ and $j$ communicate** if they are

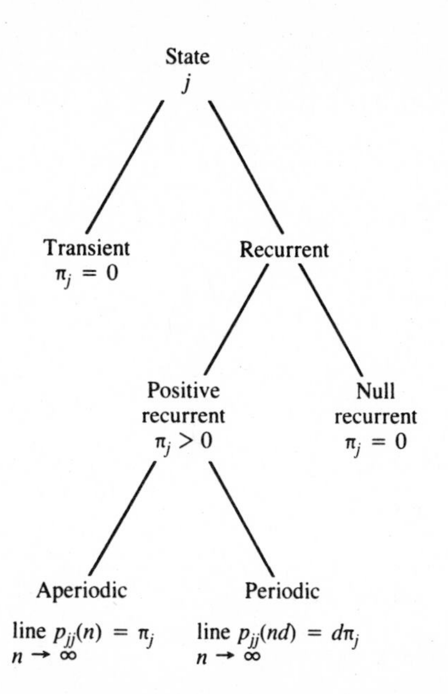


**FIGURE 8.6** Classification of states and associated long-term behavior. The proportion of time spent in state $j$ is denoted by $\pi_j$.

accessible to each other; we then write $i \leftrightarrow j$. Note that a state communicates with itself since $p_{ii}(0) = 1$.

If state $i$ communicates with state $j$ and state $j$ communicates with state $k$, that is $i \leftrightarrow j$ and $j \leftrightarrow k$, then state $i$ communicates with $k$. To see this note that $i \leftrightarrow j$ implies that there is a nonzero-probability path from $i$ to $j$ and $j \leftrightarrow k$ implies that there is a subsequent nonzero-probability path from $j$ to $k$. The combined paths form a nonzero-probability path from $i$ to $k$. A nonzero-probability path in the reverse direction exists for the same reasons.

We say that two states belong to the same **class** if they communicate with each other. Note that two different classes of states must be disjoint since having a state in common would imply that the states from both classes communicate with each other. Thus *the states of a Markov chain consist of one or more disjoint communication classes.* A Markov chain that consists of a single class is said to be **irreducible**.

## ■■ Example 8.20

Figure 8.7(a) shows the state transition diagram for a Markov chain with three classes: $\{0\}$, $\{1,2\}$, and $\{3\}$. ■■

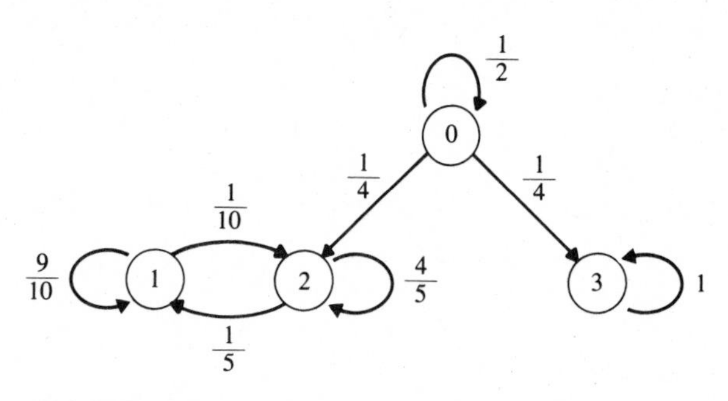

**FIGURE 8.7a** A three-class Markov chain.

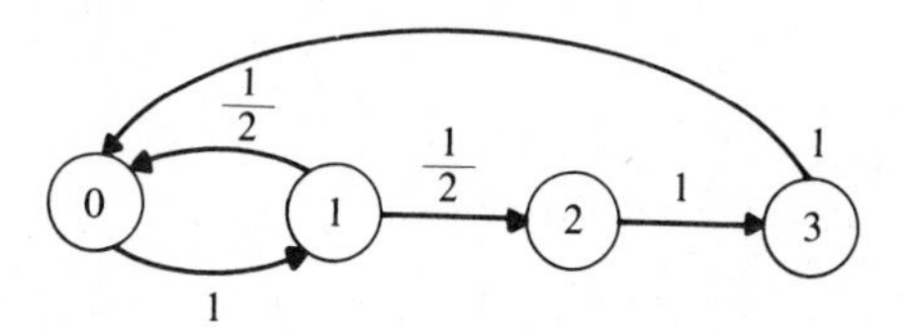


**FIGURE 8.7b** A periodic Markov chain.

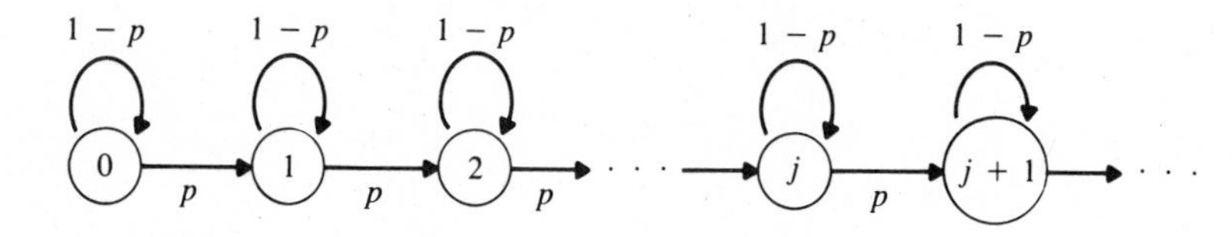

**FIGURE 8.7c** A binomial counting process.

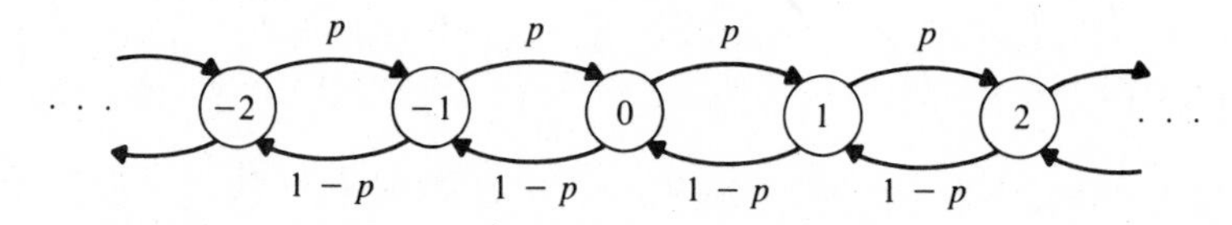

**FIGURE 8.7d** The random walk process.

**■■ Example 8.21**

Figure 8.7(b) shows the state transition diagram for a Markov chain with one class: $\{0, 1, 2, 3\}$. Thus the chain is irreducible. ■■

**■■ Example 8.22**

Figure 8.7(c) shows the state transition diagram for a binomial counting process. It can be seen that the classes are: $\{0\}$, $\{1\}$, $\{2\}, \ldots$. ■■

**■■ Example 8.23**

Figure 8.7(d) shows the state transition diagram for the random walk process. If $p > 0$, then the process has only one class, $\{0, \pm 1, \pm 2, \ldots\}$, so it is irreducible. ■■

### Recurrence Properties

Suppose we start a Markov chain in state $i$. State $i$ is said to be **recurrent** if the process returns to the state with probability one, that is,

$$f_i = P[\text{ever returning to state } i] = 1. \tag{8.38a}$$

State $i$ is said to be **transient** if

$$f_i < 1. \tag{8.38b}$$

If we start the Markov chain in a recurrent state $i$, then the state reoccurs an infinite number of times. If we start the Markov chain in a transient state, the state does not reoccur after some finite number of returns. Each reoccurrence of the state can be viewed as a success in a Bernoulli trial. The probability of success is $f_i$. Thus the number of returns to state $i$ is a geometric random variable with mean $(1 - f_i^{-1})$. If $f_i < 1$, then the probability of an infinite number of successes is zero. Therefore a transient state reoccurs only a finite number of times.

Let $X_n$ denote the Markov chain with initial state $i$, $X_0 = i$. Let $I_i(X)$ be the indicator function for state $i$, that is, $I_i(X)$ is equal to 1 if $X = i$ and equal to 0 otherwise. The expected number of returns to state $i$ is then

$$E\left[\sum_{n=1}^{\infty} I_i(X_n) \,|\, X_0 = i\right] = \sum_{n=1}^{\infty} E[I_i(X_n) \,|\, X_0 = i] = \sum_{n=1}^{\infty} p_{ii}(n) \tag{8.39}$$

since by Example 3.34

$$E[I_i(X_n) \,|\, X_0 = i] = P[X_n = i \,|\, X_0 = i] = p_{ii}(n).$$

A state is recurrent if and only if it reoccurs an infinite number of times, thus from Eq. (8.39) *state i is recurrent if and only if*

$$\sum_{n=1}^{\infty} p_{ii}(n) = \infty. \tag{8.40}$$

Similarly, *state i is transient if and only if*

$$\sum_{n=1}^{\infty} p_{ii}(n) < \infty. \tag{8.41}$$

### ■■ Example 8.24

In Example 8.20, state 0 is transient since $p_{00}(n) = (1/2)^n$, so

$$\sum_{n=1}^{\infty} p_{00}(n) = \frac{1}{2} + \left(\frac{1}{2}\right)^2 + \left(\frac{1}{2}\right)^3 + \cdots = 1 < \infty.$$

On the other hand, if the process were started in state 1, we would have the two-state process discussed in Example 8.9. For such a process we found that

$$p_{11}(n) = \frac{\beta + \alpha(1 - \alpha - \beta)^n}{\alpha + \beta} = \frac{1/2 + 1/4(7/10)^n}{3/4}$$

so that

$$\sum_{n=1}^{\infty} p_{11}(n) = \sum_{n=1}^{\infty} \left(\frac{2}{3} + \frac{(7/10)^n}{3}\right) = \infty.$$

Therefore state 1 is recurrent. ■■

### ■■ Example 8.25

In the binomial counting process all the states are transient since $p_{ii}(n) = (1 - p)^n$ so that for $p > 0$

$$\sum_{n=1}^{\infty} p_{ii}(n) = \sum_{n=1}^{\infty} (1 - p)^n = \frac{1 - p}{p} < \infty.$$

■■

### ■■ Example 8.26

Consider state zero in the random walk process in Fig. 8.7(d). The state reoccurs in $2n$ steps if and only if $n$ + 1s and $n$ − 1s occur during the $2n$ steps. This occurs with probability

$$p_{00}(2n) = \binom{2n}{n} p^n (1 - p)^n.$$

Stirling's formula for $n!$ can be used to show that

$$\binom{2n}{n}p^n(1-p)^n \sim \frac{(4p(1-p))^n}{\sqrt{\pi n}}.$$

where $a_n \sim b_n$ when $\lim_{n\to\infty} a_n/b_n = 1$.

Thus Eq. (8.39) for state 0 is

$$\sum_{n=1}^{\infty} p_{00}(2n) \sim \sum_{n=1}^{\infty} \frac{(4p(1-p))^n}{\sqrt{\pi n}}.$$

If $p = 1/2$, then $4p(1-p) = 1$ and the series diverges. It then follows that state 0 is recurrent. If $p \neq 1/2$, then $(4p(1-p)) < 1$, and the above series converges. This implies that state 0 is transient. ■■

If state $i$ is recurrent then all states in its class will be visited eventually as the process returns to $i$ over and over again. Indeed all other states in its class appear an infinite number of times. Thus *recurrence is a class property,* that is, if state $i$ is recurrent and $i \leftrightarrow j$, then state $j$ is also recurrent. Similarly, *transience is a class property.*

If a Markov chain is irreducible, that is, if it consists of a single communication class, then either all its states are transient or all its states are recurrent. If the number of the states in the chain is finite, it is impossible for the all of the states to be transient. Thus, *the states of a finite-state, irreducible Markov chain are all recurrent.*

The information about when state $i$ can reoccur is contained in $p_{ii}(n)$, the $n$-step transition probability from state $i$ to state $i$. We say that state $i$ has **period** $d$ if it can only reoccur at times that are multiples of $d$, that is $p_{ii}(n) = 0$ unless $n$ is a multiple of $d$, where $d$ is the largest integer with this property. It can be shown that all the states in a class have the same period. An irreducible Markov chain is said to be **aperiodic** if the states in its single class have period one.

### ■■ Example 8.27

In Example 8.20, all the states have the property that $p_{ii}(n) > 0$ for $n = 1, 2, \ldots$ Therefore all three classes in the Markov chain have period 1. ■■

### ■■ Example 8.28

In the Markov chain in Fig. 8.7(b) the states 0 and 1 can reoccur at times $2, 4, 6, \ldots$ and states 2 and 3 at times $4, 6, 8, \ldots$. Therefore the Markov chain has period 2. ■■

### ■■ Example 8.29

In the random walk process in Fig. 8.7(d), a state reoccurs when the number of successes (+1s) equals the number of failures (−1s). This can only happen after an even number of steps. The process therefore has period 2. ■■

### Limiting Probabilities

If all the states in a Markov chain are transient then all the state probabilities approach zero as $n \to \infty$. If a Markov chain has some transient classes and some recurrent classes, as in Fig. 8.7(a), then eventually the process falls and remains thereafter in one of the recurrent classes. Therefore we can concentrate on individual recurrent classes when studying the limiting probabilities of a chain. For this reason we assume in this section that we are dealing with an irreducible Markov chain.

Suppose we start a Markov chain in a *recurrent* state $i$ at time $n = 0$. Let $T_i(1)$, $T_i(1) + T_i(2), \ldots$ be the times when the process returns to state $i$, where $T_i(k)$ is the time that elapses between the $(k-1)$th and $k$th returns (see Fig. 8.8). The $T_i$ form an iid sequence since each return time is independent of previous return times.

The proportion of time spent in state $i$ after $k$ returns to $i$ is

$$\text{proportion of time in state } i = \frac{k}{T_i(1) + T_i(2) + \cdots + T_i(k)}. \tag{8.42}$$

Since the state is recurrent, the process returns to state $i$ an infinite number of times. Thus the law of large numbers implies that, with probability one, the reciprocal of the above expression approaches the

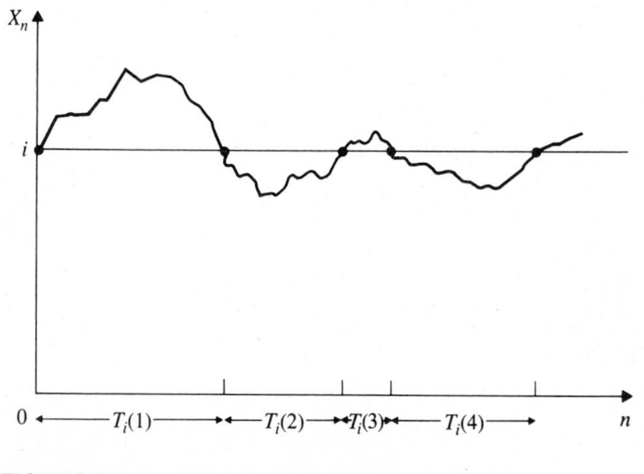


**FIGURE 8.8** Recurrence times for state $i$.

**mean recurrence time** $E[T_i]$ so the long-term proportion of time spent in state $i$ approaches

$$\text{proportion of time in state } i \rightarrow \frac{1}{E[T_i]} = \pi_i, \tag{8.43}$$

where $\pi_i$ *is the long-term proportion of time spent in state i.*

If $E[T_i] < \infty$, then we say that state $i$ is **positive recurrent.** Equation (8.43) then implies that

$$\pi_i > 0 \qquad \text{if state } i \text{ positive recurrent.}$$

If $E[T_i] = \infty$, then we say that state $i$ is **null recurrent.** Equation (8.43) then implies that

$$\pi_i = 0 \qquad \text{if state } i \text{ is null recurrent.}$$

It can be shown that positive and null recurrence are class properties.

Positive recurrent, aperiodic states are called **ergodic.** An **ergodic Markov chain** is defined as an irreducible, aperiodic, positive recurrent Markov chain.

### ■■ Example 8.30

The process in Figure 8.7(b) returns to state 0 in two steps with probability 1/2 and in four steps with probability 1/2. Therefore the mean recurrence time for state 0 is

$$E[T_0] = \frac{1}{2}2 + \frac{1}{2}4 = 3.$$

Therefore state 0 is positive recurrent and the long-term proportion of time spent in state 0 is

$$\pi_0 = \frac{1}{3}.$$

■■

### ■■ Example 8.31

In Example 8.26 it was shown that the random walk process is recurrent if $p = 1/2$. However, the mean recurrence time can be shown to be infinite when $p = 1/2$. (Feller 1968 p. 314]. Thus all the states in the chain are null-recurrent. ■■

The $\pi_j$'s in Eq. (8.43) satisfy the equations that define the stationary state pmf:

$$\pi_j = \sum_i \pi_i P_{ij} \qquad \text{for all } j \tag{8.44a}$$

and

$$1 = \sum_i \pi_i. \tag{8.44b}$$

To see this, note that since $\pi_i$ is the proportion of time spent in state $i$, then $\pi_i P_{ij}$ is the proportion of time in which state $j$ follows $i$. If we sum over all $i$, we then obtain the long-term proportion of time in state $j$, $\pi_j$.

### ■■ Example 8.32

The stationary state pmf for the periodic Markov chain in Fig. 8.7(b), is found from Eqs. (8.44a) and (8.44b):

$$\pi_0 = \frac{1}{2}\pi_1 + \pi_3$$

$$\pi_1 = \pi_0$$

$$\pi_2 = \frac{1}{2}\pi_1$$

$$\pi_3 = \pi_2.$$

These equations imply that $\pi_1 = \pi_0$ and $\pi_2 = \pi_3 = \pi_0/2$. Using the fact that the probabilities must add to one, we obtain

$$\pi_1 = \pi_0 = \frac{1}{3} \quad \text{and} \quad \pi_2 = \pi_3 = \frac{1}{6}.$$

Note that $\pi_0 = 1/3$ was obtained for the mean recurrence time in Example 8.30. ■■

In Section 8.2 we found that for Markov chains that exhibit stationary behavior, the $n$-step transition matrix approaches a fixed matrix of equal rows as $n \to \infty$ [see Eq. (8.17)]. We also saw that the rows of this limiting matrix consisted of a pmf that satisfied Eqs. (8.44a) and (8.44b). We are now ready to state under what conditions this occurs.

**THEOREM 1**

For an irreducible, aperiodic, and positive recurrent Markov chain

$$\lim_{n\to\infty} p_{ij}(n) = \pi_j \qquad \text{for all } j, \tag{8.45}$$

where $\pi_j$ is the unique nonnegative solution of Eqs. (8.44a) and (8.44b).

Theorem 1 states that for irreducible, aperiodic, and positive recurrent Markov chains, the state probabilities approach steady state values that are independent of the initial condition. These steady state probabilities correspond to the stationary probabilities obtained from Eqs. (8.44a) and (8.44b) and thus correspond to the long-term proportion of time spent in the given state. This is the reason why irreducible, aperiodic, and positive recurrent Markov chains are called "ergodic".

For periodic processes, we have the following result.

**THEOREM 2**

For an irreducible, periodic, and positive recurrent Markov chain with period $d$

$$\lim_{n\to\infty} p_{jj}(nd) = d\pi_j \qquad \text{for all } j, \tag{8.46}$$

where $\pi_j$ is the unique nonnegative solution of Eqs. (8.44a) and (8.44b).

As before, $\pi_j$ represents the proportion of time spent in state $j$. However, the fact that the occurrence of state $j$ is constrained to occur at multiples of $d$ steps implies that the probability of occurrence of the state $j$ is $d$ times greater at the allowable times and zero elsewhere.

### ■■ Example 8.33

In Examples 8.30 and 8.32 we found that the long-term proportion of time spent in state 0 is $\pi_0 = 1/3$. If we start in state 0, then only even states can occur at even time instants. Thus at these even time instants the probability of state 0 is 2/3 and of state 2 is 1/3. At odd time instants, the probabilities of states 0 and 2 are zero. ■■

## Limiting Probabilities for Continuous-Time Markov Chains

We saw in Section 8.3 that a continuous-time Markov chain $X(t)$ can be viewed as consisting of a sequence of states determined by some discrete-time Markov chain $X_n$ with transition probabilities $\tilde{q}_{ij}$ and a corresponding sequence of exponentially distributed state occupancy times. In this section, we show that if the associated discrete-time chain is irreducible and positive recurrent with stationary pmf $\pi_j$, then the long term proportion of time spent by $X(t)$ in state $i$ is

$$p_i = \frac{\pi_i/v_i}{\sum_j \pi_j/v_j},$$

where $1/v_i$ is the mean occupancy time in state $i$. Furthermore we show that the $p_i$ are the unique solution to the global balance equations, Eqs. (8.31b) and (8.31c).

Suppose that the embedded Markov chain $X_n$ is irreducible and positive recurrent, so that Eq. (8.43) holds. Let $N_i(n)$ denote the number of times state $i$ occurs in the first $n$ transitions, and let $T_i(j)$ denote the occupancy time the $j$th time state $i$ occurs. The proportion of time spent in state $i$ after the first $n$ transitions is

$$\frac{\text{time spent in state } i}{\text{time spent in all states}} = \frac{\sum_{j=1}^{N_i(n)} T_i(j)}{\sum_i \sum_{j=1}^{N_i(n)} T_i(j)} = \frac{\frac{N_i(n)}{n} \frac{1}{N_i(n)} \sum_{j=1}^{N_i(n)} T_i(j)}{\sum_i \frac{N_i(n)}{n} \frac{1}{N_i(n)} \sum_{j=1}^{N_i(n)} T_i(j)}. \tag{8.47}$$

As $n \to \infty$, by Eqs. (8.43), (8.44a), and (8.44b), with probability one,

$$\frac{N_i(n)}{n} \to \pi_i, \tag{8.48}$$

the stationary pmf of the embedded Markov chain. In addition, we also have that $N_i(n) \to \infty$ as $n \to \infty$, so that by the strong law of large numbers, with probability one,

$$\frac{1}{N_i(n)} \sum_{j=1}^{N_i(n)} T_i(j) \to E[T_i] = 1/v_i, \tag{8.49}$$

where we have used the fact that the state occupancy time in state $i$ has mean $1/v_i$. Equations (8.48) and (8.49) when applied to Eq. (8.47) imply that, with probability one, the long-term proportion of time spent in state $i$ approaches

$$p_i = \frac{\pi_i/v_i}{\sum_j \pi_j/v_j} = c\pi_i/v_i, \tag{8.50}$$

where $\pi_j$ is the unique pmf solution to

$$\pi_j = \sum_i \pi_i \tilde{q}_{ij} \qquad \text{for all } j. \tag{8.51}$$

and $c$ is a normalization constant.

We obtain the global balance equation, Eq. (8.31b), by substituting $\pi_i = v_i p_i/c$ from Eq. 8.50 and $\tilde{q}_{ij} = \gamma_{ij}/v_i$ into Eq. (8.51):

$$v_j p_j = \sum_{i \neq j} p_i \gamma_{ij} \qquad \text{for all } j.$$

Thus the $p_i$'s are the unique solution of the global balance equations.

### ■■ Example 8.34

In the two-state system in Example 8.15,

$$[\tilde{q}_{ij}] = \begin{bmatrix} 0 & 1 \\ 1 & 0 \end{bmatrix}.$$

The equation $\boldsymbol{\pi} = \boldsymbol{\pi}[\tilde{q}_{ij}]$ implies that

$$\pi_0 = \pi_1 = \frac{1}{2}.$$

In addition, $v_0 = \alpha$ and $v_1 = \beta$. Thus

$$p_0 = \frac{1/2(1/\alpha)}{1/2(1/\alpha + 1/\beta)} = \frac{\beta}{\alpha + \beta}$$

and

$$p_1 = \frac{\alpha}{\alpha + \beta}.$$

■■

## *8.5 TIME-REVERSED MARKOV CHAINS

We now consider the random process that results when we play a Markov chain backwards in time. We will see that the resulting process is also a Markov chain and develop another method for obtaining the stationary probabilities of the forward and reverse processes. The insights gained by looking at the reverse process will prove useful in developing certain results in queueing theory in Chapter 9.

Let $X_n$ be a stationary ergodic Markov chain[3] with one-step transition probability matrix $P = \{p_{ij}\}$ and stationary state pmf $\{\pi_j\}$. Consider the dependence of $X_{n-1}$, the "future" in the reverse process, on $X_n$, $X_{n+1}, \ldots, X_{n+k}$, the "present and past":

$$\begin{aligned} P[X_{n-1} = j \mid X_n = i, X_{n+1} = i_1, \ldots, X_{n+k} = i_k] &= \frac{P[X_{n-1} = j, X_n = i, X_{n+1} = i_1, \ldots, X_{n+k} = i_k]}{P[X_n = i, X_{n+1} = i_1, \ldots, X_{n+k} = i_k]} \\ &= \frac{\pi_j p_{ji} p_{i,i_1} \cdots p_{i_{k-1},i_k}}{\pi_i p_{i,i_1} \cdots p_{i_{k-1},i_k}} \\ &= \frac{\pi_j p_{ji}}{\pi_i} \\ &= P[X_{n-1} = j \mid X_n = i]. \end{aligned} \quad (8.52)$$

3. That is, let it be an irreducible, aperiodic, stationary Markov chain.

The above equations show that *the time-reversed process is also a Markov chain with one-step transition probabilities*

$$P[X_{n-1} = j \mid X_n = i] = q_{ij} = \frac{\pi_j p_{ji}}{\pi_i}. \tag{8.53}$$

Since $X_n$ is irreducible and aperiodic, its stationary state probabilities $\pi_j$ represent the proportion of time that the state is in state $j$. This proportion of time does not depend on whether one goes forward or backward in time, thus $\pi_j$ must also be the stationary pmf for the reverse process. Thus *the forward and reverse process must have the same stationary pmf.*

### ■■ Example 8.35

Suppose that a new light bulb is put in use at day $n = 0$, and suppose that each time a light bulb fails it is replaced on the next day. Let $X_n$ be the age of the light bulb (in days) at the end of day $n$. If $a_i$ is the probability that the lifetime $L$ of a light bulb is $i$ days, then the probability that the light bulb fails in day $j$ given that it has not failed up to then is

$$b_j = \frac{P[L = j]}{P[L \geq j]} = \frac{a_j}{\sum_{k=j}^{\infty} a_k} \qquad j = 1, 2, \ldots.$$

Thus the transition probabilities for $X_n$ are

$$\begin{aligned} p_{i,i+1} &= 1 - b_i && i = 1, 2, \ldots \\ p_{i1} &= b_i && i = 1, 2, \ldots \\ p_{ij} &= 0 && \text{otherwise.} \end{aligned}$$

Figure 8.9(a) shows the state transition diagram of $X_n$, and Fig. 8.10(a) shows a typical sample function that consists of a sawtooth-shaped function that increases linearly and then falls abruptly to one when a light bulb fails.

Figure 8.10(b) shows a sample function of the reverse process from which we deduce that the state transition diagram must be as shown in Fig. 8.9(b). The transition probabilities for the reverse process are obtained from Eq. (8.53):

$$\begin{aligned} q_{i,i-1} &= \frac{\pi_{i-1}}{\pi_i}(1 - b_{i-1}) && i = 2, 3, 4, \ldots \\ q_{1,i} &= \frac{\pi_i}{\pi_1} b_i && i = 1, 2, \ldots \\ q_{i,j} &= 0 && \text{otherwise.} \end{aligned}$$

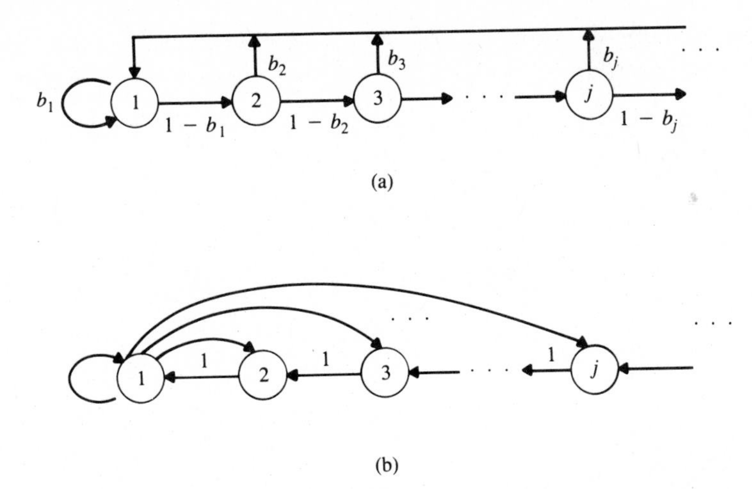


**FIGURE 8.9** (a) Transition diagram for age of a renewal process. (b) Transition diagram for a time-reversed process.

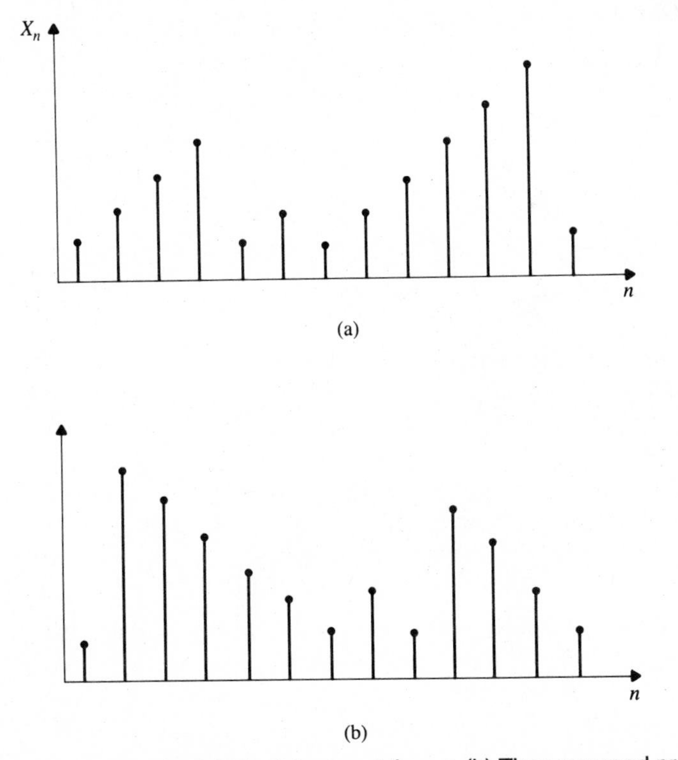


**FIGURE 8.10** (a) Age of a light bulb in use at time $n$. (b) Time-reversed process of $X_n$.

For now we defer the problem of finding the stationary state probabilities $\pi_j$. ■■

---

Example 8.35 shows that Eq. (8.53) provides us with conditions that must be satisfied by the stationary probabilities $\pi_j$. Suppose we were able to *guess* a pmf $\{\pi_j\}$ so that Eq. (8.53) holds, that is,

$$\pi_i q_{ij} = \pi_j p_{ji} \qquad \text{for all } i, j. \tag{8.54}$$

It then follows that $\{\pi_j\}$ is the stationary pmf. To see that this is so, sum Eq. (8.54) over all $j$, then

$$\sum_j \pi_j p_{ji} = \pi_i \sum_j q_{ij} = \pi_i \qquad \text{for all } i. \tag{8.55}$$

But Eq. (8.55) is the condition for $\pi_j$ to be the stationary pmf for the forward process, thus $\pi_j$ is the stationary pmf. Equation (8.54) thus provides us with another method for finding the stationary pmf of a discrete-time Markov chain.

## ■■ Example 8.36

The sample function of the reverse process in Example 8.35 suggests that for $i > 1$, the process moves from state $i$ to state $i - 1$ with probability one, that is,

$$q_{i,i-1} = \frac{\pi_{i-1}(1 - b_{i-1})}{\pi_i} = 1,$$

which implies that

$$\begin{aligned} \pi_i &= (1 - b_{i-1})\pi_{i-1} \qquad i = 2, 3, \ldots \\ &= (1 - b_{i-1})(1 - b_{i-2}) \cdots (1 - b_1)\pi_1, \end{aligned} \tag{8.56}$$

but from Example 8.35 for $i \geq 2$

$$(1 - b_{i-1}) = 1 - \frac{a_{i-1}}{\sum_{k=i-1}^{\infty} a_k} = \frac{\sum_{k=i}^{\infty} a_k}{\sum_{k=i-1}^{\infty} a_k},$$

so in Equation (8.56), the denominator of $(1 - b_{i-1})$ cancels the numerator of $(1 - b_{i-2})$, the denominator of $(1 - b_{i-2})$ cancels the numerator of $(1 - b_{i-3})$, and so on. Thus

$$\pi_i = \left\{\sum_{k=i}^{\infty} a_k\right\}\pi_1 = P[L \geq i]\pi_1 \qquad i = 2, 3, \ldots.$$

We obtain $\pi_1$ by using the fact that the probabilities sum to one:

$$1 = \pi_1 \sum_{i=1}^{\infty} P[L \geq i] = \pi_1 E[L],$$

where we have used Eq. (3.60) for $E[L]$. Thus

$$\pi_i = \frac{P[L \geq i]}{E[L]} \qquad i = 1, 2, \ldots. \tag{8.57}$$

■■

---

### Time Reversible Markov Chains

A stationary ergodic Markov chain is said to be **reversible** if the one-step transition probability matrix of the forward and reverse processes are the same, that is, if

$$q_{ij} = p_{ij} \qquad \text{for all } i, j. \tag{8.58}$$

Equations (8.53) and (8.58) together imply that a Markov chain is reversible if and only if

$$\pi_i p_{ij} = \pi_j p_{ji} \qquad \text{for all } i, j. \tag{8.59}$$

Since $\pi_i$ and $\pi_j$ are the long-term proportion of transitions out of states $i$ and $j$, respectively, Eq. (8.59) implies that a chain is reversible if the proportion of transitions from $i$ to $j$ is equal to the proportion of transitions from $j$ to $i$.

### ■■ Example 8.37
Discrete-Time Birth-and-Death Process

Figure 8.11 shows the state transition diagram for a discrete-time birth-and-death process with transition probabilities

$$\begin{aligned} p_{00} &= 0 & p_{01} &= 1 = a_0 & & \\ p_{i,i+1} &= a_i & p_{i,i-1} &= 1 - a_i & i &= 1, 2, \ldots \\ p_{ij} &= 0 & &\text{otherwise.} & & \end{aligned}$$

For any sample path, the number of transitions from $i$ to $i + 1$ can differ

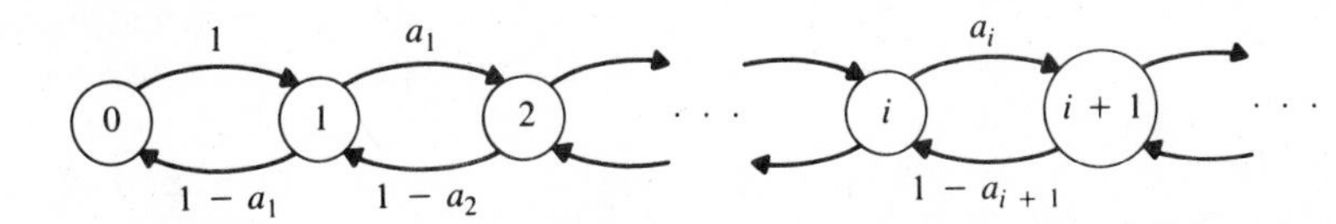


**FIGURE 8.11** Transition diagram for a discrete-time birth-and-death process.

by at most 1 from the number of transitions from $i + 1$ to $i$ since the only way to return to $i$ is through $i + 1$. Thus the long term proportion of transitions from $i$ to $i + 1$ is equal to that from $i + 1$ to $i$. Since these are the only possible transitions, it follows that birth-and-death processes are reversible.

Equation (8.59) implies that

$$a_j\pi_j = (1 - a_{j+1})\pi_{j+1} \qquad j = 0, 1, 2, \ldots,$$

which allows us to write all the $\pi_j$'s in terms of $\pi_0$:

$$\begin{aligned} \pi_j &= \left(\frac{a_{j-1}}{1 - a_j}\right)\cdots\left(\frac{a_0}{1 - a_1}\right)\pi_0 \\ &= \frac{a_{j-1}\cdots a_0}{(1 - a_j)\cdots(1 - a_1)}\pi_0 \\ &\triangleq R_j\pi_0. \end{aligned} \tag{8.60}$$

The probability $\pi_0$ is found from

$$1 = \pi_0 \sum_{j=0}^{\infty} R_j. \tag{8.61}$$

The series in Eq. (8.61) must converge in order for $\pi_j$ to exist. ■■

## Time Reversible Continuous-Time Markov Chains

Now consider a stationary, continuous-time Markov chain played backward in time. If $X(t) = i$ (i.e., the process is in state $i$ at time $t$), then the probability that the reverse process remains in state $i$ for an additional $s$ seconds is

$$\begin{aligned} P[X(t') = i, \quad t - s < t' < t \,|\, X(t) = i] &= \frac{P[X(t - s) = i, T_i > s]}{P[X(t) = i]} \\ &= \frac{P[X(t - s) = i]P[T_i > s]}{P[X(t) = i]}, \\ &= P[T_i > s] = e^{-v_i s} \end{aligned} \tag{8.62}$$

where $P[X(t - s) = i] = P[X(t) = i]$ because $X(t)$ is a stationary process, and where $T_i$ is the time spent in state $i$ for the forward process. Thus *the reverse process also spends an exponentially distributed amount of time with rate $v_i$ in state $i$.*

The jumps in the forward process $X(t)$ are determined by the embedded Markov chain $\tilde{q}_{ij}$, so the jumps in the reverse process are determined by the discrete-time Markov chain corresponding to the time-reversed embedded Markov chain given by Eq. (8.53):

$$q_{ij} = \frac{\pi_j \tilde{q}_{ji}}{\pi_i} \tag{8.63}$$

It follows that the transition rates for the time-reversed continuous-time process are given by

$$\gamma'_{ij} = \nu_i q_{ij} = \frac{\pi_j \nu_i \tilde{q}_{ji}}{\pi_i} \tag{8.64}$$

$$= \frac{\nu_i \pi_j \gamma_{ji}}{\pi_i \nu_j} = \frac{p_j \gamma_{ji}}{p_i},$$

where we used the fact that $\tilde{q}_{ji} = \gamma_{ji}/\nu_j$ and $p_j = c\pi_j/\nu_j$. By comparing Eq. (8.64) to Eq. (8.53), note that the transition rates $\gamma'_{ij}$ have simply replaced the transition probabilities $q_{ij}$ in going from the discrete-time to the continuous-time case.

The discussion that led to Eq. (8.54) provides us with another method for determining the stationary pmf $p_j$ of $X(t)$. If we can find a pmf $p_j$ such that

$$p_i \gamma'_{ij} = p_j \gamma_{ji} \qquad \text{for all } i, j, \tag{8.65}$$

then $p_i$ is the stationary pmf for $X(t)$ and for its time-reversed version.

Since the state occupancy times in the forward and reverse processes are exponential random variables with the same mean, the continuous-time Markov chain $X(t)$ is reversible if and only if its embedded Markov chain is reversible. Equation (8.59) implies that the following condition must be satisfied:

$$\pi_i \tilde{q}_{ij} = \pi_j \tilde{q}_{ji} \qquad \text{for all } i, j, \tag{8.66}$$

where $\pi_j$ is the stationary pmf of the embedded Markov chain. Recall from Eq. (8.50) that $\pi_j = c\nu_j p_j$, where $p_j$ is the stationary pmf of $X(t)$. Substituting into Eq. (8.66), we obtain

$$p_i \nu_i \tilde{q}_{ij} = p_j \nu_j \tilde{q}_{ji},$$

which is equivalent to

$$p_i \gamma_{ij} = p_j \gamma_{ji}. \tag{8.67}$$

Thus we conclude that $X(t)$ *is reversible if and only if Eq.* (8.67) *is satisfied*. As in the discrete-time case, Eq. (8.67) can be interpreted as stating that the rate at which $X(t)$ goes from state $i$ to state $j$ is equal to the rate at which $X(t)$ goes from state $j$ to state $i$.

### ■■ Example 8.38
Continuous-Time Birth-and-Death Process

Consider the general continuous-time birth-and-death process introduced in Example 8.19. The embedded Markov chain in this process is a discrete-time birth-and-death process of the type discussed in Example 8.37. It therefore follows that all continuous-time birth-and-death processes are time reversible. ■■

## SUMMARY

- A random process is said to be Markov if the future of the process, given the present, is independent of the past.
- A Markov chain is an integer-valued Markov process.
- The joint pmf for a Markov chain at several time instants is equal to product of the probability of the state at the first time instant and the probabilities of the subsequent state transitions [(Eq. (8.3)].
- For discrete-time Markov chains: (1) the $n$-step transition probability matrix $P(n)$ is equal to $P^n$, where $P$ is the one-step transition probability; (2) the state probability after $n$ steps $\mathbf{p}(n)$ is equal $\mathbf{p}(0)P^n$, where $\mathbf{p}(0)$ is the initial state probability; and (3) $P^n$ approaches a constant matrix as $n \to \infty$ for Markov chains that settle into steady state.
- A continuous-time Markov chain can be viewed as consisting of a discrete-time embedded Markov chain that determines the state transitions and of exponentially distributed state occupancy times.
- For continuous-time Markov chains: (1) the state probabilities and the transition probability matrix can be found by solving Eq. (8.30); (2) the steady state probabilities can be found by solving the global balance Eqs. (8.31b) or (8.31c).
- The states of a discrete-time Markov chain can be divided into disjoint classes. The long-term behavior of a Markov chain is determined by the properties of its classes. In particular, for ergodic Markov chains the stationary state probabilities represent the long-term proportion of time spent in each state.
- A continuous-time Markov chain has a steady state if its embedded Markov chain is irreducible and positive recurrent with unique stationary pmf given by the solution of the global balance equations.
- The time-reversed version of a Markov chain is also a Markov chain. A discrete-time (continuous-time) irreducible, stationary ergodic Markov chain is reversible if the transition probability matrix (transition rate matrix) for the forward and reverse processes are the same.

## CHECKLIST OF IMPORTANT TERMS

Birth-and-death process
Global balance equations
Chapman-Kolmogorov equations
Communication class
Embedded Markov chain
Ergodic Markov chain
Irreducible Markov chain
Markov chain
Markov process
Markov property
Null recurrent state
Period of a state/class
Positive recurrent state
Recurrent state/class
Reversible Markov chain
State occupancy time
State probabilities
Stationary probabilities
Time-reversed Markov chain
Transient state/class
Transition probability matrix

## ANNOTATED REFERENCES

References [1] and [2] contain very good discussions of discrete-time Markov chains. Reference [3] provides an introduction to discrete-time and continuous-time Markov chains at about the same level as this chapter. References [4] and [5] give a more rigorous and complete coverage of Markov chains and processes. A proof of Theorem 1 is given by Ross [4, pp. 107–114].

1. K. L. Chung, *Elementary Probability Theory with Stochastic Processes,* Springer-Verlag, New York, 1975.
2. W. Feller, *An Introduction to Probability Theory and Its Applications,* Volume 1, Wiley, New York, 1968.
3. S. M. Ross, *Introduction to Probability Models,* Academic Press, Orlando, Florida, 1985.
4. S. M. Ross, *Stochastic Processes,* Wiley, New York 1983.
5. D. R. Cox and H. D. Miller, *The Theory of Stochastic Processes,* Chapman and Hall, London, 1972.
6. H. Anton, *Elementary Linear Algebra,* Wiley and Sons, New York, 1981.

## PROBLEMS

### Section 8.1
### Markov Processes

1. Let $M_n$ denote the sequence of sample means from an iid random process $X_n$:

$$M_n = \frac{X_1 + X_2 + \cdots + X_n}{n}.$$

a. Is $M_n$ a Markov process?
b. If the answer to Part a is yes, find the following state transition pdf: $f_{M_n}(x \mid M_{n-1} = y)$.

2. An urn initially contains five black balls and five white balls. The following experiment is repeated indefinitely: A ball is drawn from the urn; if the ball is white it is put back in the urn, otherwise it is left out. Let $X_n$ be the number of black balls remaining in the urn after $n$ draws from the urn.
   a. Is $X_n$ a Markov process? If so, find the appropriate transition probabilities.
   b. Do the transition probabilities depend on $n$?
3. Show that if a random process $X(t)$ has independent increments then it is also a Markov process.
4. The result from Problem 3 implies that the Brownian motion process is a Markov process. Find the state transition pdf.
5. Let $X_n$ be the Bernoulli iid process, and let $Y_n$ be given by

$$Y_n = X_n + X_{n-1}.$$

   It was shown in Example 8.2 that $Y_n$ is not a Markov process. Consider the vector process defined by $\mathbf{Z}_n = (X_n, X_{n-1})$.
   a. Show that $\mathbf{Z}_n$ is a Markov process.
   b. Find the state transition diagram for $\mathbf{Z}_n$.
6. Show that the following autoregressive process is a Markov process:

$$Y_n = rY_{n-1} + X_n \qquad Y_0 = 0,$$

   where $X_n$ is an iid process.

## Section 8.2
## Discrete-Time Markov Chains

7. Let $X_n$ be an iid random process. Show that $X_n$ is a Markov process and give its one-step transition probability matrix.
8. Let $X_n$ be the Markov chain defined in Problem 2.
   a. Find the one-step transition probability matrix $P$ for $X_n$.
   b. Find the two-step transition probability matrix $P^2$ by matrix multiplication. Check your answer by computing $p_{54}(2)$ and comparing it to the corresponding entry in $P^2$.
   c. What happens to $X_n$ as $n$ approaches infinity? Use your answer to guess the limit of $P^n$ as $n \to \infty$.
9. The vector process $\mathbf{Z}_n$ in Problem 5 has four possible states, so in effect it is equivalent to a Markov chain with states $\{0, 1, 2, 3\}$.
   a. Find the one-step transition probability matrix $P$.

b. Find $P^2$ and check your answer by computing the probability of going from state $(0, 1)$ to state $(0, 1)$ in two steps.
c. Show that $P^n = P^2$ for all $n > 2$. Give an intuitive justification for why this is true for this random process.
d. Find the steady state probabilities for the process.

10. Show that if $P^k$ has identical rows, then $P^j$ has identical rows for all $j \geq k$.

11. Two gamblers play the following game. A fair coin is flipped; if the outcome is heads, player $A$ pays player $B$ \$1, and if the outcome is tails player $B$ pays player $A$ \$1. The game is continued until one of the players goes broke. Suppose that initially player $A$ has \$1 and player $B$ has \$2, so a total of \$3 are up for grabs. Let $X_n$ denote the number of dollars held by player $A$ after $n$ trials.

a. Show that $X_n$ is a Markov chain.
b. Sketch the state transition diagram for $X_n$ and give the one-step transition probability matrix $P$.
c. Use the state transition diagram to help you show that for $n$ even (i.e., $n = 2k$),

$$p_{ii}(n) = \left(\frac{1}{2}\right)^n \qquad \text{for } i = 1, 2$$

$$p_{10}(n) = \frac{2}{3}\left(1 - \left(\frac{1}{4}\right)^k\right) = p_{23}(n).$$

d. Find the $n$-step transition probability matrix for $n$ even using Part c.
e. Find the limit of $P^n$ as $n \to \infty$.
f. Find the probability that player 1 eventually wins.

12. A certain part of a machine can be in two states: working or undergoing repair. A working part fails during the course of a day with probability $a$. A part undergoing repair is put into working order during the course of a day with probability $b$. Let $X_n$ be the state of the part.

a. Show that $X_n$ is a two-state Markov chain and give its one-step transition probability matrix $P$.
b. Find the $n$-step transition probability matrix $P^n$.
c. Find the steady state probability for each of the two states.

13. A machine consists of two parts that fail and are repaired independently. A working part fails during any given day with probability $a$. A part that is not working is repaired by the next day with probability $b$. Let $X_n$ be the number of working parts in day $n$.

a. Show that $X_n$ is a three-state Markov chain and give its one-step transition probability matrix $P$.

b. Show that the steady state pmf $\pi$ is binomial with parameter $p = b/(a + b)$.
c. What do you expect is steady state pmf for a machine that consists of $n$ parts?

## Section 8.3
## Continuous-Time Markov Chains

14. Consider the simple queueing system discussed in Example 8.15.
    a. Use the results in Example 8.15 to find the state transition probability matrix.
    b. Find the following probabilities:

$$P[X(1.5) = 1, X(3) = 1 \mid X(0) = 0]$$
$$P[X(1.5) = 1, X(3) = 1].$$

15. A critical part of a machine has an exponentially distributed lifetime with parameter $\alpha$. Suppose that $n$ spare parts are initially in stock, and let $N(t)$ be the number of spares left at time $t$.
    a. Find $p_{ij}(t) = P[N(s + t) = j \mid N(s) = i]$.
    b. Find the transition probability matrix.
    c. Find $p_j(t)$.
16. A machine shop initially has $n$ identical machines in operation. Assume that the time until breakdown for each machine is an exponentially distributed random variable with parameter $\alpha$. Let $N(t)$ denote the number of machines in working order at time $t$. Repeat parts a, b, and c of Problem 15 and show that $N(t)$ is a binomial random variable with parameter $p = e^{-\alpha t}$.
17. A shop has $n$ machines and one technician to repair them. A machine remains in the working state for an exponentially distributed time with parameter $\mu$. The technician works on one machine at a time, and it takes him an exponentially distributed time of rate $\alpha$ to repair each machine. Let $X(t)$ be the number of working machines at time $t$.
    a. Show that if $X(t) = k$, then the time until the next machine breakdown is an exponentially distributed random variable with rate $k\mu$.
    b. Find the transition rate matrix $[\gamma_{ij}]$ and sketch the transition rate diagram for $X(t)$.
    c. Write the global balance equations and find the steady state probabilities for $X(t)$.
18. A speaker alternates between periods of speech activity and periods of silence. Suppose that the former are exponentially distributed with parameter $\alpha$ and the latter exponentially distributed with parameter

β. Consider a group of $n$ independent speakers and let $N(t)$ denote the number of speakers in speech activity at time $t$.

a. Find the transition rate diagram and the transition rate matrix for this system.

b. Write the global balance equations and find the steady state pmf. Why is the solution not surprising?

19. Consider the single-server queueing system in Example 8.18. Suppose that at most $K$ customers can be in the system at any time. Let $N(t)$ be the number of customers in the system at time $t$. Find the steady state probabilities for $N(t)$.

## Section 8.4
## Classes of States, Recurrence Properties, and Limiting Probabilities

20. Sketch the state transition diagrams for the Markov chains with the following transition probability matrices. Specify the classes of the Markov chains and classify them as recurrent or transient.

a. $$\begin{bmatrix} 0 & 1 & 0 \\ 1/2 & 0 & 1/2 \\ 1 & 0 & 0 \end{bmatrix}$$

b. $$\begin{bmatrix} 1 & 0 & 0 \\ 0 & 0 & 1 \\ 0 & 1 & 0 \end{bmatrix}$$

c. $$\begin{bmatrix} 0 & 1/2 & 1/2 & 0 \\ 0 & 0 & 1 & 0 \\ 0 & 0 & 1 & 0 \\ 1 & 0 & 0 & 0 \end{bmatrix}$$

d. $$\begin{bmatrix} 1/2 & 1/2 & 0 & 0 \\ 1 & 0 & 0 & 0 \\ 1/2 & 0 & 1/4 & 1/4 \\ 0 & 1/4 & 1/4 & 1/2 \end{bmatrix}$$

21. Characterize the long-term behavior of the Markov chains in the previous problem. Find the long-term proportion of time spent in each state, and the stationary pmf where applicable.

22. Consider a random walk in the set $\{0, 1, \ldots, M\}$ with transition probabilities

$$p_{01} = 1, \qquad p_{M,M-1} = 1, \qquad \text{and} \qquad p_{i,i-1} = q \qquad p_{i,i+1} = p \qquad i = 1, \ldots, M-1.$$

Find the long-term proportion of time spent in each state, and the limit of $p_{ii}(n)$ as $n \to \infty$.

23. Repeat the previous problem if the random walk in Problem 22 is modified so that $p_{01} = p$, $p_{00} = q$, $p_{M,M-1} = q$, and $p_{M,M} = p$.
24. Find the embedded Markov chain for the process described in Example 8.18. Characterize the long-term probabilities of the process using Eq. 8.50.
25. Find the embedded Markov chain for the process described in Example 8.19. Characterize the long-term probabilities of the process using Eq. 8.50.

### Section *8.5
### Time-Reversed Markov Chains

26. $N$ balls are distributed in two urns. At time $n$, a ball is selected at random and it is removed from its present urn and placed in the other urn. Let $X_n$ denote the number of balls in urn 1.
    a. Find the transition probabilities for $X_n$.
    b. Argue that the process is time-reversible and then obtain the steady state probabilities for $X_n$.
27. A point moves in the unit circle in jumps of $\pm 90°$. Suppose that the process is initially at $0°$, and that the probability of $+90°$ is $p$.
    a. Find the transition probabilities for the resulting Markov chain and obtain the steady state probabilities.
    b. Is the process reversible? Why?
28. Find the transition probabilities for the time-reversed version of the random walk discussed in Problem 23. Is the process revesible?
29. a. Specify the time-reversed version of the process defined in Problem 19. Is the process reversible?
    b. Find the steady state probabilities of the process using Eq. 8.67.

CHAPTER 9

# Introduction to Queueing Theory

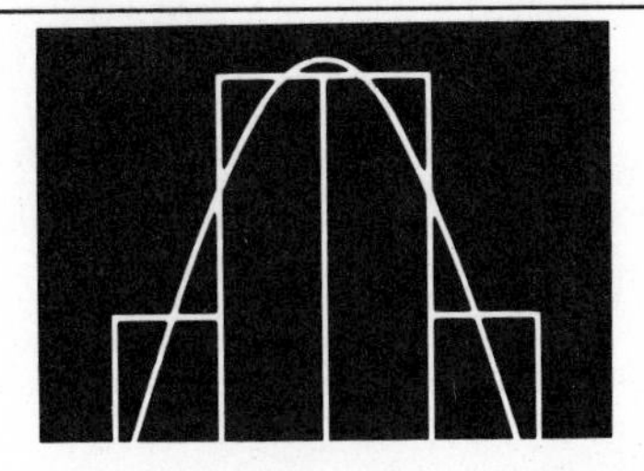

In many applications, expensive resources such as computers and communication lines are shared among a community of users. Users place demands for these resources at random times and they require use of these resources for time periods whose durations are random. Inevitably requests for the resource arrive while the resource is occupied, and a mechanism to provide an orderly access to the resource is required. The most common access control mechanism is to file user requests in a waiting line or "queue" such as might be formed at a bank by customers waiting to be serviced.

Queueing theory deals with the study of waiting lines. The random nature of the demand behavior of customers implies that probabilistic measures such as average delay, average throughput, and delay percentiles are required to assess the performance of such systems. Queueing theory provides us with the probability tools needed to evaluate these measures.

## 9.1 THE ELEMENTS OF A QUEUEING SYSTEM

Figure 9.1(a) shows a typical queueing system and Fig. 9.1(b) shows the elements of a queueing system model. Customers from some population arrive at the system at the random **arrival times** $S_1, S_2, S_3, \ldots, S_i, \ldots$ where $S_i$ denotes the arrival time of the $i$th customer. We denote the customer arrival rate by $\lambda$.

The queueing system has one or more identical servers as shown in

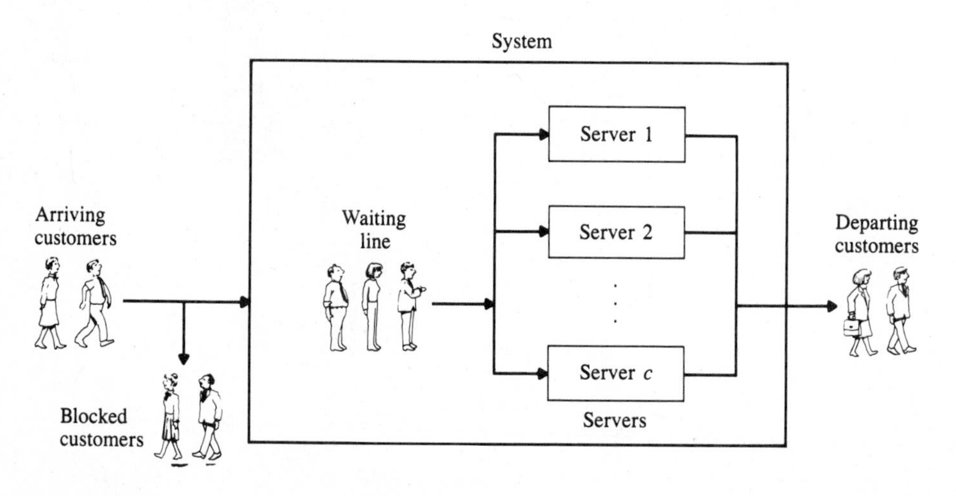

**FIGURE 9.1a** Elements of a queueing system.

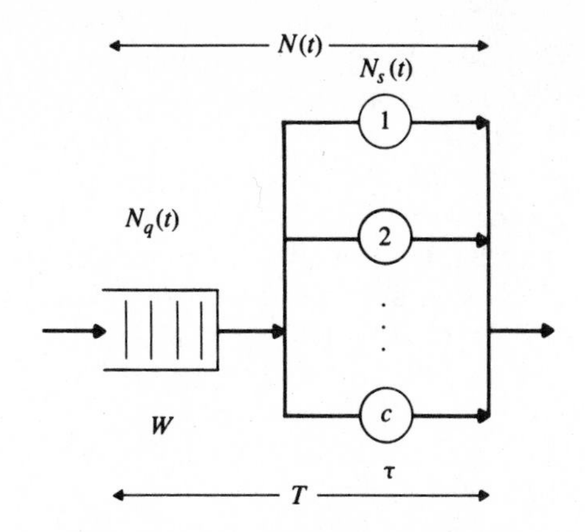


**FIGURE 9.1b** Elements of a queueing system model: $N(t)$, number in system; $N_q(t)$, number in queue; $N_S(t)$, number in service; $W$, waiting time in queue; $\tau$, service time; and $T$, total time in the system.

Fig. 9.1(a). The $i$th customer arrives at the system seeking a service that will require $\tau_i$ seconds of **service time** from one server. If all the servers are busy, then the arriving customer joins a queue where he remains until a server becomes available. Sometimes, only a limited number of waiting spaces are available so customers that arrive when there is no room are turned away. Such customers are called "blocked" and we will denote the rate at which customers are turned away by $\lambda_b$.

The **queue discipline** specifies the order in which customers are selected from the queue and allowed into service. For example, some common queueing disciplines are "first-come, first-serve," and "last-come, first-serve." The queueing discipline affects the **waiting time** $W_i$ that elapses from the arrival time of the $i$th customer until the time when it enters service. The **total delay** $T_i$ of the $i$th customer in the system is the sum of its waiting time and service time:

$$T_i = W_i + \tau_i. \tag{9.1}$$

From the customer's point of view, the performance of the system is given by the statistics of the waiting time $W$ and the total delay $T$, and the proportion of customers that are blocked, $\lambda_b/\lambda$. From a resource allocation point of view, the performance of the system is measured by the proportion of time that each server is utilized and the rate at which customers are serviced by the system, $\lambda_d = \lambda - \lambda_b$. These quantities are a function of $N(t)$, the number of customers in the system at time $t$, and $N_q(t)$, the number of customers in queue at time $t$.

The notation $a/b/m/K$ is used to describe a queueing system, where $a$ specifies the type of arrival process, $b$ denotes the service time distribution, $m$ specifies the number of servers, and $K$ denotes the maximum number of customers allowed in the system at any time. If $a$ is given by M, then the arrival process is Poisson and the interarrival times are independent, identically distributed (iid) exponential random variables. If $b$ is given by M, then the service times are iid exponential random variables. If $b$ is given by G, then the service times are iid according to some general distribution. For example, in this chapter we will deal with M/M/1, M/M/1/$K$, M/M/$c$, M/M/$c$/$c$ and M/G/1 queues.

## 9.2

## LITTLE'S FORMULA

We now develop **Little's formula,** which states that, for systems that reach steady state, the average number of customers in a system is equal to the product of the average arrival rate and the average time spent in the system:

$$E[N] = \lambda E[T]. \tag{9.2}$$

This formula is valid under very general conditions so it is applicable in an amazing number of situations.

Consider the queueing system shown in Fig. 9.2. The system begins empty at time $t = 0$, and the customer arrival times are denoted by $S_1, S_2, \ldots$. Let $A(t)$ be the number of customer arrivals up to time $t$. The $i$th customer spends time $T_i$ in the system and then departs at time $D_i = S_i + T_i$. We will let $D(t)$ be the number of customer departures up to time $t$. The **number of customers in the system** at time $t$ is the number of arrivals that have not yet left the system

$$N(t) = A(t) - D(t). \tag{9.3}$$

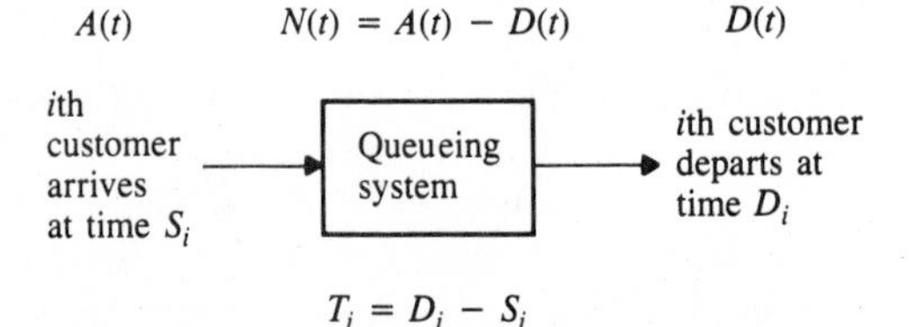


**FIGURE 9.2** Time in system is departure time minus arrival time. Number in system at time $t$ is number of arrivals minus number of departures.

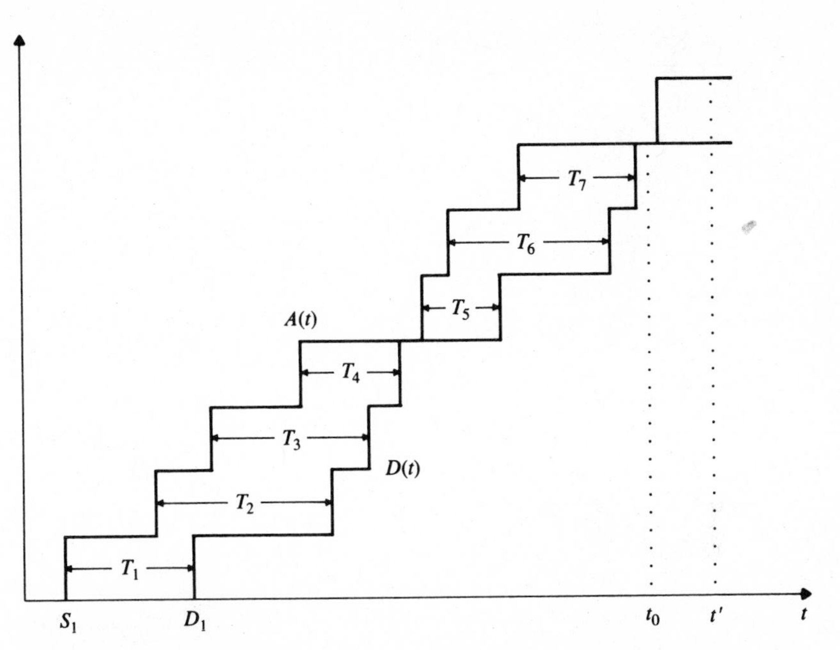


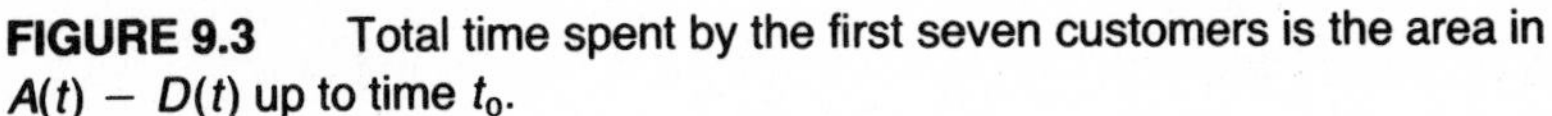

**FIGURE 9.3** Total time spent by the first seven customers is the area in $A(t) - D(t)$ up to time $t_0$.

Figure 9.3 shows a possible sample path for $A(t)$, $D(t)$, and $N(t)$ in a queueing system with "first-come, first-serve" service discipline.

Consider the time average of the number of customers in the system $N(t)$ during the interval $(0, t]$:

$$\langle N \rangle_t = \frac{1}{t} \int_0^t N(t')\, dt'. \tag{9.4}$$

In Fig. 9.3, $N(t)$ is the shaded region between $A(t)$ and $D(t)$, so the above integral is given by the area of the shaded region up to time $t$. It can be seen that each customer who has departed the system by time $t$ contributes $T_i$ to the integral, and thus the integral is simply the total time all customers have spent in the system up to time $t$.

Consider, for now, a time instant $t = t_0$ for which $N(t) = 0$ as in Fig. 9.3, then the integral is exactly given by the sum of the $T_i$ of the first $A(t)$ customers:

$$\langle N \rangle_t = \frac{1}{t} \sum_{i=1}^{A(t)} T_i. \tag{9.5}$$

The average arrival rate up to time $t$ is given by

$$\langle \lambda \rangle_t = \frac{A(t)}{t}. \tag{9.6}$$

If we solve Eq. (9.6) for $t$ and substitute into Eq. (9.5), we obtain

$$\langle N \rangle_t = \langle \lambda \rangle_t \frac{1}{A(t)} \sum_{i=1}^{A(t)} T_i. \tag{9.7}$$

Let $\langle T \rangle_t$ be the average of the times spent in the system by the first $A(t)$ customers, then

$$\langle T \rangle_t = \frac{1}{A(t)} \sum_{i=1}^{A(t)} T_i. \tag{9.8}$$

Comparing Eqs. (9.7) and (9.8), we conclude that

$$\langle N \rangle_t = \langle \lambda \rangle_t \langle T \rangle_t \tag{9.9}$$

Finally, we assume that as $t \to \infty$, with probability one, the above time averages converge to the expected value of the corresponding steady state random processes, that is,

$$\begin{aligned} &\langle N \rangle_t \to E[N] \\ &\langle \lambda \rangle_t \to \lambda \\ &\langle T \rangle_t \to E[T]. \end{aligned} \tag{9.10}$$

Equations (9.9) and (9.10) then imply Little's formula:

$$E[N] = \lambda E[T]. \tag{9.11}$$

The restriction of $t$ to instants $t_0$ where $N(t_0) = 0$ is not necessary. The time average of $N(t)$ up to an arbitrary time $t'$ as shown in Fig. 9.3 is given by the average up to time $t_0$ plus a contribution from the interval from $t_0$ to $t'$. If $E[N] < \infty$, then as $t$ becomes large, this contribution becomes negligible.

The assumption of first-come, first-serve service discipline is not necessary. It turns out that Little's formula holds for many service disciplines. See Problem 2 for examples. In addition, Little's formula holds for systems with an arbitrary number of servers.

Up to this point we have implicitly assumed that the "system" is the entire queueing system, so $N$ is the number in the queueing system and $T$ is the time spent in the queueing system. However, Little's formula is so general that it applies to many interpretations of "system." The Examples 9.1 and 9.2 show other designations for "system."

### ■■ Example 9.1
Mean Number in Queue

Let $N_q(t)$ be the number of customers waiting in the queue for the server to become available, and let the random variable $W$ denote the waiting time. If we designate the queue to be the "system," then Little's formula becomes

$$E[N_q] = \lambda E[W]. \tag{9.12}$$

■■

---

**■■ Example 9.2**
Server Utilization

Let $N_s(t)$ be the number of customers that are being served at time $t$, and let $\tau$ denote the service time. If we designate the set of servers to be the "system," then Little's formula becomes

$$E[N_s] = \lambda E[\tau]. \tag{9.13}$$

$E[N_s]$ is the average number of busy servers for a system in steady state.

For single server systems, $N_s(t)$ can only be 0 or 1, so $E[N_s]$ represents the proportion of time that the server is busy. If $p_0 = P[N(t) = 0]$ denotes the steady state probability that the system is empty, then we must have that

$$1 - p_0 = E[N_s] = \lambda E[\tau] \tag{9.14}$$

or

$$p_0 = 1 - \lambda E[\tau], \tag{9.15}$$

since $1 - p_0$ is the proportion of time that the server is busy. For this reason, the **utilization of a single server system** is defined by

$$\rho = \lambda E[\tau]. \tag{9.16}$$

We will similarly define **utilization of a $c$-server system** by

$$\rho = \frac{\lambda E[\tau]}{c}. \tag{9.17}$$

From Eq. (9.13) $\rho$ represents the average fraction of busy servers. ■■

---

## 9.3 THE M/M/1 QUEUE

Consider a single server system in which customers arrive according to a Poisson process of rate $\lambda$ so the interarrival times are iid exponential random variables with mean $1/\lambda$. Assume that the service times are iid exponential random variables with mean $1/\mu$, and that the interarrival and service times are independent. In addition, assume that the system can accommodate an unlimited number of customers. The resulting system is an M/M/1 queueing system. In this section we find the steady state pmf of $N(t)$, the number of customers in the system, and the pdf of $T$, the total customer delay in the system.

### Distribution of Number in the System

The number of customers $N(t)$ in an M/M/1 system is a continuous-time Markov chain. To see why, suppose that we are given that $N(t) = k$, and consider the next possible change in the number in the system. The time until the next arrival is an exponential random variable that is independent of the service times of customers already in the system. The memoryless property of the exponential random variable implies that this interarrival time is independent of the present and past history of $N(t)$. If the system is nonempty (i.e., $N(t) > 0$) the time until the next departure is also an exponential random variable. The memoryless property implies that the time until the next departure is independent of the time already spent in service. Thus if we know that $N(t) = k$, then the past history of the system is irrelevant as far as the probabilities of future states are concerned. This is the property required of a Markov chain.

To find the transition rates for $N(t)$, consider the probabilities of the various ways in which $N(t)$ can change.

i. Since $A(t)$, the number of arrivals in an interval of length $t$, is a Poisson process, the probability of one arrival in an interval of length $\delta$ is

$$P[A(\delta) = 1] = \frac{\lambda\delta}{1!}e^{-\lambda\delta} = \lambda\delta\left\{1 - \frac{\lambda\delta}{1!} + \frac{(\lambda\delta)^2}{2!} - \cdots\right\}$$
$$= \lambda\delta + o(\delta). \tag{9.18}$$

ii. Similarly, the probability of more than one arrival is

$$P[A(\delta) \geq 2] = o(\delta). \tag{9.19}$$

iii. Since the service time is an exponential random variable $\tau$, the time a customer has spent in service is independent of how much longer he will remain in service because of the memoryless property of $\tau$. In particular, the probability of a customer in service completing his service in the next $\delta$ seconds is

$$P[\tau \leq \delta] = 1 - e^{-\mu\delta} = \mu\delta + o(\delta). \tag{9.20}$$

iv. Since service times and the arrival process are independent, the probability of one arrival and one departure in an interval of length $\delta$ is

$$P[A(\delta) = 1, \tau \leq \delta] = P[A(\delta) = 1]P[\tau \leq \delta] = o(\delta) \tag{9.21}$$

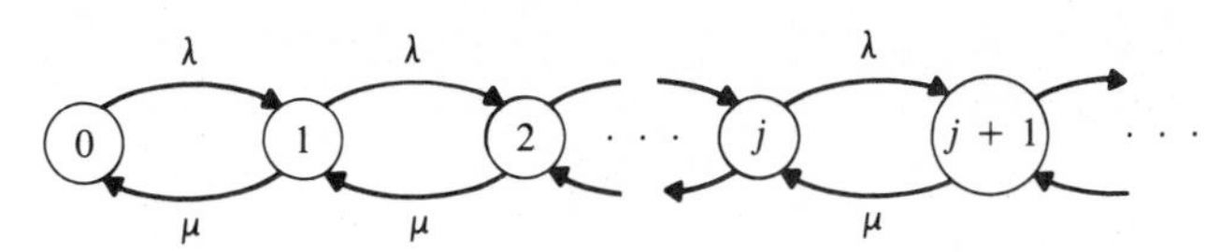


**FIGURE 9.4** Transition rate diagram for M/M/1 system.

from Eqs. (9.18) and (9.20). Similarly, the probability of any change that involves more than a single arrival or a single departure is $o(\delta)$.

Properties (i) through (iv) imply that $N(t)$ has the transition rate diagram shown in Fig. 9.4. The global balance equations for the steady state probabilities are

$$\lambda p_0 = \mu p_1$$
$$(\lambda + \mu)p_j = \lambda p_{j-1} + \mu p_{j+1} \qquad j = 1, 2, \ldots. \tag{9.22}$$

In Example 8.18, we saw that a steady state solution exists when $\rho = \lambda/\mu < 1$:

$$P[N(t) = j] = (1 - \rho)\rho^j \qquad j = 0, 1, 2, \ldots. \tag{9.23}$$

The condition $\rho = \lambda/\mu < 1$ must be met if the system is to be stable in the sense that $N(t)$ does not grow without bound. Since $\mu$ is the maximum rate at which the server can process customers, the condition $\rho < 1$ is equivalent to

$$\text{arrival rate} = \lambda < \mu = \text{maximum service rate}. \tag{9.24}$$

If the inequality is violated, we have customers arriving at the system faster than they can be processed and sent out. This is an unstable situation in which the number in the queue will grow steadily without bound.

The mean number of customers in the system is given by

$$E[N] = \sum_{j=0}^{\infty} jP[N(t) = j] = \frac{\rho}{1 - \rho}, \tag{9.25}$$

where we have used the fact that $N$ has a geometric distribution (see Table 3.1).

The mean total customer delay in the system is found from Eq. (9.25) and Little's formula:

$$E[T] = \frac{E[N]}{\lambda} = \frac{\rho/\lambda}{1 - \rho}$$
$$= \frac{1/\mu}{1 - \rho} = \frac{E[\tau]}{1 - \rho} = \frac{1}{\mu - \lambda}. \tag{9.26}$$

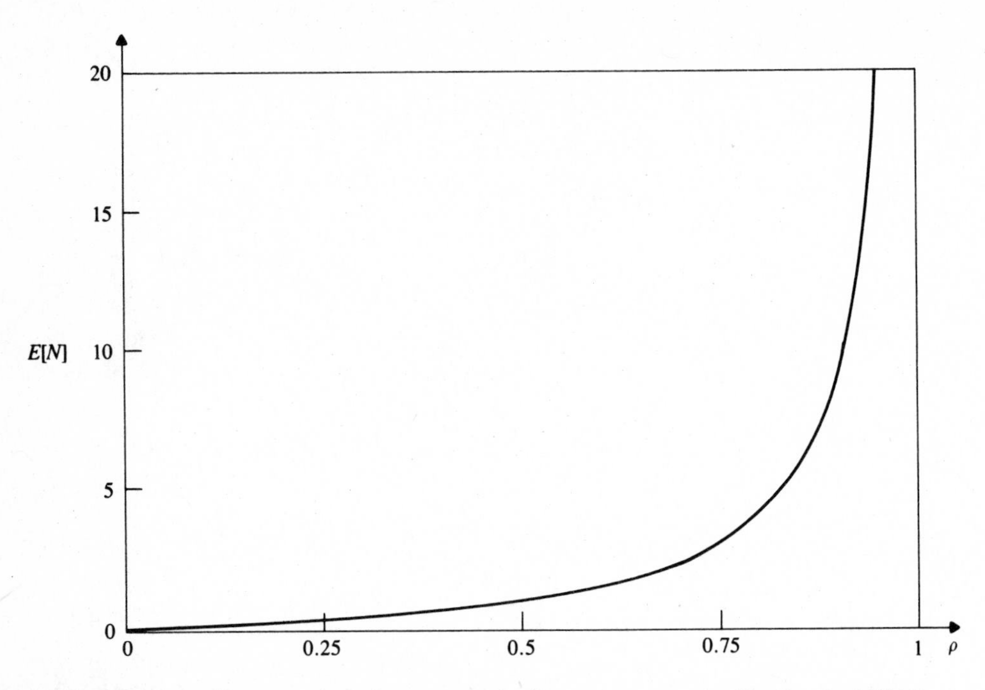


**FIGURE 9.5** Mean number of customers in the system versus utilization for M/M/1 queue.

The mean waiting time in queue is given by the mean of the total time in the system minus the service time:

$$
\begin{aligned}
E[W] &= E[T] - E[\tau] \\
&= \frac{E[\tau]}{1-\rho} - E[\tau] \\
&= \frac{\rho}{1-\rho} E[\tau]. \qquad (9.27)
\end{aligned}
$$

Little's formula then gives the mean number in queue:

$$
\begin{aligned}
E[N_q] &= \lambda E[W] \\
&= \frac{\rho^2}{1-\rho}. \qquad (9.28)
\end{aligned}
$$

The server utilization (defined in Example 9.2) is given by

$$1 - p_0 = 1 - (1 - \rho) = \rho = \frac{\lambda}{\mu}. \qquad (9.29)$$

Figures 9.5 and 9.6 show $E[N]$ and $E[T]$ versus $\rho$. It can be seen that as $\rho$

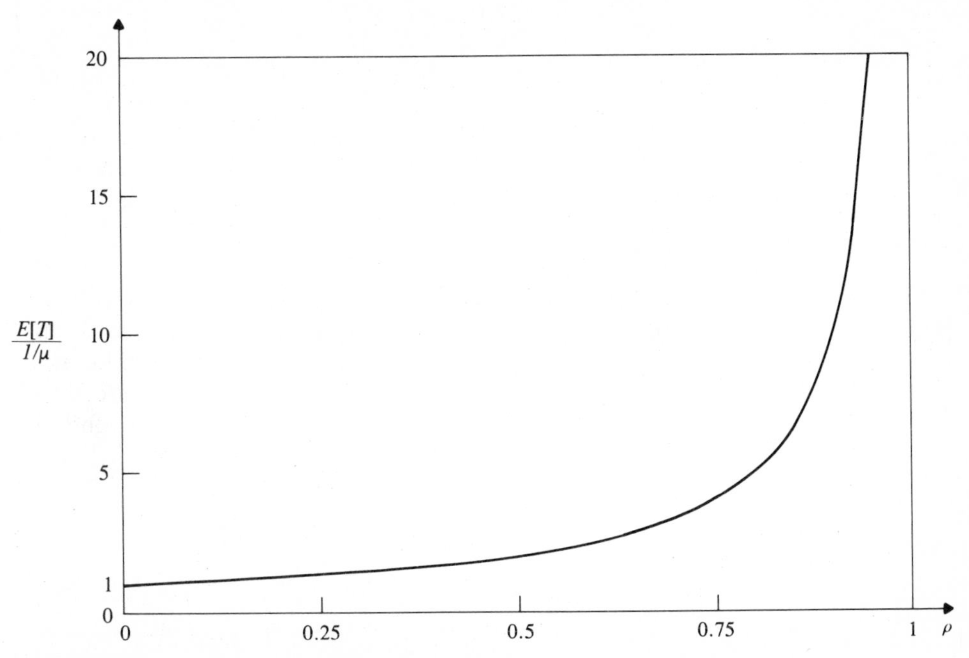


**FIGURE 9.6** Mean total customer delay versus utilization for M/M/1 system. The delay is expressed in multiples of mean service times.

approaches one, the mean number in the system and the system delay become arbitrarily large.

## ■■ Example 9.3

A concentrator receives messages from a group of terminals and transmits them over a single transmission line. Suppose that messages arrive according to a Poisson process at a rate of one message every 4 milliseconds, and suppose that message transmission times are exponentially distributed with mean 3 ms. Find the mean number of messages in the system and the mean total delay in the system. What percentage increase in arrival rate results in a doubling of the above mean total delay?

The arrival rate is 1/4 messages/ms and the mean service time is 3 ms. The utilization is therefore

$$\rho = \frac{1}{4}(3) = \frac{3}{4}.$$

The mean number of customers in the system is then

$$E[N] = \frac{\rho}{1-\rho} = 3.$$

The mean time in the system is

$$E[T] = \frac{E[N]}{\lambda} = \frac{3}{1/4} = 12 \text{ ms}.$$

The mean time in the system will be doubled to 24 ms when

$$24 = \frac{E[\tau]}{1 - \rho'} = \frac{3}{1 - \rho'}$$

The resulting utilization is $\rho' = 7/8$ and the corresponding arrival rate is $\lambda' = \rho'\mu = 7/24$. The original arrival rate was 6/24. Thus an increase in arrival rate of $1/6 = 17\%$ leads to a 100% increase in mean system delay.

The point of this example is that *the onset of congestion is swift.* The mean delay increases rapidly once the utilization increases beyond a certain point. ■■

---

### ■■ Example 9.4
Concentration and Effect of Scale

A large processor handles transactions at a rate of $K\mu$ transactions per second. Suppose transactions arrive according to a Poisson process of rate $K\lambda$ transactions/second, and that transactions require an exponentially distributed amount of processing time. Suppose that a proposal is made to eliminate the large processor and to replace it with $K$ processors each with processing rate of $\mu$ transactions per second and with arrival rates $\lambda$. Compare the mean delay performance of the existing and the proposed systems.

The large processor system is an M/M/1 queue with arrival rate $K\lambda$, service rate $K\mu$, and utilization $\rho = K\lambda/K\mu = \lambda/\mu$. The mean delay is given by Eq. (9.26):

$$E[T] = \frac{E[\tau]}{1 - \rho} = \frac{1/K\mu}{1 - \rho}.$$

Each of the small processors is an M/M/1 system with arrival rate $\lambda$, service rate $\mu$, and utilization $\rho = \lambda/\mu$. The mean delay is

$$E[T'] = \frac{E[\tau']}{1 - \rho} = \frac{1/\mu}{1 - \rho} = KE[T].$$

Thus, the system with the single large processor with processing rate $K\mu$ has a smaller mean delay than the system with $K$ small processors each of rate $\mu$. In other words, the concentration of customer demand into a single system results in significant delay performance improvement. ■■

---

### Delay Distribution in M/M/1 System and Arriving Customer's Distribution

Let $N_a$ denote the number of customers found in the system by a customer arrival. We call $P[N_a = k]$ the **arriving customer's distribution.** We now show that if arrivals are Poisson and independent of the system state and customer service times, then the arriving customer's distribution is equal to the steady state distribution for the number in the system. A customer that arrives at time $t + \delta$ finds $k$ in the system if $N(t) = k$, thus

$$\begin{aligned} P[N_a(t) = k] &= \lim_{\delta \to 0} P[N(t) = k \mid A(t + \delta) - A(t) = 1] \\ &= \lim_{\delta \to 0} \frac{P[N(t) = k, A(t + \delta) - A(t) = 1]}{P[A(t + \delta) - A(t) = 1]} \\ &= \lim_{\delta \to 0} \frac{P[A(t + \delta) - A(t) = 1 \mid N(t) = k] P[N(t) = k]}{P[A(t + \delta) - A(t) = 1]}, \end{aligned}$$

where we have used the definition of conditional probability. The probability of an arrival in the interval $(t, t + \delta]$ is independent of $N(t)$, thus

$$\begin{aligned} P[N_a(t) = k] &= \lim_{\delta \to 0} \frac{P[A(t + \delta) - A(t) = 1] P[N(t) = k]}{P[A(t + \delta) - A(t) = 1]} \\ &= P[N(t) = k]. \end{aligned}$$

Thus the probability that $N_a = k$ is simply the proportion of time during which the system has $k$ customers in the system. For the M/M/1 queueing system under consideration we have

$$P[N_a = k] = P[N(t) = k] = (1 - \rho)\rho^k. \tag{9.30}$$

We are now ready to compute the distribution for the total time $T$ that a customer spends in an M/M/1 system. Suppose that an arriving customer finds $k$ in the system, that is $N_a = k$. If the service discipline is "first-come, first-served," then $T$ is the residual service time of the customer found in service, the service times of the $k - 1$ customers found in queue, and the service time of the arriving customer. The memoryless property of the exponential service time implies that the residual service time of the customer found in service has the same distribution as a full service time. Thus $T$ is the sum of $k + 1$ iid exponential random variables. In Example 5.5 we saw that this sum has the Gamma pdf

$$f_T(x \mid N_a = k) = \frac{(\mu x)^k}{k!} \mu e^{-\mu x} \qquad x > 0. \tag{9.31}$$

The pdf of $T$ is found by averaging over the probability of an arriving customer finding $k$ messages in the system, $P[N_a = k]$. Thus the pdf of $T$

is

$$
\begin{aligned}
f_T(x) &= \sum_{k=0}^{\infty} \frac{(\mu x)^k}{k!} \mu e^{-\mu x} P[N(t) = k] \\
&= \sum_{k=0}^{\infty} \frac{(\mu x)^k}{k!} \mu e^{-\mu x}(1-\rho)\rho^k \\
&= (1-\rho)\mu e^{-\mu x} \sum_{k=0}^{\infty} \frac{(\mu \rho x)^k}{k!} \\
&= (1-\rho)\mu e^{-\mu x} e^{\mu \rho x} \\
&= (\mu - \lambda) e^{-(\mu-\lambda)x} \qquad x > 0.
\end{aligned} \tag{9.32}
$$

Thus $T$ is an exponential random variable with mean $1/(\mu - \lambda)$. Note that this is in agreement with Eq. (9.26) for the mean of $T$ obtained through Little's formula.

We can similarly show that the pdf for the waiting time is

$$
f_W(x) = (1-\rho)\delta(x) + \lambda(1-\rho)e^{-\mu(1-\rho)x} \qquad x > 0 \tag{9.33}
$$

### ■■ Example 9.5

Find the 95% percentile of the total delay.

The $p$th percentile of $T$ is that value of $x$ for which

$$
\begin{aligned}
p &= P[T \le x] \\
&= \int_0^x (\mu - \lambda) e^{-(\mu-\lambda)y}\, dy = 1 - e^{-(\mu-\lambda)x},
\end{aligned}
$$

which yields

$$
x = \frac{1}{\mu - \lambda} \ln \frac{1}{1-p} = -E[T] \ln(1-p). \tag{9.34}
$$

The 95% percentile is obtained by substituting $p = .95$ above. The result is $x \simeq 3.0\, E[T]$. ■■

---

## The M/M/1 System with Finite Capacity

Real systems can only accommodate a finite number of customers, but the assumption of infinite capacity is convenient when the probability of having a full system is negligible. Consider the M/M/1/$K$ queueing system that is identical to the M/M/1 system with the exception that it can only hold a maximum of $K$ customers in the system. Customers that arrive when the system is full are turned away.

The process $N(t)$ for this system is a continuous-time Markov chain

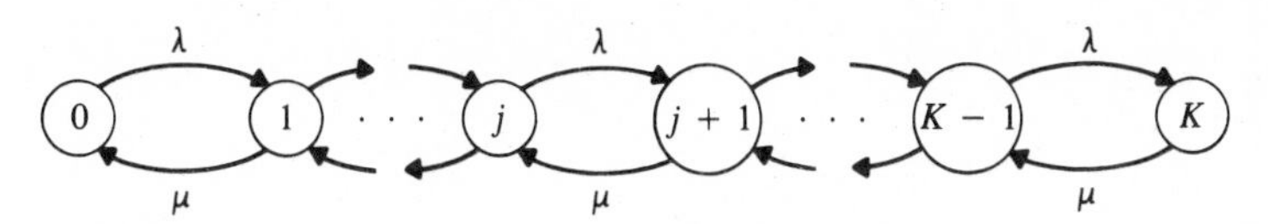


**FIGURE 9.7** Transition rate diagram for M/M/1/*K* system.

that takes on values from the set $\{0, 1, \ldots, K\}$ with transition rate diagram as shown in Fig. 9.7. It can be seen that the arrival rate *into* the system is now zero when $N(t) = K$. The transition rates from the other states are the same as for the M/M/1 system.

The global balance equations are now

$$\begin{aligned} \lambda p_0 &= \mu p_1 \\ (\lambda + \mu)p_j &= \lambda p_{j-1} + \mu p_{j+1} \qquad j = 1, 2, \ldots, K - 1 \\ \mu p_K &= \lambda p_{K-1}. \end{aligned} \tag{9.35}$$

Let $\rho = \lambda/\mu$. It can be readily shown (see Problem 14) that the steady state probabilities are

$$P[N = j] = \frac{(1 - \rho)\rho^j}{1 - \rho^{K+1}} \qquad j = 0, 1, 2, \ldots, K, \tag{9.36}$$

for $\rho < 1$ or $\rho > 1$. When $\rho = 1$ all the states are equiprobable. Figure 9.8 shows the steady state probabilities for various values of $\rho$.

The mean number of customers in the system is given by

$$\begin{aligned} E[N] &= \sum_{j=0}^{K} jP[N(t) = j] \\ &= \begin{cases} \dfrac{\rho}{1 - \rho} - \dfrac{(K + 1)^{K+1}}{1 - \rho^{K+1}} & \text{for } p \neq 1 \\ \dfrac{K}{2} & \text{for } \rho = 1. \end{cases} \end{aligned} \tag{9.37}$$

The mean total time spent by customers in the system is found from Eq. (9.37) by using Little's formula with $\lambda_a$, the rate of arrivals that actually enter the system. The proportion of time when the system turns away customers is $P[N(t) = K] = p_K$. Thus the system turns away customers at the rate

$$\lambda_b = \lambda p_K, \tag{9.38}$$

and the actual arrival rate *into* the system is

$$\lambda_a = \lambda(1 - p_K). \tag{9.39}$$

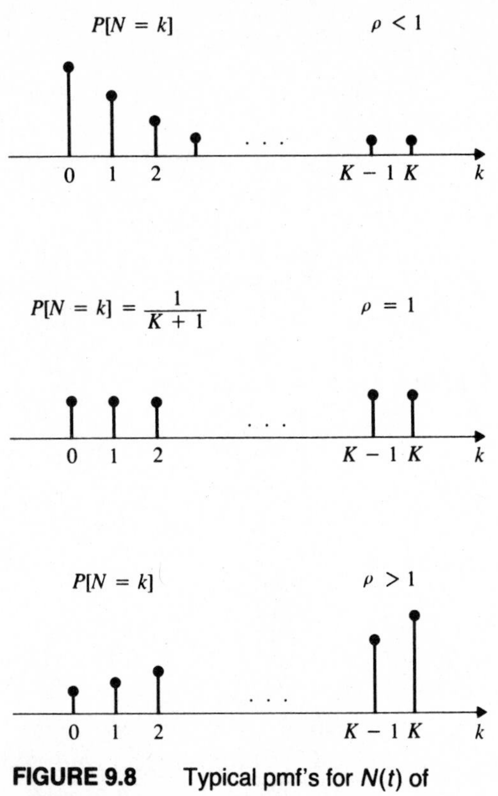


**FIGURE 9.8** Typical pmf's for $N(t)$ of M/M/1/$K$ system.

Applying Little's formula to Eq. (9.37) we obtain

$$E[T] = \frac{E[N]}{\lambda_a} = \frac{E[N]}{\lambda(1 - p_K)}. \tag{9.40}$$

In finite capacity systems, it is necessary to distinguish between the traffic load offered to a system and the actual load carried by the system. The **offered load** or **traffic intensity** is a measure of the demand made on the system and is defined as

$$\lambda \frac{\text{customers}}{\text{second}} \times E[\tau] \frac{\text{seconds of service}}{\text{customer}} \tag{9.41}$$

The **carried load** is the actual demand met by the system:

$$\lambda_a \frac{\text{customers}}{\text{second}} \times E[\tau] \frac{\text{seconds of service}}{\text{customer}}. \tag{9.42}$$

### ■■ Example 9.6
Mean Delay and Carried Load Versus $K$

Figure 9.9(a) gives a comparison of the carried load versus the offered load $\rho$ for two values of $K$. It can be seen that increasing the capacity $K$ results in an increase in carried load since more customers are allowed into the system. Figure 9.9(b) gives the corresponding values for the mean delay. We see that increasing $K$ results in increased delays, again because more customers are allowed into the system. ■■

### ■■ Example 9.7

Suppose that an M/M/1 model is used for a system that has capacity $K$, and that the probability of rejecting customers is approximated by $P[N = K]$. Compare this approximation to the exact probability given by the M/M/1/$K$ model.

For the M/M/1 system the above probability is given by

$$P[N = K] = (1 - \rho)\rho^K.$$

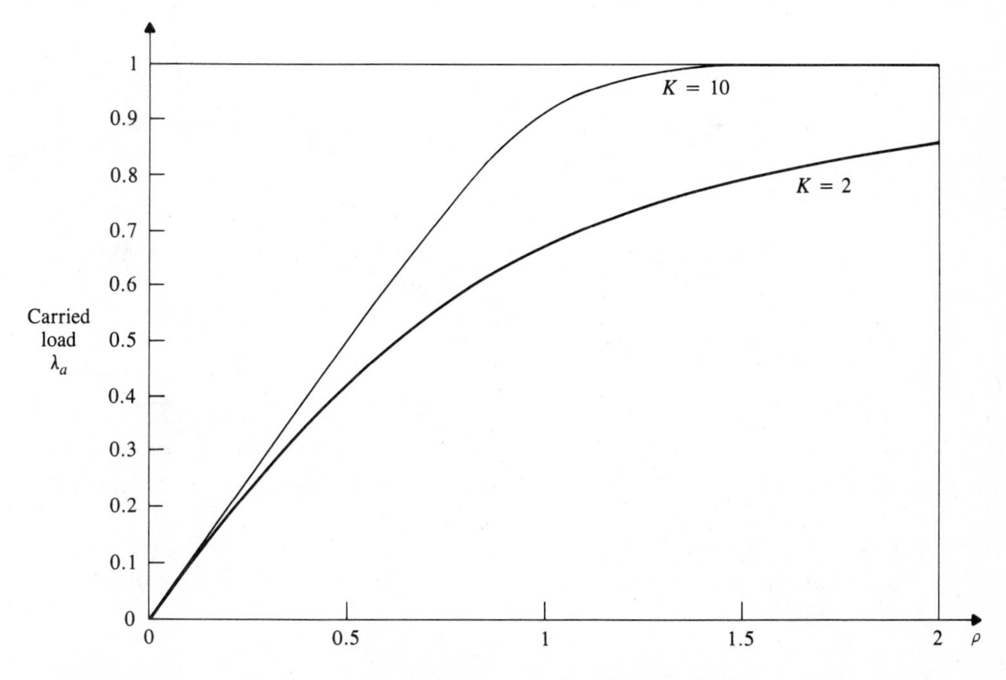


**FIGURE 9.9a** Carried load versus offered load for M/M/1/$K$ system with $K = 2$ and $K = 10$.

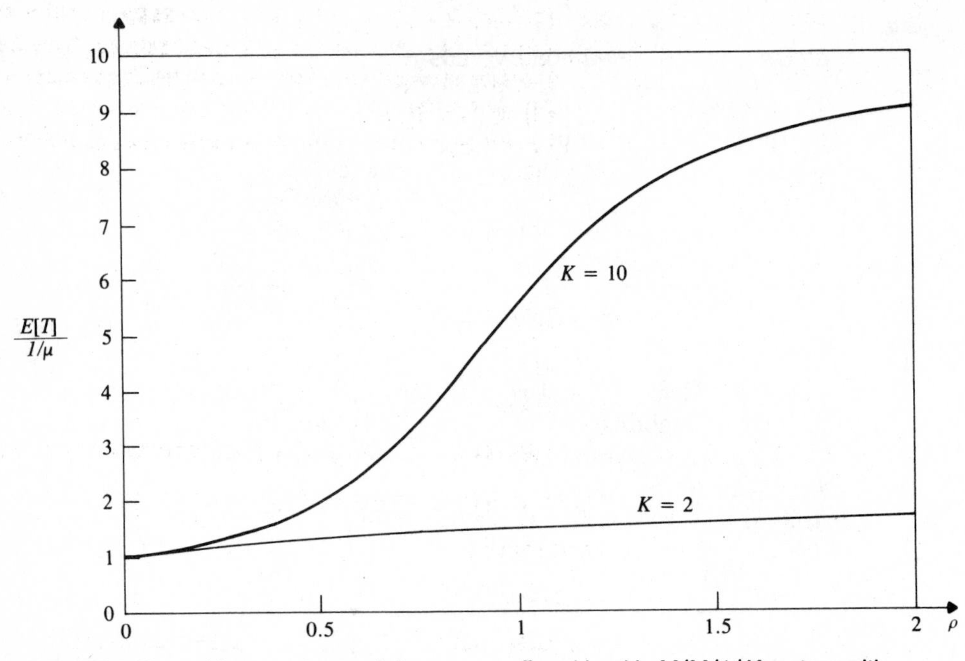


**FIGURE 9.9b** Mean customer delay versus offered load in M/M/1/*K* system with *K* = 2 and *K* = 10.

For $\rho < 1$, the M/M/1/$K$ system the probability of rejecting a customer is

$$P[N' = K] = \frac{(1-\rho)\rho^K}{1-\rho^{K+1}} = (1-\rho)\rho^K\{1 + \rho^{K+1} + (\rho^{K+1})^2 + \cdots\}.$$

For $\rho < 1$ and $K$ large, $P[N = k] \simeq P[N' = K]$. For $\rho > 1$, the M/M/1 approximation breaks down and gives a negative probability. ■■

## 9.4

## MULTI-SERVER SYSTEMS: M/M/*c*, M/M/*c*/*c*, AND M/M/∞

We now modify the M/M/1 system to consider queueing systems with multiple servers. In particular, we consider systems with iid exponential interarrival times and iid exponential service times. As in the case of the M/M/1 system, the resulting systems can be modeled by continuous-time Markov chains.

### Distribution of Number in the M/M/*c* System

The transition rate diagram for an M/M/$c$ system is shown in Fig. 9.10. As before, arrivals occur at a rate $\lambda$. The difference now is that the departure

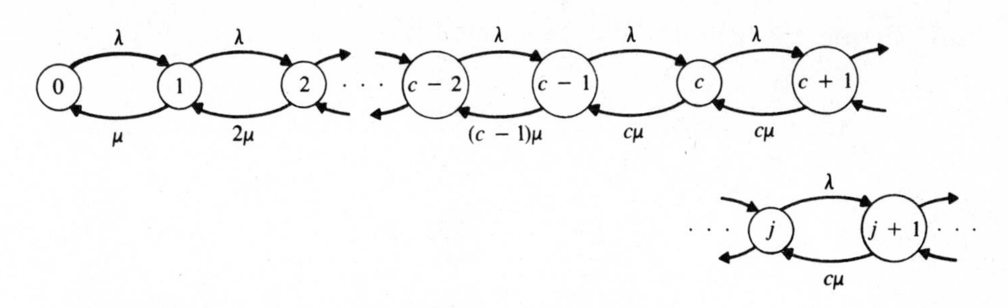

**FIGURE 9.10** Transition rate diagram for M/M/$c$ system.

rate is $k\mu$ when $k$ servers are busy. To see why, suppose that $k$ of the servers are busy, then the time until the next departure is given by

$$X = \min(\tau_1, \tau_2, \ldots, \tau_k),$$

where $\tau_i$ are iid exponential random variables with parameter $\mu$. The complementary cdf of this random variable is

$$\begin{aligned} P[X > t] &= P[\min(\tau_1, \tau_2, \ldots, \tau_k) > t] \\ &= P[\tau_1 > t, \tau_2 > t, \ldots, \tau_k > t] \\ &= P[\tau_1 > t]P[\tau_2 > t]\ldots P[\tau_k > t] \\ &= e^{-\mu t}e^{-\mu t}\ldots e^{-\mu t} \\ &= e^{-k\mu t}. \end{aligned} \tag{9.43}$$

Thus the time until the next departure is an exponential random variable with mean $1/k\mu$. Thus when $k$ servers are busy, customers depart at rate $k\mu$. When the number of customers in the system is greater than $c$, all $c$ servers are busy and the departure rate is $c\mu$.

We obtain the steady state probabilities for the M/M/$c$ system from the general solution for birth-and-death processes found in Example 8.19. The probabilities of the first $c$ states are obtained from the following recursion (see Eq. 8.36):

$$p_j = \frac{\lambda}{j\mu}p_{j-1} \qquad j = 1, \ldots, c,$$

which leads to

$$p_j = \frac{a^j}{j!}p_0 \qquad j = 0, 1, \ldots, c, \tag{9.44}$$

where

$$a = \frac{\lambda}{\mu}. \tag{9.45}$$

The probabilities for states equal to or greater than $c$ are obtained from the following recursion:

$$p_j = \frac{\lambda}{c\mu} p_{j-1} \qquad j = c, c+1, c+2, \ldots,$$

which leads to

$$p_j = \rho^{j-c} p_c \qquad j = c, c+1, c+2, \ldots \tag{9.46a}$$

$$= \frac{\rho^{j-c} a^c}{c!} p_0, \tag{9.46b}$$

where we have used Eq. (9.44) with $j = c$ and where

$$\rho = \frac{\lambda}{c\mu}. \tag{9.47}$$

Finally $p_0$ is obtained from the normalization condition:

$$1 = \sum_{j=0}^{\infty} p_j = p_0 \left\{ \sum_{j=0}^{c-1} \frac{a^j}{j!} + \frac{a^c}{c!} \sum_{j=c}^{\infty} \rho^{j-c} \right\}.$$

The system is stable and has a steady state if the term inside the brackets is finite. This is the case if the second series converges, which in turn requires that $\rho < 1$ or equivalently

$$\lambda < c\mu. \tag{9.48}$$

In other words, the system is stable if the customer arrival rate is less than the total rate at which the $c$ servers can process customers. The final form for $p_0$ is

$$p_0 = \left\{ \sum_{j=0}^{c-1} \frac{a^j}{j!} + \frac{a^c}{c!} \frac{1}{1-\rho} \right\}^{-1}. \tag{9.49}$$

The probability that an arriving customer finds all servers busy and has to wait in queue is an important parameter of the M/M/$c$ system:

$$P[W > 0] = P[N \geq c] = \sum_{j=c}^{\infty} \rho^{j-c} p_c = \frac{p_c}{1-\rho}. \tag{9.50}$$

This probability is called the **Erlang *C* formula** and is denoted by $C(c, a)$:

$$C(c, a) = \frac{p_c}{1-\rho} = P[W > 0]. \tag{9.51}$$

The mean number of customers in queue is given by

$$E[N_q] = \sum_{j=c}^{\infty} (j - c)\rho^{j-c}p_c = p_c \sum_{j'=0}^{\infty} j'\rho^{j'}$$

$$= \frac{\rho}{(1-\rho)^2} p_c$$

$$= \frac{\rho}{1-\rho} C(c, a). \tag{9.52}$$

The mean waiting time is found from Little's formula:

$$E[W] = \frac{E[N_q]}{\lambda}$$

$$= \frac{1/\mu}{c(1-\rho)} C(c, a). \tag{9.53}$$

The mean total time in the system is

$$E[T] = E[W] + E[\tau] = E[W] + \frac{1}{\mu}. \tag{9.54}$$

Finally, the mean number in the system is found from Little's formula:

$$E[N] = \lambda E[T] = E[N_q] + a, \tag{9.55}$$

where we have used Equation (9.54).

### ■■ Example 9.8

A company has a system with four private telephone lines connecting two of its sites. Suppose that requests for these lines arrive according to a Poisson process at a rate of one call every 2 minutes, and suppose that call durations are exponentially distributed with mean 4 minutes. When all lines are busy, the system delays (i.e., queues) call requests until a line becomes available. Find the probability of having to wait for a line.

First we need to compute $p_0$. Since $\lambda = 1/2$ and $1/\mu = 4$, $a = \lambda/\mu = 2$ and $\rho = a/c = 1/2$. Therefore

$$p_0 = \left\{1 + 2 + \frac{2^2}{2} + \frac{2^3}{6} + \frac{16}{24}\left(\frac{1}{1 - 1/2}\right)\right\}^{-1} = \frac{3}{23}.$$

The probability of having to wait is then

$$C(4, 2) = \frac{2^4/4!}{1 - 1/2}\frac{3}{23} = \frac{4}{23} \simeq .17.$$

■■

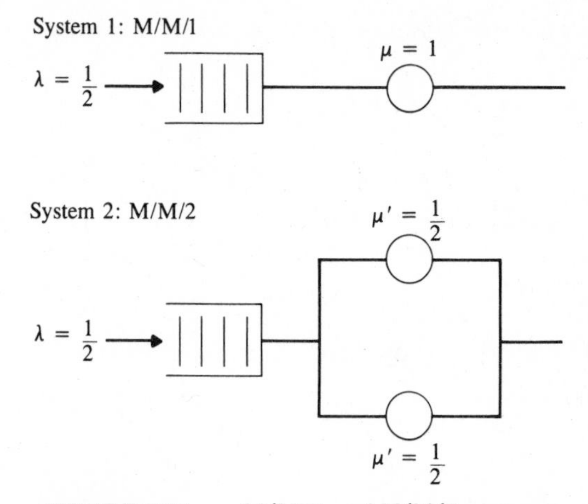


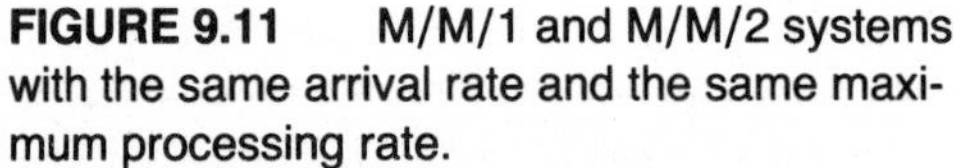

**FIGURE 9.11** M/M/1 and M/M/2 systems with the same arrival rate and the same maximum processing rate.

### ■■ Example 9.9
M/M/1 Versus M/M/c

Compare the mean delay and mean waiting time performance of the two systems shown in Fig. 9.11. Note that both systems have the same processing rate.

For the M/M/1 system, $\rho = \lambda/\mu = (1/2)/1 = 1/2$, so the mean waiting time is

$$E[W] = \frac{\rho/\mu}{1-\rho} = 1 \text{ second,}$$

and the mean total delay is

$$E[T] = \frac{1/\mu}{1-\rho} = 2 \text{ seconds.}$$

For the M/M/2 system, $a = \lambda/\mu' = 1$, and $\rho = \lambda/2\mu' = 1/2$. The probability of an empty system is

$$p_0 = \left\{1 + a + \frac{a^2/2}{1-1/2}\right\}^{-1} = \frac{1}{3}.$$

Erlang's $C$ formula is

$$C(2,1) = \frac{a^2/2}{1-\rho} p_0 = \frac{1}{3}.$$

The mean waiting time is then

$$E[W'] = \frac{1/\mu'}{2(1-\rho)} C(2,1) = \frac{2}{3},$$

and the mean delay is

$$E[T'] = \frac{2}{3} + \frac{1}{\mu'} = \frac{8}{3}.$$

Thus the M/M/1 system has a smaller total delay but a larger waiting time than the M/M/2. In general increasing the number of servers decreases the waiting time but increases the total delay. ■■

### Waiting Time Distribution for M/M/$c$

Before we compute the pdf of the waiting time, consider the conditional probability that there are $j - c > 0$ customers in queue given that all servers are busy (i.e., $N(t) \geq c$):

$$P[N(t) = j \mid N(t) \geq c] = \frac{P[N(t) = j, N(t) \geq c]}{P[N(t) \geq c]} = \frac{P[N(t) = j]}{P[N(t) \geq c]} \qquad j \geq c$$

$$= \frac{\rho^{j-c} p_c}{p_c/(1-\rho)} = (1-\rho)\rho^{j-c} \qquad j \geq c. \tag{9.56}$$

This geometric pmf suggests that when all the servers are busy, the M/M/$c$ system behaves like an M/M/1 system. We use this fact to compute the cdf of $W$.

Suppose that a customer arrives when there are $k$ customers in queue. There must be $k + 1$ service completions before our customer enters service. From Eq. (9.43), each service completion is exponentially distributed with rate $c\mu$. Thus the waiting time for our customer is the sum of $k + 1$ iid exponential random variables with parameter $c\mu$, which we know is a gamma random variable with parameter $c\mu$:

$$f_W(x \mid N = c + k) = \frac{(c\mu x)^k}{k!} c\mu e^{-c\mu x}. \tag{9.57}$$

The cdf for $W$ given that $W > 0$, or equivalently $N \geq c$, is obtained by combining Eqs. (9.56) and (9.57):

$$\begin{aligned}
F_W(x \mid W > 0) &= \sum_{k=0}^{\infty} F_W(x \mid N = c + k) P[N = c + k \mid N \geq c] \\
&= \sum_{k=0}^{\infty} \int_0^x \frac{(c\mu y)^k}{k!} c\mu e^{-c\mu y}\, dy (1-\rho)\rho^k \\
&= (1-\rho) \int_0^x \sum_{k=0}^{\infty} \frac{(c\mu y)^k}{k!} \rho^k c\mu e^{-c\mu y}\, dy \\
&= (1-\rho) c\mu \int_0^x e^{-c\mu(1-\rho)y}\, dy \\
&= 1 - e^{-c\mu(1-\rho)x}.
\end{aligned}$$

The cdf of $W$ is then:

$$\begin{aligned} P[W \le x] &= P[W = 0] + F_W(x \mid W > 0)P[W > 0] \qquad x > 0 \\ &= (1 - C(c,a)) + (1 - e^{-c\mu(1-\rho)x})C(c,a) \\ &= 1 - C(c,a)e^{-c\mu(1-\rho)x}. \end{aligned} \tag{9.58}$$

Since $T = W + \tau$, where $W$ and $\tau$ are independent random variables, it is easy to show that if $a \neq c - 1$ the cdf of $T$ is

$$P[T \le x] = 1 + \frac{a - c + P[W = 0]}{c - 1 - a} e^{-\mu x} + \frac{C(c,a)}{c - 1 - a} e^{-c\mu(1-\rho)x}. \tag{9.59}$$

**■■ Example 9.10**

What is the probability that a request for a telephone line has to wait more than one minute in the system discussed in Example 9.8.

In Example 9.8 we found that $p_0 = 3/23$ and that the probability of having to wait is

$$C(4,2) = \frac{4}{23} \simeq .17.$$

The probability of having to wait more than one minute is

$$\begin{aligned} P[W > 1] &= 1 - P[W \le 1] \\ &= C(c,a)e^{-c\mu(1-\rho)1} \\ &= \frac{4}{23} e^{-4(1/4)(1-1/2)1} \simeq .11. \end{aligned}$$

■■

---

### The M/M/$c$/$c$ Queueing System

The M/M/$c$/$c$ queueing system has $c$ servers but no waiting room. Calls that arrive when all servers are busy are turned away. The transition rate diagram for this system is shown in Fig. 9.12 where it can be seen that the arrival rate is zero when $N(t) = c$.

The steady state probabilities for this system have the same form as

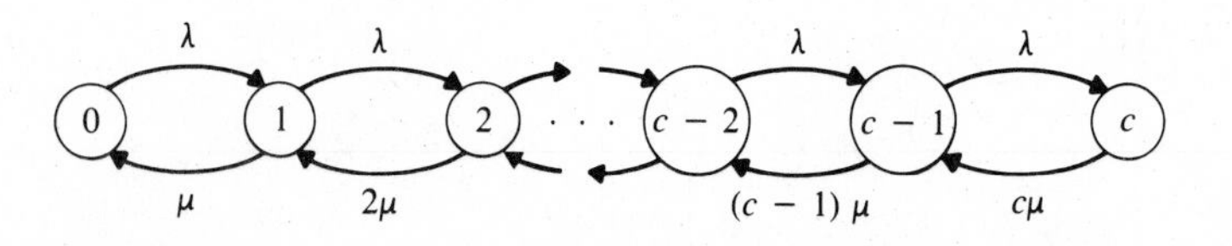

**FIGURE 9.12** Transition rate diagram for M/M/$c$/$c$ system.

those for states $0, \ldots, c$ in the M/M/$c$ system:

$$p_j = \frac{a^j}{j!} p_0 \qquad j = 0, \ldots, c, \tag{9.60}$$

where

$$a = \frac{\lambda}{\mu}, \tag{9.61}$$

is the offered load and

$$p_0 = \left\{ \sum_{j=0}^{c} \frac{a^j}{j!} \right\}^{-1}. \tag{9.62}$$

The **Erlang** ***B*** formula is defined as the probability that all servers are busy:

$$B(c, a) = P[N = c] = p_c = \frac{a^c/c!}{1 + a + a^2/2! + \cdots + a^c/c!}. \tag{9.63}$$

The actual arrival rate *into* the system is then

$$\lambda_a = \lambda(1 - B(c, a)). \tag{9.64}$$

The average number in the system is obtained from Little's formula:

$$E[N] = \lambda_a E[\tau] = \frac{\lambda}{\mu}(1 - B(c, a)). \tag{9.65}$$

Note that $E[N]$ is also equal to the carried load as defined by Eq. (9.42).

The Erlang $B$ formula depends only on the arrival rate $\lambda$, the mean service time $E[\tau] = 1/\mu$, and the number of servers $c$. It turns out that Eq. (9.63) also gives the probability of blocking for M/G/$c$/$c$ systems (see Ross 1983).

## ■■ Example 9.11

Consider the telephone system discussed in Example 9.8, but now assume that call requests that arrive when all four lines are busy are automatically redirected to public telephone lines. What proportion of calls are redirected?

The Erlang $B$ formula gives the proportion of calls that are redirected:

$$B(4, 2) = \frac{2^4/4!}{1 + 2 + 4/2 + 8/6 + 16/24} \simeq 9.5\%.$$

■■

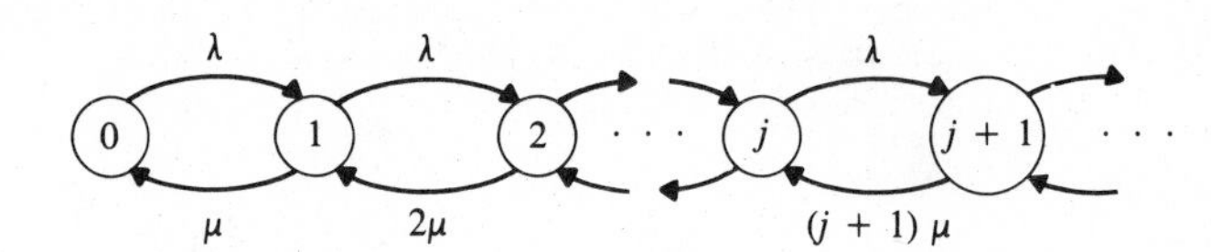


**FIGURE 9.13** Transition rate diagram for M/M/∞ system.

### The M/M/∞ Queueing System

Consider a system with Poisson arrivals, exponential service times, and suppose that the number of servers is so large that arriving customers always find a server available. In effect we have a system with an infinite number of servers. If we allow $c$ to approach infinity for the M/M/$c$/$c$ system, we obtain the *M/M/∞* system with the transition rate diagram shown in Fig. 9.13.

The steady state probabilities are also found by letting $c$ approach infinity in the equations for the M/M/$c$/$c$ system:

$$p_j = \frac{a^j}{j!}e^{-a} \qquad j = 0, 1, 2, \ldots, \tag{9.66}$$

where $a = \lambda/\mu$. Thus the number of customers in the system is a Poisson random variable. The mean number of customers in the system is

$$E[N] = a.$$

### ■■ Example 9.12

Subscribers connect to a computer system according to a Poisson process of rate $\lambda = 1$ customer/6 minutes. Suppose that subscribers remain in the system for an exponentially distributed amount of time with $1/\mu = 1$ hr. Find the probability that, at steady state, there are five or fewer subscribers in the system.

The number in the system has a Poisson distribution with parameter $a = (1/6)(60) = 10$. The probability of 5 or less in the system is then

$$P[N \le 5] = \left[1 + \frac{10}{1!} + \frac{10^2}{2!} + \frac{10^3}{3!} + \frac{10^4}{4!} + \frac{10^5}{5!}\right]e^{-10}$$

$$= .067.$$

■■

## 9.5

## FINITE-SOURCE QUEUEING SYSTEMS

Consider a single-server queueing system that serves $K$ sources as shown in Fig. 9.14(a). Each source can be in one of two states: In the first state,

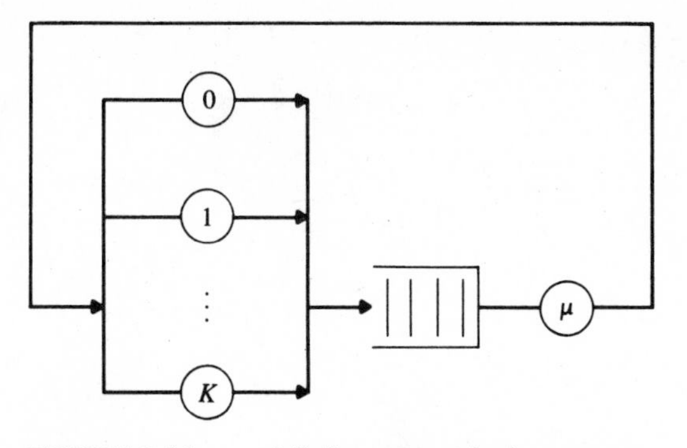


**FIGURE 9.14a** A finite-source single-server system.

the source is preparing a request for service from the server; in the second state, the source has generated a request that is either waiting in queue or being served. For example, the sources could represent $K$ machines and the server could represent a repairman that repairs machines when they break down. In another example, the $K$ sources could represent terminals that generate jobs for a computer system as shown in Fig. 9.14(b).

We assume that each source spends an exponentially distributed amount of time with mean $1/\alpha$ preparing each service request. Thus when idle, a source generates a request for service in the interval $(t, t + \delta)$ with probability $\alpha\delta + o(\delta)$. If the state of the system is $N(t) = k$, then the number of idle sources is $K - k$, so the rate at which service requests are generated is $(K - k)\alpha$. We also assume that the time required to service each request is an exponentially distributed amount of time with mean $1/\mu$. $N(t)$ is then the continuous-time Markov chain with the transition rate diagram shown in Fig. 9.15.

The steady state probabilities are found using the results obtained in

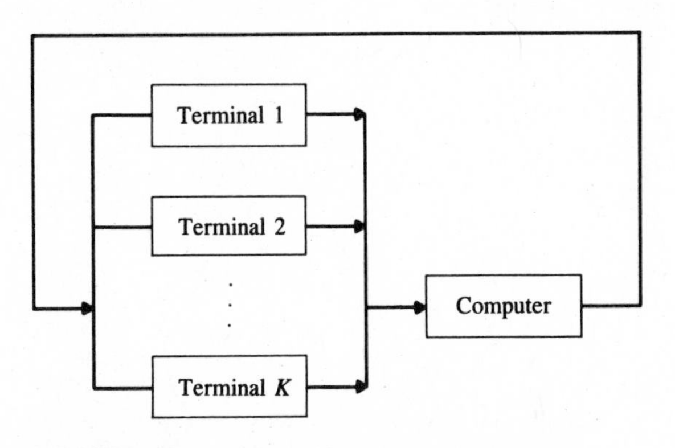

**FIGURE 9.14b** A multiuser computer system.

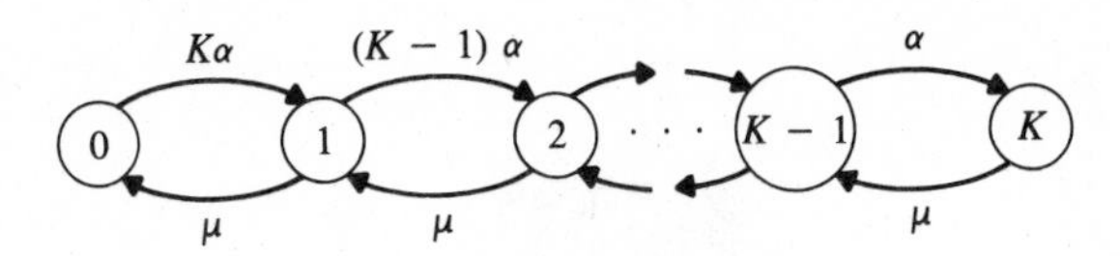


**FIGURE 9.15** Transition rate diagram for a finite-source single-server system.

Example 8.19:

$$p_k = \frac{K!}{(K-k)!}\left(\frac{\alpha}{\mu}\right)^k p_0 \qquad k = 0, 1, \ldots, K, \tag{9.67}$$

where

$$p_0 = \left\{\sum_{k=0}^{K} \frac{K!}{(K-k)!}\left(\frac{\alpha}{\mu}\right)^k\right\}^{-1}. \tag{9.68}$$

The server utilization $\rho$ is the proportion of time when the system is busy, thus

$$\rho = 1 - p_0, \tag{9.69}$$

where $p_0$ is given by Eq. (9.68). The mean arrival rate to the queue can then be found from Little's formula with "system" defined as the server:

$$\lambda E[\tau] = \rho = 1 - p_0,$$

which implies

$$\lambda = \frac{\rho}{E[\tau]} = \mu\rho = \mu(1 - p_0). \tag{9.70}$$

A source takes an average time of $1/\alpha$ to generate a request and then spends time $E[T]$ having it serviced in the queueing system. Thus each source generates a request at the rate $(1/\alpha + E[T])^{-1}$ requests per second. Since the actual arrival rate must equal the rate at which the $K$ sources generate requests, we have

$$\lambda = \frac{K}{1/\alpha + E[T]}. \tag{9.71}$$

The mean delay in the system for each request is found by solving for $E[T]$:

$$E[T] = \frac{K}{\lambda} - \frac{1}{\alpha}. \tag{9.72}$$

Finally, we can apply Little's formula to Eq. (9.72) to obtain the mean number in the system:

$$E[N] = \lambda E[T] = K - \frac{\lambda}{\alpha}. \tag{9.73}$$

Note that this implies that $\lambda/\alpha$ is the mean number of idle sources. The mean waiting time is obtained by subtracting the mean service time from $E[T]$:

$$E[W] = E[T] - \frac{1}{\mu}. \tag{9.74}$$

The proportion of time that a source spends waiting for the completion of a service request is the ratio of the time spent in the system to the mean cycle time:

$$P[\text{source busy}] = \frac{E[T]}{E[T] + 1/\alpha}. \tag{9.75}$$

### ■■ Example 9.13
Interactive Computer System

Suppose that a set of $K$ terminals generate transactions for processing by a computer. Each terminal spends an exponentially distributed "think" time preparing a transaction request, and the computer takes an exponentially distributed time processing each request. The "throughput" of the computer is defined as the rate at which it completes transactions. The response time is the total time a transaction spends in the computer. Find expressions for the throughput and response time for two extreme cases: $K$ small and $K$ large.

When $K$ is sufficiently small, there is no waiting in queue so

$$E[T] \simeq \frac{1}{\mu} \qquad \text{for } K \text{ small}, \tag{9.76}$$

and by Eq. (9.71)

$$\lambda = \frac{K}{1/\alpha + 1/\mu} \qquad \text{for } K \text{ small}. \tag{9.77}$$

Thus $\lambda$ grows linearly with $K$. As $K$ increases, the computer will eventually become fully utilized, and then outputs transactions at its maximum rate, namely $\mu$ transactions per second. Thus

$$\lambda \simeq \mu \qquad \text{for } K \text{ large}, \tag{9.78}$$

and Eq. (9.72) becomes

$$E[T] = \frac{K}{\mu} - \frac{1}{\alpha} \qquad \text{for } K \text{ large}. \tag{9.79}$$

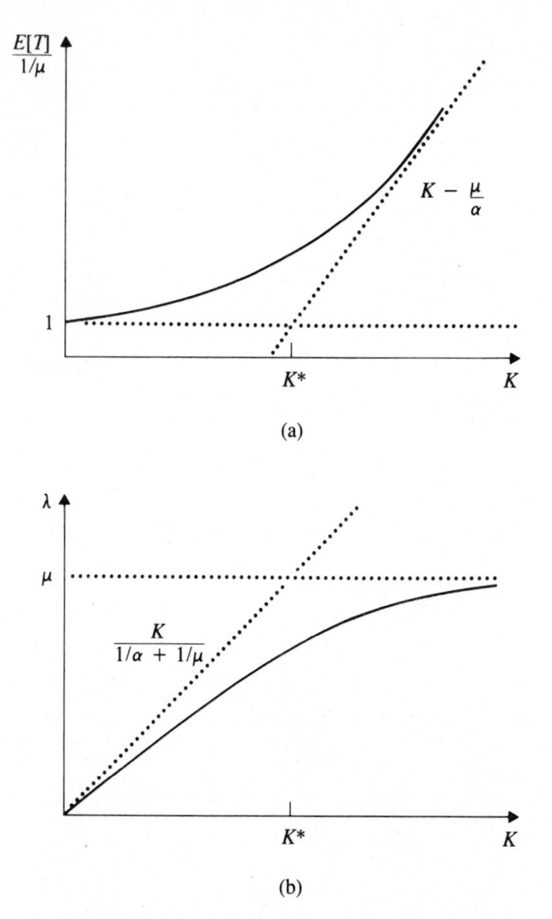


**FIGURE 9.16** Delay and throughput for finite source system as a function of number of sources. Dashed lines show small K and large K asymptotes.

These asymptotic expressions for the throughput and response time are shown in Figs. 9.16(a) and 9.16(b), which were previewed in Chapter 1 (Figs. 1.10 and 1.11). The value of $K$ where the two asymptotes for $E[T]$ intersect is called the **system saturation point,**

$$K^* = \frac{1/\mu + 1/\alpha}{1/\mu}. \tag{9.80}$$

When $K$ becomes larger than $K^*$, the requests from the terminals are certain to interfere with each other and the response time increases accordingly. ■■

### *Arriving Customer's Distribution

In the above discussion, we found $\lambda$, E[N] and E[T] in a roundabout way (see Eqs. [9.70], [9.71], and [9.72]). To calculate $E[T]$ directly, we argue as follows. If we assume a first-come, first-serve service discipline, then a customer that arrives when there are $N_a = k$ requests in the queueing system spends a total time in the system equal to the sum of 1 residual service time, $k - 1$ service times, and his own service time. Since all of these times are iid exponential random variables with mean $1/\mu$, then the mean time in the system for our request is

$$E[T \mid N_a = k] = \frac{k + 1}{\mu}.$$

The mean time in the system is then found by averaging over $N_a$:

$$E[T] = \frac{1}{\mu} \sum_{k=0}^{K-1} (k + 1) P[N_a = k]. \tag{9.81}$$

The difficulty with the above equation is that arrivals are not Poisson—remember that the arrival rate is $(K - N(t))\alpha$, and thus depends on the state of the system. Consequently, the distribution of states seen by an arriving customer is not the same as $P[N = k]$, the proportion of time that there are $k$ requests in the queueing system. For example, a service request cannot be generated when all sources have requests in the system, that is, $N(t) = K$, so $P[N_a = K] = 0$. However, $P[N = K]$ is nonzero since it is possible for all sources to simultaneously have requests in the queueing system.

To find $P[N_a = k]$ we need to find the long-term proportion of time that arriving customers find $k$ customers in the system. Since $p_k = P[N(t) = k]$ is the long-term proportion of time the system is in state $k$, then in a very long time interval of duration $T'$ approximately $p_k T'$ seconds are spent in state $k$. The arrival rate when $N(t) = k$ is $(K - k)\alpha$ requests/second, so the number of arrivals that find $k$ requests is approximately

$$(K - k)\alpha \text{ customers/second} \times p_k T' \text{ seconds in state } k. \tag{9.82}$$

The total number of arrivals in time $T'$ is obtained by summing over all states:

$$\sum_{j=0}^{K} (K - j)\alpha p_j T'. \tag{9.83}$$

Thus *the proportion of arrivals that find k requests in the system is*

$$P[N_a = k] = \frac{(K-k)\alpha p_k T'}{\sum_{j=0}^{K} (K-j)\alpha p_j T'} = \frac{(K-k)p_k}{\sum_{j=0}^{K}(K-j)p_j}$$

$$= \frac{(K-k)[K!/(K-k)!](\alpha/\mu)^k p_0}{\sum_{j=0}^{K}(K-j)[K!/(K-j)!](\alpha/\mu)^j p_0}$$

$$= \frac{[(K-1)!/(K-k-1)!](\alpha/\mu)^k}{\sum_{j=0}^{K-1}[(K-1)!/(K-j-1)!](\alpha/\mu)^j} \qquad 0 \le k \le K-1. \tag{9.84}$$

If we compare Equation (9.84) to Eq. (9.67), we see that Eq. (9.84) is the steady-state probability of having $k$ customers in a system with $K - 1$ sources. In other words, *a source when placing a request "sees" a queueing system that behaves as if the source were not present at all*!

We leave it up to you in Problem 31 to show the Eqs. (9.84) and (9.81) give $E[T]$ as given in Eq. (9.72). Indeed, this same approach can be used to find the pdf of $T$.

## 9.6 M/G/1 QUEUEING SYSTEMS

We now consider single-server queueing systems in which the arrivals follow a Poisson process but in which the service times need not be exponentially distributed. We assume that the service times are independent, identically distributed random variables with general pdf $f_\tau(x)$. The resulting queueing system is denoted by M/G/1.

The number of customers $N(t)$ in an M/G/1 system is a continuous-time random process. Recall that the "state" of the system is the information about the past history of the system that is relevant to the probabilities of future events. In the preceding sections, customer interarrival times and service times were exponential distributions, so $N(t)$ was always the state of the system. This is no longer the case for M/G/1 systems. For example, if service times are constant, then knowledge about when a customer began service specifies the customer's future departure time. Thus the state of an M/G/1 system at time $t$ is specified by $N(t)$ together with the remaining ("residual") service time of the customer being served at time $t$.

In this section we present a simple approach based on Little's formula that gives the mean waiting time and mean delay in an M/G/1 system. We also use this simple approach to find the mean waiting times in M/G/1 systems that have priority classes.

### The Residual Service Time

Suppose that an arriving customer finds the server busy, and consider the residual time of the customer found in service. Let $\tau_1, \tau_2, \ldots$ be the iid sequence of service times of customers in this M/G/1 system, and suppose we divide the positive time axis into segments of length $\tau_1, \tau_2, \ldots$ as shown in Fig. 9.17. We can then view customers that arrive when the server is busy as picking a point at random on this time axis. The residual service time is then the remainder of time in the segment that is intercepted as shown in Fig. 9.17.

In Example 5.23 we showed that the long-term proportion of time that the residual service time exceeds $x$ is given by

$$\frac{1}{E[\tau]}\int_x^\infty (1 - F_\tau(y))\,dy. \tag{9.85}$$

Since the arrival times of Poisson customers are independent of the system state, Eq. (9.85) is also the probability that the residual service time $R$ of a customer found in service exceeds $x$, that is,

$$P[R > x] = \frac{1}{E[\tau]}\int_x^\infty (1 - F_\tau(y))\,dy. \tag{9.86}$$

The pdf of $R$ is then

$$f_R(x) = -\frac{d}{dx}P[R > x] = \frac{1 - F_\tau(x)}{E[\tau]}. \tag{9.87}$$

The mean residual time is

$$E[R] = \int_0^\infty x\frac{1 - F_\tau(x)}{E[\tau]}\,dx.$$

Integrating by parts with $u = (1 - F_\tau(x))/E[\tau]$ and $dv = x\,dx$, we obtain

$$\begin{aligned} E[R] &= (1 - F_\tau(x))\frac{x^2}{2E[\tau]}\Big|_0^\infty + \frac{1}{2E[\tau]}\int_0^\infty x^2 f_\tau(x)\,dx \\ &= \frac{E[\tau^2]}{2E[\tau]}. \end{aligned} \tag{9.88}$$

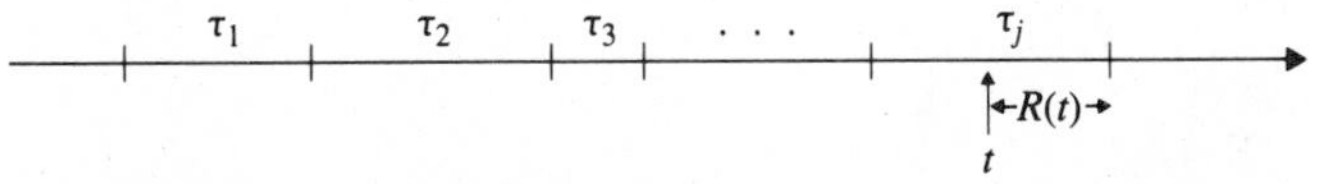

**FIGURE 9.17** Sequence of service times and a residual service time.

### ■■ Example 9.14

Compare the residual service times of two systems with exponential service times of mean $m$ and of constant service times of mean $m$, respectively.

For an exponential service time of mean $m$, the second moment is $2m^2$, thus the mean residual service time is, from Eq. (9.88),

$$E[R_{\text{exp}}] = \frac{2m^2}{2m} = m.$$

Thus, the mean residual time is the same as the full service time of a customer. This is consistent with the memoryless property of the exponential random variable.

The second moment of a constant random variable of value $m$ is $m^2$. Thus the mean residual service time is

$$E[R_{\text{const}}] = \frac{m^2}{2m} = \frac{m}{2},$$

which is what one would expect; on the average we expect to wait half a service time. ■■

---

### Mean Delay in M/G/1 Systems

Consider the time $W$ spent by a customer waiting for service in an M/G/1 system. If the service discipline is first-come-first-serve, then $W$ is the sum of the residual service time $R'$ of the customer (if any) found in service and the $N_q(t) = k - 1$ service times of the customers (if any) found in queue. Thus the mean waiting time is then

$$E[W] = E[R'] + E[N_q(t)]E[\tau], \tag{9.89}$$

since the service times are iid with mean $E[\tau]$ (see Eq. [5.13]). From Little's formula we have that $E[N_q(t)] = \lambda E[W]$, so

$$E[W] = E[R'] + \lambda E[W]E[\tau] = E[R'] + \rho E[W]. \tag{9.90}$$

The residual service time $R'$ encountered by an arriving customer is zero when the system is found empty, and $R$, as defined in the previous section, when a customer is found in service. Thus

$$\begin{aligned} E[R'] &= 0P[N(t) = 0] + E[R](1 - P[N(t) = 0]) \\ &= \frac{E[\tau^2]}{2E[\tau]}\lambda E[\tau] \\ &= \frac{\lambda E[\tau^2]}{2}, \end{aligned} \tag{9.91}$$

where we have used Eq. (9.88) for $E[R]$ and Eq. (9.14) for the fact that $1 - P[N(t) = 0] = \rho = \lambda E[\tau]$.

The **mean waiting time $E[W]$** of a customer in an M/G/1 system is found by substituting Eq. (9.91) into Eq. (9.90) and solving for $E[W]$:

$$E[W] = \frac{\lambda E[\tau^2]}{2(1-\rho)}. \tag{9.92}$$

We can obtain another expression for $E[W]$ by noting that $E[\tau^2] = \sigma_\tau^2 + E[\tau]^2$:

$$E[W] = \frac{\lambda(\sigma_\tau^2 + E[\tau]^2)}{2(1-\rho)} = \lambda E[\tau]^2 \frac{(1 + C_\tau^2)}{2(1-\rho)}$$

$$= \frac{\rho(1 + C_\tau^2)}{2(1-\rho)} E[\tau], \tag{9.93}$$

where $C_\tau^2 = \sigma_\tau^2/E[\tau]^2$ is the coefficient of variation of the service time. Equation (9.93) is called the **Pollaczek-Khinchin mean value formula.**

The **mean delay $E[T]$** is found by adding the mean service time to $E[W]$:

$$E[T] = E[\tau] + E[\tau]\frac{\rho(1 + C_\tau^2)}{2(1-\rho)}. \tag{9.94}$$

From Eqs. (9.93) and (9.94) we can see that the mean waiting time and mean delay time are affected not only by the mean service time and the server utilization but also by the coefficient of variation of the service time. Thus the degree of randomness of the service times as measured by $C_\tau^2$ affects these delays.[1]

### ■■ Example 9.15

Compare $E[W]$ for the M/M/1 and M/D/1 systems. The second moments of the exponential and constant random variables were found in Example 9.14. The exponential service time has a coefficient of variation equal to one. Thus Eq. (9.93) implies

$$E[W_{M/M/1}] = \frac{\rho}{(1-\rho)} E[\tau]. \tag{9.95}$$

The constant service time has zero variance so its coefficient of variation is

1. On the other hand, it is rather surprising that only the first two moments of the distribution of the service time affect $E[W]$ and $E[T]$.

zero. Thus

$$E[W_{M/D/1}] = \frac{\rho}{2(1-\rho)}E[\tau]. \tag{9.96}$$

Thus we see that the waiting time in an M/D/1 is half that of an M/M/1 system. ■■

---

### Mean Delay in M/G/1 Systems with Priority Service Discipline

Consider a queueing system that handles $K$ priority classes of customers. Type $k$ customers arrive according to a Poisson process of rate $\lambda_k$ and have service times with pdf $f_{\tau_k}(x)$ and mean $E[\tau_k]$. A separate queue is kept for each priority class, and each time the server becomes available it selects the next customer from the highest-priority nonempty queue. This service discipline is often referred to as **"head-of-line priority service."** We assume that customers cannot be preempted once their service has begun.

The server utilization from type $k$ customers is

$$\rho_k = \lambda_k E[\tau_k].$$

We assume that the total server utilization is less than 1:

$$\rho = \rho_1 + \cdots + \rho_K < 1. \tag{9.97}$$

If this is not the case, one or more of the lower priority queues become unstable, that is, grow without bound.

Consider the mean waiting time $W_1$ of the highest priority (type 1) customer. If an arriving type 1 customer finds $N_{q_1}(t) = k_1$ type 1 customers in queue and if the service discipline is first-come-first-serve within each class, then $W_1$ is the sum of the residual service time $R''$ of the customer (if any) found in service and the $N_{q_1}(t) = k_1$ service times of the type 1 customers (if any) found in queue. Thus

$$E[W_1] = E[R''] + E[N_{q_1}]E[\tau_1].$$

Following the same development that followed Equation (9.89) in the previous section, we arrive at the following expression for the **mean waiting time for type-1 customers:**

$$E[W_1] = \frac{E[R'']}{1-\rho_1}. \tag{9.98}$$

If an arriving type-2 customer finds $N_{q_1}(t) = k_1$ type 1 and $N_{q_2}(t) = k_2$ type 2 customers waiting in queue, then $W_2$ is the sum of the residual service time $R''$ of the customer (if any) found in service, the $k_1$ service times of the type 1 customers (if any) found in queue, the service times of the $k_2$ type 2 customers found in queue, *and* the service times of the higher priority type 1 customers that arrive while our customer is waiting in queue. Thus

$$E[W_2] = E[R''] + E[N_{q_1}]E[\tau_1] + E[N_{q_2}]E[\tau_2] + E[M_1]E[\tau_1], \tag{9.99}$$

where $M_1$ denotes the number of type 1 arrivals during our customer's waiting time. By Little's formula we have $E[N_{q_1}] = \lambda_1 E[W_1]$ and $E[N_{q_2}] = \lambda_2 E[W_2]$. In addition, the mean number of type 1 arrivals during $E[W_2]$ seconds is $E[M_1] = \lambda_1 E[W_2]$. Substituting these expressions in Eq. (9.99) gives

$$E[W_2] = E[R''] + \rho_1 E[W_1] + \rho_2 E[W_2] + \rho_1 E[W_2].$$

Solving for $E[W_2]$:

$$\begin{aligned} E[W_2] &= \frac{E[R''] + \rho_1 E[W_1]}{1 - \rho_1 - \rho_2} \\ &= \frac{E[R'']}{(1-\rho_1)(1-\rho_1-\rho_2)}, \end{aligned} \tag{9.100}$$

where we have used Eq. (9.98) for $E[W_1]$.

If there are more than two classes of customers, the above method can be used to show that the mean waiting time for a type $k$ customer is

$$E[W_k] = \frac{E[R'']}{(1-\rho_1-\cdots-\rho_{k-1})(1-\rho_1-\cdots-\rho_k)}. \tag{9.101}$$

The customer found in service by an arriving customer can be of any type, so $R''$ is the residual service time of customers of all types:

$$E[R''] = \frac{\lambda E[\tau^2]}{2}, \tag{9.102}$$

where $\lambda$ is the total arrival rate

$$\lambda = \lambda_1 + \cdots + \lambda_K, \tag{9.103}$$

and $E[\tau^2]$ is the second moment of the service time of customers of all types. The fraction of customers that are type $k$ is $\lambda_k/\lambda$, thus

$$E[\tau^2] = \frac{\lambda_1}{\lambda}E[\tau_1^2] + \cdots + \frac{\lambda_K}{\lambda}E[\tau_K^2]. \tag{9.104}$$

We finally arrive at the following expression for the **mean waiting time of type *k* customers:**

$$E[W_k] = \frac{\sum_{j=1}^{K} \lambda_j E[\tau_j^2]}{2(1-\rho_1-\cdots-\rho_{k-1})(1-\rho_1-\cdots-\rho_k)}. \tag{9.105}$$

The **mean delay for type $k$ customers** is then

$$E[T_k] = E[W_k] + E[\tau_k]. \tag{9.106}$$

Equation (9.105) reveals the effect of the priority classes on each other. Class $k$ customers are affected by lower priority customers only through the residual service time term in the numerator. On the other hand, if the server utilization of the first $k - 1$ classes exceeds one, then the queue for class $k$ customers is unstable.

### ■■ Example 9.16

A computer handles two types of jobs. Type 1 jobs require a constant service time of 1 ms, and type 2 jobs require an exponentially distributed amount of time with mean 10 ms. Find the mean waiting time if the system operates as follows: (1) an ordinary M/G/1 system and (2) a two-priority M/G/1 system with priority given to type 1 jobs. Assume that the arrival rates of the two classes are Poisson with the same rate.

The first two moments of the service time are

$$E[\tau] = \frac{1}{2}E[\tau_1] + \frac{1}{2}E[\tau_2] = 5.5$$

$$E[\tau^2] = \frac{1}{2}E[\tau_1^2] + \frac{1}{2}E[\tau_2^2] = \frac{1}{2}(1^2 + 2(10^2)) = 100.5$$

The traffic intensity for each class and the total traffic intensity are

$$\rho_1 = 1\frac{\lambda}{2}, \qquad \rho_2 = 10\frac{\lambda}{2}, \quad \text{and}$$

$$\rho = \lambda E[\tau] = 5.5\lambda,$$

where $\lambda$ is the total arrival rate. The mean residual service time is then

$$E[R] = \frac{\lambda E[\tau^2]}{2} = 50.25\lambda.$$

From Eq. (9.92), the mean waiting time for an M/G/1 system is

$$E[W] = \frac{E[R]}{1 - \rho} = \frac{50.25\lambda}{1 - 5.5\lambda}. \tag{9.107}$$

For the priority system we have

$$E[W_1] = \frac{E[R]}{1 - \rho_1} = \frac{50.25\lambda}{1 - 0.5\lambda} \tag{9.108}$$

and

$$E[W_2] = \frac{E[R]}{(1 - \rho_1)(1 - \rho)} = \frac{50.25\lambda}{(1 - 0.5\lambda)(1 - 5.5\lambda)} \tag{9.109}$$

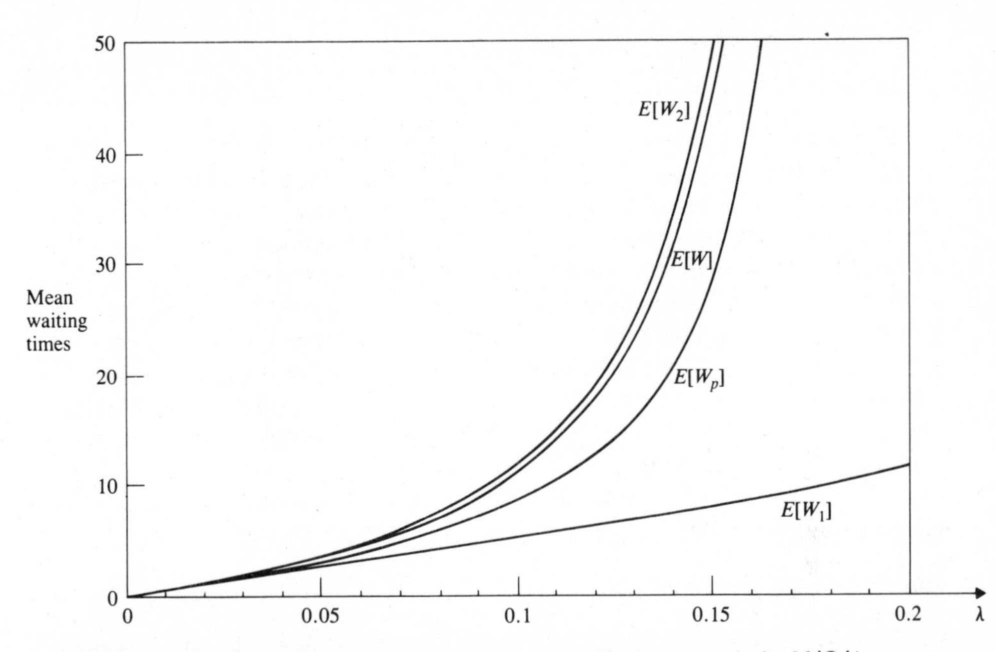


**FIGURE 9.18** Relative mean waiting times in priority and nonpriority M/G/1 systems: $E[W]$, mean waiting time in M/G/1 system; $E[W_1]$, $E[W_2]$, mean waiting time for class 1 and class 2 customers in priority system; $E[W_p]$, overall mean waiting time in priority system.

Comparison of Eqs. (9.108) and (9.109) to Eq. (9.107) shows that the waiting time of type 1 customers is improved by a factor of $(1 - \rho)/(1 - \rho_1)$ and that of type 2 is worsened by the factor $1/(1 - \rho_1)$.

The overall mean waiting for the priority system is

$$\begin{aligned} E[W_p] &= \frac{1}{2}E[W_1] + \frac{1}{2}E[W_2] = \frac{1}{2}\left(\frac{E[R]}{1 - \rho_1}\right)\left(1 + \frac{1}{1 - \rho}\right) \\ &= \left(\frac{1 - \rho/2}{1 - \rho_1}\right)\left(\frac{E[R]}{1 - \rho}\right) \\ &= \frac{1 - 2.75\lambda}{1 - 0.5\lambda}E[W], \end{aligned}$$

where $E[W]$ is the mean waiting time of the M/G/1 system without priorities. Figure 9.18 shows $E[W]$, $E[W_p]$, $E[W_1]$, and $E[W_2]$. It can be seen that the discipline "short-job class first" used here improves the average waiting time. The graphs for $E[W_1]$ and $E[W_2]$ also show that at $\lambda = 2/11$ the lower-priority queue becomes unstable but the higher priority remains stable up to $\lambda = 2$. ■■

## *9.7

## M/G/1 ANALYSIS USING EMBEDDED MARKOV CHAIN

In the previous section we noted that the state of an M/G/1 queueing system is given by the number of customers in the system $N(t)$ and the residual service time of the customer in service. Suppose we observe $N(t)$ at the instants when the residual service time becomes zero (i.e., at the instants $D_j$ when the $j$th service completion occurs), then all of the information relevant to the probability of future events is embodied in $N_j = N(D_j)$, *the number of customers left behind by the jth departing customer.* We will show that the sequence $N_j$ is a discrete-time Markov chain and that the steady state pmf at customer departure instants is equal to the steady state pmf of the system at arbitrary time instants. Thus we can find the steady state pmf of $N(t)$ if we can find the steady state pmf for the chain $N_j$.

### The Embedded Markov Chain

First, we show that the sequence $N_j = N(D_j)$ is a Markov chain. Consider the relation between $N_j$ and $N_{j-1}$. If $N_{j-1} \geq 1$, then a customer enters service immediately at time $D_j$ as shown in Fig. 9.19(a), and $N_j$ equals $N_{j-1}$, minus the customer that is served in between, plus the number of customers $M_j$ that arrive during the service time of the $j$th customer:

$$N_j = N_{j-1} - 1 + M_j \qquad \text{if } N_{j-1} \geq 1. \tag{9.110a}$$

If $N_{j-1} = 0$, then as shown in Fig. 9.19(b), there are no departures until

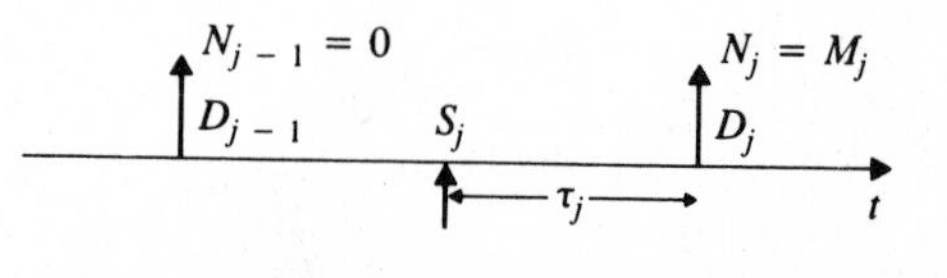


(a)

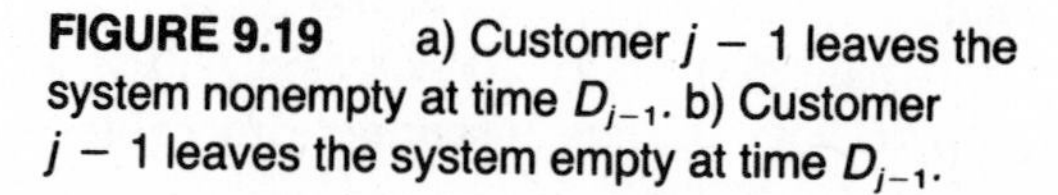


(b)

**FIGURE 9.19** a) Customer $j - 1$ leaves the system nonempty at time $D_{j-1}$. b) Customer $j - 1$ leaves the system empty at time $D_{j-1}$.

the $j$th customer arrives and completes its service; $N_j$ then is the number of customers that arrive during this service time:

$$N_j = M_j \qquad \text{if } N_{j-1} = 0. \tag{9.110b}$$

Thus we see that $N_j$ depends on the past only through $N_{j-1}$ and $M_j$. The $M_j$ form an iid sequence because the service times are iid and because of the memoryless property of Poisson arrivals. Thus $N_j$ depends on the past of the system only through $N_{j-1}$. We therefore conclude that the sequence $N_j$ is a Markov chain.

Next we need to show that the steady state pmf of $N(t)$ is the same as the steady state pmf of $N_j$. We do so in two steps: first, we show that in M/G/1 systems, the distribution of customers found by arriving customers is the same as that left behind by departing customers; second, we show that in M/G/1 systems, the distribution of customers found by arriving customers is the same as the steady state distribution of $N(t)$. It then follows that the steady state pmf of $N(t)$ and $N_j$ are the same.

First we need to show that *for systems in which customers arrive one at a time and depart one at a time (i.e., M/G/1 systems) the distribution found by arriving customers is the same as that left behind by departing customers.* Let $U_n(t)$ be the number of times that the system goes from $n$ to $n+1$ in the interval $(0, t)$; then $U_n(t)$ is the number of times an arriving customer finds $n$ customers in the system. Similarly, let $V_n(t)$ be the number of times that the system goes from $n+1$ to $n$; then $V_n(t)$ is the number of times a departing customer leaves $n$. Note that the transition $n$ to $n+1$ cannot reoccur until after the number in the system drops to $n$ once more (i.e., until after the transition $n+1$ to $n$ reoccurs). Thus $U_n(t)$ and $V_n(t)$ can differ by at most 1. As $t$ becomes large, both of these transitions occur a large number of times, so the rate of transitions from $n$ to $n+1$ equals the rate from $n+1$ to $n$. Thus the rate at which customer arrivals find $n$ in the system equals the rate at which departures leave $n$ in the system. It then follows that the probability that an arrival finds $n$ in the system is equal to the probability that a departure leaves $n$ behind.

Since the arrivals in an M/G/1 system are Poisson and independent of the customer service times, the customer arrival times are independent of the state of the system. Thus the probability that an arrival finds $n$ customers in the system is equal to the proportion of time the system has $n$ customers, that is, the steady state probability $P[N(t) = n]$. Thus *the distribution of states seen by arriving customers is the same as the steady state distribution.*

By combining the results from the two previous paragraphs, we have that for an M/G/1 system, the pmf of $N_j$, the state at customer departure points, is the same as the steady state pmf of $N(t)$. In the next section, we find the generating function of $N_j$ and thus of $N(t)$.

### The Number of Customers in an M/G/1 System

We now find the generating function for the steady state pmf of $N_j$. The transition probabilities for $N_j$ can be deduced from Eqs. (9.110a) and (9.110b).

$$p_{ik} = P[N_j = k \mid N_{j-1} = i] = P[M_j = k - i + 1] \qquad i > 0 \tag{9.111a}$$

$$p_{0k} = P[N_j = k \mid N_{j-1} = 0] = P[M_j = k]. \tag{9.111b}$$

Note that $p_{ik} = 0$ for $k - i + 1 < 0$. The probability that there are $N_j = k$ customers in the system at time $j$ is

$$\begin{aligned} P[N_j = k] &= \sum_{i=0}^{\infty} P[N_{j-1} = i]p_{ik} \\ &= P[N_{j-1} = 0]P[M_j = k] \\ &\quad + \sum_{i=1}^{k+1} P[N_{j-1} = i]P[M_j = k + 1 - i] \qquad (9.112a) \\ &= P[N_{j-1} = 0]P[M_j = k] \\ &\quad + \sum_{i=1}^{\infty} P[N_{j-1} = i]P[M_j = k + 1 - i], \qquad (9.112b) \end{aligned}$$

where we have used the fact that $P[M_j = k + 1 - i] = 0$ for $i > k + 1$.

If the process $N_j$ reaches a steady state as $j \to \infty$, then $P[N_j = k] \to P[N_d = k]$ and the above equation becomes

$$\begin{aligned} P[N_d = k] &= P[N_d = 0]P[M = k] \\ &\quad + \sum_{i=1}^{\infty} P[N_d = i]P[M = k + 1 - i]. \end{aligned} \tag{9.113}$$

where $N_d$ denotes the number of customers left behind by a departing customer.

Since the steady state pmf of $N_j$ is equal to that of $N(t)$, Eq. (9.113) also holds for the steady state pmf of $N(t)$. Equation (9.113) is readily solved for the generating function of $N(t)$ by using the probability generating function. The generating functions for $N$ and for $M$ are given by

$$G_N(z) = \sum_{k=0}^{\infty} P[N = k]z^k \qquad \text{and} \qquad G_M(z) = \sum_{k=0}^{\infty} P[M = k]z^k.$$

Multiply both sides of Equation (9.113) (with $N_d$ replaced by $N$) by $z^k$ and sum from 0 to infinity:

$$\begin{aligned} \sum_{k=0}^{\infty} P[N = k]z^k &= \sum_{k=0}^{\infty} P[N = 0]P[M = k]z^k \\ &\quad + \sum_{k=0}^{\infty}\sum_{i=1}^{\infty} P[N = i]P[M = k + 1 - i]z^k. \end{aligned} \tag{9.114}$$

The generating functions for $N$ and $M$ are immediately recognizable in the first two summations:

$$G_N(z) = P[N = 0]G_M(z) + z^{-1}\sum_{i=1}^{\infty} P[N = i]z^i \sum_{k=0}^{\infty} P[M = k + 1 - i]z^{k+1-i}.$$

The first summation is the generating function for $N$ with the $i = 0$ term missing. Let $k' = k + 1 - i$ in the second summation and note that $P[M = k'] = 0$ for $k' < 0$, then

$$\begin{aligned} G_N(z) &= P[N = 0]G_M(z) + z^{-1}\{G_N(z) - P[N = 0]\}\left\{\sum_{k'=0}^{\infty} P[M = k']z^{k'}\right\} \\ &= P[N = 0]G_M(z) + z^{-1}(G_N(z) - P[N = 0])G_M(z). \end{aligned} \tag{9.115}$$

The generating function for $N$ is found by solving for $G_N(z)$:

$$G_N(z) = \frac{P[N = 0](z - 1)G_M(z)}{z - G_M(z)}. \tag{9.116}$$

We can find $P[N = 0]$ by noting that as $z \to 1$, we must have

$$G_N(z) = \sum_{k=0}^{\infty} P[N = k]z^k \to 1. \tag{9.117}$$

When we take the limit $z \to 1$ in Eq. (9.116) we obtain zero for the numerator and the denominator. By applying L'Hopital's rule, we obtain

$$1 = P[N = 0]\left.\frac{G_M(z) + (z - 1)G'_M(z)}{1 - G'_M(z)}\right|_{z=1} = \frac{P[N = 0]}{1 - E[M]}. \tag{9.118}$$

Thus

$$P[N = 0] = 1 - E[M] \tag{9.119}$$

and

$$G_N(z) = \frac{(1 - E[M])(z - 1)G_M(z)}{z - G_M(z)}. \tag{9.120}$$

Note from Eq. (9.119) that we must have $E[M] < 1$ since $P[N = 0] \geq 0$. This stability condition makes sense since it implies that on the average less than one customer should arrive during the time it takes to service a customer.

We now determine $G_M(z)$, the generating function for the number of arrivals during a service time:

$$\begin{aligned} G_M(z) &= \sum_{k=0}^{\infty} P[M = k]z^k \\ &= \sum_{k=0}^{\infty} \int_0^{\infty} P[M = k \mid \tau = t]f_\tau(t)\,dt\, z^k. \end{aligned} \tag{9.121a}$$

Noting that the number of arrivals in $t$ seconds is a Poisson random variable:

$$
\begin{aligned}
G_M(z) &= \sum_{k=0}^{\infty} \int_0^{\infty} \frac{(\lambda t)^k}{k!} e^{-\lambda t} f_\tau(t)\, dt\, z^k \\
&= \int_0^{\infty} e^{-\lambda t} f_\tau(t) \sum_{k=0}^{\infty} \frac{(\lambda t)^k}{k!} z^k\, dt \\
&= \int_0^{\infty} e^{-\lambda t} f_\tau(t) e^{\lambda t z}\, dt \\
&= \int_0^{\infty} e^{-\lambda(1-z)t} f_\tau(t)\, dt \\
&= \hat{\tau}(\lambda(1-z)),
\end{aligned}
\tag{9.121b}
$$

where $\hat{\tau}(s)$ is the Laplace transform of the pdf of $\tau$:

$$
\hat{\tau}(s) = \int_0^{\infty} e^{-st} f_\tau(t)\, dt. \tag{9.122}
$$

We can obtain the moments of $M$ by taking derivatives of $G_M(z)$:

$$
\begin{aligned}
E[M] &= \frac{d}{dz} G_M(z) \bigg|_{z=1} = \frac{d}{du} \hat{\tau}(u) \frac{d}{dz} \lambda(1-z) \bigg|_{z=1} \\
&= \hat{\tau}'(\lambda(1-z))(-\lambda) \bigg|_{z=1} \\
&= -\lambda \hat{\tau}'(0) = \lambda E[\tau] = \rho,
\end{aligned}
\tag{9.123}
$$

where we used the chain rule in the second equality. Similarly,

$$
E[M(M-1)] = \lambda^2 \hat{\tau}''(0) = \lambda^2 E[\tau^2].
$$

Thus

$$
\begin{aligned}
\sigma_M^2 &= E[M^2] - E[M]^2 = \lambda^2 E[\tau^2] + \lambda E[\tau] - (\lambda E[\tau])^2 \\
&= \lambda^2 \sigma_\tau^2 + \lambda E[\tau].
\end{aligned}
\tag{9.124}
$$

If we substitute Eqs. (9.123) and (9.121b) into Eq. (9.120), we obtain the **Pollaczek-Khinchin transform equation**

$$
G_N(z) = \frac{(1-\rho)(z-1)\hat{\tau}(\lambda(1-z))}{z - \hat{\tau}(\lambda(1-z))}. \tag{9.125}
$$

Note that $G_N(z)$ depends on the utilization $\rho$, the arrival rate $\lambda$, and the Laplace transform of the service time pdf.

### ■■ Example 9.17
M/M/1 System

Use the Pollaczek-Khinchin formula to find the pmf for $N(t)$ for an $M/M/1$ system.

The Laplace transform for the pdf of an exponential service of mean $1/\mu$ is

$$\hat{\tau}(s) = \frac{\mu}{s+\mu}.$$

Thus the Pollaczek-Khinchin transform formula is

$$G_N(z) = \frac{(1-\rho)(z-1)[\mu/(\lambda(1-z)+\mu)]}{z - [\mu/(\lambda(1-z)+\mu)]}$$

$$= \frac{(1-\rho)(z-1)\mu}{(\lambda - \lambda z + \mu)z - \mu} = \frac{1-\rho}{1-\rho z},$$

where we cancelled the $z - 1$ term from the numerator and denominator and noted that $\rho = \lambda/\mu$. By expanding $G_N(z)$ in a power series, we have

$$G_N(z) = \sum_{k=0}^{\infty} (1-\rho)\rho^k z^k = \sum_{k=0}^{\infty} P[N=k]z^k,$$

which implies that the steady state pmf is

$$P[N=k] = (1-\rho)\rho^k \qquad k = 0, 1, 2, \ldots,$$

which is in agreement with our previous results for the M/M/1 system. ■■

---

### ■■ Example 9.18
M/H$_2$/1 System

Find the pmf for the number of customers in an M/G/1 system that has arrivals of rate $\lambda$ and where the service times are hyperexponential random variables of degree two as shown in Fig. 9.20. In other words, with probability 1/9 the service time is exponentially distributed with mean $1/\lambda$, and with probability 8/9 the service time is exponentially distributed with mean $1/2\lambda$.

In order to find $\hat{\tau}(s)$ we note that the pdf of $\tau$ is

$$f_\tau(x) = \frac{1}{9}\lambda e^{-\lambda x} + \frac{8}{9}2\lambda e^{-2\lambda x} \qquad x > 0.$$

Thus the mean service time is

$$E[\tau] = \frac{1}{9\lambda} + \frac{8}{9(2\lambda)} = \frac{5}{9\lambda},$$

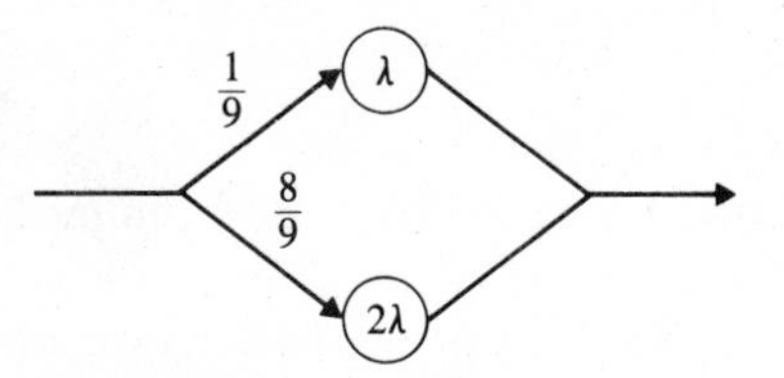


**FIGURE 9.20** A hyperexponential service time results if we select an exponential service time of rate λ with probability 1/9 and an exponential service time of rate 2λ with probability 8/9.

and the server utilization is $\rho = \lambda E[\tau] = 5/9$. The Laplace transform of $f_\tau(x)$ is

$$\hat{\tau}(s) = \frac{1}{9}\frac{\lambda}{s+\lambda} + \frac{8}{9}\frac{2\lambda}{s+2\lambda} = \frac{18\lambda^2 + 17\lambda s}{9(s+\lambda)(s+2\lambda)}$$

Substitution of $\hat{\tau}(\lambda(1-z))$ into Eq. (9.125) gives

$$G_N(z) = \frac{(1-\rho)(z-1)(18\lambda^2 + 17\lambda^2(1-z))}{9(\lambda - \lambda z + \lambda)(\lambda - \lambda z + 2\lambda)z - (18\lambda^2 + 17\lambda^2(1-z))}$$

$$= \frac{(1-\rho)(z-1)(35-17z)}{9(2-z)(3-z)z - (35-17z)},$$

where we have cancelled $\lambda^2$ from the numerator and denominator. If we factor the denominator we obtain

$$G_N(z) = \frac{(1-\rho)(35-17z)(z-1)}{9(z-1)(z-7/3)(z-5/3)}$$

$$= (1-\rho)\left\{\frac{1/3}{1-3z/7} + \frac{2/3}{1-3z/5}\right\},$$

where we have carried out a partial fraction expansion. Finally we note that since $G_N(z)$ converges for $|z| < 1$, we can expand $G_N(z)$ as follows:

$$G_N(z) = (1-\rho)\left\{\frac{1}{3}\sum_{k=0}^{\infty}\left(\frac{3}{7}\right)^k z^k + \frac{2}{3}\sum_{k=0}^{\infty}\left(\frac{3}{5}\right)^k z^k\right\}.$$

Since the coefficient of $z^k$ is $P[N = k]$, we finally have that

$$P[N = k] = \frac{4}{27}\left(\frac{3}{7}\right)^k + \frac{8}{27}\left(\frac{3}{5}\right)^k \qquad k = 0, 1, \ldots,$$

where we used the fact that $\rho = 5/9$. ■■

### Delay and Waiting Time Distribution in M/G/1 System

We now find the delay and waiting time distributions for an M/G/1 system with first-come-first-serve service discipline. If a customer spends $T_j$ seconds in the queueing system, then the number of customers $N_d$ it leaves behind in the system is the number of customers that arrive during these $T$ seconds since customers are served in order of arrival. An expression for the generating function for $N_d$ is found by proceeding as in Eq. (9.121a):

$$G_{N_d}(z) = \sum_{k=0}^{\infty} \int_0^{\infty} P[N_d = k \mid T = t] f_T(t)\, dt z^k$$
$$= \hat{T}(\lambda(1 - z)), \tag{9.126}$$

where $\hat{T}(s)$ is the Laplace transform of the pdf of $T$, the total delay in the system. Since the steady state distributions of $N_d(t)$ and $N(t)$ are equal, we have that $G_N(z) = G_{N_d}(z)$ and thus combining Eqs. (9.125) and (9.126):

$$\hat{T}(\lambda(1 - z)) = \frac{(1 - \rho)(z - 1)\hat{\tau}(\lambda(1 - z))}{z - \hat{\tau}(\lambda(1 - z))} \tag{9.127}$$

If we let $s = \lambda(1 - z)$, Eq. (9.127) yields an expression for $\hat{T}(s)$:

$$\hat{T}(s) = \frac{(1 - \rho)s\hat{\tau}(s)}{s - \lambda + \lambda\hat{\tau}(s)}. \tag{9.128}$$

The pdf of $T$ is found from the inverse transform of $\hat{T}(s)$ either analytically or numerically.

Since $T = W + \tau$, where $W$ and $\tau$ are independent random variables, we have that

$$\hat{T}(s) = \hat{W}(s)\hat{\tau}(s). \tag{9.129}$$

Equations (9.128) and (9.129) can then be solved for the Laplace transform of the waiting time pdf:

$$\hat{W}(s) = \frac{(1 - \rho)s}{s - \lambda + \lambda\hat{\tau}(s)}. \tag{9.130}$$

Equations (9.128) and (9.129) are also referred to as the **Pollaczek-Khinchin transform equations.**

#### ■■ Example 9.19
M/M/1

Find the pdf's of $W$ and $T$ for an M/M/1 system. Substituting $\hat{\tau}(s) = \mu/(s + \mu)$ into Eq. (9.128) gives

$$\hat{T}(s) = \frac{(1 - \rho)s\mu}{(s + \mu)(s - \lambda) + \lambda\mu} = \frac{(1 - \rho)\mu}{s - (\lambda - \mu)}, \tag{9.131}$$

which is readily inverted to obtain

$$f_T(x) = \mu(1-\rho)e^{-\mu(1-\rho)x} \qquad x > 0. \tag{9.132}$$

Similarly Eq. (9.130) gives

$$\hat{W}(s) = \frac{(1-\rho)s}{s - \lambda + \lambda\mu/(s+\mu)} = (1-\rho)\frac{s+\mu}{s+\mu-\lambda}.$$

In order to invert this expression, the numerator polynomial must have order lower than the denominator polynomial. We achieve this by dividing the denominator into the numerator

$$\hat{W}(s) = (1-\rho)\frac{s+\mu-\lambda+\lambda}{s+\mu-\lambda} = (1-\rho)\left\{1 + \frac{\lambda}{s+\mu-\lambda}\right\}. \tag{9.133}$$

We then obtain

$$f_W(x) = (1-\rho)\delta(x) + \lambda(1-\rho)e^{-\mu(1-\rho)x} \qquad x > 0. \tag{9.134}$$

The delta function at zero corresponds to the fact that a customer has zero wait with probability $(1-\rho)$. Equations (9.132) and (9.134) were previously obtained as Eq. (9.32) and (9.33) in Section 9.3 by a different method. ■■

## ■■ Example 9.20
M/$H_2$/1

Find the pdf of the waiting time in the M/$H_2$/1 system discussed in Example 9.18.

Substitution of $\hat{\tau}(s)$ from Example 9.18 into Eq. (9.130) gives

$$\begin{aligned}
\hat{W}(s) &= \frac{9s(1-\rho)(s+\lambda)(s+2\lambda)}{9(s-\lambda)(s+\lambda)(s+2\lambda) + \lambda(18\lambda^2 + 17\lambda s)} \\
&= \frac{(1-\rho)(s+\lambda)(s+2\lambda)}{s^2 + 2\lambda s + 8\lambda^2/9} \\
&= (1-\rho)\frac{9s^2 + 27\lambda s + 18\lambda^2}{9s^2 + 18\lambda s + 8\lambda^2} \\
&= (1-\rho)\left\{1 + \frac{9\lambda s + 10\lambda^2}{9s^2 + 18\lambda s + 8\lambda^2}\right\} \\
&= (1-\rho)\left\{1 + \frac{2\lambda/3}{s + 2\lambda/3} + \frac{\lambda/3}{s + 4\lambda/3}\right\},
\end{aligned}$$

where we have followed the same sequence of steps as in Example 9.18 and then done a partial fraction expansion.

The inverse Laplace transform then yields

$$f_W(x) = \frac{4}{9}\left\{\delta(x) + \frac{2\lambda}{3}e^{-2\lambda x/3} + \frac{1}{4}\frac{4\lambda}{3}e^{-4\lambda x/3}\right\} \qquad x > 0.$$

■■

---

Examples 9.18 and 9.19 demonstrate that the Pollaczek-Khinchin transform equations can be used to obtain closed form expressions for the pmf of $N(t)$ and the pdf's of $W$ and $T$ when the Laplace transform of the service time pdf is a rational function of $s$, that is, a ratio of polynomials in $s$. This result is particularly important because it can be shown that the Laplace transform of any service time pdf can be approximated arbitrarily closely by a rational function of $s$. Thus in principle we can obtain exact expressions for the pmf of $N(t)$ and pdf's of $W$ and $T$.

In addition it should be noted that *the Pollaczek-Khinchin transform expressions can always be inverted numerically* using fast Fourier transform methods such as those discussed in Chapter 5. This numerical approach does not require that the Laplace transform of the pdf be a rational function of $s$.

## 9.8

## BURKE'S THEOREM: DEPARTURES FROM M/M/c SYSTEMS

In many problems, a customer requires service from several service stations before a task is completed. These problems require that we consider a *network* of queueing systems. In such networks, the departures from some queues become the arrivals to other queues. This is the reason why we are interested in the statistical properties of the departure process from a queue.

Consider two queues in tandem as shown in Fig. 9.21, where the departures from the first queue become the arrivals at the second queue. Assume that the arrivals to the first queue are Poisson with rate $\lambda$ and that the service time at queue 1 is exponentially distributed with rate $\mu_1 > \lambda$. Assume that the service time in queue 2 is also exponentially distributed with rate $\mu_2 > \lambda$.

The state of this system is specified by the number of customers in the

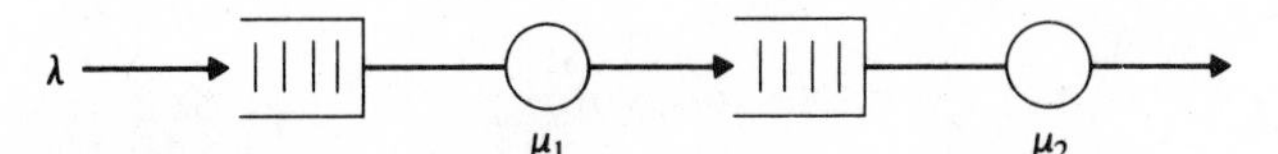


**FIGURE 9.21** Two tandem exponential queues with Poisson input.

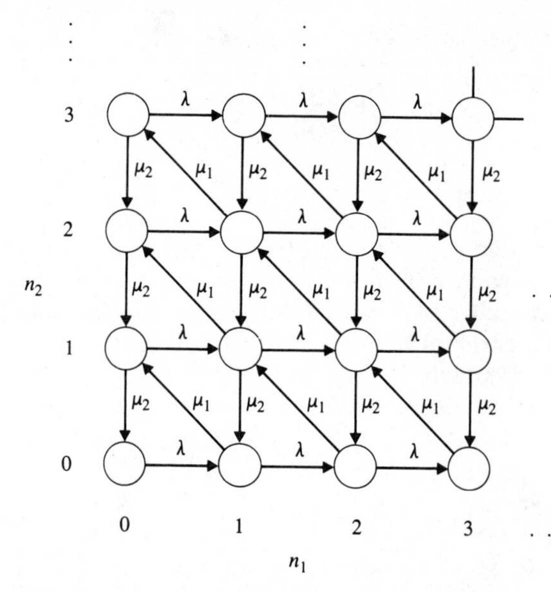


**FIGURE 9.22** Transition rate diagram for two tandem exponential queues with Poisson input.

two queues, $(N_1(t), N_2(t))$. This state vector forms a Markov process with the transition rate diagram shown in Fig. 9.22, and global balance equations:

$$\lambda P[N_1 = 0, N_2 = 0] = \mu_2 P[N_1 = 0, N_2 = 1] \tag{9.135a}$$

$$(\lambda + \mu_1)P[N_1 = n, N_2 = 0] = \mu_2 P[N_1 = n, N_2 = 1] + \lambda P[N_1 = n - 1, N_2 = 0] \qquad n > 0 \tag{9.135b}$$

$$(\lambda + \mu_2)P[N_1 = 0, N_2 = m] = \mu_2 P[N_1 = 0, N_2 = m + 1] + \mu_1 P[N_1 = 1, N_2 = m - 1] \qquad m > 0 \tag{9.135c}$$

$$\begin{aligned}(\lambda + \mu_1 + \mu_2)P[N_1 = n, N_2 = m] = {} & \mu_2 P[N_1 = n, N_2 = m + 1] \\ & + \mu_1 P[N_1 = n + 1, N_2 = m - 1] \\ & + \lambda P[N_1 = n - 1, N_2 = m] \\ & n > 0, m > 0.\end{aligned} \tag{9.135d}$$

It is easy to verify that the following joint state pmf satisfies Eqs. (9.135a) through (9.135d):

$$P[N_1 = n, N_2 = m] = (1 - \rho_1)\rho_1^n(1 - \rho_2)\rho_2^m \qquad n \geq 0, m \geq 0, \tag{9.136}$$

where $\rho_i = \lambda/\mu_i$. We know that the first queue is an M/M/1 system, so

$$P[N_1 = n] = (1 - \rho_1)\rho_1^n \qquad n = 0, 1, \ldots. \tag{9.137}$$

By summing Eq. (9.136) over all $n$, we obtain the marginal state pmf of the second queue

$$P[N_2 = m] = (1 - \rho_2)\rho_2^m \qquad m \geq 0. \tag{9.138}$$

Equations (9.136) through (9.138) imply that

$$P[N_1 = n, N_2 = m] = P[N_1 = n]P[N_2 = m] \qquad \text{for all } n, m. \tag{9.139}$$

In words, *the number of customers at queue 1 and the number at queue 2 at the same time instant are independent random variables*. Furthermore, *the steady state pmf at the second queue is that of an M/M/1 system with Poisson arrival rate $\lambda$ and exponential service time $\mu_2$.*

We say that a network of queues has a **product-form** solution when the joint pmf of the vector of numbers of customers at the various queues is equal to the product of the marginal pmf's of the number in the individual queues. We now discuss Burke's Theorem, which states the fundamental result underlying the product-form solution in Eq. (9.139).

**BURKE'S THEOREM**

Consider an M/M/1, M/M/$c$, or M/M/$\infty$ queueing system at steady state with arrival rate $\lambda$, then

1. The departure process is Poisson with rate $\lambda$.
2. At each time $t$, the number of customers in the system $N(t)$ is independent of the sequence of departure times prior to $t$.

The product-form solution for the two tandem queues follows from Burke's theorem. Queue 1 is an M/M/1 queue, so from part 1 of the theorem the departures from queue 1 form a Poisson process. Thus the arrivals to queue 2 are a Poisson process, so the second queue is also an M/M/1 system with steady state pmf given by Eq. (9.138). It remains to show that the number of customers in the two queues at the same time instant are independent random variables.

The arrivals to queue 2 prior to time $t$ are the departures from queue 1 prior to time $t$. By part 2 of Burke's theorem the departures from queue 1, and hence the arrivals to queue 2, prior to time $t$ are independent of $N_1(t)$. Since $N_2(t)$ is determined by the sequence of arrivals from queue 1 prior to time $t$ and the independent sequence of service times, it then follows that $N_1(t)$ and $N_2(t)$ are independent. Equation (9.139) then follows.

Burke's theorem implies that the generalization of Eq. (9.139) holds for the tandem combination of any number of M/M/1, M/M/$c$, or M/M/$\infty$ queues. Indeed, the result holds for any "feedforward" network of queues in which a customer cannot visit any queue more than once.

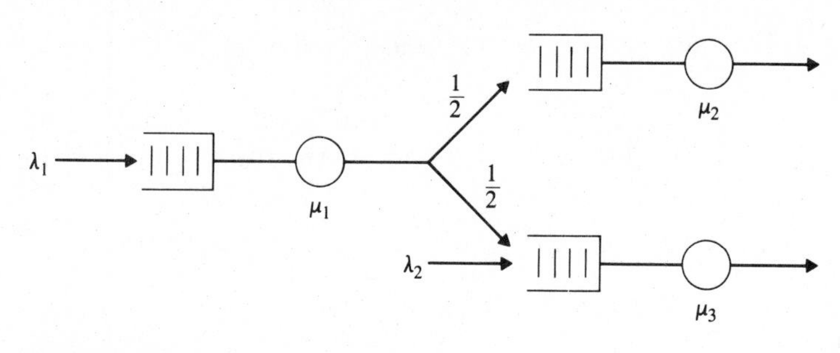


FIGURE 9.23 A feedforward network of queues.

### ■■ Example 9.21

Find the joint state pmf for the network of queues shown in Fig. 9.23 where queue 1 is driven by a Poisson process of rate $\lambda_1$, where the departures from queue 1 are randomly routed to queues 2 and 3, and where queue 3 also has an additional independent Poisson arrival stream of rate $\lambda_2$.

From Burke's theorem $N_1(t)$ and $N_2(t)$ are independent, as are $N_1(t)$ and $N_3(t)$. Since the random split of a Poisson process yields independent Poisson processes, we have that the inputs to queues 2 and 3 are independent. The input to queue 2 is Poisson with rate $\lambda_1/2$. The input to queue 3 is Poisson of rate $\lambda_1/2 + \lambda_2$ since the merge of two independent Poisson processes is also Poisson. Thus

$$P[N_1(t) = k, N_2(t) = m, N_3(t) = n]$$
$$= (1 - \rho_1)\rho_1^k(1 - \rho_2)\rho_2^m(1 - \rho_3)\rho_3^n \qquad k, m, n \geq 0,$$

where $\rho_1 = \lambda_1/\mu_1$, $\rho_2 = \lambda_1/2\mu_2$, and $\rho_3 = (\lambda_1/2 + \lambda_2)/\mu_3$, and where we have assumed that all of the queues are stable. ■■

### *Proof of Burke's Theorem Using Time Reversibility

Consider the sample path of an M/M/1, M/M/$c$, or M/M/$\infty$ system as shown in Fig. 9.24(a). Note that the arrivals in the forward process correspond to the departures in the time-reversed process. In Section 8.5, we showed that birth-and-death Markov chains in steady state are time-reversible processes, that is, the sample functions of the process played backward in time have the same statistics as the forward process. Since M/M/1, M/M/$c$, and M/M/$\infty$ systems are birth-and-death Markov chains, we have that their states are reversible processes. Thus the sample functions of these systems played backward in time correspond to the sample functions of queueing systems of the same type. It then follows that the arrival process of the time-reversed system is a Poisson process.

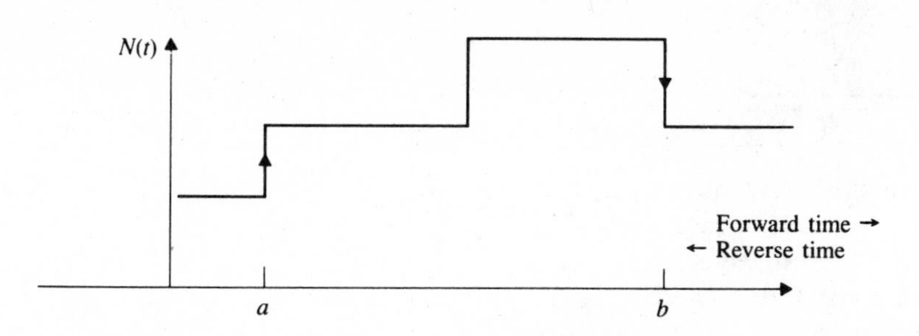


**FIGURE 9.24a** Time instant $a$ is an arrival time in the forward process and a departure time in the reverse process. Time instant $b$ in a departure in the forward process and an arrival in the reverse process.

To prove part 1 of Burke's theorem, we note that the interdeparture times of the forward-time system are the interarrival times of the time-reversed system. Since the arrival process of the time-reversed system is Poisson, it then follows that the departure process of the forward system is also Poisson. Thus we have shown that the departure process of an M/M/1, M/M/$c$, or M/M/$\infty$ system is Poisson.

To prove part 2 of Burke's theorem, fix a time $t$ as shown in Fig. 9.24(b). The departures before time $t$ from the forward system are the arrivals after time $t$ in the reverse system. In the reverse system, the arrivals are Poisson and thus the arrival times after time $t$ do not depend on $N(t)$. These arrival instants of the reverse process are exactly the departure instants before $t$ in the forward process. It then follows that $N(t)$ and the departure instants prior to $t$ are independent, so part 2 is proved.

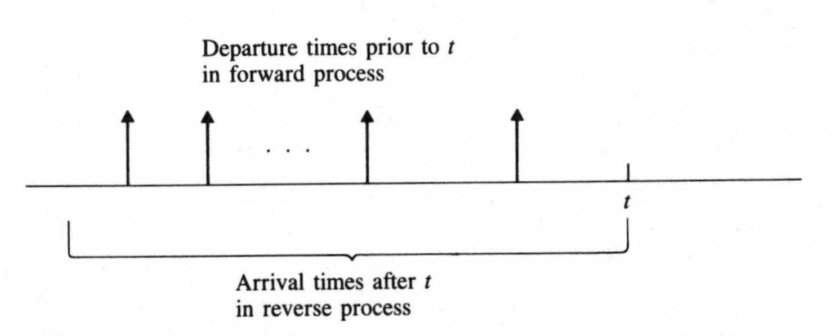


**FIGURE 9.24b** The departure times prior to time $t$ in the forward process correspond exactly to the arrival times after time $t$ in the reverse process.

## 9.9

## NETWORKS OF QUEUES: JACKSON'S THEOREM

In many queueing networks, a customer is allowed to visit a particular queue more than once. Burke's theorem does not hold for such systems. In this section we discuss Jackson's theorem, which extends the product-form solution for the steady-state pmf to a broader class of queueing networks.

If a customer is allowed to visit a queue more than once, then the arrival process at that queue will not be Poisson. For example, consider the simple M/M/1 queue with feedback shown in Fig. 9.25 where external customers arrive according to a Poisson process of rate $\lambda$ and where departures are instantaneously fed back into the system with probability .9. If the arrival rate is much less than the departure rate, then we have that the net arrival process (i.e., external and feedback arrivals) typically consists of isolated external arrivals followed by a burst of feedback arrivals. Thus the arrival process does not have independent increments and so it is not Poisson.

### Open Networks of Queues

Consider a network of $K$ queues in which customers arrive from outside the network to queue $k$ according to independent Poisson processes of rate $\alpha_k$. We assume that the service time of a customer at queue $k$ is exponentially distributed with rate $\mu_k$ and independent of all other service times and arrival processes. We also suppose that queue $k$ has $c_k$ servers. After completion of service at queue $k$, a customer proceeds to queue $i$ with probability $P_{ki}$ and it exits the network with probability

$$1 - \sum_{i=1}^{K} P_{ki}.$$

The total arrival rate $\lambda_k$ into queue $k$ is the sum of the external arrival rate and the internal arrival rates:

$$\lambda_k = \alpha_k + \sum_{j=1}^{K} \lambda_j P_{jk} \qquad k = 1, \ldots, K. \tag{9.140}$$

It can be shown that Eq. (9.140) has a unique solution if no customer

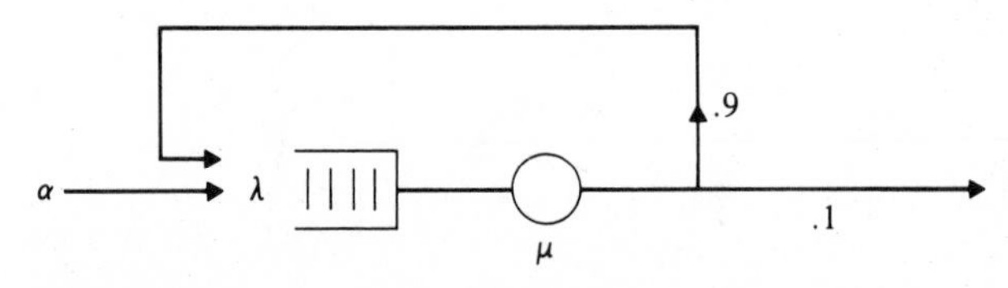


FIGURE 9.25 A queue with feedback.

remains in the network indefinitely. We call such networks **open queueing networks.**

The vector of the number of customers at all the queues

$$\mathbf{N}(t) = (N_1(t), N_2(t), \ldots, N_K(t))$$

is a Markov process. Jackson's theorem gives the steady state pmf for $\mathbf{N}(t)$.

**JACKSON'S THEOREM**

If $\lambda_k < c_k\mu_k$, then for any possible state $\mathbf{n} = (n_1, n_2, \ldots, n_K)$

$$P[\mathbf{N}(t) = \mathbf{n}] = P[N_1 = n_1]P[N_2 = n_2] \ldots P[N_K = n_K], \tag{9.141}$$

where $P[N_k = n_k]$ is the steady state pmf of an M/M/$c_k$ system with arrival rate $\lambda_k$ and service rate $\mu_k$.

Jackson's theorem states that the number of customers in the queues at time $t$ are *independent* random variables. In addition, it states that the steady state probabilities of the individual queues are those of an M/M/$c_k$ system. This is an amazing result because in general the input process to a queue is not Poisson as was demonstrated in the simple queue with feedback discussed in the beginning of this section.

### ■■ Example 9.22

Messages arrive to a concentrator according to a Poisson process of rate $\alpha$. The time required to transmit a message and receive an acknowledgement is exponentially distributed with mean $1/\mu$. Suppose that a message needs to be retransmitted with probability $p$. Find the steady state pmf for the number of messages in the concentrator.

The overall system can be represented by the simple queue with feedback shown in Fig. 9.25. The net arrival rate into the queue is $\lambda = \alpha + \lambda p$, that is,

$$\lambda = \frac{\alpha}{1 - p}.$$

Thus, the pmf for the number of messages in the concentrator is

$$P[N = n] = (1 - \rho)\rho^n \qquad n = 0, 1, \ldots,$$

where $\rho = \lambda/\mu = \alpha/(1 - p)\mu$. ■■

### ■■ Example 9.23

New programs arrive at a CPU according to a Poisson process of rate $\alpha$ as shown in Fig. 9.26. A program spends an exponentially distributed execution time of mean $1/\mu_1$ in the CPU. At the end of this service time,

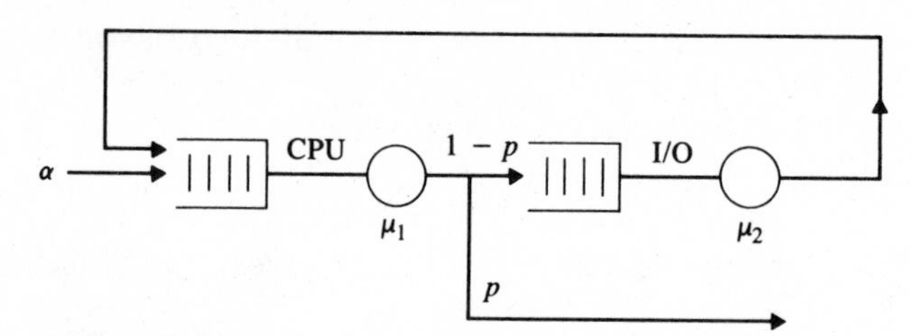


**FIGURE 9.26** An open queueing network model for a computer system.

the program execution is complete with probability $p$ or it requires retrieving additional information from secondary storage with probability $1 - p$. Suppose that the retrieval of information from secondary storage requires an exponentially distributed amount of time with mean $1/\mu_2$. Find the mean time that each program spends in the system.

The net arrival rates into the two queues are

$$\lambda_1 = \alpha + \lambda_2 \quad \text{and} \quad \lambda_2 = (1 - p)\lambda_1.$$

Thus

$$\lambda_1 = \frac{\alpha}{p} \quad \text{and} \quad \lambda_2 = \frac{(1-p)\alpha}{p}.$$

Each queue behaves like an $M/M/1$ system, so

$$E[N_1] = \frac{\rho_1}{1 - \rho_1} \quad \text{and} \quad E[N_2] = \frac{\rho_2}{1 - \rho_2},$$

where $\rho_1 = \lambda_1/\mu_1$ and $\rho_2 = \lambda_2/\mu_2$. Little's formula then gives the mean for the total time spent in the system

$$E[T] = \frac{E[N_1 + N_2]}{\alpha} = \frac{1}{\alpha}\left[\frac{\rho_1}{1 - \rho_1} + \frac{\rho_2}{1 - \rho_2}\right].$$

■■

---

### *Proof of Jackson's Theorem

Jackson's theorem can be proved by writing the global balance equations for the queueing network and verifying that the solution is given by Eq. (9.141). We present an alternative proof of the theorem using a result from time-reversed Markov chains. For notational simplicity we consider only the case of a network of single-server queues.

Let $\mathbf{n}$ and $\mathbf{n}'$ be two possible states of the network, and let $v_{\mathbf{n},\mathbf{n}'}$ denote the transition rate from $\mathbf{n}$ to $\mathbf{n}'$. In Section 8.5, we found that if we can guess a state pmf $P[\mathbf{n}]$ and a set of transition rates $\hat{v}_{\mathbf{n}',\mathbf{n}}$ for the reverse process such that (Eq. [8.65])

$$P[\mathbf{n}]v_{\mathbf{n},\mathbf{n}'} = P[\mathbf{n}']\hat{v}_{\mathbf{n}',\mathbf{n}} \tag{9.142a}$$

and such that the total rate out of state **n** is the same in the forward and reverse processes (Eq. [8.64] summed over $j$)

$$\sum_{\mathbf{m}} v_{\mathbf{n},\mathbf{m}} = \sum_{\mathbf{m}} \hat{v}_{\mathbf{n},\mathbf{m}}, \tag{9.142b}$$

then $P[\mathbf{n}]$ is the steady-state pmf of the process.

For the case under consideration our guess for the pmf is

$$P[\mathbf{n}] = \prod_{j=1}^{K} (1 - \rho_j)\rho_j^{n_j} \tag{9.143}$$

so the proof reduces to finding a consistent set of transition rates for the reverse process that satisfy Eqs. (9.142a) and (9.142b). Noting that $v_{\mathbf{n},\mathbf{n}'}$ is known and that $P[\mathbf{n}]$ and $P[\mathbf{n}']$ are specified by Eq. (9.143), Eq. (9.142a) can be solved for the transition rates of the reverse process:

$$\hat{v}_{\mathbf{n}',\mathbf{n}} = \frac{P[\mathbf{n}]v_{\mathbf{n},\mathbf{n}'}}{P[\mathbf{n}']} \tag{9.144}$$

Let $\mathbf{n} = (n_1, \ldots, n_k)$ denote a state for the network, and let $\mathbf{e}_k = (0, \ldots, 0, 1, 0, \ldots, 0)$ where the 1 is located in the $k$th component. Only three types of transitions in the state of the queueing network have nonzero probabilities. In the first type of transition, an external arrival to queue $k$ takes the state from $\mathbf{n}$ to $\mathbf{n} + \mathbf{e}_k$. In the second type of transition, a departure from queue $k$ exits the network and takes the state from $\mathbf{n}$ to $\mathbf{n} - \mathbf{e}_k$, where $n_k > 0$. In the third type of transition, a customer leaves queue $k$ and proceeds to queue $j$, thus taking the state from $\mathbf{n}$ to $\mathbf{n} - \mathbf{e}_k + \mathbf{e}_j$, where $n_k > 0$. Table 9.1 shows three types of transitions and their corresponding rates for the forward process.

A consistent set of transition rates for the reverse process are obtained by solving Eq. (9.144) for the three types of transitions possible. For example, if we let $\mathbf{n}' = \mathbf{n} + \mathbf{e}_k$, then the transition $\mathbf{n} \to \mathbf{n} + \mathbf{e}_k$ in the forward process corresponds to the transition $\mathbf{n} + \mathbf{e}_k \to \mathbf{n}$ in the reverse process. Equation (9.144) gives

$$\hat{v}_{\mathbf{n}',\mathbf{n}} = \frac{\alpha_k \prod_{j=1}^{K} (1 - \rho_j)\rho_j^{n_j}}{\rho_k \prod_{j=1}^{K} (1 - \rho_j)\rho_j^{n_j}}$$

$$= \frac{\alpha_k}{\rho_k} = \frac{\alpha_k}{\lambda_k/\mu_k} = \frac{\alpha_k \mu_k}{\lambda_k}.$$

**TABLE 9.1** Allowable Transitions in Jackson Network and Their Corresponding Rates in the Forward and Reverse Processes

| | | |
|---|---|---|
| **Forward Process** | | |
| *Transition* | *Rate* | |
| $\mathbf{n} \to \mathbf{n} + \mathbf{e}_k$ | $\alpha_k$ | all $k$ |
| $\mathbf{n} \to \mathbf{n} - \mathbf{e}_k$ | $\mu_k\left(1 - \sum_{j=1}^{K} P_{kj}\right)$ | all $k$: $n_k > 0$ |
| $\mathbf{n} \to \mathbf{n} - \mathbf{e}_k + \mathbf{e}_j$ | $\mu_k P_{kj}$ | all $k$: $n_k > 0$, all $j$ |
| **Reverse Process** | | |
| *Transition* | *Rate* | |
| $\mathbf{n} \to \mathbf{n} + \mathbf{e}_k$ | $\lambda_k\left(1 - \sum_{j} P_{kj}\right)$ | all $k$ |
| $\mathbf{n} \to \mathbf{n} - \mathbf{e}_k$ | $\frac{\alpha_k \mu_k}{\lambda_k}$ | all $k$: $n_k > 0$ |
| $\mathbf{n} \to \mathbf{n} - \mathbf{e}_k + \mathbf{e}_j$ | $\frac{\lambda_j P_{jk} \mu_k}{\lambda_k}$ | all $k$: $n_k > 0$, all $j$ |

The other reverse process transition rates are found in similar manner. Table 9.1 shows the results for the transition rates of the reverse process that are implied by Eq. (9.144).

The proof that the pmf in Eq. (9.143) gives the steady state pmf of the network of queues is completed by showing that the total transition rate out of any state **n** are the same in the forward and in the reverse process, that is, Eq. (9.142b) holds. In the forward process, the total transition rate out of state **n** is obtained by adding the entries for the forward process in Table 9.1:

$$\sum_{\mathbf{m}} v_{\mathbf{n},\mathbf{m}} = \sum_{k} \alpha_k + \sum_{k:n_k>0} \mu_k. \tag{9.145a}$$

For the reverse process, we have from Table 9.1 that

$$\sum_{\mathbf{m}} \hat{v}_{\mathbf{n},\mathbf{m}} = \sum_{k} \lambda_k\left(1 - \sum_{j} P_{kj}\right) + \sum_{k:n_k>0} \left\{\frac{\alpha_k \mu_k}{\lambda_k} + \sum_{j} \frac{\lambda_j P_{jk} \mu_k}{\lambda_k}\right\}. \tag{9.145b}$$

We need to show that the right-hand sides of Eq. (9.145a) and (9.145b) are equal. First, note that Eq. (9.140) implies that

$$\lambda_k - \alpha_k = \sum_{j=1}^{K} \lambda_j P_{jk}.$$

The right-hand side of Eq. (9.145b) then becomes

$$\left(\sum_k \lambda_k - \sum_j \sum_k \lambda_k P_{kj}\right) + \sum_{k:n_k>0} \left\{\frac{\alpha_k \mu_k}{\lambda_k} + \frac{\mu_k}{\lambda_k} \sum_j \lambda_j P_{jk}\right\}$$

$$= \sum_k \lambda_k - \sum_j (\lambda_j - \alpha_j) + \sum_{k:n_k>0} \left\{\frac{\alpha_k \mu_k}{\lambda_k} + \frac{\mu_k}{\lambda_k}(\lambda_k - \alpha_k)\right\}$$

$$= \sum_k \alpha_k + \sum_{k:n_k>0} \mu_k.$$

Thus the right-hand side of Eqs. (9.145a) and (9.145b) are equal and thus Eq. (9.143) is the steady state pmf of the network of queues. This completes the proof of Jackson's theorem for a network of single server queues.

### Closed Networks of Queues

In some problems, a *fixed* number of customers, say $I$, circulate endlessly in a network of queues. For example, some computer system models assume that at any time a fixed number of programs use the CPU and input/output (I/O) resources of a computer as shown in Fig. 9.27. We now consider queueing networks that are identical to the above open networks in every respect except that the external arrival rates are zero and the networks always contain a fixed number of customers $I$. We show that the steady state pmf for such systems is product form but that the states of the queues are no longer independent.

The net arrival rate into queue $k$ is now given by

$$\lambda_k = \sum_{j=1}^{K} \lambda_j P_{jk} \qquad k = 1, \ldots, K. \tag{9.146}$$

Note that these equations have the same form as the set of equations that define the stationary pmf for a discrete-time Markov chain with transition probabilities $P_{jk}$. The only difference is that the sum of the $\lambda_k$'s need not be one. Thus the solution vector to Eq. (9.146) must be proportional to the

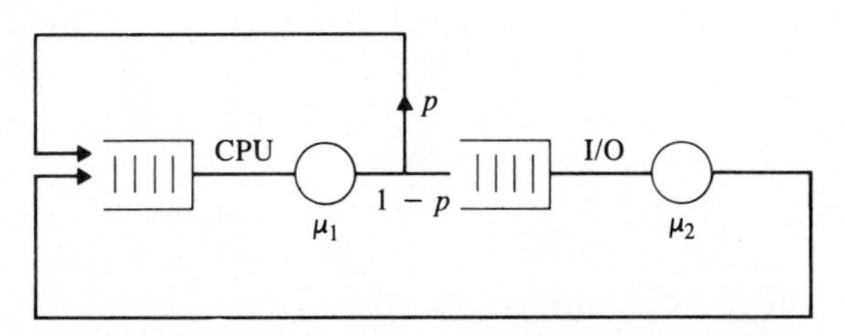


**FIGURE 9.27** A closed queueing network model for a computer system.

stationary pmf $\{\pi_j\}$ corresponding to $\{P_{jk}\}$:

$$\lambda_k = \lambda(I)\pi_k, \tag{9.147}$$

where

$$\pi_k = \sum_{j=1}^{K} \pi_j P_{jk} \tag{9.148}$$

and where $\lambda(I)$ is a constant that depends on $I$, the number of customers in the queueing network. If we sum both sides of Eq. (9.147) over $k$, we see that $\lambda(I)$ is the sum of the arrival rates in all the queues in the network, and $\pi_k = \lambda_k/\lambda(I)$ is the fraction of total arrivals to queue $k$.

**THEOREM**

Let $\lambda_k = \lambda(I)\pi_k$ be a solution to Eq. (9.146), and let $\mathbf{n} = (n_1, n_2, \ldots, n_K)$ be any state of the network for which $n_1, \ldots, n_K \geq 0$ and

$$n_1 + n_2 + \cdots + n_K = I, \tag{9.149}$$

then

$$P[\mathbf{N}(t) = \mathbf{n}] = \frac{P[N_1 = n_1]P[N_2 = n_2]\ldots P[N_K = n_K]}{S(I)}, \tag{9.150}$$

where $P[N_k = n_k]$ is the steady state pmf of an M/M/$c_k$ system with arrival rate $\lambda_k$ and service rate $\mu_k$, and where $S(I)$ is the normalization constant given by

$$S(I) = \sum_{\mathbf{n}:n_1+\cdots+n_K=I} P[N_1 = n_1]P[N_2 = n_2]\ldots P[N_K = n_K]. \tag{9.151}$$

Equation (9.150) states that $P[\mathbf{N}(t) = \mathbf{n}]$ has a product form. However, $P[\mathbf{N}(t) = \mathbf{n}]$ is no longer equal to the product of the marginal pmf's because of the normalization constant $S(I)$. This constant arises because the fact that there are always $I$ customers in the network implies that the allowable states $\mathbf{n}$ must satisfy Eq. (9.149). The theorem can be proved using the approach used to prove Jackson's theorem above.

### ■■ Example 9.24

Suppose that the computer system in Example 9.23 is operated so that there are always $I$ programs in the system. The resulting network of queues is shown in Fig. 9.27. Note that the feedback loop around the CPU signifies the completion of one job and its instantaneous replacement by another one. Find the steady state pmf of the system. Find the rate at which programs are completed.

The stationary probabilities associated with Eq. (9.146) are found by solving

$$\pi_1 = p\pi_1 + \pi_2, \qquad \pi_2 = (1-p)\pi_1, \qquad \text{and} \qquad \pi_1 + \pi_2 = 1.$$

The stationary probabilities are then

$$\pi_1 = \frac{1}{2-p} \qquad \text{and} \qquad \pi_2 = \frac{1-p}{2-p} \tag{9.152}$$

and the arrival rates are

$$\lambda_1 = \lambda(I)\pi_1 = \frac{\lambda(I)}{2-p} \qquad \text{and} \qquad \lambda_2 = \frac{(1-p)\lambda(I)}{2-p}. \tag{9.153}$$

The stationary pmf for the network is then

$$P[N_1 = i, N_2 = I - i] = \frac{(1-\rho_1)\rho_1^i(1-\rho_2)\rho_2^{I-i}}{S(I)} \qquad 0 \le i \le I \tag{9.154}$$

where $\rho_1 = \lambda_1/\mu_1$ and $\rho_2 = \lambda_2/\mu_2$, and where we have used the fact that if $N_1 = i$ then $N_2 = I - i$. The normalization constant is then

$$\begin{aligned} S(I) &= (1-\rho_1)(1-\rho_2)\sum_{i=0}^{I} \rho_1^i \rho_2^{I-i} \\ &= (1-\rho_1)(1-\rho_2)\rho_2^I \frac{1-(\rho_1/\rho_2)^{I+1}}{1-(\rho_1/\rho_2)} \end{aligned} \tag{9.155}$$

Substitution of Eq. (9.155) into Eq. (9.154) gives

$$P[N_1 = i, N_2 = I - i] = \frac{1-\beta}{1-\beta^{I+1}}\beta^i \qquad 0 \le i \le I, \tag{9.156}$$

where

$$\beta = \frac{\rho_1}{\rho_2} = \frac{\pi_1\mu_2}{\pi_2\mu_1} = \frac{\mu_2}{(1-p)\mu_1}. \tag{9.157}$$

Note that the form of Eq. (9.156) suggests that queue 1 behaves like an M/M/1/$K$ queue. The apparent load to this queue is $\beta$, which is proportional to the ratio of I/O to CPU service rates and inversely proportional to the probability of having to go to I/O.

The rate at which programs are completed is $p\lambda_1$. We find $\lambda_1$ from the relation between server utilization and probability of empty system:

$$1 - \lambda_1/\mu_1 = P[N_1 = 0] = \frac{1-\beta}{1-\beta^{I+1}},$$

which implies that

$$p\lambda_1 = p\mu_1 \frac{\beta(1-\beta^I)}{1-\beta^{I+1}}.$$

■■

---

## ■■ Example 9.25

A transmitter (queue 1 in Fig. 9.28) has 2 permits for message transmission. As long as the transmitter has a permit ($N_1 > 0$) it generates messages with exponential interarrival times of rate $\lambda$. The messages enter the transmission system and require an exponential service time at station 2. As soon as a message arrives at the other side of the transmission system, the corresponding permit is sent back via station 3. Thus the transmitter can have at most 2 messages outstanding in the network at any given time. Find the steady state pmf for the network of queues. Find the rate at which messages enter the transmission system.

We can view the 2 permits as 2 customers circulating the queueing network. Since $P_{1,2} = P_{2,3} = P_{3,1} = 1$, we have that $\pi_1 = \pi_2 = \pi_3 = 1/3$ and thus

$$\lambda_1 = \lambda_2 = \lambda_3 = \frac{\lambda(2)}{3}.$$

The steady state pmf for the network is

$$P[N_1 = i, N_2 = j, N_3 = 2-i-j] = \frac{(1-\rho_1)\rho_1^i(1-\rho_2)\rho_2^j(1-\rho_3)\rho^{I-i-j}}{S(2)}$$

$$\text{for } 0 \le i \le 2,\ 0 \le j \le 2-i,$$

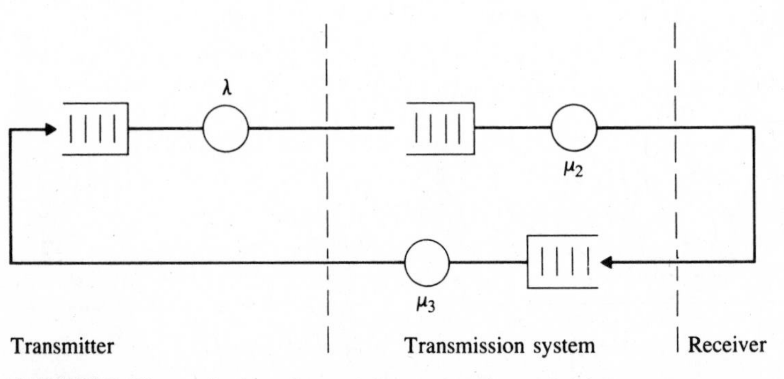


**FIGURE 9.28** A closed queueing network model for a message transmission system.

where $\rho_1 = \lambda(2)/3\lambda$ and $\rho_2 = \rho_3 = \lambda(2)/3\mu$. The normalization constant $S(2)$ is obtained by summing the above joint pmf over all possible states and equating the result to one. There are six possible network states: $(2,0,0)$, $(0,2,0)$, $(0,0,2)$, $(1,1,0)$, $(1,0,1)$, $(0,1,1)$. Thus the normalization constant is given by

$$\begin{aligned} S(2) &= (1-\rho_1)(1-\rho_2)(1-\rho_3)\{\rho_1^2 + \rho_2^2 + \rho_3^2 + \rho_1\rho_2 + \rho_1\rho_3 + \rho_2\rho_3\} \\ &= (1-\rho_1)(1-\rho_2)^2\{\rho_1^2 + 2\rho_2^2 + 2\rho_1\rho_2 + \rho_2^2\}, \end{aligned}$$

where we have used the fact that $\rho_2 = \rho_3$.

The rate at which messages enter the system is

$$\lambda_1 = \lambda(1 - P[N_1 = 0]),$$

where

$$\begin{aligned} P[N_1 = 0] &= P[\mathbf{N} = (0,2,0)] + P[\mathbf{N} = (0,0,2)] + P[\mathbf{N} = (0,1,1)] \\ &= \frac{3\rho_2^2}{\rho_1^2 + 2\rho_1\rho_2 + 3\rho_2^2} = \frac{3/\mu^2}{1/\lambda^2 + 2/\lambda\mu + 3/\mu^2} \end{aligned}$$

■■

### Mean Value Analysis

Example 9.25 shows that the evaluation of the normalization constant is the fundamental difficulty with closed queueing networks. A number of efficient algorithms have been devised to evaluate this constant (Lavenberg, 1983). Nevertheless, the evaluation of this constant poses severe difficulties as the size of a network grows. Fortunately, a method has been developed for obtaining certain average quantities of interest without having to evaluate this constant (Reiser and Lavenberg, 1980). This **mean value analysis** method is based on the following theorem.

**ARRIVAL THEOREM**
In a closed queueing network with $I$ customers, the system as seen by a customer arrival to queue $j$ is the steady state pmf of the same network with one fewer customer.

We have already encountered this result in the discussion of finite-source queueing systems in Section 9.5. We prove the result in the last part of this section. We now use the result to develop the mean value analysis method.

Let $E[N_j(I)]$ be the mean number of customers in the $j$th queue for a network that has $I$ customers, let $E[T_j(I)]$ denote the mean time spent by a customer in queue $j$, and let $\lambda_j(I)$ denote the average customer arrival rate at queue $j$. The mean time spent by a customer in queue $j$ is its service time plus the service times of the customers it finds in the queue upon arrival:

$$\begin{aligned} E[T_j(I)] &= E[\tau_j] + E[\tau_j] \times \text{mean number found upon arrival} \\ &= E[\tau_j] + E[\tau_j]E[N_j(I-1)] \\ &= \frac{1 + E[N_j(I-1)]}{\mu_j}, \end{aligned} \tag{9.158}$$

where $E[N_j(I-1)]$ is the mean number found upon arrival by the arrival theorem. By Little's formula, the mean number of customers in queue $j$ when there are $I$ in the network is

$$E[N_j(I)] = \lambda_j(I)E[T_j(I)] = \lambda(I)\pi_j E[T_j(I)]. \tag{9.159}$$

Since the sum of the customers in all queues is $I$ in the previous equation, we have that

$$I = \sum_{j=1}^{K} E[N_j(I)] = \lambda(I)\sum_{j=1}^{K} \pi_j E[T_j(I)]. \tag{9.160}$$

Thus

$$\lambda(I) = \frac{I}{\sum_{j=1}^{K} \pi_j E[T_j(I)]}. \tag{9.161}$$

The *mean value analysis* method combines Eqs. (9.158) through (9.161) in the following way. First compute $\pi_j$ by solving Eq. (9.148), then For $I = 0$:

$$E[N_j(0)] = 0 \qquad \text{for } j = 1, \ldots, K.$$

For $I = 1, 2, \ldots$:

$$E[T_j(I)] = \frac{1}{\mu_j} + \frac{E[N_j(I-1)]}{\mu_j} \qquad j = 1, \ldots, K \tag{9.158}$$

$$\lambda(I) = \frac{I}{\sum_{i=0}^{K} \pi_i E[T_j(I)]} \tag{9.161}$$

$$E[N_j(I)] = \lambda(I)\pi_j E[T_j(I)] \qquad j = 1, \ldots, K, \tag{9.159}$$

Thus the mean value algorithm begins with an empty system and by use of the above three equations builds up to a network with the desired number of customers. This method has considerably simplified the numerical solution of closed queueing networks and extended the range of network sizes that can be analyzed.

### Example 9.26

In Example 9.24, let $\mu_1 = \mu_2 = 1$, and $p = 1/2$. Find the rate at which programs are completed if $I = 2$.

It was already indicated in Example 9.24 that the rate of program completion is $p\lambda_1(2) = p\pi_1\lambda(2)$. From Eq. (9.152), we have that $\pi_1 = 1/(2 - p) = 2/3$. Thus we only need to find $\lambda(2)$, the total arrival rate of the network with $I = 2$.

Starting the mean value method with $I = 1$, we have

$$E[T_1(1)] = \frac{1}{\mu_1} = 1 \qquad E[T_2(1)] = \frac{1}{\mu_2} = 1$$

$$\lambda(1) = \frac{1}{\pi_1 T_1(1) + \pi_2 T_2(1)} = 1$$

$$E[N_1(1)] = \lambda(1)\pi_1 E[T_1(1)] = \frac{2}{3}$$

$$E[N_2(1)] = \lambda(1)\pi_2 E[T_2(1)] = \frac{1}{3}.$$

Continuing with $I = 2$, we have

$$E[T_1(2)] = \frac{1}{\mu_1} + \frac{E[N_1(1)]}{\mu_1} = \frac{5}{3}$$

$$E[T_2(2)] = \frac{1}{\mu_2} + \frac{E[N_2(1)]}{\mu_2} = \frac{4}{3}$$

$$\lambda(2) = \frac{2}{\pi_1 E[T_1(2)] + \pi_2 E[T_2(2)]} = \frac{9}{7}.$$

Thus the program completion rate is

$$p\pi_1\lambda(2) = \frac{3}{7}.$$

You should verify that this is consistent with the results of Example 9.24.

---

### Example 9.27

In Example 9.25, let $1/\lambda = a$ and $\mu = 1$. Find the rate at which messages enter the system when $I = 2$.

We previously found that $\pi_1 = \pi_2 = \pi_3 = 1/3$ and

$$\lambda_1(I) = \lambda_2(I) = \lambda_3(I) = \frac{\lambda(I)}{3}.$$

Starting the mean value method with $I = 1$, we have

$$E[T_1(1)] = a \qquad E[T_2(1)] = E[T_3(1)] = 1$$

$$\lambda(1) = \frac{1}{\pi_1 E[T_1(1)] + \pi_2 E[T_2(1)] + \pi_3 E[T_3(1)]} = \frac{3}{a+2}$$

$$E[N_1(1)] = \lambda(1)\pi_1 E[T_1(1)] = \frac{a}{a+2}$$

$$E[N_2(1)] = \lambda(1)\pi_2 E[T_2(1)] = \frac{1}{a+2} = E[N_3(1)].$$

Continuing with $I = 2$, we have

$$E[T_1(2)] = a + aE[N_1(1)] = \frac{2a^2 + 2a}{a+2}$$

$$E[T_2(2)] = 1 + 1E[N_2(1)] = \frac{a+3}{a+2} = E[T_3(2)]$$

$$\lambda_2(2) = \frac{2}{(1/3)\{(2a^2 + 2a)/(a+2) + [2(a+3)/(a+2)]\}}$$

$$= \frac{3(a+2)}{a^2 + 2a + 3}.$$

Finally, messages enter the transmission network at a rate $\lambda_1(2) = \lambda(2)/3$, so

$$\lambda_1(2) = \frac{a+2}{a^2 + 2a + 3}.$$

You should verify that this is consistent with the results obtained in Example 9.25. ■■

### *Proof of the Arrival Theorem

Consider the instant when a customer leaves queue $j$ and is proceeding to queue $k$. We are interested in the pmf of the system state at these arrival instants. Suppose that at this instant, with the customer removed from the system, the customer sees the network in state $\mathbf{n} = (n_1, \ldots, n_K)$. This occurs only when the network state goes from the state $\mathbf{n}' = (n_1, \ldots, n_j + 1, \ldots, n_K)$ to the state $\mathbf{n}'' = (n_1, \ldots, n_j, \ldots, n_k + 1, \ldots, n_K)$. Thus:

$$P[\text{customer sees } \mathbf{n} \mid \text{customer goes from } j \text{ to } k]$$

$$= \frac{P[\text{customer sees } \mathbf{n}, \text{ customer goes from } j \text{ to } k]}{P[\text{customer goes from } j \text{ to } k]}$$

$$= \frac{P[\text{customer goes from } j \text{ to } k \mid \text{state is } \mathbf{n}']P[\mathbf{N}(I) = \mathbf{n}']}{P[\text{customer goes from } j \text{ to } k]}$$

$$= \frac{\mu_j P_{jk} P[\mathbf{N}(I) = \mathbf{n}']}{\mu_j P_{jk} P[N_j(I) > 0]}$$

$$= \frac{P[\mathbf{N}(I) = \mathbf{n}']}{P[N_j(I) > 0]}. \tag{9.162}$$

To simplify the notation, let us assume that we are dealing with a network of $M/M/1$ queues, then

$$P[\mathbf{N}(I) = \mathbf{n}'] = \frac{P[N_1 = n_1] \ldots P[N_j = n_j + 1] \ldots P[N_K = n_K]}{S(I)}$$

$$= \rho_j \prod_{m=1}^{K} \frac{\rho_m^{n_m}}{S'(I)}, \tag{9.163}$$

where $S'(I)$ absorbs all the constants associated with the $P[N_m = n_m]$:

$$S'(I) = \sum_{\mathbf{n}: n_1 + \cdots + n_K = I} \prod_{m=1}^{K} \rho_m^{n_m}. \tag{9.164}$$

Next, consider the probability that queue $j$ is not empty:

$$P[N_j(I) > 0] = \sum_{\mathbf{n}: n_1 + \cdots + n_K = I-1} P[N_1 = n_1] \ldots P[N_j = n_j + 1] \ldots P[N_K = n_k]$$

$$= \sum_{\mathbf{n}: n_1 + \cdots + n_K = I-1} \rho_j \frac{\prod_{m=1}^{K} \rho_m^{n_m}}{S'(I)}$$

$$= \frac{\rho_j}{S'(I)} \sum_{\mathbf{n}: n_1 + \cdots + n_K = I-1} \prod_{m=1}^{K} \rho_m^{n_m}$$

$$= \frac{\rho_j S'(I-1)}{S'(I)}, \tag{9.165}$$

where we have noted that the above summation is the normalization constant for a network with $I - 1$ customers $S'(I - 1)$.

Finally, we substitute Eqs. (9.165) and (9.163) into Eq. (9.162):

$$P[\text{customer sees } \mathbf{n} \mid \text{customer goes from } j \text{ to } k]$$

$$= \frac{\rho_j \prod_{m=1}^{K} \rho_m^{n_m} / S'(I)}{[\rho_j S'(I-1)]/S'(I)}$$

$$= \prod_{m=1}^{K} \frac{\rho_m^{n_m}}{S'(I-1)}$$

$$= P[\mathbf{N}(I-1) = \mathbf{n}],$$

which is the steady state probability for **n** in a network with $I - 1$ customers. This completes the proof of the arrival theorem.

## SUMMARY

- A queueing system is specified by the arrival process, the service time distribution, the number of servers, the waiting room, and the queue discipline. Kendall's notation is used to specify the first four properties.
- Little's formula states that under very general conditions: The mean number in a system is equal to the product of the mean arrival rate and the mean time spent in the system.
- In M/M/1, M/M/1/$K$, M/M/$c$, M/M/$c$/$c$, and M/M/$\infty$ queueing systems, the number of customers in the system is a continuous-time Markov chain. The steady state distribution for the number in the system is found by solving the global balance equations for the Markov chain. The waiting time and delay distribution when the service discipline is first-come-first-serve is found by using the arriving customer's distribution.
- If the arrival process in a queueing system is a Poisson process and if the customer interarrival times are independent of the service times, then the arriving customer's distribution is the same as the steady state distribution of the queueing system.
- In M/G/1 queueing systems the arriving customer's distribution and the departing customer's distribution are both equal to the steady state distribution of the queueing system. The steady state distribution for the number of customers in an $M/G/1$ system can be found by embedding a discrete-time Markov chain at the customer departure instants.
- Burke's theorem states that the output process of M/M/1, M/M/$c$, and M/M/$\infty$ systems at steady state are Poisson processes, and that the departure instants prior to time $t$ are independent of the state of the system at time $t$. As a result, feedforward combinations of queueing systems with exponential service times have a product-form solution.
- Jackson's theorem states that for networks of queueing systems with exponential service times and external Poisson input processes, the joint state pmf is of product form. If the network of queues is open, the marginal state pmf of each queue is the same as that of a queue in isolation that has Poisson arrivals of the same rate. If the network of queues is closed, finding the joint state pmf requires finding a normalization constant. The mean-value analysis method allows us to find the mean number in each queue, the mean time spent at each queue, and the arrival rate in each queue in a closed network of queues.

## CHECKLIST OF IMPORTANT TERMS

A/B/$m$/$K$
Arrival rate
Arriving customer's distribution
Burke's theorem
Carried load
Closed networks
Delay
Departing customer's distribution
Embedded Markov chain
Erlang $B$ formula
Erlang $C$ formula
Finite source queueing system
First-come-first-serve
Head-of-line priority
Interarrival times
Jackson's formula
Little's formula
Mean-value analysis
M/G/1 queueing system
M/M/$c$ queueing system
M/M/$c$/$c$ queueing system
M/M/1 queueing system
M/M/1/$K$ queueing system
Offered load
Open networks
Product-form solution
Queue discipline
Server utilization
Service discipline
Service time
Total delay
Waiting time

## ANNOTATED REFERENCES

References [1] through [3] provide an introduction to queueing theory at a level slightly higher than that given here. Reference [2] is an invaluable source of classical queueing theory results in telephony problems. Reference [3] is an excellent introduction to the application of queueing theory to computer system performance evaluation. Reference [4] demonstrates the application of queueing theory to data communication networks. Reference [5] discusses techniques for simulating queueing systems and for analyzing the resulting data.

1. L. Kleinrock, *Queueing Systems,* Volume 1, Wiley, New York, 1975.
2. R. B. Cooper, *Introduction to Queueing Theory,* Macmillan, New York, 1972.
3. H. Kobayashi, *Modeling and Analysis: An Introduction to System Performance Evaluation,* Addison-Wesley, Reading, Mass., 1978.
4. D. Bertsekas and R. Gallager, *Data Networks,* Prentice-Hall, Englewood Cliffs, N.J., 1987.
5. A. M. Law and W. D. Kelton, *Simulation, Modeling, and Analysis,* McGraw-Hill, New York, 1982.
6. S. M. Ross, *Stochastic Processes,* Wiley, New York, 1983.
7. M. Reiser and S. S. Lavenberg, "Mean-value Analysis of Closed

Multichain Queueing Networks," *J. Assoc. Comput. Mach.* 27: 313–322, 1980.

8. S. S. Lavenberg, *Computer Performance Modeling Handbook*, Academic Press, New York, 1983.

## PROBLEMS

### Sections 9.1–9.2
### The Elements of a Queueing Network and Little's Formula

1. Describe the following queueing systems: M/M/1, M/D/1/$K$, M/G/3, D/M/2, G/D/1, D/D/2.
2. Suppose that a queueing system is empty at time $t = 0$, and let the arrival times of the first six customers be 1, 3, 4, 7, 8, 15 and let their respective service times be 3.5, 4, 2, 1, 1.5, 4. Find $S_i$, $\tau_i$, $D_i$, $W_i$, and $T_i$ for $i = 1, \ldots, 5$; sketch $N(t)$ versus $t$; and check Little's formula by computing $\langle N \rangle_t$, $\langle \lambda \rangle_t$, and $\langle T \rangle_t$ for each of the following three service disciplines:
   a. First-come-first served.
   b. Last-come-first-served.
   c. Shortest-job first (assume that the precise service time of each job is known before it enters service).
3. A data communication line delivers a block of information every 10 microseconds. A decoder checks each block for errors and corrects the errors if necessary. It takes 1 microsecond to determine whether a block has any errors. If the block has one error, it takes 5 microseconds to correct it, and if it has more than one error it takes 20 microseconds to correct the error. Blocks wait in a queue when the decoder falls behind. Suppose that the decoder is initially empty and that the number of errors in the first ten blocks are: 0, 1, 3, 1, 0, 4, 0, 1, 0, 0.
   a. Plot the number of blocks in the decoder as a function of time.
   b. Find the mean number of blocks in the decoder.
   c. What percent of the time is the decoder empty?
4. Three queues are arranged in a loop as shown in Fig. P9.1. Assume that the mean service time in queue $i$ is $m_i = 1/\mu_i$.
   a. Suppose the queue has a single customer circulating in the loop. Find the mean time $E[T]$ it takes the customer to cycle around the loop. Deduce from $E[T]$ the mean arrival rate $\lambda$ at each of the queues. Verify that Little's formula holds for these two quantities.

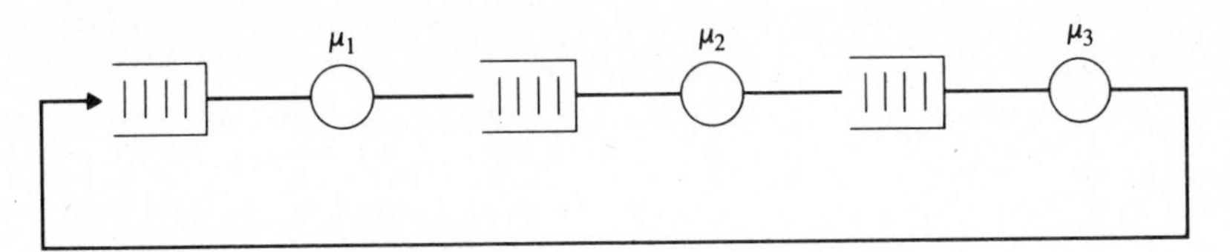


**FIGURE P9.1**

b. If there are $N$ customers circulating in the loop, how are the mean arrival rate and the mean cycle time related?

5. A very popular barber shop is always full. The shop has two barbers and three chairs for waiting and as soon as a customer completes his service and leaves the shop, another enters the shop. Assume the mean service time is $m$.
   a. Use Little's formula to relate the arrival rate and the mean time spent in the shop.
   b. Use Little's formula to relate the arrival rate and the mean time spent in service.
   c. Use the above formulae to find an expression for the mean time spent in the system in terms of the mean service time.

6. In Problem 3, suppose that the probabilities of zero, one, and more than one errors are $p_0$, $p_1$, and $p_3$, respectively. Use Little's formula to find the mean number of blocks in the decoder.

7. A communication network receives messages from $R$ sources with mean arrival rates $\lambda_1, \ldots, \lambda_R$. On the average there are $E[N_i]$ messages from source $i$ in the network.
   a. Use Little's formula to find the average time $E[T_i]$ spent by type $i$ customers in the network.
   b. Let $\lambda$ denote the total arrival rate into the network. Use Little's formula to find an expression for the mean time $E[T]$ spent by customers (of all types) in the network in terms of the $E[N_i]$.
   c. Combine the results of Part a and Part b to obtain an expression for $E[T]$ in terms of $E[T_i]$. Can you justify this expression in another way?

## Section 9.3
## The M/M/1 Queue

8. a. Find $P[N \geq n]$ for an M/M/1 system.
   b. What is the maximum allowable arrival rate in a system with service rate $\mu$, if we require that $P[N \geq 10] = 10^{-3}$?

9. A decision to purchase one of two machines is to be made. Machine 1 has a processing rate of $\mu$ transactions/hour and it costs $B$ dollars/hour to operate; machine 2 is twice as fast but costs twice as much to operate. Suppose that transactions arrive at the system according to a Poisson process of rate $\lambda$ and that the transaction processing times are exponentially distributed. The total cost of the system is the operation cost plus a cost of $A$ dollars for each hour a customer has to wait.
   a. Find expressions for the total cost per hour for each of the systems. Plot this cost versus the arrival rate.
   b. If $A = B/10$, for what range of arrival rates is machine 1 cheaper? Repeat for $A = 10B$.
10. Consider an M/M/1 queueing system in which each customer arrival brings in a profit of \$5 but in which each unit time of delay costs the system \$1. Find the range of arrival rates for which the system makes a net profit.
11. Consider an M/M/1 queueing system with arrival rate $\lambda$ customers/second.
   a. Find the service rate required so that the average queue is five customers (i.e., $E[N_q] = 5$).
   b. Find the service rate required so that the queue that forms from time to time has mean 5 (i.e., $E[N_q \mid N_q > 0] = 5$).
   c. Which of the two criteria, $E[N_q]$ or $E[N_q \mid N_q > 0]$, do you consider the more appropriate?
12. Show that the $p$th percentile of the waiting time for an $M/M/1$ system is given by

$$x = \frac{1/\mu}{1-\rho} \ln\left(\frac{\rho}{1-p}\right).$$

13. Consider an M/M/1 queueing system with service rate two customers per second.
   a. Find the maximum allowable arrival rate if 90% of customers should not have a delay of more than 3 seconds.
   b. Find the maximum allowable arrival rate if 90% of customers should not have to wait for service for more than 2 seconds. *Hint*: Use the result from Problem 12, and then find $\lambda$ by trial and error.
14. Verify Eq. (9.36) for the steady state pmf of an M/M/1/$K$ system.
15. Consider an $M/M/1/2$ queueing system in which each customer accepted into the system brings in a profit of \$5 and in which each customer rejection results in a loss of \$1. Find the arrival rate at which the system breaks even.

16. For an $M/M/1/K$ system show that

$$P[N = k \mid N < K] = \frac{P[N = k]}{1 - P[N = K]} \qquad 0 \le k < K.$$

Why does this probability represent the proportion of arriving customers who actually enter the system and find exactly $k$ customers in the system?

17. Suppose that two types of customers arrive at queueing system according to independent Poisson process of rate $\lambda/2$. Both types of customers require exponentially distributed service times of rate $\mu$. Type 1 customers are always accepted into the system, but type 2 customers are turned away when the total number of customers in the system exceeds $K$.
   a. Sketch the transition rate diagram for $N(t)$, the total number of customers in the system.
   b. Find the steady state pmf of $N(t)$.

### Section 9.4
### Multi-Server Systems: M/M/*c*, M/M/*c*/*c*, and M/M/∞

18. Find $P[N \ge c + k]$ for an M/M/$c$ system.

19. Customers arrive at a shop according to a Poisson process of rate 12 customers per hour. The shop has two clerks to attend to the customers. Suppose that it takes a clerk an exponentially distributed amount of time with mean 5 minutes to service one customer.
   a. What is the probability that an arriving customer must wait to be served?
   b. Find the mean number of customers in the system and the mean time spent in the system.
   c. Find the probability that there are more than 4 customers in the system.

20. Little's formula applied to the servers implies that the mean number of busy servers is $\lambda E[\tau]$. Verify this by explicit calculation of the mean number of busy servers in an M/M/$c$ system.

21. Inquiries arrive at an information center according to a Poisson process of rate 10 inquiries per second. It takes a server 2 seconds to answer each query.
   a. How many servers are needed if we require that the mean total delay for each inquiry should not exceed 4 seconds, and 90% of all queries should wait less than 8 seconds.
   b. What is the resulting probability that all servers are busy? idle?

22. Consider a queueing system in which the maximum processing rate is $c\mu$ customers per second. Let $k$ be the number of customers in the

system. When $k \geq c$, $c$ customers are served at a rate $\mu$ each. When $0 < k \leq c$, these $k$ customers are served at a rate $c\mu/k$ each. Assume Poisson arrivals of rate $\lambda$ and exponentially distributed times.

a. Find the transition rate diagram for this system.
b. Find the steady state pmf for the number in the system.
c. Find $E[W]$ and $E[T]$.
d. For $c = 2$, compare $E[W]$ and $E[T]$ for this system to those of M/M/1 and M/M/2 systems of the same maximum processing rate.

23. Show that the Erlang $B$ formula satisfies the following recursive equation:

$$B(c, a) = \frac{aB(c-1, a)}{c + aB(c-1, a)},$$

where $a = \lambda E[\tau]$.

24. Consider an M/M/5/5 system in which the arrival rate is 10 customers per minute and the mean service time is 1/2 minute.

a. Find the probability of blocking a customer. *Hint*: Use the result from the previous problem.
b. How many more servers are required to reduce the blocking probability to 10%?

25. A tool rental shop has four floor sanders. Customers for floor sanders arrive according to a Poisson process at a rate of one customer every two days. The average rental time is exponentially distributed with mean two days. If the shop has no floor sanders available, the customers go to the shop across the street.

a. Find the proportion of customers that go to the shop across the street.
b. What is the mean number of floor sanders rented out?
c. What is the increase in lost customers if one of the sanders breaks down and is not replaced?

26. a. Show that the Erlang $C$ formula is related to the Erlang $B$ formula by

$$C(c, a) = \frac{cB(c, a)}{c - a\{1 - B(c, a)\}} \qquad \text{for } c > a.$$

b. Show that this implies that $C(c, a) > B(c, a)$.

27. Suppose that department "A" in a certain company has three private telephone lines connecting two sites. Calls arrive according to a Poisson process of rate 1 call/minute, and have an exponentially distributed holding time of 2 minutes. Calls that arrive when the

three lines are busy are automatically redirected to public telephone lines. Suppose that department "B" also has three private telephone lines connecting the same sites, and that it has the same arrival and service statistics.

a. Find the proportion of calls that are redirected to public lines.
b. Suppose we consolidate the telephone traffic from the two departments and allow all calls to share the six lines. What proportion of calls are redirected to public lines?

28. Suppose we use $P[N = c]$ from an M/M/$\infty$ system to approximate $B(c, a)$ in selecting the number of servers in an M/M/$c$/$c$ system. Is the resulting design optimistic or pessimistic?

## Section 9.5
## Finite-Source Queueing Systems

29. A computer is shared by 15 users as shown in Fig. 9.14(b). Suppose that the mean service time is 2 seconds and the mean think time is 30 seconds, and that both of these times are exponentially distributed.
    a. Find the mean delay and mean throughput of the system.
    b. What is the system saturation point $K^*$ for this system?
    c. Repeat Part a if 5 users are added to the system.
30. Find the transition rate diagram and steady state pmf for a two-server finite source queueing system.
31. Verify that Eqs. (9.84) and (9.81) give $E[T]$ as given in Eq. (9.72).
32. Consider a $c$-server, finite source queueing system that allows no queueing for service. Requests that arrive when all servers are busy are turned away, and the corresponding source immediately returns to the "think state," and spends another exponentially distributed think time before submitting another request for service.
    a. Find the transition rate diagram and show that the steady state pmf for the state of the system is

$$P_K[N = j] = \frac{\binom{K}{j} p^j (1-p)^{K-j}}{\sum_{i=0}^{c} \binom{K}{i} p^i (1-p)^{K-i}} \qquad i = 0, \ldots, c,$$

    where $c$ is the number of servers, $K$ is the number of sources, and

$$p = \frac{\alpha/\mu}{1 + \alpha/\mu}.$$

    b. Find the probability that all servers are busy.

c. Use the fact that arriving customers "see" the steady state pmf of a system with one fewer source to show that the fraction of arrivals that are turned away is given by $P_{K-1}(c)$. The resulting expression is called the Engset formula.

## Section 9.6
## M/G/1 Queueing Systems

33. Find the mean waiting time and mean delay in an M/G/1 system in which the service time is a $k$-Erlang random variable (see Table 3.2) with mean $1/\mu$. Compare the results to M/M/1 and M/D/1 systems.
34. A $k = 2$ hyperexponential random variable is obtained by selecting a service time at random from one of two exponential random variables as shown in Fig. P9.2. Find the mean delay in an M/G/1 system with this hyperexponential service time distribution.

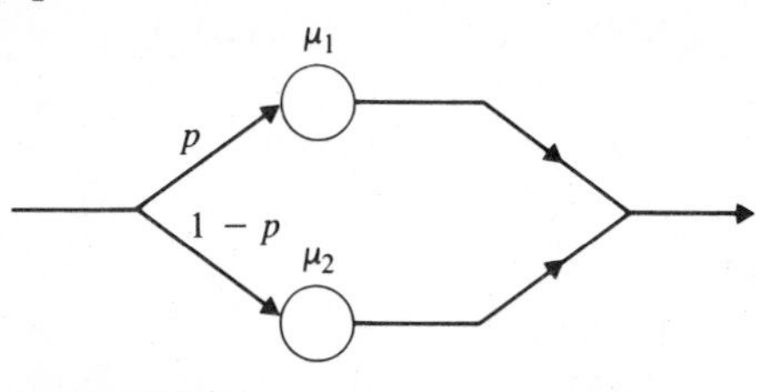


**FIGURE P9.2**

35. Customers arrive at a queueing system according to a Poisson process of rate $\lambda$. A fraction $\alpha$ of the customers require a fixed service time $d$, and a fraction $1 - \alpha$ require an exponential service time of mean $1/\mu$. Find the mean waiting time and mean delay in the resulting M/G/1 system.
36. Find the mean waiting time and mean delay in an M/G/1 system in which the service time consists of a fixed time $d$ plus an exponentially distributed time of mean $1/\mu$.
37. Fixed-length messages arrive at a transmitter according to a Poisson process of rate $\lambda$. The time required to transmit a message and to receive an acknowledgement is $d$ seconds. If a message is acknowledged as having been received correctly, then the transmitter proceeds with the next message. If the message is acknowledged as having been received in error, the transmitter retransmits the message. Assume that a message undergoes errors in transmission with probability $p$, and that transmission errors are independent.
    a. Find the mean and variance of the effective message service time.
    b. Find the mean message delay.
38. Jobs arrive at a machine according to a Poisson process of rate $\lambda$. The

service time for the jobs are exponentially distributed with mean $1/\mu$. The machine has a tendency to break down while it is serving customers; if a particular service time is $t$, then the probability that it will break down $k$ times during this service time is a Poisson random variable with mean $\alpha t$. It takes an exponentially distributed time with mean $1/\beta$ to repair the machine. Assume a machine is always working when it begins a job.

a. Find the mean and variance of the total time required to complete a job. *Hint*: Use conditional expectation.
b. Find the mean job delay for this system.

39. Consider a two-class nonpreemptive priority queueing system, and suppose that the lower-priority class is saturated (i.e., $\lambda_1 E[\tau_1] + \lambda_2 E[\tau_2] > 1$).

a. Show that the rate of low-priority customers that are served by the system is $\lambda_2' = (1 - \lambda_1 E[\tau_1])/E[\tau_2]$. *Hint:* What proportion of time is the server busy with class two customers?
b. Show that the mean waiting time for class 1 customers is

$$E[W_1] = \frac{(1/2)\lambda_1 E[\tau_1^2]}{1 - \lambda_1 E[\tau_1]} + \frac{E[\tau_2^2]}{2E[\tau_2]}.$$

40. Consider an M/G/1 system in which the server goes on vacations (becomes unavailable) whenever it empties the queue. If upon returning from vacation, the system is still empty, the server takes another vacation, and so on until it finds customers in the system. Suppose that vacation times are independent of each other and of the other variables in the system. Show that the mean waiting time for customers in this system is

$$E[W] = \frac{(1/2)\lambda E[\tau_2]}{1 - \lambda E[\tau]} + \frac{E[V^2]}{2E[V]},$$

where $V$ is the vacation time. *Hint:* Show that this system is equivalent to a nonpreemptive priority system and use the result of Problem 39.

41. Fixed-length packets arrive at a concentrator that feeds a synchronous transmission system. The packets arrive according to a Poisson process of rate $\lambda$, but the transmission system will only begin packet transmissions at times $id$, $i = 1, 2, \ldots$, where $d$ is the transmission time for a single packet. Find the mean packet waiting time. *Hint*: Show that this is an M/D/1 queue with vacations as in Problem 40.

42. A queueing system handles two types of traffic. Type $i$ traffic arrives according to a Poisson process and has exponentially distributed service times with mean $1/\mu_i$ for $i = 1, 2$. Suppose that type 1

customers are given nonpreemptive priority. Plot the overall and per-class mean waiting time vs. $\lambda$ if $\lambda_1 = \lambda_2 = \lambda$, $\mu_1 = 1$, $\mu_2 = 1/10$.

43. Consider a two-class priority M/G/1 system in which high-priority customer arrivals preempt low-priority customers that are found in service. Preempted low-priority customers are placed at the head of their queue and they resume service when the server again becomes available to low-priority customers.

    a. What is the mean waiting time and the mean delay for the high-priority customers?

    b. Show that the time required to service all customers found by a type 2 arrival to the system is

    $$\frac{R_2}{1 - \rho_1 - \rho_2},$$

    where $\rho_j = \lambda_j E[\tau_j]$, and

    $$R_2 = \frac{1}{2} \sum_{j=1}^{2} \lambda_j E[\tau_j^2].$$

    c. Show that the time required to service all type 1 customers that arrive during the time a type 2 customer spends in the system is $\rho_1 E[T_2]$.

    d. Use Parts b and c to show that

    $$E[T_2] = \frac{(1 - \rho_1 - \rho_2)/\mu_2 + R_2}{(1 - \rho_1)(1 - \rho_1 - \rho_2)}.$$

44. Evaluate the formulas developed in Problem 43 using the two traffic classes described in Problem 42.

## Section *9.7
## M/G/1 Analysis Using Embedded Markov Chain

45. The service time in an M/G/1 system has a $k = 2$ Erlang distribution with mean $1/\mu$ and $\lambda = \mu/2$.

    a. Find $G_N(z)$ and $P[N = j]$.

    b. Find $\hat{W}(s)$ and $\hat{T}(s)$ and the corresponding pdf's.

46. a. In Problem 38, show that the Laplace transform of the pdf for the total time $\tau$ required to complete the service of a customer is

    $$\hat{\tau}(s) = \frac{\mu(s + \beta)}{(s + \beta)(s + \mu) + \alpha s}.$$

    *Hint:* Use conditional expectation in evaluating $E[e^{-s\tau}]$, and note that the number of breakdowns depends on the service time of the customer.

    b. Find $\hat{W}(s)$ and $\hat{T}(s)$ and the corresponding pdf's.

47. a. Show that Eqs. (9.110a) and (9.110b) can be written as

$$D_j = D_{j-1} - U(D_{j-1}) + M_j, \tag{9.166}$$

where

$$U(x) = \begin{cases} 1 & x > 0 \\ 0 & x \leq 0. \end{cases}$$

b. Take the expected value of both sides of Eq. (9.166) to obtain an expression for $P[N > 0]$.

c. Square both sides of Eq. (9.166) and take the expected value to obtain the Pollaczek-Khinchin formula for $E[N]$.

48. a. Show that for an M/D/1 system

$$G_N(z) = \frac{(1 - \rho)(1 - z)}{1 - ze^{\rho(1-z)}}$$

b. Expand the denominator in a geometric series, and then identify the coefficient of $z^k$ to obtain

$$P[N = k] = (1 - \rho)\sum_{j=0}^{\infty} \frac{(-j\rho)^{k-j-1}(-j\rho - k + 1)e^{j\rho}}{(k - j)!}$$

49. a. Show that Eq. (9.130) can be rewritten as

$$\hat{W}(s) = \frac{1 - \rho}{1 - \rho\hat{R}(s)}, \tag{9.167}$$

where

$$\hat{R}(s) = \frac{1 - \hat{\tau}(s)}{sE[\tau]}$$

is the Laplace transform of the pdf of the residual service time.

b. Expand the denominator of Eq. (9.167) in a geometric series and invert the resulting transform expression to show that

$$f_W(x) = \sum_{k=0}^{\infty} (1 - \rho)\rho^k f^{(k)}(x), \tag{9.168}$$

where $f^{(k)}(x)$ is the $k$th order convolution of the residual service time.

50. Approximate $f_W(x)$ for an M/D/1 system using the $k = 0, 1, 2$ terms of Eq. (9.168). Sketch the resulting pdf for $\rho = 1/2$.

### Section 9.8
### Burke's Theorem: Departures from M/M/*c* Systems

51. Consider the interdeparture times from a stable M/M/1 system in steady state.

a. Show that if a departure leaves the system nonempty, then the time to the next departure is an exponential random variable with mean $1/\mu$.
b. Show that if a departure leaves the system empty, then the time to the next departure is the sum of two independent exponential random variables of means $1/\lambda$ and $1/\mu$, respectively.
c. Combine the results of Parts a and b to show that the interdeparture times are exponential random variables with mean $1/\lambda$.

52. Verify that Eqs. (9.137) through (9.139) satisfy Eqs. (9.135).
53. Find the joint pmf for the number of customers in the queues in the network shown in Fig. P9.3.

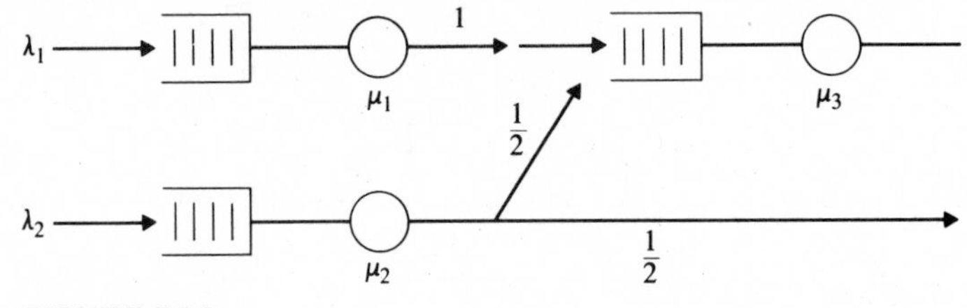


**FIGURE P9.3**

54. Write the balance equations for the feedforward network shown in Fig. P9.4 and verify that the joint state pmf is of product form.

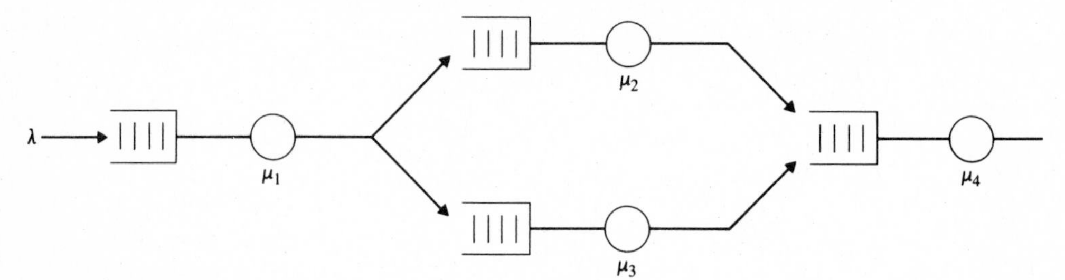


**FIGURE P9.4**

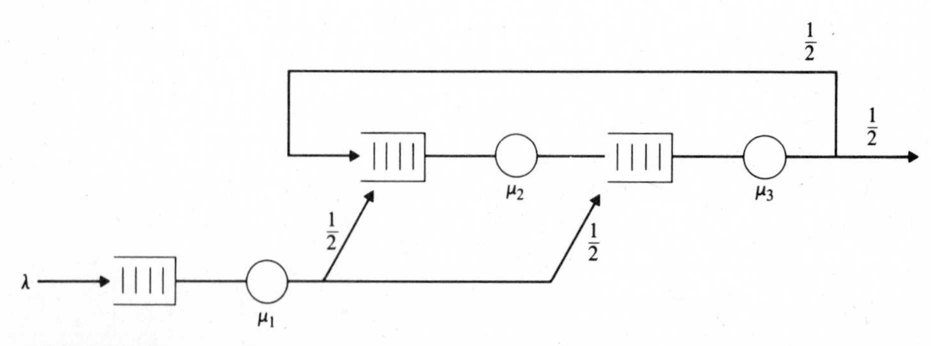


**FIGURE P9.5**

## Section 9.9
## Networks of Queues: Jackson's Theorem

55. Find the joint state pmf for the open network of queues shown in Fig. P9.5.

56. A computer system model has three programs circulating in the network of queues shown in Fig. P9.6.

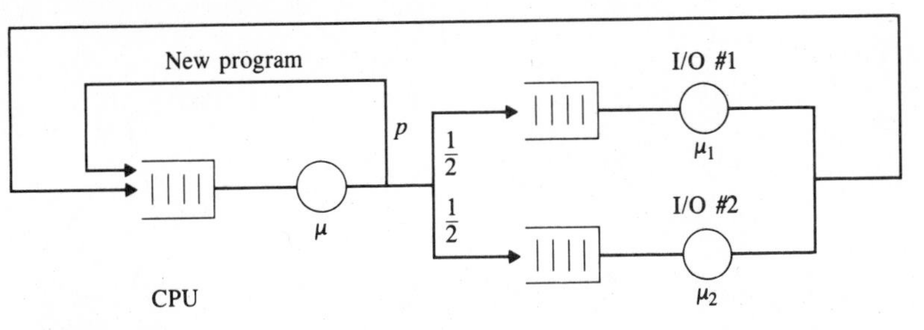


**FIGURE P9.6**

a. Find the joint state pmf of the system.
b. Find the average program completion rate.

57. Use the mean value analysis algorithm to answer Problem 56, Part b.

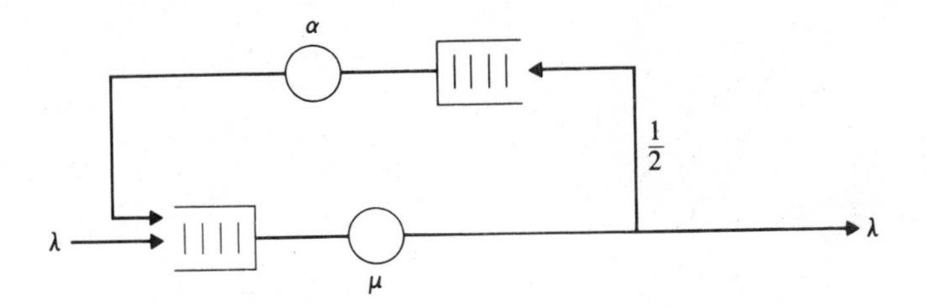

# APPENDIX A

# Mathematical Tables

## A. Trigonometric Identities

$\sin^2 \alpha + \cos^2 \alpha = 1$

$\sin(\alpha + \beta) = \sin \alpha \cos \beta + \cos \alpha \sin \beta$

$\sin(\alpha - \beta) = \sin \alpha \cos \beta - \cos \alpha \sin \beta$

$\cos(\alpha + \beta) = \cos \alpha \cos \beta - \sin \alpha \sin \beta$

$\cos(\alpha - \beta) = \cos \alpha \cos \beta + \sin \alpha \sin \beta$

$\sin 2\alpha = 2 \sin \alpha \cos \alpha$

$\cos 2\alpha = \cos^2 \alpha - \sin^2 \alpha = 2 \cos^2 \alpha - 1 = 1 - 2 \sin^2 \alpha$

$$\sin \alpha \sin \beta = \frac{1}{2} \cos(\alpha - \beta) - \frac{1}{2} \cos(\alpha + \beta)$$

$$\cos \alpha \cos \beta = \frac{1}{2} \cos(\alpha - \beta) + \frac{1}{2} \cos(\alpha + \beta)$$

$$\sin \alpha \cos \beta = \frac{1}{2} \sin(\alpha + \beta) + \frac{1}{2} \sin(\alpha - \beta)$$

$$\cos \alpha \sin \beta = \frac{1}{2} \sin(\alpha + \beta) - \frac{1}{2} \sin(\alpha - \beta)$$

$$\sin^2 \alpha = \frac{1}{2}(1 - \cos 2\alpha)$$

$$\cos^2 \alpha = \frac{1}{2}(1 + \cos 2\alpha)$$

$e^{j\alpha} = \cos \alpha + j \sin \alpha$

$\cos \alpha = (e^{j\alpha} + e^{-j\alpha})/2$

$\sin \alpha = (e^{j\alpha} - e^{-j\alpha})/2j$

$\sin \alpha = \cos(\alpha - \pi/2)$

## B. Indefinite Integrals

$\int u\, dv = uv - \int v\, du$ where $u$ and $v$ are functions of $x$

$$\int x^n\,dx = x^{n+1}/(n+1) \qquad \text{except for } n = -1$$

$$\int x^{-1}\,dx = \ln x$$

$$\int e^{ax}\,dx = e^{ax}/a$$

$$\int \ln x\,dx = x\ln x - x$$

$$\int (a^2 + x^2)^{-1}\,dx = (1/a)\tan^{-1}(x/a)$$

$$\int (\ln x)^n/x\,dx = (1/(n+1))(\ln x)^{n+1}$$

$$\int x^n \ln ax\,dx = (x^{n+1}/(n+1))\ln ax - x^{n+1}/(n+1)^2$$

$$\int xe^{ax}\,dx = e^{ax}(ax - 1)/a^2$$

$$\int x^2e^{ax}\,dx = e^{ax}(a^2x^2 - 2ax + 2)/a^3$$

$$\int \sin ax\,dx = -(1/a)\cos ax$$

$$\int \cos ax\,dx = (1/a)\sin ax$$

$$\int \sin^2 ax\,dx = x/2 - \sin(2ax)/4a$$

$$\int x\sin ax\,dx = (1/a^2)(\sin ax - ax\cos ax)$$

$$\int x^2 \sin ax\,dx = \{2ax\sin ax + 2\cos ax - a^2x^2\sin ax\}/a^3$$

$$\int \cos^2 ax\,dx = x/2 + \sin(2ax)/4a$$

$$\int x\cos ax\,dx = (1/a^2)(\cos ax + ax\sin ax)$$

$$\int x^2\cos ax\,dx = (1/a^3)\{2ax\cos ax - 2\sin ax + a^2x^2\sin ax\}$$

## C. Definite Integrals

$$\int_0^\infty t^{n-1} e^{-(a+1)t}\, dt = \frac{\Gamma(n)}{(a+1)^n} \qquad n > 0, a > -1$$

$\Gamma(n) = (n - 1)!$ if $n$ is an integer, $n > 0$

$$\Gamma\left(\frac{1}{2}\right) = \sqrt{\pi}$$

$$\Gamma\left(n + \frac{1}{2}\right) = \frac{1 \cdot 3 \cdot 5 \cdots (2n - 1)}{2^n} \sqrt{\pi} \qquad n = 1, 2, 3, \ldots$$

$$\int_0^\infty e^{-\alpha^2 x^2}\, dx = \sqrt{\pi}/2\alpha$$

$$\int_0^\infty x e^{-\alpha^2 x^2}\, dx = \sqrt{\pi}/2\alpha^2$$

$$\int_0^\infty x^2 e^{-\alpha^2 x^2}\, dx = \sqrt{\pi}/2\alpha^3$$

$$\int_0^\infty x^n e^{-\alpha^2 x^2}\, dx = \Gamma((n + 1)/2)/(2\alpha^{n+1})$$

$$\int_0^\infty a/(a^2 + x^2)\, dx = \pi/2 \qquad \text{if } a > 0$$

$$\int_0^\infty \frac{\sin^2 ax}{x^2}\, dx = |a|\, \pi/2 \qquad \text{if } a > 0$$

# APPENDIX B

# Tables of Fourier Transforms

## A. Fourier Transform Definition

$$G(f) = \mathscr{F}\{g(t)\} = \int_{-\infty}^{\infty} g(t)e^{-j2\pi ft}\,dt$$

$$g(t) = \mathscr{F}^{-1}\{G(f)\} = \int_{-\infty}^{\infty} G(f)e^{j2\pi ft}\,df$$

## B. Properties

| | |
|---|---|
| Linearity: | $\mathscr{F}\{ag_1(t) + bg_2(t)\} = aG_1(f) + bG_2(f)$ |
| Time scaling: | $\mathscr{F}\{g(at)\} = G(f/a)/\lvert a\rvert$ |
| Duality: | If $\mathscr{F}\{g(t)\} = G(f)$, then $\mathscr{F}\{G(t)\} = g(-f)$ |
| Time shifting: | $\mathscr{F}\{g(t - t_0)\} = G(f)e^{-j2\pi ft_0}$ |
| Frequency shifting: | $\mathscr{F}\{g(t)e^{j2\pi f_0 t}\} = G(f - f_0)$ |
| Differentiation: | $\mathscr{F}\{g'(t)\} = j2\pi fG(f)$ |
| Integration: | $\mathscr{F}\left\{\int_{-\infty}^{t} g(s)\,ds\right\} = G(f)/(j2\pi f) + (G(0)/2)\,\delta(f)$ |
| Multiplication in time: | $\mathscr{F}\{g_1(t)g_2(t)\} = G_1(f) * G_2(f)$ |
| Convolution in time: | $\mathscr{F}\{g_1(t) * g_2(t)\} = G_1(f)G_2(f)$ |

## C. Transform Pairs

$g(t)$ $\qquad$ $G(f)$

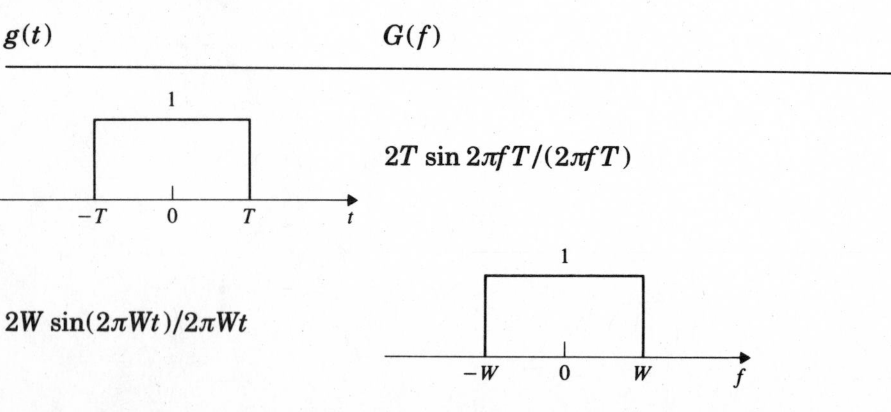

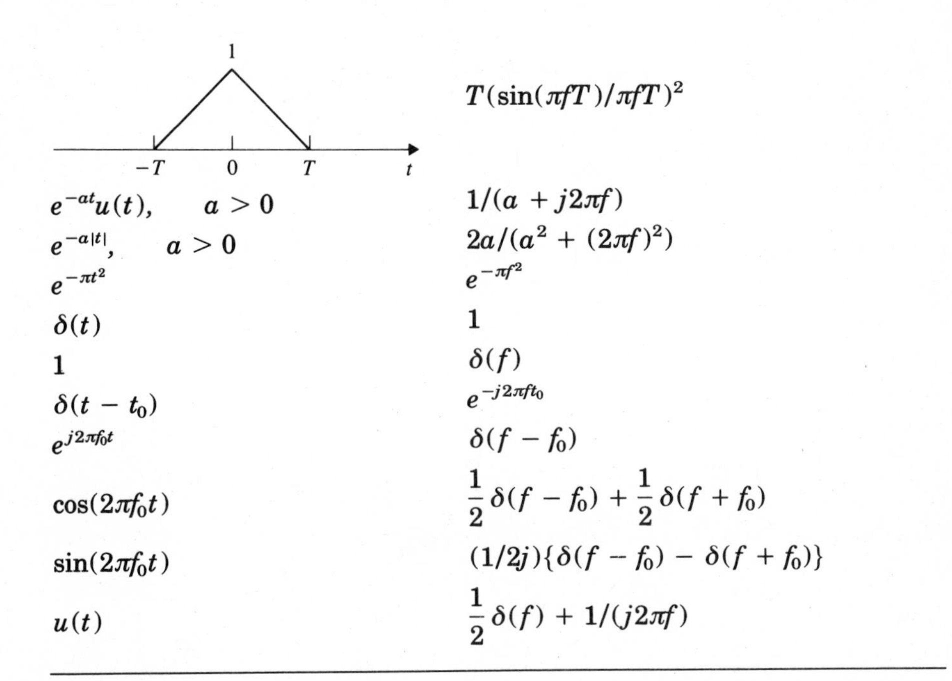

| | |
|---|---|
| (triangle: 1, $-T$, 0, $T$, $t$) | $T(\sin(\pi f T)/\pi f T)^2$ |
| $e^{-at}u(t), \quad a > 0$ | $1/(a + j2\pi f)$ |
| $e^{-a\|t\|}, \quad a > 0$ | $2a/(a^2 + (2\pi f)^2)$ |
| $e^{-\pi t^2}$ | $e^{-\pi f^2}$ |
| $\delta(t)$ | $1$ |
| $1$ | $\delta(f)$ |
| $\delta(t - t_0)$ | $e^{-j2\pi f t_0}$ |
| $e^{j2\pi f_0 t}$ | $\delta(f - f_0)$ |
| $\cos(2\pi f_0 t)$ | $\frac{1}{2}\delta(f - f_0) + \frac{1}{2}\delta(f + f_0)$ |
| $\sin(2\pi f_0 t)$ | $(1/2j)\{\delta(f - f_0) - \delta(f + f_0)\}$ |
| $u(t)$ | $\frac{1}{2}\delta(f) + 1/(j2\pi f)$ |

# Answers to Selected Problems

## Chapter 1

1. a. $S = \{bb, bw, wb\}$     *b.* $S = \{bb, bw, wb, ww\}$
   c. 0 in Part a and 1/4 in Part b
   d. In Part a the outcome of the first draw affects the probabilities of the outcomes of the second draw.
6. a. $\sum k^2 f_k(n)$

## Chapter 2

2. d. $B$ implies $A$     f. 0
3. c. $\{\text{total} = 4\} = \{(1,3), (2,2), (3,1)\}$
5. a. $\{g, bg, bbg, bbbg\}$     b. $\{1, 2, 3, 4\}$
7. b. $A_1 = \{(1, 2, 3), (1, 3, 2)\}$
9. $A \cap B = B, A \cap C = \varnothing, A \cup B = A$
12. $(A \cap B^c) \cup (A^c \cap B)$
16. a. all outcomes have probability 1/6
    b. $p_1 = 2/7$ and the other outcomes have probability 1/7
24. a. $P[A] = 1/2, P[B] = 3/4, P[A \cap B] = 1/4$
28. 1/4
30. 624
34. 10!, 12!/2!
36. 20!, $2(10!)^2/20!$
38. $$\frac{\binom{k}{m}\binom{100-k}{M-m}}{\binom{100}{M}}$$
41. 1260
47. $P[A \mid B] = 1 \qquad P[B \mid A] = 1/2$

49. a. .7696, .5838
52. 1/30, 1/10
54. b. 19/120
60. a. $P[A] + P[B] - P[A]P[B]$  b. $P[A] + P[B]$
64. Compare $f_{A \cap B}(n)$ to $f_A(n) f_B(n)$
66. $1.5(10^{-4})$
68. 11 chips have probability .898 of yielding 10 good chips.
74. $P[k] = p(1-p)^k$ for $k < m$, $P[m] = (1-p)^m$
76. $P[A] = (k-1)p^2(1-p)^{k-2}$
80. $Y_n = (b-a)U_n + a$
81. b. If $U_n \leq (1-p)$ then $B_n = 0$; otherwise $B_n = 1$.
84. Use the result of Problem 81, Part b, to generate Bernoulli trials and count the number of successes.

## Chapter 3

1. a. $S = \{\xi_1, \xi_2, \ldots, \xi_{100}\}$ $P[\{\xi_k\}] = 1/100$
   b. $S_X = \{1, 5, 50\}$ $p_1 = .9$, $p_5 = .09$, $p_{50} = .01$
3. c. $F_Z(z) = \begin{cases} z^2/2b^2 & 0 < z < b \\ 1 - (2b-z)^2/2b^2 & b \leq z \leq 2b \end{cases}$
11. b. $P[X < 0] = 0$,  $P[X \leq 0] = 1/4$
13. c. $c = 1/2$
17. a. $c = 6$  b. 0.34375
21. $f_X(x) = (1/4)\,\delta(x) + (1/4)[u(x) - u(x-1)] + (1/2)\,\delta(x-1)$
24. a. $1/n$
28. $[(1-p)^{k-1}p]/[1-(1-p)^m]$  $1 \leq k \leq m$
29. a. $e^{-15}$  b. .8815
33. a. $\pi(90) \approx 2.3/\lambda$,  $\pi(95) \approx 3.0/\lambda$,  $\pi(99) \approx 4.6/\lambda$
37. Chip 2 preferred for 20 Khrs. Chip 1 preferred for 24 Khrs.
38. .7149, .09963
45. a. $F_Y(y) = F_X(x) - F_X(-x)$
46. b. $f_Y(y) = F_X(-a)\,\delta(x+a) + f_X(y) + (1 - F_X(a))\,\delta(x-a)$ for $-a \leq y \leq a$, and zero elsewhere
50. 1.85
51. 15/8
61. $E[Y] = -aF_X(-a) + \int_{-a}^{a} x f_X(x)\,dx + a(1 - F_X(a^-))$

64. $E[X^n] = 1/(n + 1)$

65. a. $P[|X| > c] = 1 - c/b$ for $0 < c < b$, Chebychev bound gives $P[|X| > c] \le b^2/3c^2$

66. Case 1: $D^2 = 130 > 21.7$, so reject hypothesis
Case 2: $D^2 = 83 > 18.5$, so reject hypothesis

77. $X^*(s) = (\lambda/(s + \lambda))^\alpha$

79. $f_X(x) = (\alpha\beta/(\alpha + \beta))[e^{-\beta x} - e^{-\alpha x}]$ for $x > 0$

80. c. $T_0 + 0.01/\lambda$

85. a. $3e^{-2t} - 2e^{-3t}$

88. $n = -\ln(100)/\ln(1 - e^{-3t})$

89. $X = -a + a\sqrt{(2U)}$ for $0 < U < 1/2$ and $X = a - a\sqrt{(2 - 2U)}$ for $1/2 < U < 1$

93. $E[N] = 2$

96. $F_Z(z) = z^\lambda/(z^\lambda + \alpha^\lambda)$ for $z > 0$

## Chapter 4

3. b. $\{1 - F_X(5)\}\{F_Y(0) - P[Y = 0]\}P[Z = 1]$

5. a. In all three cases: $P[X = -1] = P[X = 0] = P[X = 1] = 1/3$ and $P[Y = -1] = P[Y = 0] = P[Y = 1] = 1/3$.

8. $$F_X(x, y) = \begin{cases} (1 - e^{-2y}) - 2e^{-x}(1 - e^{-y}) & \text{for } 0 \le y \le x \\ 1 - 2e^{-x} + e^{-2x} & \text{for } 0 < x < y \end{cases}$$

10. c. $f_X(x) = x + \frac{1}{2}$ for $0 \le x \le 1$ and $f_Y(y) = y + \frac{1}{2}$ for $0 \le y \le 1$

16. $$f_Y(y) = \begin{cases} 1/8 & \text{for } 1 < |y| < 3 \\ 1/4 & \text{for } |y| < 1 \end{cases}$$

19. i. No, ii. Yes, iii. No

21. a. $1/2\sqrt{2}$ b. 1

25. $f_Y(y \mid x) = (x + y)/\left(x + \frac{1}{2}\right)$ for $0 \le y \le 1$

28. $$f_Y(y \mid x) = \frac{1}{2}\delta(y + \sqrt{(1 - x^2)}) + \frac{1}{2}\delta(y - \sqrt{(1 - x^2)})$$

30. $$P[K = k] = \frac{(1 - p)(1 - a)}{1 - a(1 - p)}\left(\frac{p}{1 - a(1 - p)}\right)^k$$

31. $$P[K = k] = \frac{\Gamma(k + \alpha)}{\Gamma(\alpha)k!}\frac{\lambda^\alpha}{(1 + \lambda)^{k+\alpha}}$$

33. $f_{X_n}(x) = (-1)^{n-1}(\ln(x))^{n-1}/(n-1)!$ for $0 < x < 1$
35. b. $X_1$, $X_2$, and $X_3$ are independent Poisson random variables with parameters $p_1\alpha$, $p_2\alpha$, and $p_3\alpha$, respectively.
38. $T$ is Rayleigh with parameter $\alpha/\sqrt{n}$.
43. $f_Z(z) = (2\alpha^2/\beta^2)z/(z^2 + \alpha^2/\beta^2)^2$ for $z > 0$
44. a. $f_{X,Y,Z}(u, v - u, w - v)$
45. $f_{X_1,X_2}(m + \sqrt{v/2}, m - \sqrt{v/2})$
46. $Z$ is uniformly distributed in $[0, 1]$.
49. $f_Z(z) = \sqrt{(2/\pi)}\, z^2 e^{-z^2/2\sigma^2}/\sigma^3$ for $z > 0$.
51. 1
56. $\rho_{X,Y} = (1 + \sigma_N/\sigma_X)^{-1/2}$
57. a. i. $\rho_{X,Y} = 0$ ii. $\rho_{X,Y} = 0$ iii. $\rho_{X,Y} = -1$
59. b. Linear estimator: $\hat{Y} = (1/11)(X - 7/12) + 7/12$
    Maximum likelihood estimator: $\hat{Y} = 1$
    Min. MSE estimator: $\hat{Y} = (X/2 + 1/3)/(X + 1/2)$
66. $\begin{bmatrix} 1 & 1/\sqrt{2} & 0 \\ 1/\sqrt{2} & 1 & 0 \\ 0 & 0 & 1 \end{bmatrix}$
67. $\sqrt{2}\ln(2)$
70. $Z_1 = \sigma_1 X_1 + m_1$ $Z_2 = \sigma_2\rho X_1 + \sigma_2\sqrt{(1 - \rho^2)}X_2$

## Chapter 5

1. a. $\text{VAR}[Z] = 3$ b. $\text{VAR}[Z] = 3$
2. $\text{VAR}[S_n] = n\sigma^2 + 2(n - 1)\rho\sigma^2$
4. a. $\Phi_z(\omega) = e^{-(\alpha+\beta)|\omega|}$
7. b. $((\beta - \alpha)/\alpha\beta)(e^{-\alpha t} - e^{-\beta t})$
12. a. $\text{VAR}[S] = \text{VAR}[X]E[N] + \text{VAR}[N]E[X]^2$
15. $P[|N(t)/t - \lambda| \geq \varepsilon] \leq \lambda/\varepsilon^2 t$
16. $n = 4500$
19. yes
22. $1 - 2.54(10^{-10})$, $0.5 - 1.27(10^{-10})$
25. 0.5692
27. $P[S = k] \approx Q((k - 1/2 - n\lambda)/\sqrt{n\lambda}) - Q((k + 1/2 - n\lambda)/\sqrt{n\lambda})$
31. (98.87, 101.13)
33. (221.7, 224.3)
34. (29.2, 36.8)
40. $1/mT$

44. $m_1/(m_1 + m_2)$

53. a. $2T/3$ b. $T$

55. $c_0 = 1, c_1 = 0, c_2 = 0$

## Chapter 6

2. b. $P[X_n = \pm 1] = 1/2$
   c. $P[X_n = \pm 1, X_{n+k} = \pm 1] = 1/2, \quad P[X_n = \pm 1 \neq X_{n+k}] = 0$

5. $P[X(t) = \pm 1] = 1/2$ for $0 < t < 1$, $P[X(t) = 0] = 1$ otherwise

6. $P[X(t) = 0] = 1 - t$ for $0 < t < 1$, $P[X(t) = 0] = t - 1$ for $1 < t < 2$

10. c. $P[H(t) = 1 - F_X(0^-), E[H(t)] = 1 - 2F_X(0^-)$

11. $E[(X(t_2) - X(t_1))^2] = R_X(t_2, t_2) + R_X(t_1, t_1) - 2R_X(t_1, t_2)$

18. a. $C_Z(t_1, t_2) = C(t_1, t_2)\cos\omega(t_1 - t_2)$

23. c. $C_Y(k) = \begin{cases} (1/2)(p - p^2) & k = 0 \\ (1/4)(p - p^2) & k = 1 \\ 0 & k > 1 \end{cases}$

24. b. $E[W_n] = (2^n - 1)E[X], E[Z_n] = 2(1 - (1/2)^{n-1})E[X]$

28. $P[S_n = j, S_{n+d} = k] = \dfrac{(n\alpha)^j}{j!} e^{-n\alpha} \dfrac{(d\alpha)^{k-j}}{(k-j)!} e^{-d\alpha}$

31. $e^{-5}$

34. a. $P[X_1 < X_2] = \lambda_1/(\lambda_1 + \lambda_2)$
    b. $P[\min(X_1, X_2) > x] = e^{-(\lambda_1+\lambda_2)x}$
    c. $N(t)$ is Poisson with rate $\lambda_1 + \lambda_2$.

36. $P[N = k] = (\beta/(\lambda + \beta))(\lambda/(\lambda + \beta))^k, \quad k = 0, 1, \ldots$

40. $Y(t)$ is a random telegraph process with transition rate $p\alpha$.

44. $\text{VAR}[X(t)] = \lambda \int_0^t h^2(u)\,du$

50. No, No

51. Yes, Yes

54. a. Yes b. Yes

58. a. Yes

62. Yes

66. Yes

## Chapter 7

1. a. $S_X(f) = AT(\sin(WT/2)/(WT/2))^2$
   b. $R_X(\tau) = AW(\sin(W\tau/2)/(W\tau/2)^2$

7. a. $R_{X,Y}(\tau) = R_X(\tau) - R_X(\tau - d) \qquad S_{X,Y}(f) = S_X(f)[1 - e^{-j2\pi fd}]$
   b. $R_Y(\tau) = 2R_X(\tau) - R_X(\tau - d) - R_X(\tau + d)$
   $S_Y(f) = 2S_X(f)[1 - \cos 2\pi fd]$

9. $S_X(f) = 3/[5 - 4\cos 2\pi f] + 240/[17 - 8\cos 2\pi f]$

14. a. $R_X(k) = (\sin 2\pi f_c k)/\pi k$

17. a. $S_Y(f) = 4\pi^2 f^2 S_X(f) \qquad R_Y(\tau) = -d^2/d\tau^2 R_Y(\tau)$

18. b. $R_Y(0) = 4\pi^2 N_0 W^3/3$

19. a. $h(t) = 1$ for $0 < t < 1$; $h(t) = 0$ otherwise

21. a. $S_{Y,X}(f) = (N_0/2)/(1 + j2\pi f)$ c. $R_Y(0) = N_0/4$

29. a. $S_Y(f) = 2N_0/(5 - 4\cos 2\pi f)$ $S_z(f) = 16S_Y(f)/(17 - 8\cos 2\pi f)$
   b. $R_{W,Z}(k) = 2(1/2)^k - (1/4)^k$ for $k \geq 0$

30. a. $R_{Y,X}(k) = R_X(k) + \beta R_X(k - 1)$ c. $\alpha = -\beta$

34. c. $R_Y(k) = (16/21)(1/2)^{|k|} - (32/105)(1/4)^{|k|}$

40. $\text{SNR} = \sigma_X^2/(4f_c^2 W + 2W^3/3)$

44. a. $h_0 = (1 + \Gamma - \rho^2)/((1 + \Gamma)^2 - \rho^2) \qquad h_1 = \Gamma/((1 + \Gamma)^2 - \rho^2)$
   b. $E[e_n^2] = \sigma^2\Gamma(\Gamma + 1 - \rho)/((1 + \Gamma)^2 - \rho^2)$

48. b. $h_1 = 1/2 \quad h_2 = 0$

49. a. $a = R_X(t - t_1)/(R_X(0)^2 - R_X(t_1 - t_2)^2)$
   $b = R_X(t - t_2)/(R_X(0)^2 - R_X(t_1 - t_2)^2)$

53. $E[e(t)^2] = [1 + 2/\alpha N_0]^{-1/2}$

## Chapter 8

1. a. Yes b. $f_{M_n}(x \mid M_{n-1} = y) = f_X(n - x - (n - 1)y)$

2. a. for $k = 1, \ldots, 5$: $P[k - 1 \mid k] = k/(5 + k) = 1 - P[k \mid k]$,
   $P[0 \mid 0] = 1$,
   b. No

8. a. 
$$\begin{bmatrix} 1 & 0 & 0 & 0 & 0 & 0 \\ 1/6 & 5/6 & 0 & 0 & 0 & 0 \\ 0 & 2/7 & 5/7 & 0 & 0 & 0 \\ 0 & 0 & 3/8 & 5/8 & 0 & 0 \\ 0 & 0 & 0 & 4/9 & 5/9 & 0 \\ 0 & 0 & 0 & 0 & 1/2 & 1/2 \end{bmatrix}$$

11. b. 
$$\begin{bmatrix} 1 & 0 & 0 & 0 \\ 1/2 & 0 & 1/2 & 0 \\ 0 & 1/2 & 0 & 1/2 \\ 0 & 0 & 0 & 1 \end{bmatrix}$$
f. 2/3

14. a. 
$$\frac{1}{\alpha + \beta}\begin{bmatrix} \beta + \alpha e^{-(\alpha+\beta)t} & \alpha(1 - e^{-(\alpha+\beta)t}) \\ \beta(1 - e^{-(\alpha+\beta)t}) & \alpha + \beta e^{-(\alpha+\beta)t} \end{bmatrix}$$

15. a. $p_{i,j}(t) = (\alpha t)^{i-j}/(i-j)!$ for $i > j$,
$p_{i,j}(t) = 0$, otherwise
b. $p_j(t) = p_{n,j}(t)$

17. b. $\tau_{i,i+1} = \alpha$ for $i = 0, \ldots, n-1$
$\tau_{i,i-1} = i\mu$ for $i = 1, \ldots, n$
$\tau_{0,0} = -\alpha \quad \tau_{n,n} = -n\mu$
$\tau_{i,1} = -\alpha - k\mu \quad i = 1, \ldots, n-1$
c. $p_j = (\alpha/\mu)^j p_0/j! \quad j = 1, \ldots, n$

20. a. $\{1, 2, 3\}$ recurrent
b. $\{1\}$ recurrent, $\{2, 3\}$ recurrent
c. $\{0\}$, $\{1\}$, $\{3\}$ are transient classes, $\{2\}$ recurrent
d. $\{0, 1\}$ recurrent, $\{2, 3\}$ transient

21. a. $(2/5, 2/5, 1/5)$
b. if system starts in $\{1\}$, then $(1, 0, 0)$
if system starts in $\{2, 3\}$, then $(0, 1/2, 1/2)$
c. $(0, 0, 1, 0)$ d. $(2/3, 1/3, 0, 0)$

23. $\pi_j = (1 - (p/q))(p/q)^j/[1 - (p/q)^{N+1}] \quad j = 0, 1, \ldots, N$

24. $p_{0,1} = 1 \quad p_{i,i+1} = \lambda/(\lambda + \mu) \quad p_{i,i-1} = \mu/(\lambda + \mu)$

26. $\pi_j = \binom{N}{j}\left(\frac{1}{2}\right)^j \quad j = 0, \ldots, N$

28. Yes

## Chapter 9

4. a. $\lambda = (m_1 + m_2 + m_3)^{-1} \quad E[N_i] = m_i/(m_1 + m_2 + m_3)$
b. $\lambda = N/E[C]$

5. a. $\lambda E[T] = 5$ b. $\lambda m = 2$ c. $E[T] = 5m/2$

8. a. $P[N \geq n] = \rho^n$

10. $0 < \lambda < \mu - 5$

11. a. $\rho = (-5 + \sqrt{45})/2$ b. $\rho = 4/5$

15. $\lambda = \mu(5 + \sqrt{45})/2$

18. $P[N \geq c + k] = a^c \rho^k/(c!(1 - \rho))$

21. a. $c = 6 \quad p_0 = .004512 \quad p_6 = .0979$

24. $B(5, 5) = .2849 \quad B(8, 5) = .07004$

25. a. $B(4, 1) = 1/65 = 1.54\%$
b. $E[N] = 64/65$
c. $B(3, 1) = 1/16 = 6.25\%$

29. b. $K^* = 16$ c. $E[t] \approx K/\mu - 1/\alpha = 10$

33. $E[W]/(1/\mu) = \rho(1 + k^{-1})/2(1 - \rho))$

36. $E[W]/[1/\mu) = \rho(1 + (1 + \mu d)^{-2})/(2(1 - \rho))$
42. $E[W_1] = 101\lambda/(1 - \lambda)$, $E[W_2] = 101\lambda/((1 - \lambda)(1 - 11\lambda))$
45. $P[N = j] = (1/2)(2/(9 + \sqrt{17}))^j + (1/2)(2/(9 - \sqrt{17}))^j$
53. $\rho_1 = \lambda_1/\mu_1, \rho_2 = \lambda_2/\mu_2, \rho_3 = (\lambda_1 + (1/2)\lambda_2)/\mu_3$
Assuming $\rho_j < 1$ for $j = 1, 2, 3$ then
$P[N_1 = l, N_2 = m, N_3 = n] = (1 - \rho_1)\rho_1^l(1 - \rho_2)\rho_2^m(1 - \rho_3)\rho_3^n.$
55. $\rho_1 = \lambda/\mu_1, \rho_2 = 1.5\lambda/\mu_2, \rho_3 = 2\lambda/\mu_3$
Assuming $\rho_j < 1$ for $j = 1, 2, 3$ then
$P[N_1 = l, N_2 = m, N_3 = n] = (1 - \rho_1)\rho_1^l(1 - \rho_2)\rho_2^m(1 - \rho_3)\rho_3^n.$

# Index

# Continuous Random Variables

**Uniform Random Variable**

$S_X = [a, b]$

$$f_X(x) = \frac{1}{b - a} \qquad a \leq x \leq b$$

$$E[X] = \frac{a + b}{2} \qquad \text{VAR[X]} = \frac{(b - a)^2}{12}$$

$$\Phi_X(\omega) = \frac{e^{j\omega b} - e^{j\omega a}}{j\omega(b - a)}$$

**Exponential Random Variable**

$S_X = [0, \infty)$

$f_X(x) = \lambda e^{-\lambda x} \qquad x \geq 0 \qquad \text{and } \lambda > 0$

$$E[X] = \frac{1}{\lambda} \qquad \text{VAR}[X] = \frac{1}{\lambda^2}$$

$$\Phi_X(\omega) = \frac{\lambda}{\lambda - j\omega}$$

*Remarks*: The exponential random variable is the only random variable with the memoryless property.

**Gaussian (Normal) Random Variable**

$S_X = (-\infty, +\infty)$

$$f_X(x) = \frac{e^{-(x-m)^2/2\sigma^2}}{\sqrt{2\pi}\,\sigma} \qquad -\infty < x < +\infty \quad \text{and} \quad \sigma > 0$$

$E[X] = m \qquad \text{VAR}[X] = \sigma^2$

$\Phi_X(\omega) = e^{jm\omega - \sigma^2\omega^2/2}$

*Remarks*: Under a wide range of conditions $X$ can be used to approximate the sum of a large number of independent random variables.

**Gamma Random Variable**

$S_X = (0, +\infty)$

$$f_X(x) = \frac{\lambda(\lambda x)^{\alpha-1}e^{-\lambda x}}{\Gamma(\alpha)} \qquad x > 0 \quad \text{and} \quad \alpha > 0, \lambda > 0$$

where $\Gamma(z)$ is the gamma function (Eq. 3.46).

$E[X] = \alpha/\lambda \qquad \text{VAR}[X] = \alpha/\lambda^2$

$$\Phi_X(\omega) = \frac{1}{(1 - j\omega/\lambda)^{\alpha}}$$